SPACE PLASMA SIMULATIONS

SPACE PLASMA SIMULATIONS

Proceedings of the Second International School for Space Simulations, Kapaa, Hawaii, February 4-15, 1985

Edited by

MAHA ASHOUR-ABDALLA and DARYL ANN DUTTON

Institute of Geophysics, University of California, Los Angeles

Reprinted from Space Science Reviews, Vol. 42, Nos. 1-4 (1985)

D. Reidel Publishing Company / Dordrecht / Boston

Library of Congress Cataloging-in-Publication Data

CIP

International School for Space Simulations (2nd: 1985: Kopaa, Hawaii)
Space plasma simulations.

Includes indexes.
1. Space plasmas--Mathematical models--Congresses. I. Ashour-Abdalla, Maha. II. Dutton, Daryl Ann. III. Title.
QC809.P5I83 1985 538'.766 85-19671
ISBN 90-277-2108-4

Published by D. Reidel Publishing Company,
P.O. Box 17, 3300 AA Dordrecht, Holland.

Sold and distributed in the U.S.A. and Canada
by Kluwer Academic Publishers,
190 Old Derby Street, Hingham, MA 02043, U.S.A.

In all other countries, sold and distributed
by Kluwer Academic Publishers Group,
P.O. Box 322, 3300 AH Dordrecht, Holland.

Printed in The Netherlands

TABLE OF CONTENTS

PREFACE vii

PART 1 - SIMULATIONS

D. W. FORSLUND / Fundamentals of Plasma Simulation 3

P. L. PRITCHETT / Electromagnetic Particle Simulation Codes 17

D. W. HEWETT / Elimination of Electromagnetic Radiation in Plasma Simulation: The Darwin or Magnetoinductive Approximation 29

H. OKUDA / Particle Simulation Models for Low Frequency Microinstabilities in a Magnetic Field 41

D. WINSKE / Hybrid Simulation Codes with Application to Shocks and Upstream Waves 53

A. B. LANGDON / Implicit Plasma Simulation 67

J. DENAVIT / Time-Implicit Simulation of Particle-Fluid Systems 85

O. BUNEMAN / Multidimensional Particle Codes: Their Capabilities and Limitations 103

V. K. DECYK / Wave-Particle Diagnostics for Plasma Simulation 113

C. H. ALDRICH / Particle Code Simulations with Injected Particles 131

C. W. BARNES / Optimizing Computational Efficiency and User Convenience in Plasma Simulation Codes 145

J. U. BRACKBILL / Fluid Modeling of Magnetized Plasmas 153

S. H. BRECHT / Global Simulations Using MHD Codes: A Few Points to Consider Before You Try One 169

J. M. DAWSON / The Future of Space Plasma Simulation 187

PART 2 - OBSERVATIONS

L. A. FRANK / Plasmas in the Earth's Magnetotail 211

F. L. SCARF / Wave Observations in Outer Planet Magnetospheres 241

S. D. SHAWHAN / The Menagerie of Geospace Plasma Waves 257

B. HULTQVIST / Observations of Low Energy Magnetospheric Plasma Outside the Plasmasphere 275

T. A. POTEMRA / Field-Aligned (Birkeland) Currents 295

F. S. MOZER, M. H. BOEHM, C. A. CATTELL, M. TEMERIN, and J. R. WYGANT / Large Electric Fields in the Magnetosphere 313

J. F. FENNELL / Plasma Observations in the Auroral and Polar Cap Region 337

B. J. FRASER / Observations of Ion Cyclotron Waves Near Synchronous Orbit and on the Ground 357

D. J. WILLIAMS / Dynamics of the Earth's Ring Current: Theory and Observation 375

PART 3 - THEORY

F. V. CORONITI / Space Plasma Turbulent Dissipation: Reality or Myth? 399

K. G. RÖNNMARK / Kinetic Theory of Plasma Waves 411

H. MATSUMOTO / Coherent Nonlinear Effects on Electromagnetic Wave-Particle Interactions 429

I. KIMURA / Whistler Mode Propagation in the Earth and Planetary Magnetospheres and Ray Tracing Techniques 449

C. T. DUM / Coupling of Macroscopic and Small Scale Phenomena 467

T. SATO / Global Energy Regulation in the Solar Wind-Magnetosphere-Ionosphere System 485

C. K. GOERTZ / Auroral Arc Formation: Kinetic and MHD Effects 499

R. GENDRIN / Consequences of Hydromagnetic Waves on Magnetospheric Particle Dynamics 515

B. McNAMARA and L. D. PEARLSTEIN / Spectral Computation of Tandem Mirror Equilibria with Finite Larmor Radius Effects 541

ISSS2 PARTICIPANTS AND CONTRIBUTORS 553

ISSS2 SUPPORTING ORGANIZATIONS 568

SUBJECT INDEX 569

AUTHOR INDEX 577

ANNOUNCEMENT 578

Preface

The emergence over the past several years of space plasma simulations as a distinct field of endeavor, rather than simply the somewhat startling offspring of plasma physics, computer simulations and space observations, has necessitated a concentrated effort at interdigitating its parent and component fields. After several years of working without the benefits of a well-defined interactive community of those working in the field, a group of those who had gained greatly from setting up joint research projects and other lines of communication, arranged to further these gains by setting up the First International School for Space Simulations, which was organized by Kyoto University and held in Kyoto, Japan in November 1982. Its unqualified success led to the organization of the second such School, this time by the University of California, Los Angeles, and held in Kapaa, Kauai, Hawaii.

The Second International School for Space Simulations drew some 175 attendees from around the world; the distribution of attendees approached the targeted equal representation by established investigators and graduate students/beginning investigators. This strong attendance by graduate students and beginning investigators was due to the generous support of a number of funding agencies from the United States and Japan as well as international scientific organizations.

As in the first School, the school nature of the ISSS-2 gathering was emphasized, with a majority of the assembly time being devoted to basic tutorials given by leaders in the field of space plasma simulations. This volume is made up of these tutorial presentations. Other significant contributions to the School were made by the evening observations speakers and by poster presentations, but as these elements of the School have ample opportunity for publication elsewhere, they are not included in the volume. Nevertheless, the School benefitted greatly from their presentation. Session chairmen, poster previewers and closing colloquium members also made real contributions to the School. Of particular note were the well-received talks of the funding agencies' representatives who showed another aspect of the future of space plasma simulations - their assessment of the probability of future funding for the supercomputer time inherent of the field. All of these elements together with the enthusiasm of the students of the School resulted in a very successful School, one that hopefully further encouraged the establishment of a true community of space plasma simulation investigators.

Particular credit must be given to the School's Organizing Committee, Drs. P.J. Coleman, Jr. (UCLA, USA), R. Gendrin (CNET/CRNS,

France), H. Matsumoto (Kyoto University, Japan), T. Obayashi (ISAS, Japan), H. Okuda (Princeton University, USA), T. Sato (Hiroshima University, Japan), for their efforts at every stage of the School's organization. The Program Committee members (listed on a previous page) were also generous with their time and ideas. Much credit for the success of the School is due to the many organizations which supported it both financially and programatically. The work of my co-editor, Mrs. Daryl Ann Dutton, was vital, both in the preparation and running of the School and the organization of this volume; Mr. Jean Berchem and Mr. Willis Livesey contributed much time and effort to the mechanics of the preparation and running of the School; Mrs. M. Ishiwata did wonders with the School's artwork. The business office, headed by Mr. K. Olwin, of the Institute of Geophysics and Planetary Physics at UCLA was very helpful in managing many of the financial aspects of the School. The UCLA postdoctoral fellows and graduate students managed the audio-visual requirements as well as the required transportation and packing of office supplies and equipment, and did so with willingness and grace. The Kauai BeachBoy Hotel staff, headed by Mr. B. Goo, provided professional and enthusiastic support at the site of the School. All of the above contributions are greatly appreciated and were indeed largely responsible for the School's occurrence as well as its ultimate success.

Maha Ashour-Abdalla
Institute of Geophysics
University of California
Los Angeles

PART 1 - SIMULATIONS

FUNDAMENTALS OF PLASMA SIMULATION

David W. Forslund
Applied Theoretical Physics
Los Alamos National Laboratory
Los Alamos, New Mexico 87545

ABSTRACT. With the increasing size and speed of modern supercomputers, the incredibly complex nonlinear properties of plasmas in the laboratory and in space are being successfully explored in increasing depth. Of particular importance have been numerical simulation techniques involving finite size particles on a discrete mesh. After discussing the importance of this means of understanding a variety of nonlinear plasma phenomena, we describe the basic elements of particle-in-cell simulation and their limitations and advantages. The differencing techniques, stability and accuracy issues, data management and optimization issues are discussed by a simple example of a particle-in-cell code. Recent advances in simulation methods allowing large space and time scales to be treated with minimal sacrifice in physics are reviewed. Various examples of nonlinear processes successfully studied by plasma simulation are given.

1. INTRODUCTION

The many degrees of freedom in plasmas leads to an incredible variety of plasma phenomena on a spectrum of space and time scales. Nonlinear phenomena in plasmas are known to have many more diverse properties than even compressible fluid dynamics. Various theoretical models have been developed to describe the observed plasma properties, but these, of necessity, require many simplifying assumptions whose validity is many times questionable and difficult to verify. Although weak turbulence theory has been successful over a limited range of parameters, frequently the nonlinearity of some plasma process is so strong that it breaks down. It is in this regime that plasma simulations are the most useful and the easiest to perform. In regimes where weak turbulence theory is reasonably accurate, particle simulation is the most expensive and difficult to perform. Thus there has been considerable complementarity between weak turbulence theory and plasma simulation, each having its own domain of importance.

For more than 15 years the simulation of plasma phenomena on large scale computers has played an important role in explaining many observed plasma phenomena and in verifying or disproving various nonlinear plasma theories. The large

Space Science Reviews **42** (1985) 3–16. 0038–6308/85.15

variation in time and space scales present in typical plasma problems has continued to present a challenge to the simulationist. Today, as computers continue to increase in size and speed the nonlinear plasma problems that can be studied on the computer have become even more realistic. As an introduction to computer simulation techniques of plasmas, we will review the basic methods and illustrate their use on problems of significance. For a more complete discussion of simulation techniques, the reader is referred to the textbook by *Birdsall and Langdon*, [*1985*].

2. BASIC EQUATIONS

2.1. Liouville Equation

The equations usually used to describe a plasma are Newton's laws and Maxwell's equations:

$$\frac{d\vec{v}_j}{dt} = \frac{q_j}{m_j}(\vec{E} + \frac{\vec{v}_j \times \vec{B}}{c})$$

$$\frac{d\vec{x}}{dt} = \vec{v}$$

$$\nabla \times \vec{E} = -\frac{1}{c}\frac{\partial \vec{B}}{\partial t}$$

$$\nabla \cdot \vec{B} = 0, \quad \nabla \cdot \vec{E} = 4\pi\rho \tag{1}$$

$$\nabla \times \vec{B} = \frac{4\pi\vec{j}}{c} + \frac{1}{c}\frac{\partial \vec{E}}{\partial t}$$

$$\rho(\vec{x}) = \sum_j q_j \delta(\vec{x} - \vec{x}_j)$$

$$\vec{j}(\vec{x}) = \sum_j q_j \vec{v}_j \delta(\vec{x} - \vec{x}_j)$$

Almost all classical plasma physics involves the solving of these equations with various approximations. The simultaneous solution of these equations for a large collection of particles is equivalent to the solution of the Liouville equation for the N body distribution function. Solutions to these equations exhibit collisional effects determined by the structure of the δ functions in the above equations. Usually one speaks of a plasma as having a large number of particles per Debye sphere, in which case only terms to the lowest order in this plasma parameter are kept.

2.2. Vlasov Equations

In this collisionless limit, we have the Vlasov equation system:

$$\frac{\partial f}{\partial t} + \vec{v} \cdot \frac{\partial f}{\partial \vec{x}} + \frac{q}{m}\left(\vec{E} + \frac{\vec{v} \times \vec{B}}{c}\right) \cdot \frac{\partial f}{\partial \vec{v}} = 0$$

$$\nabla \cdot \vec{E} = 4\pi \int q f d^3\vec{v},$$

$$\nabla \times \vec{B} = \frac{1}{c}\frac{\partial \vec{E}}{\partial t} + \frac{4\pi}{c}\int q f \vec{v} d^3\vec{v}, \tag{2}$$

$$\nabla \cdot \vec{B} = 0, \quad \nabla \times \vec{E} = -\frac{1}{c}\frac{\partial \vec{B}}{\partial t}.$$

Although much successful effort has gone into solving Eqs. (2) for the distribution function directly[*Sakanaka, et al., 1971; Denavit, 1972*], the methods have been difficult or expensive to extend to 2 dimensions. The traditional form of plasma simulation [*Buneman, 1967*] involving particles is equivalent to solving the Vlasov equation by the method of characteristics, where the particle orbits are the characteristics of the Vlasov equation. Since the particles have finite size, however, they exhibit properties of the original Liouville system from which the Vlasov equation is derived. Consequently, plasma simulation codes exhibit many kinetic properties of plasmas including bremsstrahlung, collisional excitation and damping of plasma waves. With care one can study these phenomena directly with a simulation code. However, usually the collisions are not desireable, so one must take care to reduce them .

2.3. General Methods

To solve Eqs. (2) one must represent the spatial coordinate as a discrete grid and advance the equations over discrete time intervals or steps. Stability and accuracy conditions dictate the choice of the space and time steps. Early methods in one dimension treated particles as points moving through each other. However, because one cannot have as many particles per Debye sphere in the computer as one has in most plasmas, this method results in an excessively high noise level (and hence collisional level). To reduce this problem, particles are made to be of finite size and to deposit information onto and take information from a grid. In other words the δ functions of Eqs. (1) are made finite in width [*Birdsall and Fuss, 1969; Morse and Nielson, 1969*]. This drastically decreases the noise level by significantly altering the collisional process. Particles no longer undergo large angle collisions from a close approach but experience only small angle collisions as they pass through one another with a minimum effective distance equal to the particle size. Consequently, the finite size particle simulation method is most suited for studying collisionless plasmas. In this limit collisions can be added back in a controlled manner. To the lowest order then one can view the particle-in-cell simulation technique as a Monte Carlo solution to the Vlasov equation by integration along its characteristics. This Monte Carlo aspect of the solution illustrates why problems that are sensitive to the behavior of particles in a small region of velocity space are difficult to handle. Problems that depend primarily on the bulk plasma properties, albeit high nonlinear, are most readily solved by this technique.

The solution to Eqs. (2) are obtained by integrating in a series of discrete time steps the orbits of a large number of finite size charged particles (typically $10^4 - 10^6$). The lagrangian positions of the particles are used to deposit the charge and currents onto the fixed discrete grid. The field equations are then solved on this discrete grid (typically $10^3 - 10^5$ cells) to obtain the self-consistent fields for the next time step. The discrete field points are then interpolated to form a continuous field variable to advance the particle orbit. The two important numerical aspects are the space and time differencing and the interpolation of the fields from the grid to the particle and the particle charge onto the grid (i.e. the shape of the particle). A variety of techniques have been employed successfully and I will discuss only a few.

3. A ONE DIMENSIONAL ELECTROSTATIC CODE

3.1. Basic Equations

To clarify the preceding general introduction, let us consider what is involved in writing a simple one dimensional electrostatic plasma simulation code. As an interesting generalization of the usual one dimensional code we consider the equations for a spherical plasma and include the case of a space and time varying grid on which the equations are solved. The basic equations to be solved are:

$$\frac{1}{r^2}\frac{d}{dr}r^2 E_r = 4\pi\rho \tag{3}$$

$$m_j\frac{dv_r}{dt} = q_j E_r + \frac{l^2}{m_j r^3}, \quad l = m_j r v_\theta = constant \tag{4}$$

$$\frac{dr}{dt} = v_r \tag{5}$$

$$\rho = \sum_j q_j \delta(r - r_j) \tag{6}$$

In Eq. (4) l is the angular momentum of the particle and the additional term in the momentum equation is the centrifugal force.

3.2. Space and Time Differencing

We begin by first discussing the spatial differencing for these equations. The method we describe here is a simplified one dimensional form of the one used in the new two dimensional code, CELESTE, developed by *Brackbill and Forslund*, [*1984*]. We consider a set of N mesh points, $r(1), \ldots, r(N)$, that define the cell centers. These points can be arbitrarily spaced although the rate of change should be small compared to the cell size in order for the differencing to be reasonably accurate. We first consider how to deposit charge onto this grid, as shown in Eq. (6), in order to solve Poisson's equation, Eq. (3). The simplest procedure is to consider the particles to have zero size and to deposit all of the charge onto the grid point that is closest to the particle. This procedure is known as the nearest grid point (NGP) method. Although simple, it introduces large fluctuations on the scale of the grid point spacing as the particles move from grid point to grid point. These fluctuations are a form of collisions as mentioned in Section 2. Instead we take the particles to be of finite size and deposit charge on the grid in proportion to the overlap of the particle with each cell. Although a variety of particle sizes are possible, *Langdon and Birdsall* [*1970*] have shown that a particle size comparable with the cell size introduces the least amount of numerical error. For simplicity of calculation we choose a particle to be the size of the cell in which it is located. Since the grid spacing can vary with the index i, the particle size can change as it moves through the grid. However, the charge on a particle does not change. The shape of the particle is defined by the interpolation formula to and from the grid. For a uniform mesh a gaussian shape for the particle has also been used successfully[*Lin, et al., 1974*].

We use logical or natural coordinates, x_p, to describe the position of the particles such that the cell index is given by:

$$i = integer(x_p). \tag{7}$$

and the fractional part of x_p determines where in the cell the particle is located. The weights applied to cells i and $i+1$ are given by:

$$w_{i+1} = x_p - i, \tag{8}$$

$$w_i = 1.0 - w_{i+1}. \tag{9}$$

The particle position then is converted to physical coordinates by:

$$r_p = w_i r_i + w_{i+1} r_{i+1}. \tag{10}$$

Keeping the logical position instead of the entirely equivalent physical one makes it easier to determine the cell location of the particle. The reverse process is more complicated and is done only once per time step. Similarly, the electric field, E_p, used to advance the particle velocity is interpolated with the same weight functions from the grid. The time centered acceleration and particle position advancement are given by:

$$v_r^{j+1/2} = v_r^{j-1/2} + \left(\frac{q_p}{m_p} E_p^j + \frac{l^2}{m_p^2 (r^j)^3} \right) dt \tag{11}$$

$$r_p^{j+1} = r_p^j + v_r^{j+1/2} dt \tag{12}$$

where the superscript j is the time step index. The quantity, v_θ, can be calculated from the conserved angular momentum at any time. We then use a rapid zero point search to determine the new natural coordinate of the particle. If the particle does not change cells, this can be done with only one iteration. With this new natural coordinate, new values of w_i and w_{i+1} are calculated and the charge is interpolated onto the grid:

$$\begin{aligned} q_i &= q_i + w_i q_p \\ q_{i+1} &= q_{i+1} + w_{i+1} q_p. \end{aligned} \tag{13}$$

If desired the grid can be moved before this new natural coordinate of the particle is calculated without introducing any diffusion into the system. Adding up the charge from all the particles yields the total charge at each grid point. The solution of Poisson's equation gives E^{j+1} and the time cycle is completed.

The solution of Poisson's Equation in one dimension is simple. In differenced form the equation is:

$$(r^2E)_{i+1/2} = (r^2E)_{i-1/2} + 4\pi(\rho r^2 dr)_i \tag{14}$$

The term $\rho r^2 dr$ is the total charge in a cell calculated above. The cell centered electric field then is obtained by averaging the quantity r^2E at the cell edges:

$$E_i = \left((r^2E)_{i+1/2} + (r^2E)_{i-1/2}\right) / \left(r^2_{i+1/2} + r^2_{i-1/2}\right) \tag{15}$$

where $r_{i+1/2} = (r_i + r_{i+1})/2$. Boundary conditions are applied by specifying E at the edge of the first cell. Alternatively Poisson's equation could be written in terms of the electrostatic potential and solved with a tridiagonal solver and center differenced to obtain E at the cell centers.

3.3. Boundary Conditions

An additional problem exists during the charge accumulation stage. The finite size of the particles causes some charge to be deposited outside the grid from cells at the edge of the system. An extra cell is added at each end to contain this charge. The charge accumulated in these "ghost" cells is added back onto the physical grid. In the Cartesian case one may also choose to have periodic boundary conditions, in which case the charge would be added into the first real cells at opposite ends of the grid.

Finally, the particles themselves must be checked for crossing of the system boundaries so that they may be absorbed, reflected, or recreated at some new velocity. These operations are done right after the new particle position is calculated before the new natural coordinates are calculated for redeposition of charge onto the grid.

This completes the basic description of a simple one dimensional electrostatic code. The algorithms described above are readily extendable to two dimensions. The appropriate cell size, time step, and number of particles that need to be used for any given problem is determined by numerical stability and accuracy considerations discussed below.

3.4. Particle Loading

We digress for a moment on how to load the particles with an arbitrary density distribution. Suppose we have N simulation particles which we would like to represent the initial arbitrary charge density, $\rho(r)$. The probability of a particle being at the position r is given by

$$P(r) = \frac{\int_{r_1}^{r} \rho(r')r'^2 dr'}{\int_{r_1}^{r_2} \rho(r')r'^2 dr'} \tag{16}$$

where r_2 and r_1 are the upper and lower limits of the system. We can then randomly pick a number P from $0 \rightarrow 1$ for each simulation particle and use the inverse function r(P) to give the particle position. There is less noise in the initial loading, however, if one uniformly divides the interval $0 \rightarrow 1$ into N parts. The charge associated with each of the particles is

$$q_p = \frac{1}{N} \int_{r_1}^{r_2} \rho r^2 dr \tag{17}$$

which, because of the small value of N, is much larger than the charge on an ion or electron. That is, we see that each particle here represents a collection of a large number of ions or electrons and hence is sometimes called a macroparticle. This loading technique is readily extendable to two dimensions for arbitrary two dimensional density distributions and may also be used for generating particles with some specified velocity distribution.

4. DATA MANAGEMENT

Typically for large scale plasma simulations, modern computers do not have a enough large memory to contain all the particle data. Since external storage is usually slow compared to the fast memory in a modern supercomputer, a well-tuned multi-channel input/output (I/O) package is needed to interface with this slow storage (usually disks). The fastest method is not always the best due to the environment of other users on the machine and the algorithms used to charge for computer resources. However, generally synchronous I/O (in which the I/O is going on simultaneous with the computing) on at least one channel each for input and output is required. The simplest form of I/O divides a buffer into three parts doing input into 1/3, computing on 1/3, and output on 1/3. In a multi- channel method there may be separate buffer for input and output. The package written for the WAVE code on the CRAY-X/MP computers issues multiple simultaneous I/O requests on each channel and packs the particle data in order to maximize the I/O rate for a given buffer size. An interrupt driven I/O scheme has not been found to be fast enough to obtain data from the disk without missing revolutions on the disks. Also available on some modern supercomputers is an intermediate speed memory device (the SSD on the CRAY- X/MP) which is usually more than adequate to keep up with the particle pushing. In this case a sophisticated I/O algorithm is not required.

5. OPTIMIZATION

5.1. Vectorization

Most modern supercomputers have very high speed vector performance compared with their scalar performance. It is important in the coding process to ensure that the critical sections of the code are as vectorized as possible. Within the particle mover itself most of the operations are vectorizable. For example, in the one dimensional electrostatic code described above, Eqs. (7-9), (11), and (12) are complete vectorizable. However, because the cell index is random from one particle to the next, the interpolation from and to the grid (Eqs. (10) and (12)) is not. These two parts need to be in separate do loops from the vectorizable operations to ensure maximum speed. The electric field accumulation in Eq. (13) is also not vectorizable.

On the CRAY-X/MP a new hardware instruction has been added which now allows the interpolation of Eq. (10) to be completely vectorized. However, the scatter operation of Eq. (13) remains non-vectorizable because of the accumulation. Supercomputers which have this hardware gather-scatter and a fast vector sum routine would allow complete vectorization of the particle mover.

5.2. Assembly Language

In the WAVE code the particle movers have been hand coded in assembly language to make optimal use of the CRAY-X/MP. We find that both the vector and scalar operations are about twice as fast as they are in FORTRAN. This should always be considered when the code is going to be used heavily with infrequent changes. If a careful isolation of parameters for the particle mover is done, the assembly language code will not have to be modified often even though the remainder of the code may change a great deal. With the WAVE code this modularity has allowed the same assembly language movers to be used for over 8 years on the CRAY-1 with essentially no changes, even though the remainder of the WAVE code has been almost entirely rewritten.

5.3. Multiprocessing

With the arrival of multiple processors in supercomputers it is possible to reduce the running time by processing more than one group of particles at a time. This does not reduce the CPU time of a given job (it actually increases it slightly) but may reduce the turn around time when this is most important. However, care must be taken because multiple groups of particles may try to access a shared piece of memory at the same time. In particular, the charge weighting process may require separate copies of the charge array for each task followed by a combining at the end of the cycle to avoid overwriting data. Some computer hardware provides a memory lock out when another processor is accessing the memory, but this can result in some degradation of speed. In the CELESTE code we have sought to overcome this problem by sorting the particles with linked lists. There is some additional overhead for this, but it allows for complete separation of particle pushing tasks. In addition, it allows for the inclusion of momentum and energy conserving binary collisions in a fully vectorized form.

6. NUMERICAL STABILITY

6.1. Time Step Constraints

The discretization of time and space in simulation codes introduces accuracy and stability conditions. In the usual explicit particle code in which information at time level j is used to advance the fields and particles to $j+1$ there is a limit to the size of the time step before the code becomes unstable due to time aliasing. This is usually $\omega_p \delta t < 2$. Therefore, the electron dynamics cannot be avoided in an explicit plasma simulation code by taking very large time steps. The time step must be large enough to resolve the highest frequency in the system.

For electromagnetic codes including the displacement current one has an additional constraint that can be even more severe. The timestep must satisfy the Courant condition, $c\delta t \leq \delta x$. Again this arises because of the necessity of resolving the highest frequencies in the system. Other instabilities involving Cerenkov processes have also been shown to exist[*Godfrey, 1974,1975*]

6.2. Finite Grid Instability

The aliases arising from the spatial grid are coupled together by the interpolation method in a complicated way. In general, it is found that if the Debye length, $\lambda_d \geq \delta x/2$, the numerical instabilities are almost entirely absent both for drifting and stationary plasmas because modes with the largest wavenumber, $k_{max}\delta x \sim 1$ are heavily Landau damped. For large δx the incorrect of damping on the aliases can overcome the physical Landau damping at k_{max} and hence can cause strong instability. The addition of strong smoothing at large k can reduce the growth rate somewhat.

The properties of the numerical instabilities depend on the details of the particular differencing scheme. *H. R. Lewis*[*1970*] has suggested a differencing scheme derived from the system Hamiltonian that conserves energy in the limit of $\delta t \to 0$. The only difference from the method described above that conserves momentum exactly but not energy is that E is defined at the cell edge and the particle velocities are advanced with the NGP method. As one might expect, the method is somewhat noisier than the usual method and, in practice, does not conserve energy well because of the finite δt. If, however, smoothing is introduced on the charge density, the two methods conserve energy about equally well. For the physically stable case of a Maxwellian electron plasma drifting through a fixed ion background on a uniformly space Cartesian grid with periodic boundary conditions, we compare the stability properties [*Lindman, 1970; Langdon, 1970a,1970b; Okuda, 1972*] of the two methods. The energy conserving method is stable for zero drift while the momentum conserving method is not. For drift velocities about equal to the electron thermal speed, the energy conserving method becomes unstable and the growth rate of the momentum conserving method diminishes sharply. For $\delta x = 10\lambda_d$, the maximum growth rates in both cases are about $0.02\omega_p$ at $k\lambda_d = .16$. For smaller grid size the growth rate drops off sharply.

7. TWO DIMENSIONAL ELECTROMAGNETIC CODES

7.1. Units

We now discuss the basic difference equations for electromagnetic plasma simulation in two dimensions. Before considering the difference equations, let us first discuss the system of units used in the two dimensional WAVE code[*Morse and Nielson, 1971*]. We find it useful to normalize the density to some charge density, n_o, the charge to mass ratio to some q_o/m_o and velocities to the speed of light. Time then is normalized to $\omega_p = (4\pi n_o q_o^2/m_o)^{1/2}$, the electric and magnetic fields are in units of $(4\pi n_o m_o c^2)^{1/2}$ and the potentials in units of $m_o c^2/q_o$. Maxwell's equations then have no numerical factors and are dimensionless. Several additional nontrivial advantages result from this system of units. The equations are automatically ordered so that one can compare the size of various terms. Also, the dimensionless equations imply that a given simulation is the solution to a whole class of physical problems that are scaled by the units of time and fields. This reduces the likelihood of unnecessarily running a whole series of equivalent problems. It is easy then to convert the scaled units into the physical units of a problem of interest. Doing this at the outset can shed considerable insight into the physics.

7.2. Field Equations

The interpolation techniques used in one dimension are immediately extended to two dimensions. For simplicity we consider the case of a uniformly spaced mesh in which the particle is the same size as the cell spacing. Here the amount of charge and current deposited into each cell is proportional to the area overlap in the cell, hence the term "area weighting" or bilinear interpolation.

In the WAVE code the equations are written in term of the vector and scalar potentials for historical reasons. *Langdon and Lasinski* [*1976*] have shown that an E and B based code has essentially identical properties provided the equations are differenced properly. Although an explicit wave equation solution exists in the WAVE code, we describe here the difference equations that allow for δt to exceed the Courant limit by controlling the high frequencies allowed in the system[*Nielson and Lindman, 1972*]. The equations for Ampere's Law and Faraday's Law give rise to the wave equation for $\vec{A}$:

$$\begin{aligned} &\nabla^2(\vec{A}^{3/2} + 2\vec{A}^{1/2} + \vec{A}^{-1/2})/4 \\ &- (1 + \beta D^2\nabla^2)(\vec{A}^{3/2} - 2\vec{A}^{1/2} + \vec{A}^{-1/2})/c^2(\delta t)^2 = -\vec{J}_T^{1/2} \end{aligned} \tag{18}$$

where

$$\nabla^2\vec{A} = \frac{\vec{A}_{j+1,m} - 2\vec{A}_{j,m} + \vec{A}_{j-1,m}}{(\delta x)^2} + \frac{\vec{A}_{j,m+1} - 2\vec{A}_{j,m} + \vec{A}_{j,m-1}}{(\delta y)^2} \tag{19}$$

and

$$\frac{1}{D^2} = \frac{1}{(\delta x)^2} + \frac{1}{(\delta y)^2} \tag{20}.$$

The superscripts indicate the time levels of the various terms. These equations, of course, are solved for the 3/2 time level using the earlier values as input. The parameter β is used to modify the dispersion relation at large wavenumbers. The Coulomb gauge is used so that the transverse current $\vec{J}_T$, is given by:

$$\vec{J}_T = \vec{J} - \nabla\frac{\partial\phi}{\partial t}. \tag{21}$$

In order to guarantee that $\nabla\cdot\vec{A} = 0$ and $\nabla\cdot\vec{J}_T = 0$ so that charge is conserved at each time step, we must solve the auxiliary equation:

$$\nabla^2\frac{\partial\phi}{\partial t} = \nabla\cdot\vec{J}. \tag{22}$$

where ϕ is at the whole time step and $\vec{J}$ is at the half time step and use this $\partial\phi/\partial t$ in Eq. (21). The fields $\vec{E}$ and $\vec{B}$ are computed by:

$$\begin{aligned}\vec{E}^1 &= -\nabla\phi^1 - (\vec{A}^{3/2} - \vec{A}^{1/2})/\delta t \\ \vec{B}^1 &= \nabla \times (\vec{A}^{3/2} + \vec{A}^{1/2})/2\end{aligned} \tag{23}$$

The electrostatic potential is obtained from

$$\nabla^2\phi^1 = -\rho^1 \tag{24}$$

All fields, charge and currents are defined at cell centers. Given the currents at the half time step and the charge at the whole time step the fields can be advanced to the next whole time step with the vector potential being one half time step ahead.

The field equations, Eqs. (18) and (24), are solved with a direct two dimensional Poisson solve. The method proceeds by Fast Fourier Transform in the y direction followed by Gaussian elimination in the x direction on the linearly independent Fourier components. The result is then Fourier transformed back to physical space. This allows for arbitrary boundary conditions in x. General boundary conditions in y are obtained by choosing an appropriate set of basis functions for the FFT algorithm. Boundary conditions that have been implemented in the y direction besides periodic include Neumann and Dirichlet for $\vec{A}$ and ϕ. Even more general boundary conditions are available in x.

7.3. Particle Equations

The particle velocities are known at the half time step while $\vec{x}$, $\vec{E}$, and $\vec{B}$ are known at the whole time step. If we define $h = q\delta t/m$, the centered form for the equation of motion [*Morse and Nielson, 1971*] is:

$$\vec{v}^{3/2} = \vec{v}^{1/2} + h(\vec{E}^1 + \vec{v}^1 \times \vec{B}^1) \tag{25}$$

However, we do not yet know $\vec{v}^1$. By using $\vec{v}^1 = \frac{1}{2}(\vec{v}^{3/2} + \vec{v}^{1/2})$, we can solve this equation for $\vec{v}^{3/2}$ implicitly. We do this by taking the dot product and cross product with $\vec{B}^1$. The time centered reversible equation then is

$$\begin{aligned}\vec{v}^{3/2} = \frac{1}{1+\frac{h^2}{4}\vec{B}\cdot\vec{B}}\Bigg(&\vec{v}^{1/2}(1-\frac{h^2}{4}\vec{B}\cdot\vec{B}) + h(\vec{E}+\vec{v}^{1/2}\times\vec{B}) + \frac{h^2}{2}\vec{E}\times\vec{B} \\ &+ \frac{h^2}{2}\vec{B}(\vec{v}^{1/2}\cdot\vec{B}) + \frac{h^3}{4}\vec{B}(\vec{E}\cdot\vec{B})\Bigg)\end{aligned} \tag{26}$$

If we wish to keep terms only to order $(\delta t)^2$, the Eq. (26) becomes:

$$\vec{v}^{3/2} = \vec{v}^{1/2}(1-\frac{h^2}{2}\vec{B}\cdot\vec{B}) + h(\vec{E}+\vec{v}^{1/2}\times\vec{B}) + \frac{h^2}{2}\vec{E}\times\vec{B} + \frac{h^2}{2}\vec{B}(\vec{v}^{1/2}\cdot\vec{B}) \tag{27}$$

For the relativistic equations a similar expression can be derived. The position advancement is the same as in one dimension. The velocities are interpolated onto the grid at a position corresponding to the advance of one- half time step to obtain $J^{1/2}$. The charges are interpolated on the grid at the whole time step.

8. IMPLICIT TECHNIQUES

8.1. Motivation

In studying the nonlinear properties of many plasma systems we are primarily interested in the long time and long scale length plasma behavior. For example, in collisionless shocks we frequently are interested in time scales long compared to the ion gyro frequency and long compared to the ion gyro radius. For space plasmas there may be over 10^3 Debye lengths in an ion gyro radius and over 10^4 plasma periods in an ion gyro period. To resolve all these time and space scales is prohibitive even with the next generation supercomputers. There are some problems in which the high frequencies and short scale lengths communicate effectively with the long wavelength fluctuations. If this is the case, one cannot make any significant simplifications. However, frequently the highest frequencies and shortest scale lengths are not important for the evolution of the longest time and space scales. Here several solutions are possible.

A common technique is to solve a reduced set of equations for one of the components of the plasma system. This is usually referred to as a hybrid method. There are a variety of hybrid models that have had considerable success in modeling a wide variety of plasma phenomena. In recent years, however, implicit techniques have undergone considerable development [*Brackbill and Forslund, 1982; Friedman, et al., 1982; Langdon, et al., 1983*]. By implicit we mean that the equations are solved backward in time from time level $j+1$ to time level j. This results in a decaying of the solutions that would be unstably growing in an explicit formulation. A simple example of this was given above for the wave equation solved in the two dimensional electromagnetic code WAVE. Although this technique is well known in hydrodynamics and magnetohydrodynamics, it has only recently been found to be practical in plasma simulation. The difficulty arises because there are many more simultaneous equations to be solved in the plasma system. If one had to iterate on all these equations together, the cost would be prohibitive and there would be no advantage over simpler explicit methods.

8.2. Outline of Methods

There is not enough space here to discuss the details of the implicit methods that have been developed. Other papers in this volume will deal with this in more detail as will a forthcoming book in the Computational Techniques series by Academic Press entitled "*Multiple Timescales*" to be published in 1985. With some oversimplification it can be stated that it has been found that one does not have to iterate the simultaneous set of equations together to obtain a stable accurate system. There are several approaches to this problem, but with a modified set of equations for the particles and fields one can obtain a stable system. For example, in the moment method [*Mason, 1981; Denavit, 1981*] a fluid-like description of the plasma is solved for implicitly with conventional implicit techniques. The particles then are advanced with these new fields and used to correct the moments that were obtained from the fluid-like treatment. At the same time the implicit solution of the fluid equations over a large time step removes the high frequency noise injected by the particle moments. Thus the two stages serve to correct each other. It has been shown [*Brackbill and Forslund, 1982*] that this results in a systematic elimination of the short wavelength and high frequency waves in the system allowing for large time and space steps. This method has been particularly successful in studying the self-consistent turbulence generated in magnetic shock waves[*Quest, et al., 1983; Forslund, et al., 1984*]. The direct method [*Friedman, et al., 1981; Langdon, et al., 1983*] which is derived on more formal grounds has similar properties to the moment method although all the differences have yet to be discussed in the literature. In some of the formulations of these implicit methods there are still some time and space step constraints that limit the application of the method, although it is not clear if these are in principle limitations or simply implementation limitations.

9. PROBLEMS SOLVED

The number of problems that have been solved by means of particle simulation are too numerous to cite accurately in this limited space. However the range has extended from space physics to laser fusion to magnetic confinement fusion, from collisionless shock waves to parametric instabilities to electron transport. For example, an important mechanism causing an anomalous resistivity in magnetic shock waves has been identified with two dimensional implicit calculations with the implicit code, VENUS [*Forslund, et al., 1984*]. Magnetic field induced electron transport on the surface of laser fusion targets was first discovered with plasma simulation[*Forslund and Brackbill, 1982*]. Most of the information known about how energetic electrons are produced by intense laser light have been obtained by plasma simulation[*Forslund, et al., 1977*]. Recently in the area of particle acceleration by a beating between two light waves of different frequencies, limitations due to self-focusing, filamentation, back scattering and magnetic field generation have all been identified in simulation before they could be observed experimentally[*Joshi, et al., 1984*]. Frequently the results from simulations were unexpected but are easily understood in simple terms once they are known. This has been a typical path of research in plasma simulation, a field which has only begun to develop.

Acknowledgment

This work was performed under the auspices of the United States Department of Energy.

References

Birdsall, C. K. and D. Fuss, *J. Comput. Phys.*, **3**, 494, 1969.

Birdsall, C. K. and A. B. Langdon. *Plasma Physics Via Computer Simulation,* McGraw-Hill, New York, 1985.

Brackbill, J. U. and D. W. Forslund, *J. Comput. Phys.*, **46**, 271, 1982.

Brackbill, J. U. and D. W. Forslund, *Bull. Am. Phys. Soc.*, **29**, 1208, 1984.

Brackbill, J. U., D. W. Forslund, K. B. Quest, and D. Winske, *Phys. Fluids*, **27**, 2682, 1984.
Buneman, O., *J. Comput. Phys.*, **1**, 517, 1967.
Denavit, J., *J. Comput. Phys.*, **9**, 75, 1972.
Denavit, J., *J. Comput. Phys.*, **47**, 337, 1981.
Forslund, D. W. and J. U. Brackbill, *Phys. Rev. Lett.*, **48**, 1614, 1982.
Forslund, D. W., J. M. Kindel, and K. Lee, *Phys. Rev. Lett.*, **39**, 284, 1977.
Forslund, D. W., K. B. Quest, J. U. Brackbill, and K. Lee, *J. Geophys. Res.*, **89**, 2142, 1984.
Friedman, A., A. B. Langdon, and B.I. Cohen, *Comm. Plasma Phys. Contr. Fusion*, **6**, 225, 1981.
Godfrey, B. B., *J. Comput. Phys.*, **15**, 504, 1974.
Godfrey, B. B., *J. Comput. Phys.*, **19**, 58, 1975.
Joshi, C., W. B. Mori, T. Katsouleas, J. M. Dawson, J. M. Kindel, and D. W. Forslund, *Nature*, **311**, 525, 1984.
Langdon, A. B., *J. Comput. Phys.*, **6**, 247, 1970a.
Langdon, A. B., *Proc. 4th Conf. on Num. Sim. of Plasmas*, 1970b.
Langdon, A. B., *J. Comput. Phys.*, **12**, 247, 1973.
Langdon, A. B., and C. K. Birdsall, *Phys. Fluids*, **13**, 2115, 1970.
Langdon, A. B., and B. F. Lasinski, *Methods Comput. Phys.*, **16**, 327, 1976.
Langdon, A. B., B. I. Cohen, and A. Friedman, *J. Comput. Phys.*, **51**, 107, 1983.
Lewis, H. R., *J. Comput. Phys.*, **6**, 126, 1970.
Lin, A. T., J. M. Dawson, and H. Okuda, *Phys. Fluids*, **17**, 1995, 1974.
Lindman, E. L., *J. Comput. Phys.*, **5**, 13, 1970.
Mason, R. J., *J. Comput. Phys.*, **47**, 233, 1981.
Morse, R. L. and C. W. Nielson, *Phys. Fluids*, **12**, 2418, 1969.
Morse, R. L. and C. W. Nielson, *Phys. Fluids*, **14**, 830, 1971.
Nielson, C. W. and E. L. Lindman, *Proc. 6th Conf. on Num. Sim. of Plasmas*, 1972.
Okuda, H., *J. Comput. Phys.*, **10**, 475, 1972.
Quest, K. B., D. W. Forslund, J. U. Brackbill, and K. Lee, *Geophys. Res. Lett.*, **10**, 471, 1983.
Sakanaka, P. H., et al., *Phys. Fluids*, **14**, 611, 1971.

ELECTROMAGNETIC PARTICLE SIMULATION CODES

P. L. Pritchett
Department of Physics
University of California
Los Angeles, California 90024
U.S.A.

ABSTRACT. Electromagnetic particle simulations solve the full set of Maxwell's equations. They thus include the effects of self-consistent electric and magnetic fields, magnetic induction, and electromagnetic radiation. The algorithms for an electromagnetic code which works directly with the electric and magnetic fields are described. The fields and current are separated into transverse and longitudinal components. The transverse $\underset{\sim}{E}$ and $\underset{\sim}{B}$ fields are integrated in time using a leapfrog scheme applied to the Fourier components. The particle pushing is performed via the relativistic Lorentz force equation for the particle momentum. As an example, simulation results are presented for the electron cyclotron maser instability which illustrate the importance of relativistic effects on the wave-particle resonance condition and on wave dispersion.

1. INTRODUCTION

The early particle simulation codes were concerned only with the solution of Poisson's equation to determine the self-consistent electrostatic field. The magnetic field, if any, was treated as a fixed external field. While there are many interesting problems in plasma physics for which this is an adequate approximation, there are many other problems (such as instabilities in high-beta plasmas, shocks, and generation of radiation) where the inclusion of self-consistent magnetic fields and/or electromagnetic radiation is necessary. To handle such problems, we need a more complete model.

The most fundamental model retains all of Maxwell's equations,

$$\partial \underset{\sim}{B}/\partial t = -c\, \underset{\sim}{\nabla} \times \underset{\sim}{E} \;, \tag{1}$$

$$\partial \underset{\sim}{E}/\partial t = c\, \underset{\sim}{\nabla} \times \underset{\sim}{B} - (4\pi/c)\underset{\sim}{J} \;, \tag{2}$$

$$\underset{\sim}{\nabla} \cdot \underset{\sim}{E} = 4\pi\rho \;, \tag{3}$$

$$\underset{\sim}{\nabla} \cdot \underset{\sim}{B} = 0. \tag{4}$$

Space Science Reviews **42** (1985) 17–27. 0038–6308/85.15

Particle codes which solve these equations are termed "electromagnetic." Since these codes make no approximations in the basic physical laws (the effects of relativity can be included in the particle equations of motion), they constitute the most complete and fundamental model for numerical representation of a plasma. It should be noted, however, that this exact treatment of the basic laws of physics is not always desirable from a numerical standpoint. Other particle models are discussed in this volume by Hewett and Okuda.

In this chapter we shall concentrate on the issues raised by the extension from electrostatic to electromagnetic simulations. As general references on this subject which present further details, the reader should consult Langdon and Lasinski [1976], Buneman et al. [1980], Dawson [1983], and Birdsall and Langdon [1985]. Specific problems involved in formulating multidimensional codes are discussed in the chapter by Buneman in this volume. We have chosen to discuss algorithms which involve the fields $\underset{\sim}{E}$ and $\underset{\sim}{B}$ directly. It is also possible to construct a code which integrates the scalar and vector potentials ϕ and $\underset{\sim}{A}$ in time. There seem to be no fundamental reasons to choose one approach over the other [Langdon and Lasinski, 1976].

2. DESCRIPTION OF AN ELECTROMAGNETIC CODE

2.1 Solution of the Field Equations

Our solution of (1) - (4) will be obtained in Fourier space. The transformations between coordinate space and Fourier space are carried out via fast Fourier transforms (FFT). To this end, it is useful to introduce the decomposition of a vector field into longitudinal and transverse components. For an arbitrary vector $\underset{\sim}{A}(\underset{\sim}{k})$ these are defined by

$$\underset{\sim}{k} \cdot \underset{\sim}{A}_T(\underset{\sim}{k}) = 0 , \qquad \underset{\sim}{k} \times \underset{\sim}{A}_L(\underset{\sim}{k}) = 0 , \tag{5}$$

where the subscripts T and L stand for transverse and longitudinal. Explicitly, we have

$$\underset{\sim}{A}_L(\underset{\sim}{k}) = \underset{\sim}{k}(\underset{\sim}{k}\cdot\underset{\sim}{A})/k^2 , \tag{6}$$

$$\underset{\sim}{A}_T(\underset{\sim}{k}) = \underset{\sim}{A}(\underset{\sim}{k}) - \underset{\sim}{A}_L(\underset{\sim}{k}). \tag{7}$$

From (4) it is clear that the magnetic field has only a transverse component. The longitudinal part of $\underset{\sim}{E}$ is to be obtained at any time from (3), and this can be done exactly as in an electrostatic code,

$$\underset{\sim}{E}_L(\underset{\sim}{k},t) = - 4\pi i\underset{\sim}{k}\rho(\underset{\sim}{k},t)/k^2. \tag{8}$$

The time behavior of the transverse fields $\underset{\sim}{E}_T$ and $\underset{\sim}{B}$ is obtained from (1) and (2). In Fourier space these become

$$\partial \underset{\sim}{B}(\underset{\sim}{k},t)/\partial t = - ic\underset{\sim}{k} \times \underset{\sim}{E}_T(\underset{\sim}{k},t) \ , \tag{9}$$

$$\partial \underset{\sim}{E}_T(\underset{\sim}{k},t)/\partial t = ic\underset{\sim}{k} \times \underset{\sim}{B}(\underset{\sim}{k},t) - (4\pi/c)\underset{\sim}{J}_T(\underset{\sim}{k},t) \ , \tag{10}$$

where only the transverse component of the current appears in (10). If (9) and (10) are converted to finite difference equations in time, and if $\underset{\sim}{E}_T$ and $\underset{\sim}{B}$ are defined at staggered times, say $\underset{\sim}{B}$ at $n\Delta t$ and $\underset{\sim}{E}_T$ at $(n-1/2)\Delta t$ (here n is an integer and Δt is the time step), then the transverse fields can be advanced in time using a simple leapfrog scheme,

$$\underset{\sim}{E}_T(\underset{\sim}{k},n+1/2) = \underset{\sim}{E}_T(\underset{\sim}{k},n-1/2) + ic(\Delta t)\underset{\sim}{k} \times \underset{\sim}{B}(\underset{\sim}{k},n) - $$

$$- (4\pi/c)\Delta t \underset{\sim}{J}_T(\underset{\sim}{k},n) \ , \tag{11}$$

$$\underset{\sim}{B}(\underset{\sim}{k},n+1) = \underset{\sim}{B}(\underset{\sim}{k},n) - ic(\Delta t)\underset{\sim}{k} \times \underset{\sim}{E}_T(\underset{\sim}{k},n+1/2). \tag{12}$$

What determines the choice of Δt? As an example, let us apply (11) and (12) to the case of electromagnetic waves in vacuum ($\underset{\sim}{J}_T$ = 0). We assume the fields have the form

$$\underset{\sim}{E}_T(\underset{\sim}{k},t) = \underset{\sim}{E}_o(\underset{\sim}{k}) \exp[- i\omega(n-1/2)\Delta t] \ , \tag{13}$$

$$\underset{\sim}{B}(\underset{\sim}{k},t) = \underset{\sim}{B}_o(\underset{\sim}{k}) \exp[- i\omega n\Delta t]. \tag{14}$$

In the limit $\Delta t \to 0$, we should recover $\omega = \pm\, kc$ as the dispersion relation. Substituting (13) and (14) into (11) and (12), we obtain

$$k^2c^2 = 4 \sin^2(\omega\Delta t/2)/(\Delta t)^2. \tag{15}$$

This has the correct behavior as $\Delta t \to 0$. For finite Δt we see that there are real solutions for ω only if

$$k^2c^2(\Delta t)^2 < 4. \tag{16}$$

If this inequality is violated, there are only complex solutions, and the leapfrog integration algorithm is unstable. Thus the time step is limited by the largest k mode or highest-frequency electromagnetic mode that enters the problem. This is an example of the Courant-Friedrichs-Lewy [1928] condition.

In addition to being stable, we want the integration scheme to be accurate. Thus we are also interested in the magnitude of the frequency error given by (15). For small $kc\Delta t$, ω is given approximately by

$$\omega = \pm\, kc[1 + k^2c^2(\Delta t)^2/24 + \ldots]. \tag{17}$$

A typical compromise between the competing demands of accuracy and finite computing budgets is to choose $kc\Delta t \leq 0.2$ for all modes that are expected to be important.

2.2 Time Integration of the Particles

We discuss the particle advancement using the relativistic force equation. Thus we define $\underset{\sim}{u} \equiv p/m \equiv \gamma \underset{\sim}{v}$, where $\underset{\sim}{p}$ is the relativistic momentum and γ is the usual relativistic factor, $\gamma^2 = (1 - v^2/c^2)^{-1} = 1 + u^2/c^2$. m is the particle rest mass. The particle equations of motion are

$$d\underset{\sim}{u}_i/dt = (q_i/m_i)\left[\underset{\sim}{E}_i + (\underset{\sim}{u}_i \times \underset{\sim}{B}_i)/\gamma_i c\right] , \tag{18}$$

$$d\underset{\sim}{r}_i/dt = \underset{\sim}{u}_i/\gamma_i , \tag{19}$$

where the subscript i is the particle index. It is assumed here that the fields $\underset{\sim}{E}_i$ and $\underset{\sim}{B}_i$ have been interpolated from the spatial grid to the location of particle i.

We again construct a leapfrog scheme with the particle coordinates defined at $(n - 1/2)\Delta t$ and the momenta at $n\Delta t$. This time sequencing, together with that for the fields, is indicated in Figure 1. Converting (18) and (19) to finite difference equations, we obtain the scheme

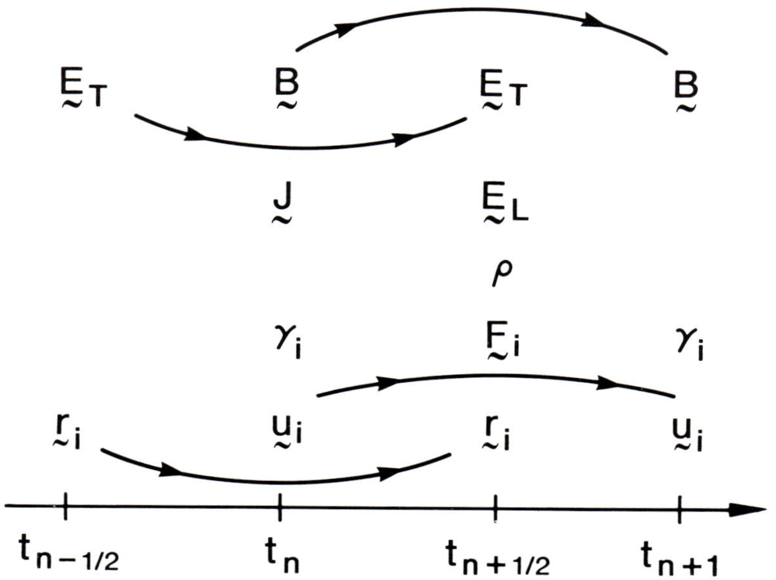

Figure 1. The time sequencing of particle and field quantities involved in advancing the electromagnetic simulation code through one cycle.

$$\mathbf{r}_i^{n+1/2} = \mathbf{r}_i^{n-1/2} + (\Delta t)\mathbf{u}_i^n/\gamma_i^n \,, \tag{20}$$

$$\mathbf{u}_i^{n+1} = \mathbf{u}_i^n + (q_i\Delta t/m_i)\left[\mathbf{E}_i^{n+1/2} + \right.$$
$$\left. + \left(\mathbf{u}_i^{n+1} + \mathbf{u}_i^n\right) \times \mathbf{B}_i^{n+1/2}/\left(2c\gamma_i^{n+1/2}\right)\right]. \tag{21}$$

In the $\mathbf{u} \times \mathbf{B}$ term we have taken $\mathbf{u}_i^{n+1/2} = \left(\mathbf{u}_i^{n+1} + \mathbf{u}_i^n\right)/2$. Likewise, $\mathbf{B}_i^{n+1/2}$ is to be computed from $\left(\mathbf{B}_i^{n+1} + \mathbf{B}_i^n\right)/2$.

The vector equations (21) constitute a set of three linear equations for the $\mathbf{u}_i^{n+1}$, and the explicit solutions can be obtained in a straightforward manner. Alternatively, a faster solution is available [Boris, 1970; see also Birdsall and Langdon, 1985]. If we define $\mathbf{u}_i^-$ and $\mathbf{u}_i^+$ by the relations

$$\mathbf{u}_i^n = \mathbf{u}_i^- - (q_i\Delta t/2m_i)\,\mathbf{E}_i^{n+1/2} \,, \tag{22}$$

$$\mathbf{u}_i^{n+1} = \mathbf{u}_i^+ + (q_i\Delta t/2m_i)\,\mathbf{E}_i^{n+1/2} \,, \tag{23}$$

and substitute these into (21), then the $\mathbf{E}_i^{n+1/2}$ terms cancel. We are left with

$$\mathbf{u}_i^+ - \mathbf{u}_i^- = \left(q_i\Delta t/2\gamma_i^{n+1/2} m_i c\right)\left(\mathbf{u}_i^+ + \mathbf{u}_i^-\right) \times \mathbf{B}_i^{n+1/2}. \tag{24}$$

This represents a rotation of the momentum about an axis parallel to $\mathbf{B}$ through an angle $\theta = -2\arctan(qB\Delta t/2\gamma mc)$. Thus the particle push can be accomplished by first adding half the electric impulse to $\mathbf{u}_i^n$ to obtain $\mathbf{u}_i^-$, rotating according to (24) to obtain $\mathbf{u}_i^+$, and adding the remaining half of the electric impulse to obtain $\mathbf{u}_i^{n+1}$.

The field and particle leapfrog schemes are time centered provided that $\mathbf{J}_T$ in (11) is known at $n\Delta t$ and γ_i in (21) is known at $(n+1/2)\Delta t$. These involve modifications to the time sequence indicated in Figure 1. Accumulation of the current on the spatial grid requires both the particle positions and velocities. The latter are known at $n\Delta t$, but the former are at $(n-1/2)\Delta t$. Since the velocity is far more important to the determination of the current than is the position, it is usually sufficient to estimate $\mathbf{r}_i^n$ as $\mathbf{r}_i^{n-1/2} + (\Delta t/2)\mathbf{v}_i^n$ for use only in calculating $\mathbf{J}_T^n$. γ_i is usually computed from $\gamma_i^2 = 1 + u_i^2/c^2$, but the momenta are defined at $n\Delta t$. $\gamma_i^{n+1/2}$ can be obtained, however, using the $\mathbf{u}_i^-$ introduced in (22). The result is correct to first order in Δt since the magnetic deflection does not change the particle's energy and hence does not change γ_i.

2.3 Summary of Basic Cycle

We summarize here the procedures for advancing the electromagnetic code through one cycle. At the beginning of the cycle one has $\underset{\sim}{r}_i^{n-1/2}$, $\underset{\sim}{u}_i^n$, $\underset{\sim}{E}_T^{n-1/2}$, $\underset{\sim}{B}^n$, and γ_i^n available.

(1) Advance $\underset{\sim}{r}_i^{n-1/2}$ to $\underset{\sim}{r}_i^n$ using $\underset{\sim}{v}_i^n = \underset{\sim}{u}_i^n/\gamma_i^n$; accumulate $\underset{\sim}{J}^n$ on grid; compute $\underset{\sim}{J}_T^n(\underset{\sim}{k})$.

(2) Advance $\underset{\sim}{r}_i^n$ to $\underset{\sim}{r}_i^{n+1/2}$ using $\underset{\sim}{v}_i^n$; accumulate $\rho^{n+1/2}$ on grid; compute $\rho^{n+1/2}(\underset{\sim}{k})$; solve for $\underset{\sim}{E}_L^{n+1/2}(\underset{\sim}{k})$.

(3) Advance $\underset{\sim}{E}_T^{n-1/2}(\underset{\sim}{k})$ to $\underset{\sim}{E}_T^{n+1/2}(\underset{\sim}{k})$ using $\underset{\sim}{B}^n(\underset{\sim}{k})$ and $\underset{\sim}{J}_T^n(\underset{\sim}{k})$.

(4) Advance $\underset{\sim}{B}^n(\underset{\sim}{k})$ to $\underset{\sim}{B}^{n+1}(\underset{\sim}{k})$ using $\underset{\sim}{E}_T^{n+1/2}(\underset{\sim}{k})$.

(5) Transform $\underset{\sim}{E}^{n+1/2}(\underset{\sim}{k})$ and $\underset{\sim}{B}^{n+1/2}(\underset{\sim}{k})$ to coordinate space.

(6) Interpolate fields onto particles; compute $\gamma_i^{n+1/2}$ from $\underset{\sim}{u}_i^n$ and $\underset{\sim}{E}_i^{n+1/2}$; form force $\underset{\sim}{F}_i^{n+1/2}$.

(7) Advance $\underset{\sim}{u}_i^n$ to $\underset{\sim}{u}_i^{n+1}$ using $\underset{\sim}{F}_i^{n+1/2}$; compute γ_i^{n+1}.

3. APPLICATION: THE ELECTRON CYCLOTRON MASER INSTABILITY

As an illustration of the use of electromagnetic particle simulations in space plasma physics, we discuss some recent work concerning the electron cyclotron maser instability. This is a process whereby electromagnetic fluctuations near the electron cyclotron frequency $\Omega_e = |e| B_o/mc$ (or its harmonics) are amplified through a gyroresonant interaction. The source of free energy is a population inversion in the electron velocity distribution function. This mechanism appears to be the most likely explanation for the generation of the earth's auroral kilometric radiation (AKR) [Melrose, 1976; Wu and Lee, 1979]. The region in which the most intense AKR is generated is characterized by a large plasma cavity with $\omega_{pe}/\Omega_e < 0.1$ [Calvert, 1981]. (Here, ω_{pe} is the plasma frequency). In this region the primary plasmasheet electrons have energies of a few keV.

The condition for a wave with frequency ω and parallel wavenumber $k_\parallel$ to resonate with an electron near Ω_e is

$$\omega - k_{\parallel} v_{\parallel} - \Omega_e/\gamma = 0. \quad (25)$$

This condition describes an ellipse in the $v_{\perp} - v_{\parallel}$ plane [Omidi and Gurnett, 1982; Melrose et al., 1982; Dusenbery and Lyons, 1982] centered at $v_{\parallel} = d$, $v_{\perp} = 0$, with

$$d/c = \omega k_{\parallel} c/(k_{\parallel}^2 c^2 + \Omega_e^2) , \quad (26)$$

with semimajor axis (in the $v_{\perp}$ direction)

$$a/c = [1 - \omega^2/(k_{\parallel}^2 c^2 + \Omega_e^2)]^{1/2} , \quad (27)$$

and with eccentricity

$$e = [k_{\parallel}^2 c^2/(k_{\parallel}^2 c^2 + \Omega_e^2)]^{1/2}. \quad (28)$$

The nonrelativistic limit ($\gamma \to 1$) of (25) defines the straight line $v_{\parallel} = (\omega - \Omega_e)/k_{\parallel}$. This limit is approached as $e \to 1$ so that the ellipse becomes highly elongated in the $v_{\perp}$ direction. Note, however, that for $k_{\parallel}^2 c^2/\omega^2 << 1$ the resonance ellipse is nearly circular. The nonrelativistic resonance condition is thus invalid for describing the gyroresonant interaction when $k_{\parallel}^2 c^2/\omega^2 << 1$ even if the velocity of the resonating electrons is small [Chu and Hirshfield, 1978]. Rather a "semi-relativistic" approximation must be used in which the v^2/c^2 terms in the expansion of γ in (25) are retained [Wu and Lee, 1979]. In this limit the growth or damping of the waves is determined by the sign of $\partial f/\partial v_{\perp}$ integrated around the resonant contour, where f is the electron distribution function. For an unstable distribution the R-X mode is expected to have larger growth rates than the L-O mode.

The R-X mode has a cutoff at

$$\omega_x = (\Omega_e/2) [1 + (1 + 4 \omega_{pe}^2/\Omega_e^2)^{1/2}] > \Omega_e. \quad (29)$$

Thus with $\omega > \omega_x$ on the R-X mode, we see from (27) that there is a minimum value of $k_{\parallel}$ in order for the resonance ellipse to exist,

$$(k_{\parallel} c)_{min} = (\omega^2 - \Omega_e^2)^{1/2}. \quad (30)$$

Thus, if cold-plasma dispersion is valid, R-X mode radiation cannot be generated by the cyclotron maser instability at exactly 90° to the magnetic field. Rather, for the conditions in the AKR generation region, the maximum emission is expected to occur at ~ 70°-75° to the field.

We have performed a number of electromagnetic simulations to investigate the growth and saturation of the maser instability. Here we shall consider one simple configuration which illustrates quite dramatically the effect of relativistic dispersion shifts. Examples which are more applicable to the generation of AKR are discussed elsewhere [Pritchett, 1984; Pritchett and Strangeway, 1985]. The initial electron distribution consists of a ring in perpendicular momentum, as illustrated in Figure 2a. The central value is $p_o/mc = 0.14$, and the thermal spread is $p_{te}/p_o = 1/8$. The ratio ω_{pe}/Ω_e is 0.05. This distribution clearly has a strong positive value of $\partial f/\partial p_{\perp}$ for $p_{\perp} \lesssim p_o$, and one would expect a strong instability. This indeed

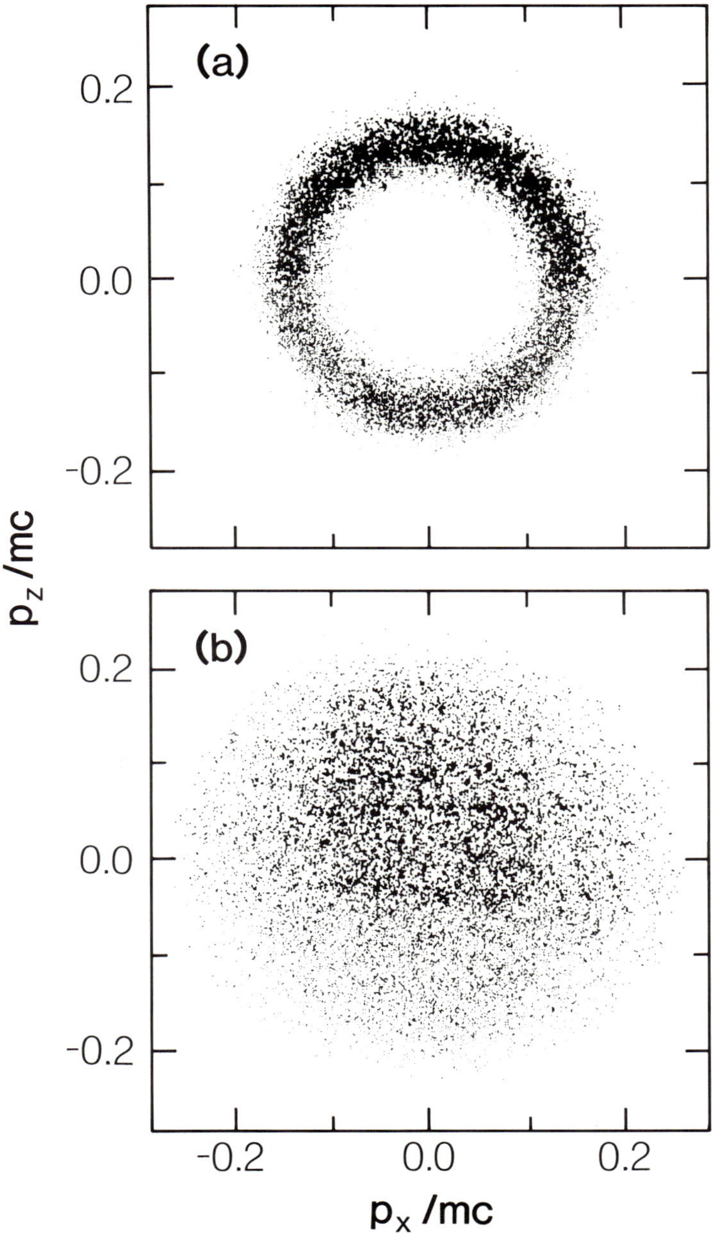

Figure 2. Electron distribution in $p_\perp$ space. (a) Initial distribution (t = 0). (b) Distribution at saturation ($\Omega_e t$ = 800).

occurs in the simulation, and Figure 2b shows the distribution at saturation. The ring distribution has completely diffused away.

A series of simulations with one spatial dimension and three velocity and field components (characterized as "one-and-two-halves" dimensional) was performed in which the angle θ between the magnetic field and the spatial axis was varied. Figure 3 shows the time history of the total electromagnetic energy for runs with $\theta = 70°$ and $\theta = 90°$. Contrary to expectations, the radiation at 90° is not negligible. In fact, the emission at 90° is the strongest of all angles. The radiation pattern is thus quite different from predictions based on cold-plasma theory. The radiation is observed, however, to have the polarization of the R-X mode. The emission has a very narrow bandwidth centered (for $\theta = 90°$) at $\omega/\Omega_e = 0.99$ and $kc/\Omega_e = 0.96$. Thus the phase velocity is greater than c. The dispersion is such that the unstable modes connect smoothly to $\omega = kc$ at larger values of k. The linear growth rate for the fastest growing mode is $\omega_i/\Omega_e = 6.5 \times 10^{-3}$, and 5% of the initial ring kinetic energy is converted into electromagnetic

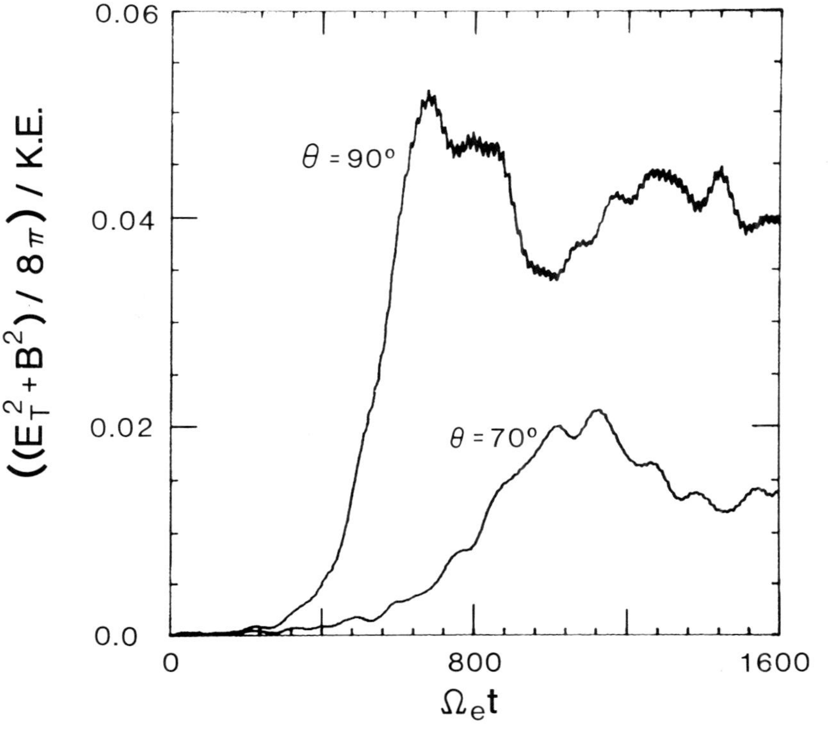

Figure 3. Time history of the total electromagnetic energy (normalized to the initial electron kinetic energy (K.E.)) in one-dimensional simulations with propagation angles of 70° and 90° to the magnetic field.

radiation. The electrostatic component of the instability is very weak. At 70° the growth rate and efficiency are about half the values at 90°.

Can these simulation results be explained in terms of the usual cyclotron maser instability? The fact that $\omega < \Omega_e$ for the unstable modes suggested that relativistic corrections to the dispersion relation might be important. When $\omega_{pe}/\Omega_e \ll 1$, the R-X cutoff approaches Ω_e,

$$\omega_x \approx \Omega_e \, (1 + \omega_{pe}^2/\Omega_e^2). \tag{31}$$

The dispersion relation for a relativistic Maxwellian distribution has been evaluated [e.g., Shkarofsky, 1966]. For $v_{te}/c \ll \omega_{pe}/\Omega_e \ll 1$, the correction to (31) is

$$\omega_x \approx \Omega_e \left[1 + \omega_{pe}^2/\Omega_e^2 - (5/2)v_{te}^2/c^2\right]. \tag{32}$$

Thus when $v_{te}/c \sim \omega_{pe}/\Omega_e$, we can expect a substantial modification of the cold-plasma cutoff. In fact, the R-X cutoff equals Ω_e for $v_{te}^2/c^2 = (2/3)\omega_{pe}^2/\Omega_e^2$ [Pritchett, 1984]. For larger values of v_{te}/c, the cutoff lies below Ω_e. Thus the simulation results are actually quite reasonable since v/c for the ring is several times ω_{pe}/Ω_e. The small shift of ω_x below Ω_e proves to be sufficient to drastically change the radiation pattern produced by the cyclotron maser instability. The frequency shift needed to satisfy the resonance condition (25) is here provided by relativistic effects rather than by a Doppler shift associated with finite $k_\parallel$.

The present discussion illustrates the role that simulations can play in revealing new phenomena in plasma physics. When confronted with an unexpected result in a simulation, it is incumbent upon the researcher to attempt to explain the result. This is the only way to distinguish between spurious results introduced by programming errors or numerical procedures and real physical effects. Research in plasma simulation is most successful when there is a mutual interplay between theory and the numerical experimentation.

ACKNOWLEDGMENTS. During the course of the present work the author has benefitted from useful discussions with M. Ashour-Abdalla, A. Baños,Jr., F.V. Coroniti, J.M. Dawson, V.K. Decyk, and R.J. Strangeway. This work was supported by NASA Solar Terrestrial Theory grant NAGW-78 and by National Science Foundation grants ATM 82-18746 and PHY 83-04641.

REFERENCES

Birdsall, C.K. and A.B. Langdon, Plasma Physics via Computer Simulation, McGraw-Hill, New York, 1985.

Boris, J.P., Relativistic plasma simulation-optimization of a hybrid code, in Proceedings of the Fourth Conference on Numerical Simulation of Plasmas, p.3, Naval Res. Lab., Washington, D.C., 1970.

Buneman, O., C.W. Barnes, J.C. Green, and D.E. Nielsen, Principles and capabilities of 3-D, E-M particle simulations, J. Comput. Phys., **38**, 1, 1980.

Calvert, W., The auroral plasma cavity, Geophys. Res. Lett., **8**, 919, 1981.

Chu, K.R., and J.L. Hirshfield, Comparative study of the axial and azimuthal bunching mechanisms in electromagnetic cyclotron instabilities, Phys. Fluids, **21**, 461, 1978.

Courant, R., K.O. Friedrichs, and H. Lewy, Über die partiellen Differenzengliechungen der mathematischen Physik, Math Ann., **100**, 32, 1928.

Dawson, J.M., Particle simulation of plasmas, Rev. Mod. Phys., **55**, 403, 1983.

Dusenbery, P.B., and L.R. Lyons, General concepts on the generation of auroral kilometric radiation, J. Geophys. Res., **87**, 7467, 1982.

Langdon, A.B., and B.F. Lasinski, Electromagnetic and relativistic plasma simulation models, in Methods in Computational Physics, Vol. 16, edited by B. Alder, S. Fernbach, M. Rotenberg, and J. Killeen, p. 327, Academic, New York, 1976.

Melrose, D.B., An interpretation of Jupiter's decametric radiation and the terrestrial kilometric radiation as direct amplified gyroemission, Astrophys. J., **207**, 651, 1976.

Omidi, N., and D.A. Gurnett, Growth rate calculations of auroral kilometric radiation using the relativistic resonance condition, J. Geophys. Res., **87**, 2377, 1982.

Pritchett, P.L., Relativistic dispersion, the cyclotron maser instability, and auroral kilometric radiation, J. Geophys. Res., **89**, 8957, 1984.

Pritchett, P.L., and R.J. Strangeway, A simulation study of kilometric radiation generation along an auroral field line, J. Geophys. Res., submitted, 1985.

Shkarosfsky, I.P., Dielectric tensor in Vlasov plasmas near cyclotron harmonics, Phys. Fluids, **9**, 561, 1966.

Wu, C.S. and L.C. Lee, A theory of the terrestrial kilometric radiation, Astrophys. J., **230**, 621, 1979.

ELIMINATION OF ELECTROMAGNETIC RADIATION IN PLASMA SIMULATION: THE DARWIN OR MAGNETOINDUCTIVE APPROXIMATION

Dennis W. Hewett
Physics Department
PO Box 5508, LLNL
Livermore, California 95540

ABSTRACT. For many astrophysical and magnetic fusion applications, the purely electromagnetic modes generated by real as well as simulation "plasma" fluctuations are a source of high frequency radiation that is often irrelevant to the physics of interest. Unfortunately, a numerical CFL stability limit prevents either making c infinite or Δt large while using the usual explicit Maxwell's equations for the fields. A modification of Maxwell's equations, which provides implicitly the field components, circumvents this problem. The solution is to neglect retardation effects so that the electromagnetic propagation speed is effectively infinite. The purely electromagnetic modes in this limit evolve "instantly" to a time-asymptotic configuration about the macroscopic plasma configuration at each new time level. The Darwin or magnetoinductive approximation effectively provides infinite propagation speeds for purely electromagnetic modes by converting Maxwell's equations from hyperbolic to elliptic in character. In practice, this is accomplished by neglecting the solenoidal part of the displacement current. The elimination of the CFL time step constraint more than offsets the substantially more complicated field solution that is required. The details of a numerical implementation of this model will be presented. Numerical examples will be given and extentions of the Darwin field solution to other plasma models also will be considered.

1. INTRODUCTION

The tremendous variety of collective behavior in plasma models with both electric and magnetic fields usually necessitates some approximations to narrow the range of phenomena treated by the model. This concentration of effort is desirable; there is no reason to follow low frequency behavior if it does not change appreciably during the time scales of interest. Similarly, high frequencies contained in full models can be so much faster that often, to good approximation, the phenomenon could have been represented by its "average" value. Numerical stability

This work was performed under the auspices of the U. S. Department of Energy by the Lawrence Livermore National Laboratory under Contract No. W-7405-Eng-48.

Space Science Reviews **42** (1985) 29–40. 0038–6308/85.15

requirements of fully explicit models frequently require small enough time steps to follow the rapid oscillations about the mean even though this behavior is of no particular interest to the physics under study.

A case in point, and the subject of this paper, is the purely electromagnetic behavior of plasmas. For many astrophysical and magnetic fusion applications the speed of light is effectively infinite; the kinetic behavior of the plasma and certainly the macroscopic movement of the plasma can be orders of magnitude slower than the transit time of light. The purely electromagnetic modes in these cases evolve "instantly" to a time-asymptotic configuration about the macroscopic plasma configuration at each new time level.

It is desirable to solve the field equations in the limit of instantaneous propagation and thus avoid retardation effects and the resulting electromagnetic modes. Unfortunately, as demonstrated in the next section, if the electric and magnetic fields are governed by the full set of Maxwell's equations, the desired limit cannot be obtained by setting the velocity of light c to infinity. If c is set to ∞ in Maxwell's equations, nothing more than electrostatics is retained. The intent of this paper is to show how to eliminate purely electromagnetic modes by neglecting retardation while properly maintaining magnetic and inductive effects.

There is an additional advantage of neglecting the pure electromagnetic modes, as Nielson and Lewis point out, (Nielson and Lewis, 1976). In simulations representing the plasma as macroscopic particles, anomalously large fluctuations result from the relatively small number of simulation compared to real particles representing the plasma. These fluctuations may drive much more than a realistic amount of energy into purely electromagnetic modes, which can obscure and distort interesting physics.

In following sections, a discussion of a plasma model which treats the field quantities in the Darwin (Kaufman and Rostler, 1971; Nielson and Lewis, 1976) or magnetoinductive (Busnardo-Neto et. al.,1977) approximation is presented. Entailing much more than a simple neglect of the displacement current, this approximation allows the fields to be obtained implicitly, effectively defeating the rather restrictive Courant, Friedrichs and Lewy (CFL) condition (Richtmeyer and Morton, 1976) which requires $c\ \Delta t \leq \Delta x$. In effect this approximation is accomplished by neglecting only the solenoidal part of the displacement current--which leads to an elliptic rather than a hyperbolic set of field equations. The cost of this improvement, while nominal compared to the price paid for CFL-limited small Δt runs, is a substantially more complicated field solution.

In the next section, the details of a numerical implementation of this model is presented. The task is shown to be a straightforward solution of the Darwin-modified model. The biggest numerical difference is the iteration needed for the inductive electric field solution, and the new conceptual difference is that more boundary conditions need to be specified. Complex geometries, such as those with internal conductors, require consideration of the interrelationship of these conditions. Extentions of the Darwin field solution to other plasma models is also considered.

2. THE DARWIN APPROXIMATION

The essence of the Darwin approximation can be distilled easily from the complete Maxwell's equations

$$\nabla \times \mathbf{B} = 4\pi/c\ \mathbf{J} + 1/c\ \partial\mathbf{E}/\partial t \tag{1a}$$

$$\nabla \cdot \mathbf{B} = 0 \tag{1b}$$

$$\nabla \times \mathbf{E} = -\ 1/c\ \partial\mathbf{B}/\partial t \tag{1c}$$

$$\nabla \cdot \mathbf{E} = 4\pi\ \rho \tag{1d}$$

in which $\mathbf{E}$ and $\mathbf{B}$ are the electric and magnetic fields and ρ and $\mathbf{J}$ are the plasma source terms which are obtained by integrating over the ensemble of moving point charges. Although instantaneous electromagnetic (EM) propagation is an important part of the Darwin model, setting the propagation speed c to infinity in Maxwell's equations eqs. 1 leaves only electrostatic fields (ES), governed by eq. 1d. From eqs. 1a–1b the $\mathbf{B}$ field can only exist as a constant in both time and space and the inductive part of $\mathbf{E}$, governed by eq. 1c, also has only the trivial solution. Thus the most straightforward technique for enforcing instantaneous propagation ($c\rightarrow\infty$) removes all nontrivial magnetic and inductive phenomena from the field equations.

This assumption is far too restrictive for most present day topics of research; even time-independent inhomogeneous magnetic configurations such as magnetic confinement or field shear cannot be self-consistently represented in this limit. A model containing more physics than the ES model but less than the full EM model is the object of this report. A commonly used approximation is to neglect all of the displacement current (the $\partial\mathbf{E}/\partial t$ term in eq. 1a). This assumption results in a low frequency model that certainly does not support purely EM modes since the coupling between the inductive $\mathbf{E}$ and $\mathbf{B}$ no longer exists. As will be shown, however, this model is inconsistent with charge continuity; serious errors can result when applying this model to nonneutral plasmas.

The Darwin model provides an avenue into the desired physics regime while avoiding the continuity problem. To introduce the Darwin model, it is convenient to consider Maxwell's equations written in terms of the scaler and vector potentials Φ and $\mathbf{A}$. These equations are, in the Coulomb guage $\nabla\cdot\mathbf{A} = 0$,

$$\nabla^2 \Phi = - 4\pi \rho \tag{2a}$$

$$\nabla^2 \underset{\sim}{A} - 1/c^2 \; \partial^2 \underset{\sim}{A}/\partial t^2 = - 4\pi/c \; \underset{\sim}{J} + 1/c \; \underset{\sim}{\nabla}\dot{\Phi} \; . \tag{2b}$$

The $\dot{}$ signifies a time derivative and

$$\underset{\sim}{B} = \underset{\sim}{\nabla} \times \underset{\sim}{A} \tag{2c}$$

and

$$\underset{\sim}{E} = - \underset{\sim}{\nabla}\Phi - 1/c \; \dot{\underset{\sim}{A}}. \tag{2d}$$

Formally the solutions can be obtained by integration and expressed by (Jackson, 1962)

$$\Phi(\underset{\sim}{r},t) = \int d\underset{\sim}{r}' \; 4\pi \; \rho(\underset{\sim}{r}',t)/|\underset{\sim}{r}-\underset{\sim}{r}'| \tag{3a}$$

$$\underset{\sim}{A}(\underset{\sim}{r},t) = 1/c \int d\underset{\sim}{r}' \; [4\pi \; \underset{\sim}{J} - \underset{\sim}{\nabla}\dot{\Phi}]_{ret}/|\underset{\sim}{r}-\underset{\sim}{r}'| \tag{3b}$$

where "ret" means that the quantity is to be evaluated at position $\underset{\sim}{r}'$ and time

$$t' = t - |\underset{\sim}{r}-\underset{\sim}{r}'|/c \; . \tag{4}$$

Eq. 3a expresses the concept that given the instantaneous charge density, the instantaneous self-consistent ES potential Φ can be obtained by integration over all space. The "propagation speed" is infinite. This is not the case in eq. 3b for the vector potential $\underset{\sim}{A}$ which, at a given position $\underset{\sim}{r}$ and time t, is determined by those currents at position $\underset{\sim}{r}'$ and retarded time t' such that their existence has just been communicated to position $\underset{\sim}{r}$ at time t. The time t' at which the current at $\underset{\sim}{r}'$ must be evaluated is retarded from t by the time it takes the signal to cross the distance $|r-r'|$.

The essence of the Darwin model is to neglect this retardation. Eq. 3b is replaced by

$$\underset{\sim}{A}(\underset{\sim}{r},t) = 1/c \int d\underset{\sim}{r}' \; [4\pi \; \underset{\sim}{J} - \underset{\sim}{\nabla}\dot{\Phi}]/|\underset{\sim}{r}-\underset{\sim}{r}'| \; . \tag{5}$$

so that, like 3a, the instantaneous current at position $\underset{\sim}{r}'$ and time t contributes to $\underset{\sim}{A}(\underset{\sim}{r},t)$. The effective propagation speed is infinite but c is finite. (Equivalently, only the c in eq. 4 is assumed infinite.) By analogy with eqs. 2a and 3a, eq. 5 is the solution of

$$\nabla^2 \underset{\sim}{A} = - 4\pi/c \ \underset{\sim}{J} + 1/c \ \underset{\sim}{\nabla}\dot{\Phi} \ . \tag{6}$$

The approximation can be expressed in a different form by noting that all vector fields can be expressed as the sum of irrotational and solenoidal parts, denoted here by subscripts ℓ and t respectively. The irrotational part can be obtained from the gradient of a scaler; the solenoidal part, from the curl of a vector. This means that irrotational vectors have no curl; solenoidal vectors have no divergence. Eq. 2d reveals that the $\underset{\sim}{\nabla}\dot{\Phi}$ term in eq. 6 is the irrotational part of the displacement current. Since $\underset{\sim}{A}$ is itself purely solenoidal (remember $\underset{\sim}{\nabla}\cdot\underset{\sim}{A} = 0$ in Coulomb guage), $\nabla^2\underset{\sim}{A} = - \underset{\sim}{\nabla}\times\underset{\sim}{\nabla}\times\underset{\sim}{A} + \underset{\sim}{\nabla}(\underset{\sim}{\nabla}\cdot\underset{\sim}{A})$ is also purely solenoidal, implying that the irrotational part of the right-hand-side of eq. 6 must vanish. This condition is also easily derivable from the continuity equation and Poisson's equation eq. 1d. Models which neglect all the displacement current force the plasma current to be purely solenoidal which is inconsistent with the continuity requirement in non-neutral plasmas. Thus, the Darwin limit of eqs. 2 can be written

$$\nabla^2 \Phi = - 4\pi \ \rho \tag{7a}$$

$$\nabla^2 \underset{\sim}{A} = - 4\pi/c \ \underset{\sim}{J}_t \ . \tag{7b}$$

The irrotational part of $\underset{\sim}{E}$ is obtained from

$$\underset{\sim}{E}_\ell = - \underset{\sim}{\nabla}\Phi \tag{7c}$$

and the solenoidal part $\underset{\sim}{E}_t$ can be obtained from eq. 2d and the time derivative of eq. 7b

$$\nabla^2 \underset{\sim}{E}_t = 4\pi/c \ \dot{\underset{\sim}{J}}_t \tag{7d}$$

The only difference between eqs. 2 and 7 is that the solenoidal part of the displacement current has been dropped on the right-hand-side of eq. 7b. Another identical statement is that the Darwin limit entails neglecting just the solenoidal part of the displacement current in Ampere's law eq. 1a. If all the displacement were neglected, the right-hand-side of 7b would retain a relatively large irrotational contribution which for consistency must go to zero. Careful numerical implementation is required to guarantee consistency in the Darwin model and is essential to avoid driving spurious energy into the magnetic field as the low frequency approximation breaks down and inductive effects become important.

In summary, the Darwin limit neglects only the solenoidal part of the displacement current in Ampere's law. This limit provides a model which is identical to the full Maxwell's equations through terms of order $(v/c)^2$. The first missing term is the v/c correction to the solenoidal part of $\underset{\sim}{J}_{ret}$ in eq. 3b. The expression for $\underset{\sim}{A}$ in eq. 5 then differs from the correct expansion 3b by terms of order $(v/c)^2$. Finally, before any fields determined by $\underset{\sim}{A}$ interact with the particles, they are multiplied again by v/c. Terms neglected by the Darwin approximation are therefore of order $(v/c)^3$ or higher.

3. NUMERICAL IMPLEMENTATION

The most important point to remember in implementing a Darwin field model is that the modified equations are no longer hyperbolic. The concept of instantaneous propagation, reflected in the elliptic eqs. 7, is evident: given the charge and current densities ρ and $\underset{\sim}{J}$, the self-consistent fields $\underset{\sim}{E}$ and $\underset{\sim}{B}$ can be found if the appropriate boundary conditions are specified. As Nielson and Lewis point out, no finite differencing in time is required or allowed.

Given an adequate Poisson solver, the issues to be dealt with in this section are largely reduced to a question of the proper boundary conditions. As is now shown, even the required decomposition of vectors is reduced to a question of proper boundary conditions. We find, for example, $\underset{\sim}{J}_t$ from $\underset{\sim}{J}$ for use in eq. 7b by solving

$$\nabla^2\psi = - \underset{\sim}{\nabla}\cdot\underset{\sim}{J} \tag{8a}$$

so that

$$\underset{\sim}{J}_t = \underset{\sim}{J} - \underset{\sim}{J}_\ell = \underset{\sim}{J} + \underset{\sim}{\nabla}\psi \ . \tag{8b}$$

The boundary conditions needed for eq. 8a take on physical significance if we realize that specifying the normal derivative of ψ on the simulation boundary is the same as specifying the normal component of the irrotational current on the boundary. Gauss's law eq. 1d gives a further constraint here because the integral of $\underset{\sim}{J}_\ell\cdot\hat{\mathbf{n}}$ ($\hat{\mathbf{n}}$ is the normal vector to the surface) around the boundary must equal the rate at which the net charge within the simulation region is changing.

With $\underset{\sim}{J}_t$ now calculated, $\underset{\sim}{A}$ can now be computed from eq. 7b with boundary conditions which are physically motivated by specifying magnetic fields or fluxes on the simulation boundary. Notice from eq. 2c that a tangential component of $\underset{\sim}{B}$ implies a normal derivative of $\underset{\sim}{A}$. After $\underset{\sim}{A}$ is computed in the simulation region, eq. 2c provides the $\underset{\sim}{B}$ field everywhere else.

The electrostatic or irrotational part of $\underset{\sim}{E}$ is obtained through eqs. 7a and 7c. This calculation is reminiscent of eq. 8a because the equations have similar form and the normal component of the gradient of the scaler potential provides the physical significant boundary condition. For consistency, $\underset{\sim}{E}_\ell\cdot\hat{\mathbf{n}}$ integrated around the simulation region must equal the net charge contained in the simulation region.

The equation for the solenoidal part of $\mathbf{E}$, eq. 7d, follows the same pattern for boundary conditions. Frequently it is most convenient to specify the functional value $\mathbf{E}_t$ on the boundary rather than a condition on normal derivatives. Again a consistency relation is given by Stoke's law relating the integral of $\mathbf{E}_t$ at the boundary to the rate at which the net magnetic flux through the simulation surface changes. Assuming we know $\dot{\mathbf{J}}_t$, the solution of eq. 7d is relatively easy.

Unfortunately, $\dot{\mathbf{J}}_t$ is not as easy to obtain as the other source terms. The temptation to express $\dot{\mathbf{J}}_t$ as a finite difference in time has been the source of much difficulty: temporal finite differencing in the field equations is inconsistent with "instantaneous propagation". Nielson was among the first to realize that if more information about each particles kinetic tensor is determined, the divergence of this tensor may be accumulated directly for each species much the same way a species density of current is obtained. Following Nielson and Lewis, we see that an expression of $\dot{\mathbf{J}}$ for each species is

$$\partial \mathbf{J}_s/\partial t = \mathbf{D}_s + q_s/m_s \; (\rho_s \; \mathbf{E} + \mathbf{J}_s/c \times \mathbf{B}) \tag{9}$$

where $\mathbf{D}_s$ is the vector that results from the divergence of the kinetic energy tensor

$$\mathbf{D}_s = - \nabla \cdot \int q_s \; \mathbf{v}_p \; \mathbf{v}_p \; f_s \; d\mathbf{v} \; . \tag{10}$$

Another expression for $\dot{\mathbf{J}}_s$ can be obtained by differentiating

$$\mathbf{J}_s = q_s \sum_p \mathbf{v}_p \; S(\mathbf{x}_g - \mathbf{x}_p) \quad , \tag{11}$$

where S is the particle shape factor, to obtain

$$\dot{\mathbf{J}}_s(\mathbf{x}_g) = - q_s \sum_p \mathbf{v}_p \mathbf{v}_p \cdot \nabla S(\mathbf{x}_g - \mathbf{x}_p)$$
$$+ q_s^2/m_s \sum_p (\mathbf{E}(\mathbf{x}_p) + \mathbf{v}_p/c \times \mathbf{B}(\mathbf{x}_p)) \; S(\mathbf{x}_g - \mathbf{x}_p) \; . \tag{12}$$

We have used the Lorentz force and Newton's laws to eliminate $\dot{\mathbf{v}}_p$. A comparison of eqs. 9 and 12 indicates how an expression for $\mathbf{D}_s$ may be obtained.

By accumulating a $\mathbf{D}_s$ for each species in the particle pusher, we can obtain an expression for $\dot{\mathbf{J}}$ by summing eq. 9 over species. We now must find the solenoidal part using the same procedure outlined for other steps in this calculation. An exception in this case compared with eq. 8 is that now we must specify the normal component of the time derivative of $\mathbf{J}_\ell$ rather than $\mathbf{J}_\ell$ itself.

A final complication is that the expression for $\dot{\mathbf{J}}_t$ contains the unknown $\mathbf{E}_t$. We cannot algebraically solve for $\mathbf{E}_t$ and thus find $\mathbf{E}_t$ directly because the right-hand-side must be reduced to just its solenoidal component. As a consequence iteration is required in which the latest value of $\underset{\sim}{E}_t$ is used in the expression for $\dot{\underset{\sim}{J}}_t$ in eq. 9 prior to its decomposition. Other techniques can be employed to speed convergence but in practice the iteration has converged quickly.

4. EXAMPLES

One use to which Darwin codes are particularly well suited is the investigation of the stability of the sheath regions of magnetically confined plasmas. The interface or sheath between the region of high density and relatively low $|\underset{\sim}{B}|$ and the region with low ρ and high $|\underset{\sim}{B}|$ typically must carry a diamagnetic current whose magnitude is inversely proportional to the thickness of the sheath. A wide variety of kinetic instabilities exist in such a region--the source of free energy being the localized drift kinetic energy. A typical approach is to initialize the Darwin code with an inhomogeneous density with a corresponding magnetic field such that pressure balance exists. The electron distribution is now constructed such that the electrons carry the required self-consistent current. Details of this method can be found in references 1 and 5 and references therein.

The Darwin code is now used to advance the equilibrium in time. Figure 1 indicates the evolution of Φ and B_z, shown by contour plots in Fig. 1a and 1b, respectively. The ripples in the two quantities are localized about the region of strong electron drift and grow to large amplitude. The instability has been identified as the lower hybrid drift (LHD) instability (Winske and Hewett, 1975; Winske and Liewer, 1978). Run parameters are ω_{pe} $\Delta t=0.2$, $N_x=40$, $N_y=128$, $0<x<8$ c/ω_{pe}, $0<y<25$ c/ω_{pe}, giving c $\Delta t/\Delta x = 1.0$. The principal advantage to the Darwin model here is the reduced EM fluctuation levels. Full EM codes need many times more particles/cell to reduce noise levels so that the small collective behavior is not destroyed by large simulation "particle" fluctuations.

A second example is a recent application (Boyd and Hewett, 1984) to the generation of relativistic electron beams. This application requires the Darwin model to be implemented in r-z geometry and as shown in Fig. 2 requires the application of internal boundary conditions to mock up the shapes of the anodes. Considerable care is required to properly set the boundary conditions required for the vector decompositions on these internal conductors. The electrons are emitted from the cathode in such a way that the surface irrotational electric field is held to zero; that is, enough electrons are "born" in cells adjacent to the cathode so that the $\underset{\sim}{E}_\ell$ at the boundary is zero.

Ultimately the relativistic electron velocities in the axial direction approach C. Gains to be realized by the Darwin model are two-fold. The mesh cell is much elongated in the axial direction to accommodate the much greater axial dimension of the problem. The time step requirement for particle motion accuracy is that $c\ \Delta t < \Delta z$; the CFL condition for a full EM model is that $c\ \Delta t < \min(\Delta r, \Delta z)$--enabling the Darwin model to run with Δt 2 to 5 times larger. The other advantage is

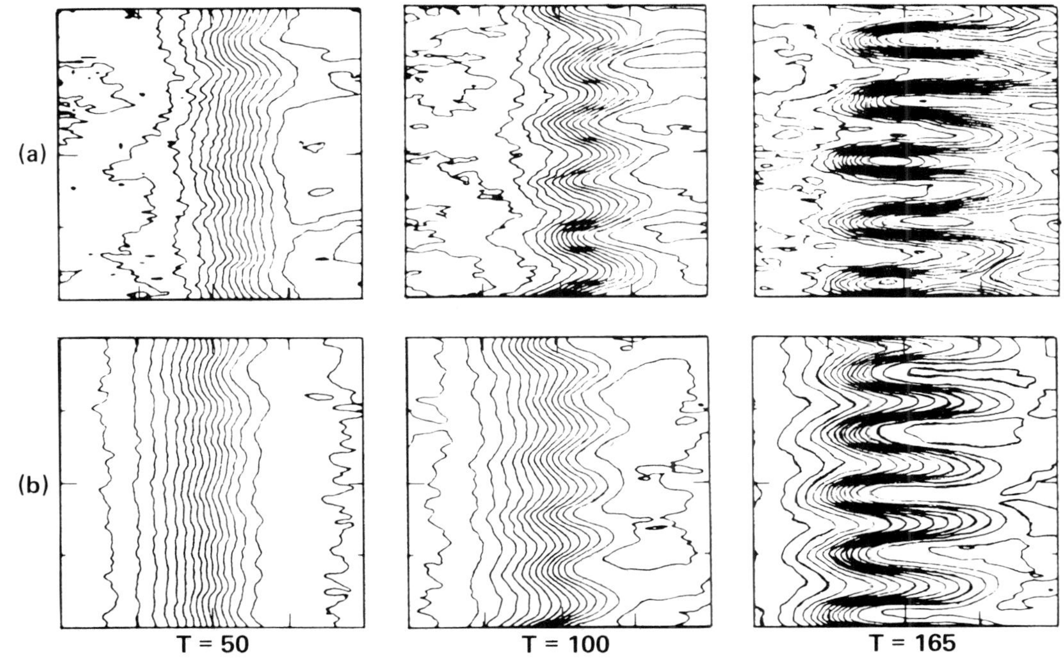

Figure 1. Results of a 2-D Darwin simulation at times 50, 100, and 165 ω_{pe}^{-1} showing (a) contours of constant electrostatic potential Φ and (b) contours of constant magnetic field B_z, both displayed in the x-y computational rectangle of dimensions 8 C/ω_{pe} wide and 25 C/ω_{pe} high (from Nielson and Lewis, 1976).

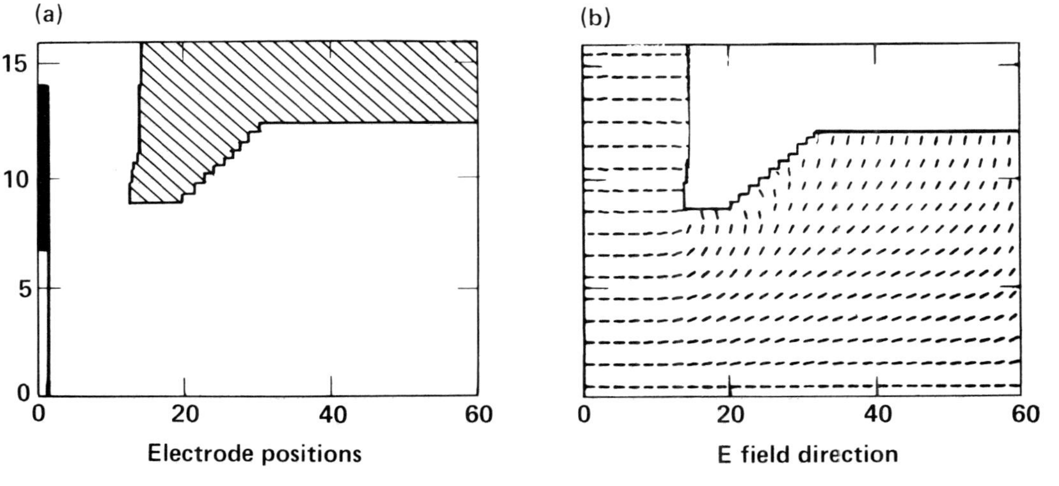

Figure 2. Axisymmetric r-z simulation region for Darwin simulation of an electron gun. (a) shows the cathode-anode placement and (b) gives the initial electric field direction. (courtesy of J.K. Boyd)

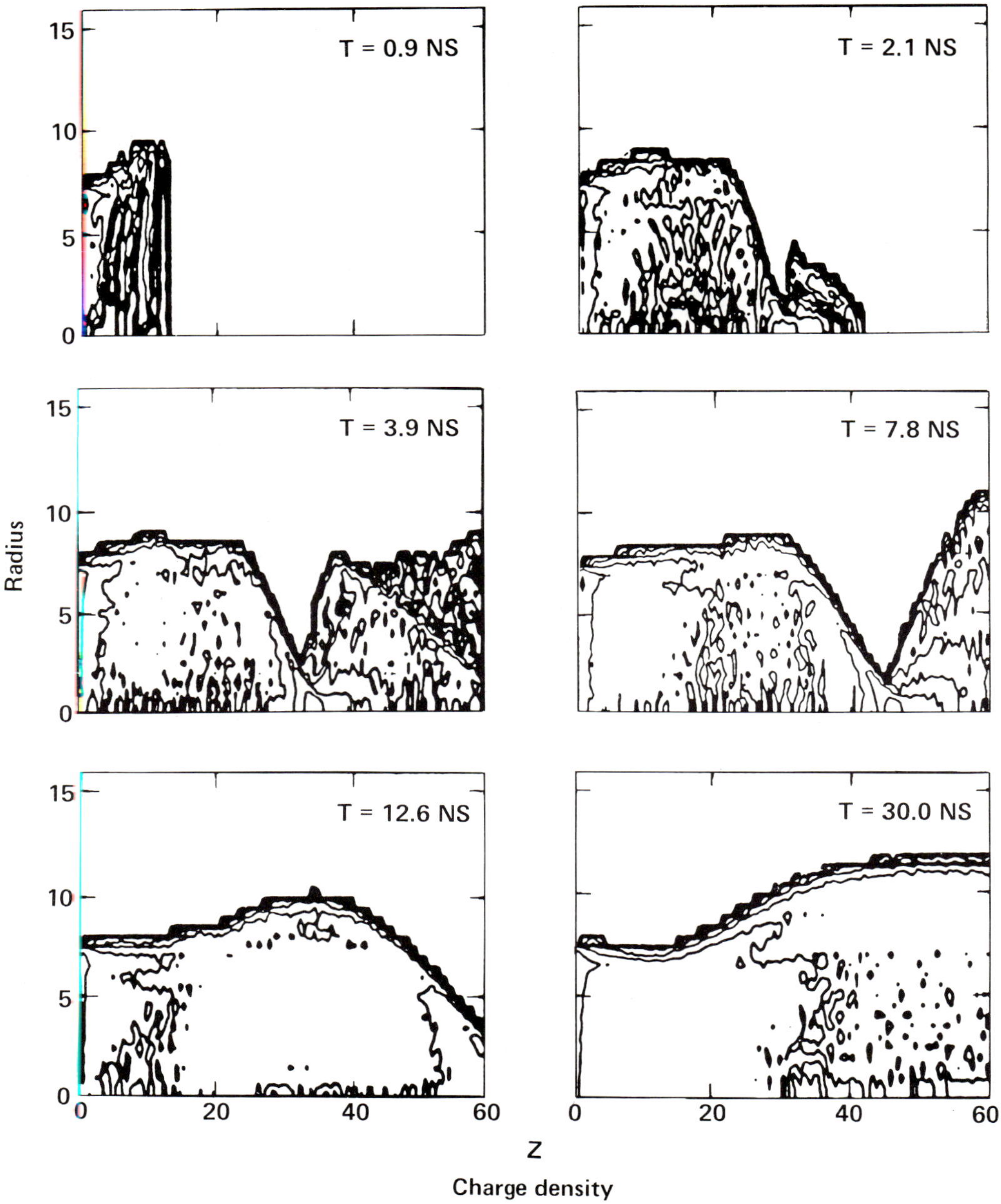

Figure 3. Time evolution of the electron charge density for the electron gun configuration shown in fig. 2. (courtesy of J.K. Boyd)

again that the Darwin model eliminates "excessive bremstrahlung" which allows significantly fewer simulation particles to be used. Figure 3 shows the electron charge density as it evolves to fill the initially-evacuated axisymmetric gun. Notice in Fig. 3b-3e the formation and subsequent evolution of a virtual cathode. This run is made with $\Delta t = 2.0 \times 10^{-11}$ seconds, $N_r=33$, $N_z=121$ and $\Delta r = \Delta z = 0.5$ cm. The applied voltage is 2.4 MeV which is turned on with a rise time of 20 ns.

5. EXTENSION TO FLUID ELECTRONS

The early simulations of the LHD instability with the Darwin model were started with exceedingly narrow sheath equilibria so that the local electron drift velocity was quite large and so, consequently, was the LHD growth rate. The overriding physics issue in these studies is the determination of the saturation level and mechanism. As the initial sheath width increases to a more physically reasonable size, the growth rate decreases as well. As the growth rate decreases simulations require more particles/cell so that the inherent simulation noise level does not disrupt the early collective behavior of the instability. Increased particles/cell requirements are directly proportional to increased computational requirements.

A solution to this dilemma (Hewett and Nielson, 1978) is to eliminate electron fluctuations by treating the electrons as a fluid. Further, as the thickness of the sheath increases, the assumption of quasi-neutrality becomes quite appropriate. As a consequence of this assumption, $\rho_e \sim -\rho_i$ and the irrotational electron current can be determined from $\underset{\sim}{\nabla}\cdot\underset{\sim}{J} \sim 0$ since $\underset{\sim}{J}_i$ is known from the PIC ion process. The solenoidal electron current $\underset{\sim}{J}_{et}$ can be advanced in time using the solenoidal part of the electron momentum eq. 9. The Darwin field equations can be solved just as before with one exception: $\underset{\sim}{E}_\ell$ can no longer be obtained from Poisson's eq. 1d in the quasi-neutral limit. Fortunately a quasi-neutral Poisson replacement can be obtained by summing eq. 9 over all components. Taking the divergence of this sum and using $\underset{\sim}{\nabla}\cdot\dot{\underset{\sim}{J}} \sim 0$, we have

$$\underset{\sim}{\nabla}\cdot \mu \underset{\sim}{\nabla}\Phi = - \underset{\sim}{\nabla}\cdot(D_i+D_e - \mu \underset{\sim}{E}_t + \underset{\sim}{\zeta} \times \underset{\sim}{B}) \qquad (13)$$

where

$$\mu = \sum_s q_s\rho_s \; (1/m_i + 1/m_e)$$

$$\zeta = \sum_s q_s/c \; (\underset{\sim}{J}_i/m_i + \underset{\sim}{J}_e/m_e) \quad .$$

Eq. 13 replaces eq. 7a in the quasi-neutral limit. A final requirement is that since Φ is now a function of $\underset{\sim}{E}_t$ and since $\underset{\sim}{E}_t$ is still a function of $\underset{\sim}{E}_\ell$ in this model, an overall iteration is required over eqs. 7d and 13. Details can be found in the literature (Hewett and Nielson, 1978) and a more recent example of the model's use (Gladd, Sgro and Hewett, 1984).

6. REFERENCES

C.W. Nielson and H.R. Lewis, Methods in Computational Physics, 16 (1976), pg. 367-388.

A.N. Kaufman and P.S. Rostler, Phys. Fluids, 14 (1971), pg. 446.

J. Busnardo-Neto, P.L. Pritchett, A.T. Lin, and J.M. Dawson, J. Comp. Phys., 23 (1977), pg. 300-312.

R.D. Richtmeyer and K.W. Morton, Difference Methods for Initial-Value Problems, Interscience Publishers (1976).

J.D. Jackson, Classical Electrodynamics, J. Wiley (1962), p. 186.

D. Winske and D.W. Hewett, Phys. Rev. Let., 35 (1975), pg. 937.

D. Winske and P. Liewer, Phys. Fluids, 21 (1978), pg. 1017.

J.K. Boyd and D.W. Hewett, Bulletin APS (1984), paper 9x14.

D.W. Hewett and C.W. Nielson, J. Comp. Phys. 29 (1978), pg. 219.

N.T. Gladd, A.G. Sgro, and D.W. Hewett, Phys. of Fluids submitted (1984).

PARTICLE SIMULATION MODELS FOR LOW FREQUENCY MICROINSTABILITIES IN A MAGNETIC FIELD

H. Okuda
Plasma Physics Laboratory, Princeton University
P.O. Box 451, Princeton, New Jersey 08544 USA

ABSTRACT. Several numerical plasma simulation models using particles are described which are appropriate for low frequency electrostatic and electromagnetic microinstabilities in a strong magnetic field. The model makes use of the guiding center drift approximations for the electrons while the ions are represented either as particles obeying the equation of motion with the full Lorentz force or a fluid which includes finite-Larmor-radius effects. These models are particularly useful for studying low frequency microinstabilities ($\omega \lesssim \omega_{pi}$, Ω_i) propagating nearly perpendicular to an external magnetic field ($k_\perp >> k_\parallel$).

1. INTRODUCTION

The use of plasma simulation models for studying the complex behavior of plasmas both in space as well as in laboratory has been established. Along with the advancement of super computers and the numerical techniques, more and more sophisticated plasma simulation models have been developed in 2- and 3-dimensions.

For studying microscopic processes in a plasma particle simulation models play an essential role in understanding nonlinear plasma processes resulting in acceleration, heating and transport of particles. Of particular importance are the low frequency microinstabilities in a strong magnetic field including ion cyclotron, $\omega \sim n\Omega_i$, lower hybrid, $\omega \sim \omega_{pi}$, pressure -gradient driven drift, $\omega \sim \omega^*$, and shear Alfvén, $\omega \sim k_\parallel v_A$, waves. One of the important characteristics of these waves is that the parallel wavelength is much longer than the perpendicular wave length ($k_\perp >> k_\parallel$) propagating nearly perpendicular to an external magnetic field. While the electron motion across magnetic field may be described within the framework of guiding center drift approximation, the motion along magnetic field must be followed exactly in order to retain the wave-particle interactions which are so essential for the stability of low frequency waves.

In the following, we shall describe plasma simulation models appropriate for studying such low frequency microinstabilities in a strong magnetic field. In Sec. II, electrostatic models using guiding center electrons and full dynamics ions are described. In Sec. III, the

Space Science Reviews **42** (1985) 41–52. 0038–6308/85.15

model is extended to include low frequency electromagnetic effects in a low β plasma. In Sec. IV, a brief description is given on the hybrid model in which ions are approximated as a fluid with the finite-Larmor-radius effects while the electrons are treated as guiding center particles.

2. ELECTROSTATIC MODELS IN 2- AND 3-DIMENSIONS

Let us first consider an electrostatic model in a uniform external magnetic field in which electrons are treated as guiding center particles while the full dynamics is used for the ions. The equations of motion for a single particle are given by

$$\frac{d\underset{\sim}{v}^i}{dt} = \frac{q_i}{m_i}\left[E(\underset{\sim}{r}^i) + \frac{1}{c}\underset{\sim}{v}^i \times \underset{\sim}{B}_o\right] \tag{1}$$

for the ions and for the electrons

$$\frac{dv_\parallel^{\,e}}{dt} = \frac{q_e}{m_e} E_\parallel(\underset{\sim}{r}^e) \tag{2}$$

$$\underset{\sim}{v}_\perp^{\,e} = \frac{c\underset{\sim}{E}(\underset{\sim}{r}^e) \times \underset{\sim}{B}_o}{B_o^2} \tag{3}$$

where $\parallel$ and $\perp$ correspond to parallel and perpendicular directions with respect to an external magnetic field. It is clear that Eq.(2) corresponds to the exact equation of motion along the magnetic field while Eq. (3) corresponds to the guiding center drift approximation in which the inertia term, $m_e\, d\underset{\sim}{v}_\perp^{\,e}/dt$, is neglected in the equation of motion. It is well-known that the drift approximation is valid for low frequency, $\omega << \Omega_e$, and long perpdicular wavelength, $k_\perp\rho_e << 1$, waves in a magnetic field where Ω_e is the electron gyrofrequency and ρ_e is the electron gyroradius. These conditions are usually met for low frequency, $\omega << \Omega_e$, microinstabilities in which the ions play dominant role in determining the propagation characteristics of the waves.

In order to study the characteristics of the electron guiding center model, let us first derive the electrostatic dispersion relation in a uniform magnetic field using full dynamics ions and electrons. Using a set of two fluid equations (Braginskii, 1965),

$$\frac{\partial n_j}{\partial t} + \nabla \cdot \left(n_j\, \underset{\sim}{v}_j\right) = 0$$

$$\frac{\partial n_j}{\partial t} + \left(\underset{\sim}{v}_j \cdot \underset{\sim}{\nabla}\right) \underset{\sim}{v}_j = \frac{q_j}{m_j}\left(\underset{\sim}{E} + \frac{1}{c}\underset{\sim}{v}_j \times \underset{\sim}{B}_o\right) \tag{4}$$

$$\nabla^2\phi = -4\pi|q|\left(n_i - n_e\right)$$

$$\underset{\sim}{E} = -\nabla\phi$$

where j = ion and electrton and the plasma is assumed cold, the dispersion relation is given by

$$\varepsilon\left(k_\perp,\, k_\parallel,\, \omega\right) = 1 + \frac{k_\perp^2}{k^2}\frac{\omega_{pi}^2}{\Omega_i^2 - \omega^2} - \frac{k_\parallel^2}{k^2}\frac{\omega_{pi}^2}{\omega^2} + \frac{k_\perp^2}{k^2}\frac{\omega_{pe}^2}{\Omega_e^2 - \omega^2} - \frac{k_\parallel^2}{k^2}\frac{\omega_{pe}^2}{\omega^2} = 0 \,. \tag{5}$$

$\underset{\sim}{k} = \left(k_\perp,\, k_\parallel\right)$ is the wave vector perpendicular and parallel to the external magnetic field.

The corresponding dispersion relation for the electron guiding center model is obtained by using $\underset{\sim}{v}_\perp^e = c\,\underset{\sim}{E} \times \underset{\sim}{B}_o/B_o^2$ in Eq. (4) and is given by

$$\varepsilon\left(k_\perp,\, k_\parallel,\, \omega\right) = 1 + \frac{k_\perp^2}{k^2}\frac{\omega_{pi}^2}{\Omega_i^2 - \omega^2} - \frac{k_\parallel^2}{k^2}\frac{\omega_{pi}^2}{\omega^2} - \frac{k_\parallel^2}{k^2}\frac{\omega_{pe}^2}{\omega^2} = 0 \tag{6}$$

which is equivalent to $\Omega_e \to \infty$ limit in Eq. (5).

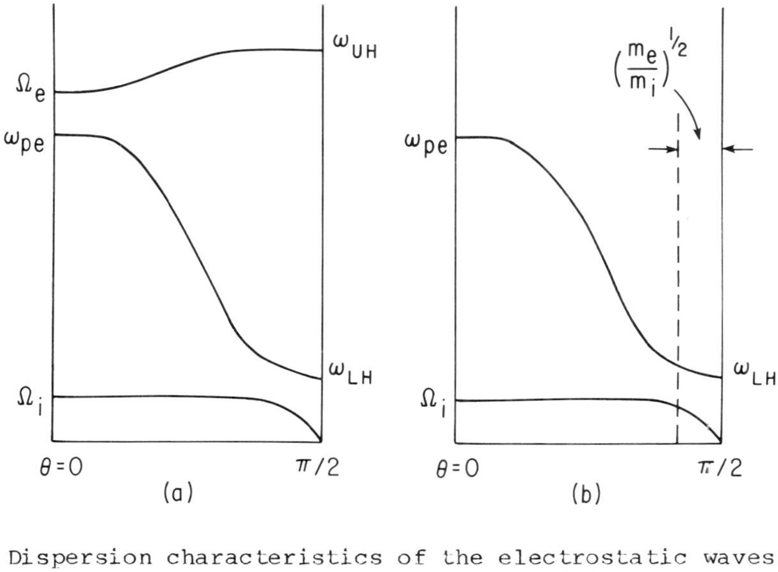

Figure 1. Dispersion characteristics of the electrostatic waves in a magnetic field using full dynamics for electrons (a) and the guiding center drift approximations for electrons (b).

A sketch of the dispersion relations corresponding to Eqs. (5) and (6) is shown in Fig. 1 (a) and (b) as a function of the angle θ between $\underset{\sim}{k}$ and $\underset{\sim}{B}_o$. For the case of full dynamics model given by Eq. (5), one finds three modes shown in Fig. 1(a), high frequency mode associated with electron gyration, electron plasma wave and the low frequency ion cyclotron wave while for the electron guiding center model, high frequency mode associated with electron gyration has been eliminated. Electron plasma waves and ion cyclotron waves are unmodified in the electron guiding center model. It is interesting to realize that for both models, the frequency of the plasma oscillation becomes smaller as the propagation angle approaches near perpendicular direction. When the propagation angle is large enough so that the electron plasma frequency for oblique propagation becomes equal to the lower hybrid frequency, $\omega_{LH} \approx \omega_{pi}$,

$$\omega_{pe} \frac{k_{\parallel}}{k} \approx \omega_{pi}$$

the electron plasma waves goes over the ion plasma waves. The angle for this to occur is given by

$$\frac{k_{\parallel}}{k} = \frac{\omega_{pi}}{\omega_{pe}} = \left(\frac{m_e}{m_i}\right)^{1/2} << 1 \tag{7}$$

which is nearly perpendicular propagation. Therefore, if one is interested in lower hybrid instabilities, for example, it is sufficient to consider the waves propagating nearly perpendicular to a magnetic field given by Eq. (7). This would then allow us to use a large time step, $\omega_{pi}\ \Delta t \approx 0.2$, much larger than the electron plasma time scale (Lee and Okuda, 1978).

While the case of lower hybrid waves is considered as an example, the model is quite general so that it can be used to simulate, for example, electrostatic ion cyclotron waves, drift waves, and shear Alfvén waves when the electromagnetic effects are included. These waves propagate nearly perpendicular to magnetic field so that only the modes propagating nearly perpendicular to magnetic field are important. Needless to say, there are situations where the plasma waves propagating parallel to the magnetic field are very important such as ion acoustic waves. Then it is necessary to use a time step determined from the electron plasma oscillation as ω_{pe} is the highest frequency in the model for such a case ($k_{\parallel} >> k_{\perp}$).

For the case of two-and-one-half dimensions, a sketch of the simulation model is given in Fig. 2. The magnetic field is taken in (y,z) plane and the angle between $\underset{\sim}{B}_o$ and $\hat{z}$ axis is arbitrarily chosen. For studying waves propagating nearly perpendicular to $\underset{\sim}{B}_o$, $\underset{\sim}{B}_o$ is taken nearly parallel to the z axis. In this case all the modes in the system are nearly perpendicular modes and no high frequency electron plasma waves appear. Typical integration time step is $\Delta t \approx 0.25\ \omega_{pi}^{-1}$ which is much larger than ω_{pe}^{-1}.

Standard time-centered leap-frog method can be used for the equation of ions (Buneman, 1967) while a predictor-corrector method must be used for the guiding center electron motion as the leap-frog method cannot be applied there (Lee and Okuda, 1978).

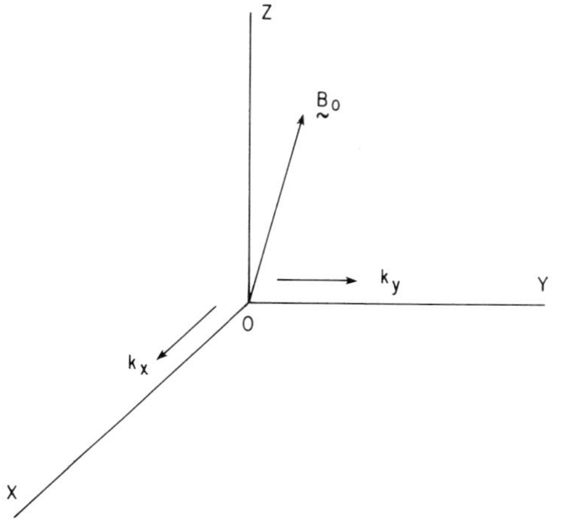

Figure 2. A sketch of a two-and-one-half dimensional model in a magnetic field $\underset{\sim}{B}_o$.

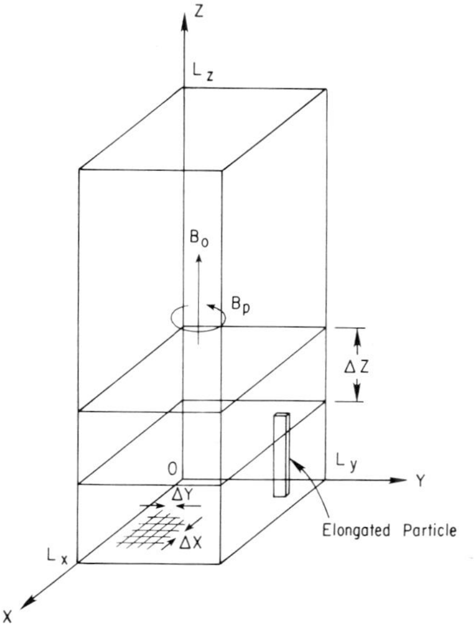

Figure 3. A sketch of a three-dimensional model elongated in the direction of the main magnetic field $B_o\hat{z}$.

It is straightforward to extend the model to full three dimensions where a wave vector $\mathbf{k} = (k_x, k_y, k_z)$ has three components. An example of such models is given in Fig. 3 where a strong uniform magnetic field $\mathbf{B}_o$ is pointing in the $\hat{z}$ direction and a weak poloidal field $\mathbf{B}_p$ is in the x-y plane satisfying $B_o >> B_p$. The system length is assumed $L_x \sim L_y << L_z$ so that $k_z \sim L_z^{-1} << k_x \sim k_y$. Since we are interested in the modes with $k_z << k_\perp$, it is necessary to keep only the long wavelengths modes in the z-direction compared with the perpendicular wavelengths in the x-y plane where short wavelengths modes must be retained. This can be done by either using a spectral method in the z-direction with a two-dimensional grid in the x-y plane (Cheng and Okuda, 1977) or a three-dimensional grid elongated in the z-direction so that

$$\Delta z >> \Delta x \sim \Delta y \tag{8}$$

is satisfied (Okuda et al., 1979). The advantage of using a spectral method is it is accurate but also slow in computing since the number of operations scales as NM_z where N is the total number of simulation particles and M_z is the number of the Fourier modes in the z-direction.

When a three dimensional grid is used, it is necessary to adopt a higher order interpolation in z-direction to avoid grid instabilities (Langdon, 1970) since the grid size Δz is much larger than the Debye length. Typically a quadratic or cubic spline is enough to suppress the numerical grid instabilities (Okuda, et al., 1979). It is also natural to use an elongated particles in the z-direction so that the size of simulation particles follows the grid size shown in Fig. 3.

While most of the modes in the three dimensional model satisfy $k_\parallel << k_\perp$ because $L_\perp >> L_\parallel$ and $\Delta z >> \Delta x \sim \Delta y$, a particular class of modes with $k_\perp = 0$, but $k_\parallel \neq 0$ do not satisfy the condition for nearly perpendicular propagation. It is straightforward, however, to remove such high frequency plasma waves from the simulations. First, for a laboratory plasma confined in a conducting wall, these high frequency modes ($k_\perp = 0$, $k_\parallel \neq 0$) do not appear since they do not satisfy conducting boundary condition where $\phi = 0$ on the wall. Second, even in the case of space plasmas where the boundary condition is more complex so that the high frequency plasma waves exist in general, one can discard these modes ($k_\perp = 0$, $k_\parallel \neq 0$) in the equation motion while retaining all the other modes ($k_\perp \neq 0$) in the system allowing the use of a large time step determined essentially from the ion plasma time scale (Cheng and Okuda, 1977; Okuda et al., 1979).

While we considered only a system of simple magnetic field geometry, it is straightforward to extend the electron guiding center model to a more complex magnetic field configuration such as a toroidal system (Cheng and Okuda, 1977) or a system involving a dipole magnetic field (Smith, 1985). For such a non-uniform magnetic field, the electron guiding center motion includes perpendicular drifts due to magnetic field gradient and curvature in addition to the $c\,\mathbf{E} \times \mathbf{B}/B^2$ drift. Furthermore the mirror force due to the magnetic field gradient along magnetic field must be added to the parallel electric field for the parallel motion of electrons. With these additions, a predictor corrector method works without any problem while allowing a use of a large time step such as $\omega_{pi}\ \Delta t \sim 0.2$ (Cheng and Okuda, 1977).

3. ELECTROMAGNETIC MODEL FOR LOW FREQUENCY WAVES IN A LOW β PLASMA

We shall now consider an electromagnetic model appropriate for a low β plasma in which the plasma pressure is much smaller than the pressure of an external magnetic field. In such a case, it is only necessary to retain the magnetic field perturbations perpendicular to the external magnetic field $B_o\,\hat{z}$ neglecting the compression of the magnetic field. Furthermore only the z component of the transverse electric field $\underset{\sim}{E}_T$ is kept while neglecting the transverse components $E_{x,y}^T$. This is justified in a low β plasma where $E_{x,y}^T \ll E_{xy}$ but $E_z^T \sim E_z^L$ are satisfied which corresponds to the one-component approximation of the vector potential A_z or current j_z generally valid in a low β plasma (Kadomtsev, 1965),

The set of field equations in this model may be written as

$$\nabla \times \underset{\sim}{B} = \frac{4\pi}{c}\,\underset{\sim}{j}^T \tag{9}$$

$$\nabla \times \underset{\sim}{E}^T = -\frac{1}{c}\frac{\partial \underset{\sim}{B}}{\partial t} \tag{10}$$

$$\nabla^2\phi = -4\pi\,|q|\,(n_i - n_e)$$

$$\underset{\sim}{E}^L = -\nabla\phi$$

which corresponds to the magnetostatic model valid for low frequency electromgnetic waves (Hasegawa and Okuda, 1968). Since we are considering only the transverse components of the magnetic field, B_x and B_y, we find

$$B_x(\underset{\sim}{k}) = i\,\frac{4\pi}{ck^2}\left(k_y j_z - k_z j_y\right) \simeq i\,\frac{4\pi}{c}\,\frac{k_y}{k^2}\,j_z$$

$$B_y(\underset{\sim}{k}) = i\,\frac{4\pi}{ck^2}\left(k_z j_x - k_x j_z\right) \simeq -i\,\frac{4\pi}{c}\,\frac{k_x}{k^2}\,j_z$$

where $k_z \ll k_x,\ k_y$ is assumed in an elongated system along the external magnetic field.

To obtain E_z^T, we first take the curl of Eq. (10) to find

$$\nabla^2 \underset{\sim}{E}^T = \frac{4\pi}{c^2}\frac{\partial}{\partial t}\underset{\sim}{J}^T.$$

$\partial j/\partial t$ is replaced by using the equation of motion (Busnardo-Neto el al., 1977; Okuda et al., 1979) as described elsewhere in this book in detail. Let us note here in our discussions that the electron guiding center drift approximations must be used in the equation of motion while the full Lorentz force is used for the ions in accordance with our model.

Such a model has been developed and used for studying low frequency electromagnetic instabilities such as shear Alfvén waves in three dimensions (Kim and Okuda, 1985). Since the nonlinear equation is solved iteratively to find E_z^T, it is usually necessary to use a higher order interpolation in the z direction such as a cubic spline where the grid size is much larger than that in the x,y-directions.

4. PARTICLE ELECTRON-FLUID ION MODEL FOR VERY LOW FREQUENCY WAVES

We have developed simulation models using guiding center drift approximations for the electrons and full dynamics ions appropriate for low frequency electrostatic and electromagnetic instabilities in a strong magnetic field. The model eliminates high frequency waves associated with the electron gyration while keeping the electron motion along magnetic field exact. The model is particularly useful for low frequency waves propagating nearly perpendicular to an external field. Since the full ion dynamics is used, lower hybrid waves, $\omega \approx \omega_{pi}$ and ion cyclotron waves, $\omega \approx \Omega_i$, can be studied using the model.

When considering very low frequency waves whose frequency is below the ion gyrofrequency $\omega << \Omega_i$, then it is possible to use guiding center fluid approximations for the motion of ions retaining the lowest order finite Larmor radius effects while using the guiding center drift approximation for the electrons as before. Since the fluid approximation is used for the ions, resonant wave-ion interactions are neglected in this model while the electron-wave interactions along magnetic field is retained. It is well-known that the resonant wave-particle interactions are very important for the stability of low frequency microinstabilities such as drift waves and shear Alfvén waves (Kadomtsev, 1965). Since we are considering the modes whose frequency is below Ω_i, we are restricting $k_\parallel/k$ is even much smaller than $(m_e/m_i)^{1/2}$ for this model more useful than the previously described model where full dynamic is used for the ion particles.

Let us consider, as an example, an electrostatic drift wave destabilized by the Landau damping of the electrons (Kadomtsev, 1965). The frequency of the wave is given by

$$\omega \sim \omega^* = k_\perp \rho_s \left(\rho_s/L_\perp\right) \Omega_i$$

where ρ_s is the ion gyroradius at the electron temperature defined by $\rho_s = (T_e/m_i)^{1/2}/\Omega_i$ and $L_\perp$ is the density scale length of the plasma perpendicular to an external magnetic field. For $k_\perp\rho_s \lesssim 1$ which is the most important wavenumbers for the drift instabilities, it is clear $\omega << \Omega_i$ since $\rho_s/L_\perp << 1$ is usually satisfied. The drift waves can become unstable in the presence of electron Landau damping and the resonance condition is given by

$$\omega \sim \omega^* \approx k_\parallel v_e.$$

Then we find that

$$\frac{k_\parallel}{k_\perp} = \left(\frac{\rho_s}{L_\perp}\right)\left(\frac{c_s}{v_e}\right) = \left(\frac{m_e}{m_i}\right)^{1/2}\left(\frac{\rho_s}{L_\perp}\right) << \left(\frac{m_e}{m_i}\right)^{1/2} \tag{11}$$

as expected. While the motion of electrons along magnetic field is followed exactly in this model, one can use a very large time step of integration valid for $\omega << \Omega_i$ and $k_\parallel/k_\perp << (m_e/m_i)^{1/2}$.

Let us consider an electrostatic drift wave in a uniform magnetic field $\underset{\sim}{B}_o$. Since the quasi-neutrality is satisfied for a low frequency wave,

$$\nabla \cdot \underset{\sim}{j} = 0 ,$$

we find

$$\nabla_\parallel j_\parallel = - \underset{\sim}{\nabla}_\perp \cdot \underset{\sim}{j}_\perp = - \underset{\sim}{\nabla}_\perp \cdot \left(\underset{\sim}{j}_p + \underset{\sim}{j}_{E\times B}\right)$$

where

$$\underset{\sim}{j}_p = \frac{m_i n}{B_o^2}\frac{dE_\perp}{dt} = \frac{m_i n_o}{B_o^2}\left[\frac{\partial}{\partial t} + \left(\frac{c\underset{\sim}{E} \times \underset{\sim}{B}_o}{B_o^2}\right) \cdot \underset{\sim}{\nabla}_\perp\right] \underset{\sim}{E}_\perp$$

$$\underset{\sim}{j}_{E\times B} = en\left(1 + \rho_i^2 \nabla^2\right)\frac{c\underset{\sim}{E} \times \underset{\sim}{B}_o}{B_o^2} - en\frac{c\underset{\sim}{E} \times \underset{\sim}{B}_o}{B_o^2}$$

$$= en_o \rho_i^2 \nabla^2 \left(\frac{c\underset{\sim}{E} \times \underset{\sim}{B}_o}{B_o^2}\right)$$

are polarization and the $c\underset{\sim}{E} \times \underset{\sim}{B}_o/B_o^2$ currents repectively (Hasegawa and Wakatani, 1983). Note we replaced n by n_o assuming $\underset{\sim}{E}$ is a small parameter. Since we are considering a plasma in a uniform magnetic field, currents due to the gradient and curvature of the magnetic field do not exist so long as we consider the electrostatic waves.

Substituting $\underset{\sim}{j}_p$ and $\underset{\sim}{J}_{E\times B}$ to $\underset{\sim}{\nabla} \cdot \underset{\sim}{j} = 0$, we find

$$\frac{\partial}{\partial t} \nabla_\perp^2\phi + \underset{\sim}{\nabla}_\perp \cdot \left(e n_o \rho_i^2 \nabla_\perp^2 \frac{c\underset{\sim}{E} \times \underset{\sim}{B}_o}{B_o^2}\right) + \underset{\sim}{\nabla}_\perp \cdot \left[\left(\frac{c\underset{\sim}{E} \times \underset{\sim}{B}_o}{B_o^2} \cdot \underset{\sim}{\nabla}_\perp\right) \underset{\sim}{\nabla}_\perp \phi\right]$$

$$= \frac{B_o^2}{m_i n_o} \nabla_\parallel j_\parallel$$

$$= \frac{B_o^2}{m_i n_o} \nabla_\parallel j_\parallel^e \tag{12}$$

where $j_\parallel = j_\parallel^e$ is assumed neglecting the field aligned current associated with the ions. As mentioned earlier $j_\parallel^e$ is determined from the electrons simulated by the guiding center drift approximations. It is clear from Eq. (12) that the ions are now represented by a fluid moving under the influence of an electric field perpendicular to magnetic field with the finite gyroradius effects given by the second term of the left hand side of Eq. (12). In additions to Eq. (12) which determines the potential ϕ, Eqs. (2) and (3) are used for the electron dynamics. Equation (12) corresponds to Poisson's equation which determines the electrostatic potential from the quasineutrality condition $\underset{\sim}{\nabla} \cdot \underset{\sim}{j} = 0$.

In order to study the characteristics of Eq. (12), it is straightforward to derive a dispersion relation for cold electrons to determine $j_\parallel^e$ using a set of electron guiding center equations. We find

$$\omega^2 = \frac{k_\parallel^2}{k_\perp^2} \Omega_e \Omega_i = \frac{k_\parallel^2}{k_\perp^2} \Omega_e^2 \frac{m_e}{m_i} \tag{13}$$

which gives the highest frequency of the system. In order the drift approximation of the ions valid, $\omega^2 << \Omega_i^2$ must be satisfied so that

$$\frac{k_\parallel}{k_\perp} << \left(\frac{m_e}{m_i}\right)^{1/2} \tag{14}$$

which is consistent with Eq. (11) valid for drift waves. For the case of warm electrons, it is well-known that Eq. (12) gives rise to a dispersion relation for drift waves

$$\omega = \frac{\omega^*}{1 + k_\perp^2 \rho_s^2}$$

which are destabilized in the presence of wave-electron resonant interactions which are included in the model.

Note, however, since the parallel currents associated with the ions are neglected, ion acoustic waves, $\omega^2 = k_\parallel^2 c_s^2$, are not retained in the model. Furthermore, in the presence of ∇B and curvature drifts, it is necessary to determine the currents perpendicular to magnetic field associated with these drift. Since those drifts include particle energy parallel and perpendicular to magnetic field, equations which determine the ion pressure $p_\perp^i$ and $p_\parallel^i$, must be supplemented (Hasegawa and Wakatani, 1983) while the pressure of the electrons, $p_\perp^e$ and $p_\parallel^e$, may be calculated from the moments of the particle distribution function determined from the electron particles.

A set of Eqs. (2), (3) and (12) are used to study drift wave instabilities in the presence of a density gradient in a uniform magnetic field. The model reproduces unstable modes associated with the drift instabilities resulting in strong particle diffusion. For more general cases where the electromagnetic waves such as Alfvén or magnetohydrodynamic waves become important, more study is clearly necessary.

ACKNOWLEDGEMENTS

The hybrid model described in Sec. 4 is developed in collaboration with Dr. Akira Hasegawa. This work is supported by the National Science Foundation Grant ATM-83-11102 and the United States Department of Energy Contract No. De-AC02-76-CHO-3073.

REFERENCES

Braginskii, S. I., in Reviews of Plasma Physics, Vol. 1, Consultants Bureau, New York, 1965 (Ed. M.A. Leontovich), p. 214.

Busnardo-Neto, J., P. L. Pritchett, A. T. Lin, and J. M. Dawson, A self-consistent magnetostatic particle code for numerical simulation of plasmas, J. Compt. Phys. 23, 300 (1977).

Buneman, O., Time reversible difference procedures, J. Compt. Phys. 1, 517 (1967).

Cheng, C. Z. and H. Okuda, New three dimensional simulation models for cylindrical and toroidal plasmas, J. Compt. Phys. 25, 133 (1977).

Hasegawa, A., and H. Okuda, One-dimensional plasma model in the presence of a magnetic field, Phys. Fluids 11, 1995 (1968).

Hasegawa, A., and M. Wakatani, Finite-larmor-radius magnetohydrodynamic equations for microturbulence, Phys. Fluids 26, 2770 (1983).

Kadomtsev, B. B., Plasma Turbulence, Academic press, New York, 1965, p. 78-106.

Kim, S. Y., and H. Okuda, Three dimensional magnetostatic particle model using guiding center electrons, To be published.

Langdon, A. B., Effects of the spatial grid in simulation plasmas, J. Compt. Phys. 6, 247 (1970).

Lee, W. W., and H. Okuda, A simulation model for studying low frequency microinstabilities, J. Compt. Phys. 26, 139 (1978).

Okuda, H., C. Z. Cheng, and W. W. Lee, Electrostatic and magnetostatic particle simulation models in three dimensions, Compt. Phys. Comm. 17, 233 (1979).

Smith, B. L., Simulations of auroral double layers, Ph.D. dissertation, Astophysical Sciences Department, Princeton University, 1985 (to be published)

HYBRID SIMULATION CODES WITH APPLICATION TO SHOCKS AND UPSTREAM WAVES

D. Winske
Los Alamos National Laboratory
Los Alamos, New Mexico, 87545
USA

ABSTRACT. Hybrid codes in which part of the plasma is represented as particles and the rest as a fluid are discussed. In the past few years such codes with particle ions and massless, fluid electrons have been applied to space plasmas, especially to collisionless shocks. All of these simulation codes are one-dimensional and similar in structure, except for how the field equations are solved. We describe in detail the various approaches that are used (resistive Ohm's law, predictor-corrector, Hamiltonian) and compare results from the various codes with examples taken from collisionless shocks and low frequency wave phenomena upstream of shocks.

1. INTRODUCTION

Plasma physics phenomena are characterized by a multitude of length and time scales, primarily due to the different responses of the light electrons and the massive ions to the imposed and self-generated electric and magnetic fields. Typically, one is interested in particular processes which occur on some of these scales and not interested in other processes that occur on shorter or longer time or distance scales. This can be accomplished in numerical simulation by treating the various plasma species in different ways, for example, as discrete particles or as fluids. Hybrid codes are defined as those numerical algorithms in which the various plasma species are treated in a different manner, as distinct from particle codes where all the plasma species are treated as particles or fluid codes where each species (or several species together) is treated as a fluid.

Various types of hybrid codes are of course possible, depending on the problem at hand. One important subclass of hybrid models are those in which there are two (or more) populations of one particular charge species, whose properties on a particular length or time scale are different. For example, consider the interaction of a small, cold electron beam with a hot background electron population (O'Neil et al., 1971). In this case the unstable waves generated by the presence of the beam strongly affect it, and thus a particle description is needed to correctly model the highly nonlinear dynamics of

Space Science Reviews **42** (1985) 53–66. 0038–6308/85.15

the beam electrons. On the other hand, the waves do not affect the background population very much. Thus, there is no need to follow the dynamics of each individual background electron; rather, the contribution of the background component can be included simply as a linear dielectric in Poisson's equation. Similar methods can be used when there are several ion populations. For example, in the study of the interaction of an ion ring velocity distribution with a background ion core, Lee and Birdsall (1979) treated the ring ions as discrete particles, with a fluid description for the background ions (and the electrons).

The most common type of hybrid code, however, occurs when the two species involved are the electrons and ions. The simplest kind of hybrid model of this type is to ignore one species entirely. For example, in the study of high frequency electron behavior, it is very common to eliminate the ions, except as a charge-neutralizing background. It is also possible to ignore the electrons entirely, as has been done for tearing mode calculations (Dickman et al. (1969) and later work to be cited in the next section). Another useful approximation that is commonly invoked in this type of hybrid model is quasineutrality, which makes use of the fact that the electron (n_e) and ion (n_i) charge densities are nearly equal. If one is interested in scale lengths much larger than the Debye length, the condition $n_e = n_i$ is imposed; for smaller systems, a Boltzmann relation between the electron charge density and the electrostatic potential may be used instead, which gives rise to a nonlinear Poisson equation. (Okuda et al., 1978). Another common approximation involves the electron mass. Depending on the frequency range of interest, the electron mass may be kept (Hewett and Nielson, 1978) or not.

A very common type of hybrid code, and the one of interest throughout the rest of this chapter, treats the electrons as a massless, charge-neutralizing fluid. In recent years this type of model has become widely used in space physics for the study of phenomena at the bow shock (Leroy et al. 1981 and 1982; Leroy and Winske, 1983, Kan and Swift, 1983; Mandt and Kan, 1985), upstream of the bow shock (Winske and Leroy 1984a; Winske et al. 1984 and 1985; Hada and Kennel, 1985), the magnetopause (Swift and Lee, 1983), the magnetotail (Swift, 1983b), and the magnetosphere (Omura et al. 1985; Tanaka and Goodrich, 1985).

While all of the calculations cited here are based on hybrid models with similar properties, there are differences in the way the models are implemented numerically, primarily in how the field equations are solved. Three different techniques, referred to hereafter as the resistive (Ohm's law) method, the predictor-corrector method, and the Hamiltonian method, have been employed. The purpose of this paper is to explain how these various methods work in some detail (Section 2) and then compare results of simulations based on each method (Section 3). The examples chosen are well studied phenomena from the earth's bow shock and the upstream region, and the discussion will emphasize numerics rather than the physics content. A short summary is given in Section 4.

2. DESCRIPTION OF HYBRID CODE MODELS

In this section the numerical schemes used in three hybrid code models are described. In each case particle-in-cell techniques are employed for the ion dynamics, while fluid equations are used for the (massless) electrons. Quasineutrality is assumed and the low frequency (Darwin) approximation is used (Nielson and Lewis, 1976). For the methods described here the one-dimensional nature of the calculation is utilized, but as will be discussed, each of the models has been generalized to two spatial dimensions.

2.1 Resistive Ohm's Law Method

This model was originally devised by Chodura (1975), and further developed by Sgro and Nielson (1976) and Hamasaki et al. (1977), for the analysis of laboratory magnetic fusion experiments and more recently for the study of the earth's bow shock by Leroy et al. (1981, 1982). The details of the method and references to earlier work are given by Winske and Leroy (1984b). In this method, as in the two to follow, a standard leapfrog scheme is used to advance the particles and fields. The velocities of the particles are known at the half time steps, while positions of the particles and the fields are known at the even time steps (time levels to be denoted by superscripts). Thus, at time step N, we know $\underline{v}^{N-1/2}$, x^N, $\underline{E}^N$, $\underline{B}^N$. To advance the particles, we solve

$$\frac{d\,\underline{v}^N}{dt} = \frac{\underline{v}^{N+1/2} - \underline{v}^{N-1/2}}{\Delta t} = \frac{q}{m}\left(\underline{E}^N + \underline{v}^N \times \underline{B}^N/c\right) \tag{1}$$

implicitly by substituting $\underline{v}^N = (\underline{v}^{N+1/2} + \underline{v}^{N-1/2})/2$ and solving to obtain

$$\underline{v}^{N+1/2} = f\,\underline{v}^{N-1/2} + \frac{\Delta t q}{m}\left(\underline{E}^N + g\underline{B}^N + \underline{v}^N \times \underline{B}^N/c\right) \tag{2}$$

where

$$\underline{v}^N = \underline{v}^{N-1/2} + \frac{\Delta t q}{2m}\underline{E}^N$$

$$f = 1 - \frac{1}{2}\left(\frac{\Delta t q}{mc}\right)^2 \underline{B}^N \cdot \underline{B}^N \tag{3}$$

$$g = \frac{\Delta t q}{mc^2}\left(\underline{v}^{N-1/2} \cdot \underline{B}^N\right)$$

and then

$$x^{N+1} = x^N + \Delta t v_x^{N+1/2} \tag{4}$$

We then solve Ampere's law

$$\nabla^2 \underline{A}_T^{N+1} = -\frac{4\pi}{c}\underline{J}_T^{N+1} \tag{5}$$

for the transverse components (y or z). The ion part of the current comes directly from collecting the ion moments, where we advance (but do not save) the particles one additional half time step

$$\underline{v}^{N+1} = \underline{v}^{N+1/2} + \frac{\Delta t q}{2m} (E^N + \underline{v}^{N+1/2} \times \underline{B}^N/c) \qquad (6)$$

where $\underline{E}^N$ and $\underline{B}^N$ are evaluated at x^{N+1}, to obtain the ion density n_i^{N+1} and velocities $\underline{V}_i^{N+1}$. Quasineutrality (in 1-D) then gives $n_e = n_i = n$ and $V_{ex} = V_{ix} = V_x$.

The electron part of the current comes from the electron momentum equation with a resistive term (i.e. Ohm's law)

$$m_e \frac{d\underline{V}_e}{dt} = 0 = -qn_e(\underline{E} + \frac{\underline{V}_e \times \underline{B}}{c}) - \hat{x} \frac{\partial}{\partial x} n_e T_e + qn_e \underset{\approx}{\eta} \cdot \underline{J} \qquad (7)$$

After solving for $\underline{A}_T^{N+1}$ (described shortly), we obtain in the usual way

$$\underline{B}_T = \underline{\nabla} \times \underline{A}_T \qquad (B_x = \text{constant})$$

$$\underline{E}_T = \frac{-1}{c} \frac{\partial A_T}{\partial t} \qquad (8)$$

Then solving an energy equation for T_e,

$$\frac{3}{2} \frac{\partial}{\partial t} nT_e + \frac{\partial}{\partial x} (\frac{3}{2} nT_e V_{ex}) + nT_e \frac{\partial V_{ex}}{\partial x} = \eta J^2 \qquad (9)$$

we can obtain the last field component, E_x, from the x component of (7)

$$E_x = \frac{-1}{c} (\underline{V}_e \times \underline{B})_x - \frac{1}{nq} \frac{\partial nT_e}{\partial x} \qquad (10)$$

The method used to solve (5) for $\underline{A}_T$ is slightly different from that presented in Winske and Leroy (1984b) and works better for oblique shocks. We assume the resistivity tensor to be diagonal (i.e. $\eta_{\parallel} = \eta_{\perp} = \eta$), which has been shown to correctly model turbulent systems, such as the z-pinch, very well (Sgro and Nielson, 1976). We solve

$$\underline{J}_T = \eta^{-1}(\underline{E} + \underline{V}_e \times \underline{B}/c)_T = qn(V_i - V_e)_T \qquad (11)$$

for $\underline{V}_{eT}$, and substitute into (5), obtaining

$$\frac{-c}{4\pi} \frac{\partial^2 A_y}{\partial x^2} = \frac{F_y + \delta F_z}{1 + \delta^2}$$

$$\frac{-c}{4\pi} \frac{\partial^2 A_z}{\partial x^2} = \frac{F_z - \delta F_y}{1 + \delta^2} \qquad (12)$$

where

$$F_y = qn(V_{iy} - \frac{cE_z}{B_x} - \frac{V_x B_y}{B_x})$$

$$F_z = qn\left(V_{iz} + \frac{cE_y}{B_x} - \frac{V_x B_z}{B_x}\right) \tag{13}$$

$$\delta = \left(\frac{n}{n_o}\right)\left(\frac{\omega_{io}{}^2 \eta}{4\pi}\right)\left(\frac{mc}{qB_x}\right)$$

and n_o=reference density, $\omega_{io}=(4\pi n_o q^2/m)^{1/2}$, and all quantities are at time level N+1. Using (8) to express $\underline{E}_T$ and $\underline{B}_T$ in terms of $\underline{A}_T$, (12) can be written in finite difference form and solved, as described in Winske and Leroy (1984b).

In addition to the references cited in Winske and Leroy (1984b), calculations based on this method have been done to study oblique shocks (Leroy and Winske, 1983), the electromagnetic ion beam instability (Winske and Leroy, 1984a), the interaction of heavy ions with the solar wind (Winske et al. 1984 and 1985) and the steepening of slow waves (Hada and Kennel, 1985). The method has been extended to two dimensions (Hewett, 1980) and has been applied to the study of magnetic reconnection in laboratory experiments (Hewett, 1984).

2.2 Predictor-Corrector Method

A second method used for hybrid code calculations employs the predictor-corrector technique. The method is described by Byers et al. (1978) and has been implemented in a one-dimensional code by Tanaka and Goodrich (1985) and Omura et al. (1985) in the study of heating of heavy ions by ion cyclotion instabilities. The method has been extended to two dimensions by Harned (1982) and used in the study of rotational instabilities in laboratory field-reversed configurations (Harned, 1983).

We will describe the method in its simplest form, where the electron temperature is kept constant. Again, we assume at time step N the $\underline{v}^{N-1/2}$, x^N, $\underline{E}^N$ and $\underline{B}^N$ are known. In this case the advance of one time step involves two steps, a predictor step and a corrector step, each of which involves going through the particle table.

In the first step, $\underline{v}^{N+1/2}$ and x^{N+1} are obtained as in the previous case using (1) - (4). In the process, however, we first advance the particles to $x^{N+1/2}=x^N+\Delta t v_x{}^{N+1/2}/2$ to collect the ion moments, $\underline{J}_i{}^{N+1/2}$ and $n_i{}^{N+1/2}$. We then compute the predictor fields (denoted by subscript p) using Faraday's law and (7):

$$\begin{aligned} \underline{B}^{N+1/2} &= \underline{B}^N - \frac{c\Delta t}{2}(\underline{\nabla}\mathrm{x}\underline{E}^N) \\ \underline{E}^{N+1/2} &= \frac{1}{qn_i{}^{N+1/2}}\left\{\left(\frac{\underline{\nabla}\mathrm{x}\underline{B}^{N+1/2}}{4\pi} - \frac{\underline{J}_i{}^{N+1/2}}{c}\right)\mathrm{x}\ \underline{B}^{N+1/2} - T_e\underline{\nabla}n_i{}^{N+1/2}\right\} \end{aligned} \tag{14}$$

to obtain

$$\begin{aligned} \underline{E}_p{}^{N+1} &= -\underline{E}^N + 2\underline{E}^{N+1/2} \\ \underline{B}_p{}^{N+1} &= \underline{B}^{N+1/2} - \frac{c\Delta t}{2}(\underline{\nabla}\mathrm{x}\underline{E}_p{}^{N+1}) \end{aligned} \tag{15}$$

We then go thru the particles again, using the predictor fields, to calculate $\underline{v}_p^{N+3/2}$ and $x_p^{N+3/2}$, again using (1) - (4), in order to collect $\underline{J}_{i,p}^{N+3/2}$ and $n_{i,p}^{N+3/2}$. Then, we calculate $\underline{B}_p^{N+3/2}$ and $\underline{E}_p^{N+3/2}$ using (14) with predictor fields $\underline{E}_p^{N+1}$, $\underline{B}_p^{N+1}$ and ion moments to obtain the new fields

$$\underline{E}^{N+1} = \frac{1}{2} (\underline{E}^{N+1/2} + \underline{E}_p^{N+3/2})$$

$$\underline{B}^{N+1} = \underline{B}^{N+1/2} - \frac{c\Delta t}{2} (\underline{\nabla} x \underline{E}^{N+1}) \tag{16}$$

Having now advanced the fields, we are thus ready to start the entire sequence again.

2.3 Hamiltonian Method

This method employs the canonical momenta of the particles $\underline{P}$ in place of the velocities. A description of the method is given, for example, by Morse and Nielson (1971), and a comparison between the Hamiltonian and Lagragian methods is discussed by Nielson and Lewis (1976). The Hamiltonian method has been used by Swift and Lee (1983) to study the rotational discontinuity at the magnetopause, Swift (1983b) to examine magnetic slow shocks, Kan and Swift (1983) and Mandt and Kan (1985) to simulate nearly parallel shocks. The method has been generalized to two spatial dimensions in the limited sense of either completely ignoring the electrons (Dickman et al.,1969; Ambrosiano et al.; 1983; Teresawa, 1981) for the study of tearing modes and ion fusion physics (Friedman et al., 1977; Mankofsky et al., 1981; Mankofsky and Sudan, 1984) or in restricting the types of perturbations allowed (Swift, 1983a).

In this case, at time step N, the canonical momenta of the particles

$$\underline{P}_T = m\underline{v}_T + q\underline{A}_T/c \tag{17}$$

are known at the half time step N-1/2. The particle equation of motion is

$$\frac{d\underline{P}_T}{dt} = \frac{q}{c} (\underline{v}_T \times B_x \hat{x}) \tag{18}$$

Note that in the case $B_x=0$ the method can be very attractive since $\underline{P}_T$ is a conserved quantity. Using (17), (18) can be written as

$$\frac{d\underline{P}_T}{dt} = \frac{qB_x}{mc} (\underline{P}_T - \frac{q\underline{A}_T}{c}) \times \hat{x} \tag{19}$$

Another nice feature is that using a complex representation ($P = P_y + iP_z$, $A = A_y + iA_z$, $\Omega = qB_x/mc$) the equations of motion can be written compactly as

$$\frac{dP^N}{dt} = -i\Omega(P^N - \frac{qA^N}{c}) \tag{20}$$

or

$$e^{-i\Omega t} \frac{d}{dt} P^N e^{i\Omega t} = \frac{i\Omega q A^N}{c} \tag{21}$$

Thus, in difference form (Swift and Lee, 1983)

$$P^{N+1/2} = P^{N-1/2} e^{-i\Omega\Delta t} + \frac{i\Delta t\Omega q}{c} A^N e^{-i\Omega\Delta t/2} \tag{22}$$

Then

$$\underline{v_T}^N = \frac{1}{2m} (\underline{P_T}^{N+1/2} + \underline{P_T}^{N-1/2}) - \frac{q\underline{A_T}^N}{mc} \tag{23}$$

is known and v_x can be advanced

$$v_x^{N+1/2} = v_x^{N-1/2} + \frac{q\Delta t}{m} (E_x^N + \underline{v_T}^N \times \underline{B_T}^N/c) \tag{24}$$

as can the particle positions, x^{N+1}, using (4). (I have used v_x here for simplicity. It is possible to write $P_x = mv_x$ (A_x=0, since $\underline{\nabla}\cdot\underline{A}$ = 0) and use P's throughout.)

Again, we need to solve Ampere's law (5). There are several different ways to difference the equations. One method is to use (22) to advance $P_T^{N+1/2}$ to $P_T^{N+3/2}$ and express it in terms of $P_T^{N+1/2}$ and A_T^{N+1}. Another (Swift and Lee, 1983) is to solve $\nabla^2\underline{A_T}^{N+1/2} = -4\pi\underline{J_T}^{N+1/2}/c$, expressing $\underline{A_T}^{N+1/2}$ as an average between $\underline{A_T}^N$ and $\underline{A_T}^{N+1}$. In either case the resulting equation is implicit in that it involves $\underline{A_T}^{N+1}$ on the right hand side. This adds a slight complication, known as double area weighting, in gathering the moments, as described by Forslund et al. (1972), Forslund (1974) and Mankofsky et al. (1981). Once $\underline{A_T}^{N+1}$ is known, $\underline{E}$ and $\underline{B}$ can be obtained using (8). An equation for T_e, such as (9), can be solved (Kan and Swift (1983) use an equation of state instead), and then E_x can be obtained as before, from (10).

Finally, it should be noted that there is an extra complication if there is an external B_y or B_z in the system. A constant B_z can be added in through an additional term $A_y = B_z x$. This contribution to P_y of the particles must be added or subtracted if the boundary conditions are such that particles exiting at one end of the system reenter at the other end.

3. NUMERICAL COMPARISON OF THE MODELS

In this section we compare results of simulations based on the three hybrid models discussed previously for two test problems. The first is an electromagnetic ion beam instability, the second is a quasiperpendicular collisionless shock. Both problems have been well studied, and the reader is referred to the literature for details.

The first problem involves the ion beam instability, which is thought to be the mechanism which produces low frequency hydromagnetic waves and diffuse ion populations upstream of the earth's bow shock (e.g., see Gary *et al*. (1981) and Winske and Leroy (1984a)). The instability results from the interaction of a weak beam of ions backstreaming from the shock and the incident solar wind. The situation to be simulated consists of an ion beam drifting relative to an ambient plasma along a uniform magnetic field B_o. The parameters are chosen to match those in Winske and Leroy (1984a) for the resonant instability: the beam is weak, 1.5% of the total ion density, with a drift speed of 10 V_A relative to the background ions. Both ion components have $\beta=8\pi nT/B_o^2=1.0$, with a cold ($\beta_e=0.01$) current-neutralizing electron background. The instability leads to the generation of low frequency waves of well defined wavelength that scatter the beam. In the nonlinear regime these regular waves break up into very nonlinear waveforms, producing a diffuse ion distribution in the process.

In Figure 1 we compare results of three simulation. The top panels correspond to the resistive code with $\eta=0$; the middle panels are from the predictor-corrector code; the bottom panels are from a code based on the Hamiltonian method. In each case we have used 10,000 particles (half to represent the beam, half for the background ions) on a grid of 256 cells with cell size $\Delta x=c/\omega_i$. In each case we use the same random numbers to initialize the particle velocities. The left side of the figure shows one component of the magnetic field, B_z, normalized in terms of the ambient field B_o at about the time when the waves have achieved their largest amptitude, $\Omega_i t=38.4$ ($\Omega_i=eB_o/m_i c$). The results from the three cases are very similar, differing only in the amount of low amptitude, short wavelength noise that is superimposed on the dominant structures. The right side of the figure shows the time histories of the fluctuating magnetic field energy density, $W_b=\int dxB_T^2/\int dxB_o^2$. Again, the overall results in each case agree very well. The peak field energy density achieved is slightly (~3%) larger in the predictor-corrector case, which may be a reflection of the better energy conservation in the predictor-corrector code ($\Delta E/E_o$~0.03%), compared with the resistive code ($\Delta E/E_o$~1.2%) and the Hamiltonian code ($\Delta E/E_o$~5.4%). The poorer energy conservation in the Hamiltonian code suggests that the differencing scheme used here for the test problem could be improved and should not be taken as an indication that this method is inherently inferior.

The second test problem involves a quasiperpendicular collisionless shock. Again, in this case the physics has been investigated in detail (Leroy *et al*, 1981 and 1982; Leroy and Winske, 1983; Forslund *et al*., 1984). The simulation is initialized with uniform upstream and downstream states related by Rankine-Hugoniot conditions and then is allowed to evolve in time. The parameters chosen for the test case are upstream Mach number $M_A=V_1/V_A=8$, upstream shock normal angle $\theta_{Bn}=60°$, and upstream $\beta_e=\beta_i=0.5$. Again, we show (Figure 2) the results of three simulations: top panels correspond to the resistive code with resistive length $L_R=(\eta\omega_i/4\pi)$

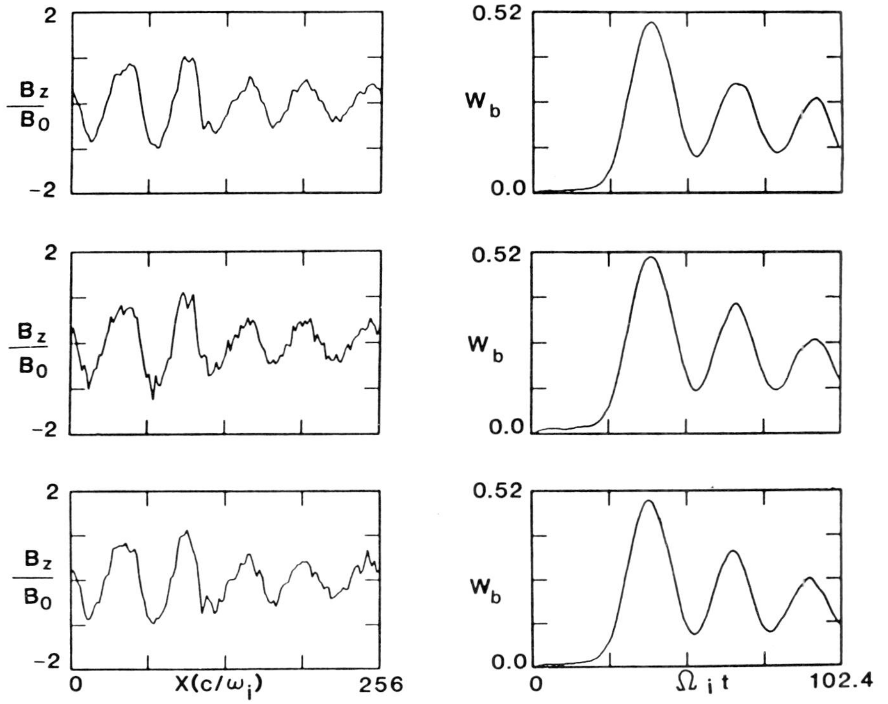

Figure 1. Results of three different hybrid simulations of the electromagnetic ion beam instability (top: resistive model; middle: predictor-corrector model; bottom: Hamiltonian model) showing: (left) one component of the fluctuating magnetic field at $\Omega_i t = 38.4$, (right) time history of the fluctuating magnetic field energy density.

$(c/V_1)(c/\omega_i)$ equal to the cell size $\Delta x=0.3\ c/\omega_i$, middle panels correspond to the results of the resistive code with $L_R=0.01\Delta x$, and bottom panels correspond to the resistance-free predictor-corrector code. In each case 10,000 particles on a grid of 200 cells with a time step $\Omega_i \Delta t=0.0125$ (where the upstream magnetic field B, is used to compute Ω_i).

The left side of Figure 2 shows one component of the transverse magnetic field B_Z at $\Omega_i t=12.5$. The upstream magnetic field is to the left in each picture; the shock occurs in the center where the magnetic field rises rapidly. The peak value of the magnetic field component B_Z is somewhat larger than the downstream value B_{Z2} (computed from the Rankine-Hugoniot relations), which is held fixed at the right hand boundary. This peak value, referred to as the overshoot, is followed by an undershoot and then several more oscillations. The middle panel shows the same snapshot of the magnetic field in the run where the resistivity is lowered. The same overall stucture is observed, except that the overshoot is larger and the oscillations behind the overshoot are somewhat less regular. In addition, there are small oscillations on the magnetic field in the upstream region and the sharp rise of the magnetic field is better resolved, consisting of a precursory smaller increase (called the foot) and a very steep ramp. The field profile obtained with the predictor-corrector code (bottom panel) shows the same general features as in the second case, except that the oscillations behind the shock are somewhat larger in amplitude and have a shorter wavelength.

The right hand panels of Figure 2 show the time histories of the magnetic overshoot for the three runs. In the case with the larger resistivity, the overshoot has a nearly constant value. For small resistivity, the overshoot oscillates in time and the average value is about 20% larger. The resistance-free predictor-corrector code gives a slightly larger average overshoot with larger, more frequent oscillations.

We conclude from these results that the three shock simulations give overall similar results, but the effect of resistivity is to damp out some of the oscillations. This naturally leads to the question of how much resistivity (if any) should be included. It should be kept in mind that the resistivity is added in to compensate for the fact that the simulations are one-dimensional and therefore do not include microinstabilities due to the cross-field current that are seen, for example, in 2-D particle simulations of shocks (Forslund et al., 1984). In the present simulations, as in Leroy et al. (1982), the resistivity is taken as a constant, although more realistic forms for the resistivity, either based on phenomenological expressions (Chodura, 1975) or microphysics (Hamasaki et al., 1977) are possible. In this regard, one has to be guided by space observations or laboratory data (which indeed show rather steady structures) to infer the proper amount of resistivity that should be used.

As a final note, the question of resistivity becomes increasingly complex as the Mach number is raised. Quest (1985) has recently carried out resistive hybrid simulations of perpendicular shocks with Mach numbers greater than 20. He finds that a shock which is fairly steady in time can be produced with resistivity such that $L_R \approx \Delta x$, but that with weaker resistivity the magnitude of the oscillations of the overshoot is comparable to its average value. In this case the shock steepens to a very narrow ($\sim\Delta x$) width, then collapses. Because the amount of resistivity that would be needed to maintain the shock steady is unphysically large at high Mach numbers, it is suggested that very high Mach number shocks indeed are oscillatory in character (Quest, 1985).

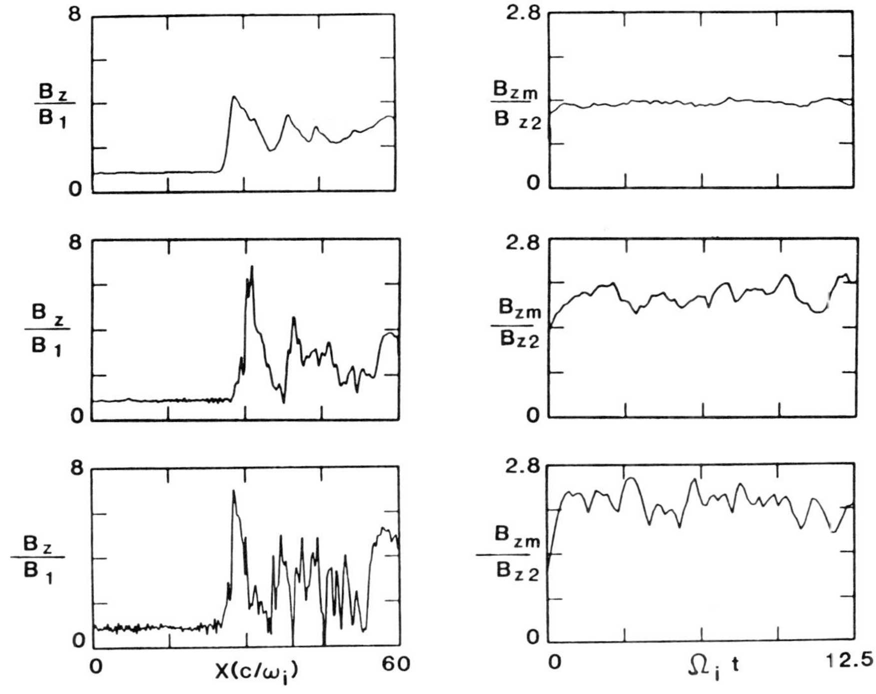

Figure 2. Results of three hybrid simulations for a quasiperpendicular shock (top: resistive code with large resistivity; middle: resistive code with smaller resistivity; bottom: predictor-corrector code) showing: (left) one component of the magnetic field at $\Omega_i t = 12.5$, (right) time history of the magnetic field overshoot.

4. SUMMARY

We have described hybrid codes in which the various plasma species are given different representations. Specializing to the important case where the ions are treated as particles and the electrons are a massless, charge-neutralizing fluid, we have discussed in some detail three ways for solving the field equations, referred to here as the resistive, predictor-corrector, and Hamiltonian methods. Using simulation codes based on each of these techniques, we have compared results for two problems of current interest in space physics: low frequency waves driven by an ion beam and quasiperpendicular collisionless shocks. For the ion beam problem all three methods give essentially the same results, with the predictor-corrector method giving better overall energy conservation. In the case of the quasiperpendicular shock, the effect of the resistivity on producing time steady solutions has been emphasized. While the use of these codes in one spatial dimension has been stressed throughout this article, all three methods discussed work in two dimensions, and will undoubtedly become widely applied in space physics, as they have already in magnetic fusion problems.

ACKNOWLEDGEMENTS. This work was supported by the NASA Solar-Terrestrial Theory Program and the Department of Energy.

REFERENCES

Ambrosiano, J.J., L.C. Lee, and D.W. Swift, 'Simulation of the ion tearing instability in the presence of a background plasma', J. Geophys. Res., 88, 7860, 1983.

Byers, J.A., B.I. Cohen, W.C. Condit, and J.D. Hanson, 'Hybrid simulations of quasineutral phenomena in magnetized plasma', J. Comp. Phys., 27, 363, 1978.

Chodura, R., 'A hybrid particle-fluid model of ion heating in high Mach number shock waves', Nucl. Fusion, 15, 55, 1975.

Dickman, D.O., R.L. Morse, and C.W. Nielson, 'Numerical simulation of axisymmetric collisionless finite-beta plasma', Phys. Fluids, 12, 1708, 1969.

Forslund, D.W., 'Plasma simulation techniques and their use', Los Alamos Scientific Laboratory Report, LA-UR 74-1333, 1974.

Forslund, D.W., E.L. Lindman, R.W. Mitchell, and R.L. Morse, 'EMI a general purpose one-dimensional electromagnetic particle simulation code', Los Alamos Scientific Laboratory Report, LA-DC 72-721, 1972.

Forslund, D.W., K.B. Quest, J.U. Brackbill, and K. Lee, 'Collisionless dissipation in quasiperpendicular shocks', J. Geophys. Res., 89, 2142, 1984.

Friedman, A., R.L. Ferch, R.N. Sudan, and A.T. Drobot, 'Numerical simulation of strong proton rings', Plasma Phys., 19, 1101, 1977.

Gary, S.P., J.T. Gosling, and D.W. Forslund, 'The electromagnetic ion beam instability upstream of the earth's bow shock', J. Geophys. Res., 86, 6691, 1981.

Hada, T., and C.F. Kennel, 'Nonlinear evolution of slow waves in the solar wind', J.Geophys. Res., 90, in press, 1985.

Hamasaki, S., N.A. Krall, C.E. Wagner, and R.N. Byrne, 'Effect of turbulence on theta pinch modeling by hybrid numerical models', Phys. Fluids, 20, 65, 1977.

Harned, D.S., 'Quasineutral hybrid simulation of macroscopic plasma phenomena', J. Comp. Phys., 47, 452, 1982.

Harned, D.S., 'Rotational instabilities in the field reversed configuration: results of hybrid simulations', Phys. Fluids, 26, 1320, 1983.

Hewett, D.W., 'A Global method for solving the electron-field equations in a zero-inertia-electron hybrid plasma simulation code', J. Comp. Phys., 38, 378, 1980.

Hewett, D.W., 'Spontaneous development of toroidal magnetic field during formation of field reversed theta pinch', Nucl. Fusion, 24, 349, 1984.
Hewett, D.W., and C.W. Nielson, ' A multidimensional quasineutral plasma simulation model', J. Comp. Phys., 29, 219, 1978.
Kan, J.R., and D.W. Swift, 'Structure of the quasiparallel bow shock: results of numerical simulations', J. Geophys. Res., 88, 6919, 1983.
Lee, J.K., and C.K. Birdsall, 'Velocity space ring plasma instability, magnetized, part II: simulations', Phys. Fluids, 22, 1315, 1979.
Leroy, M.M., C.C. Goodrich, D. Winske, C.S. Wu, and K. Papadopoulos, 'Simulation of a perpendicular bow shock', Geophys. Res. Lett., 8, 1269, 1981.
Leroy, M.M., D. Winske, C.C. Goodrich, C.S. Wu, and K. Papadopoulos, 'The structure of perpendicular bow shocks', J. Geophys. Res., 87, 5081, 1982.
Leroy, M.M., and D. Winske, 'Backstreaming ions from oblique earth bow shocks', Ann. Geophysicae, 1, 527, 1983.
Mandt, M.E., and J.R. Kan, 'Effects of electron pressure in quasiparallel collisionless shocks', J. Geophys. Res., 90, in press, 1985.
Mankofsky, A., A. Friedman, and R.N. Sudan, 'Numerical simulation of injection and resistive trapping of ion rings', Plasma Phys., 23, 521, 1981.
Mankofsky, A., and R.N. Sudan, 'Numerical simulation of ion beam propagation in Z pinch plasma channels', Nucl. Fusion, 24, 827, 1984.
Morse, R.L., and C.W. Nielson, 'Numerical simulation of the Weibel instability in one and two dimensions', Phys. Fluids, 14, 830, 1971.
Nielson, C.W., and H.R. Lewis, 'Particle code models in the nonradiative limit', in Methods of Computational Physics, Vol. 16, edited by J. Killeen, B. Alder, S. Fernbach, and M. Rotenberg, p. 367, Academic Press, New York, 1976.
Okuda, H., J.M. Dawson, A.T. Lin, and C.C. Lin, 'Quasineutral particle simulation model with application to ion wave propagation', Phys. Fluids, 21, 476, 1978.
Omura, Y., M. Ashour-Abdalla, K. Quest, and R. Gendrin, 'Heating of thermal helium in the equatorial magnetosphere: a simulation study', J. Geophys. Res., 90, in press, 1985.
O'Neil, T.M., J.H. Winfrey, and J.H. Malmberg, 'Nonlinear interaction of a small cold beam and a plasma', Phys Fluids, 14, 1204, 1971.
Quest, K.B., 'Simulations of high Mach number collsionless shocks', Phys. Rev. Lett., submitted, 1985.
Sgro, A.G., and C.W. Nielson, 'Hybrid model studies of ion dynamics and magnetic field diffusion during pinch implosions', Phys. Fluids, 19, 126, 1976.
Swift, D.W., 'A two-dimensional simulation of the interaction of the plasma sheet with the lobes of the earth's magnetotail', J.Geophys. Res., 88, 125, 1983a.
Swift, D.W., 'On the structure of the magnetic slow switch-off shock', J.Geophys. Res., 88, 5685, 1983b.

Swift, D.W., and L.C. Lee, 'Rotational discontinuities and the structure of the magnetopause', J.Geophys. Res., 88, 111, 1983.

Tanaka, M., and C.C. Goodrich, 'Simulation study of heavy ion heating by electromagnetic ion cyclotion waves induced by proton temperature anisotropy', J. Geophys. Res., 90, in press, 1985.

Teresawa, T., 'Numerical study of explosive tearing mode instability in one component plasmas', J. Geophys. Res., 86, 9007, 1981.

Winske, D., and M.M. Leroy, 'Diffuse ions produced by electromagnetic ion beam instabilities', J. Geophys. Res., 89, 2673, 1984a.

Winske, D., and M.M. Leroy, 'Hybrid simulation techniques applied to the earth's bow shock', in Computer Simulations of Space Plasmas, edited by H. Matsumoto and T. Sato, D. Reidel/Terra, Hingham, Mass., 1984b.

Winske, D., C.S. Wu, Y.Y. Li, and G.C. Zhou, 'Collective capture of released lithium ions in the solar wind', J. Geophys. Res., 89, 7327, 1984.

Winske, D., C.S. Wu, Y.Y. Li, Z.Z. Mou, and S.Y. Guo, 'Coupling of newborn ions to the solar wind by electromagnetic instabilities and their interaction with the bow shock', J. Geophys. Res., 90, in press, 1985.

IMPLICIT PLASMA SIMULATION

A. Bruce Langdon
Lawrence Livermore Laboratory
Physics Department
University of California
Livermore, California, 94550

ABSTRACT. Implicit time integration methods have been used extensively in numerical modelling of slowly varying phenomena in systems that also support rapid variation. Examples include diffusion, hydrodynamics and reaction kinetics. This article discusses implementation of implicit time integration in plasma codes of the "particle-in-cell" family, and the benefits to be gained.

1. Introduction

In plasmas, many instability, dissipation and nonlinear saturation mechanisms are kinetic in nature, i.e., fluid descriptions are inadequate. The most adaptable and reliable tools for study of complex kinetic plasma behavior are the "particle-in-cell" (PIC) codes, in which the plasma evolution is modeled by 10^3–10^7 simulation "particles", each representing a large number of plasma particles, and moving according to the classical Newton-Lorentz equations of motion in fields governed by Maxwell's equations. Characteristic time scales for collective phenomena in plasmas encompass many orders of magnitude. Already applied very successfully to studies of the nonlinear evolution of plasma phenomena on the faster time scales, there is great value in extending PIC techniques to kinetic phenomena on much longer time scales.

One approach to modeling long time-scale behavior in such systems is to alter the governing equations to eliminate uninteresting high frequency modes. Examples, such as Darwin field and hybrid PIC-fluid approximations, are described at this meeting and in chapters in the volume "Multiple Time Scales" in the series *Computational Techniques* (Academic Press, 1985). Here I discuss implicit time integration, which transforms the high-frequency modes into *damped* modes.

The divisions of this article are as follows: The remainder of this Section defines the explicit electrostatic PIC algorithm, introducing the notation to be used throughout this article, and outlines the properties of explicit differencing and an implicit scheme.

Space Science Reviews **42** (1985) 67–83. 0038–6308/85.15

Section 2 presents the "direct" and "moment" implicit methods as applied to the electrostatic field case, along with some results, and discusses remaining limitations on time step, conservation properties, and linear stability theory. In Section 3 the direct method is generalized to include the full electromagnetic field.

Langdon and Barnes (1985) provide more detail on all these subjects and others, plus many references.

1.1. Characteristic Time Scales in Weakly-Collisional Plasma

The highest frequencies are associated with the small mass of electrons and the high speed of light. The Langmuir frequency, ω_{pe}, also called simply the plasma frequency, characterizes charge-separation oscillations. Others include the cyclotron frequency, ω_{ce}, and the transit time for electrons or light to cross a characteristic distance.

In contrast, long time scales can be set by ion inertia, electromagnetic effects, and large spatial scale lengths. The ratios of electron to ion plasma and cyclotron frequencies, and of hydrodynamic to electron transit times, are determined by the small number Zm_e/m_i, where Z is the ionic charge state. Where the dominant forces are from magnetic fields due to currents in the plasma itself, the frequencies are reduced relative to ω_{pe} by at least the ratio c/v_e, where c and v_e are light and electron speeds.

1.2. PIC Electrostatic Force Calculation

Originally developed at Stanford, PIC plasma simulation models use a spatial grid to mediate the particle interactions. A charge density is formed from the particle positions onto a spatial grid. Using partial difference equations on this grid, an electric field is found. Then the particles are individually advanced in time using classical equations of motion with the acceleration found by interpolation from the electric field on the grid.

To avoid clutter, we temporarily restrict the discussion to one dimension. At the grid point located at $X_j \equiv j\Delta x$, where Δx is the grid spacing, the charge density ρ_j is formed by adding the contributions of the simulation particles at positions $\{\mathbf{x}_i\}$

$$\rho_j = \sum_i q_i S(X_j - x_i) \qquad (1.1)$$

where i is the particle index, q_i the charge, and S the particle-grid interpolation spline. Most present work is done with multilinear weighting. In one dimension, S is the linear spline, a tent-shape extending over two cells and zero elsewhere.

From ρ a potential and an electric field are derived on the same mesh. These satisfy relations such as

$$E_j = - \frac{\phi_{j+1} - \phi_{j-1}}{2\Delta x} \tag{1.2}$$

$$\frac{\phi_{j+1} - 2\phi_j + \phi_{j-1}}{\Delta x^2} = - \rho_j \tag{1.3}$$

in rationalized cgs units. Additional smoothing is often applied to ρ_j or ϕ_j.

The particle force field is obtained by an interpolation of the form

$$F_i = q_i \Delta x \sum_j E_j \, S(X_j - x_i) \tag{1.4}$$

where the sum is over grid points j, and S is the same function used in (1.1).

1.3. Explicit Time Differencing of Particle Equations of Motion

For the integration of the particle equations of motion, numerical time differencing algorithms of the elegant type discussed by Buneman (1967) are almost universally used.

$$\frac{x_{n+1} - x_n}{\Delta t} = v_{n+\frac{1}{2}}; \qquad \frac{v_{n+\frac{1}{2}} - v_{n-\frac{1}{2}}}{\Delta t} = \mathbf{a}_n + \frac{v_{n+\frac{1}{2}} + v_{n-\frac{1}{2}}}{2} \times \frac{qB_n}{mc} \tag{1.5ab}$$

$$\text{where } \mathbf{a}_n = \frac{q}{m} E_n(x_n), \tag{1.5c}$$

and subscript n denotes time level $t_n = n\Delta t$. This is usually called the centered leapfrog scheme.

Equations (1.1)–(1.5) outline the time cycle of an explicit PIC simulation.

In the simplest, unmagnetized case, longitudinal oscillations in a uniform cold plasma consist simply of harmonic oscillations. The properties of the leapfrog scheme are illustrated by considering a single particle with a linear restoring force, $a_n = -\omega_0^2 x_n$, with no magnetic field term. Setting $x_n = X \exp(-i\omega n\Delta t)$, we find for $\omega_0 \Delta t < 2$ that ω is real; the oscillations are neither growing nor damped. For $\omega_0 \Delta t > 2$, one root has Im $\omega > 0$; this numerical instability arises for *any explicit* scheme for $\omega_0 \Delta t$ above some threshold of order unity (Cohen *et al*, 1982). Instability can be avoided through the use of implicit time integration, at the expense of increased complexity.

1.4. Implicit Time Differencing of the Particle Equations of Motion

The first major issue is the choice of finite-differenced equations of motion for the particles that have the necessary stability at large time-step and are accurate enough for the low frequency phenomena to be simulated. We choose not to consider backward-biased schemes with relative errors of order Δt; it is not expensive to achieve relative error of order Δt^2, with error $\propto \Delta t^3$ in Im ω, the growth/decay rate.

Several suitable schemes for time-differencing the particles have been analyzed and applied. Here, we discuss only the "D_1" scheme (Cohen *et al*, 1982), also called the $\bar{1}$ scheme (Barnes *et al*, 1983), which can be written

$$\frac{x_{n+1} - x_n}{\Delta t} = v_{n+\frac{1}{2}}; \qquad \frac{v_{n+\frac{1}{2}} - v_{n-\frac{1}{2}}}{\Delta t} = \bar{a}_n + \frac{v_{n+\frac{1}{2}} + v_{n-\frac{1}{2}}}{2} \times \frac{qB_n}{mc} \quad (1.6ab)$$

$$\text{where } \bar{a}_n = \tfrac{1}{2}[\bar{a}_{n-1} + \tfrac{q}{m}E_{n+1}(x_{n+1})]. \quad (1.7)$$

To check the accuracy of this scheme, we can again derive and solve a dispersion relation for harmonic oscillations. For $\omega_0\Delta t \lesssim 1$, we find (Cohen *et al*, 1982) $\pm$Re $\omega/\omega_0 = 1 - (11/24)(\omega_0\Delta t)^2 + \ldots$, Im $\omega/\omega_0 = -(\omega_0\Delta t)^3/2 + \ldots$, and an extraneous damped mode with $|z| \to \frac{1}{2}$. For $\omega_0\Delta t \gg 1$, all modes are heavily damped, $|z| \to (\omega_0\Delta t)^{-2/3}$.

Equation (1.6b) can be solved exactly for $v_{n+\frac{1}{2}}$ by adding $\frac{1}{2}\bar{a}_n\Delta t$ to $v_{n-\frac{1}{2}}$, doing a rotation, and again adding $\frac{1}{2}\bar{a}_n\Delta t$. The result is

$$v_{n+\frac{1}{2}} = \tfrac{1}{2}\bar{a}_n\Delta t + R\cdot[v_{n-\frac{1}{2}} + \tfrac{1}{2}\bar{a}_n\Delta t] \quad (1.8)$$

where the operator R effects a rotation through angle $-2\tan^{-1}(\Omega\Delta t/2)$ (where $\Omega \equiv qB_n/mc$).

With explicit differencing, the time-cycle is split between advancing particles and fields; these calculations alternate and proceed independently. A price we pay for implicit differencing is that time-cycle splitting is more complicated. An implicit code must solve the coupled set of equations (1.1)–(1.4), with (1.6–1.7).

2. Direct Method With Electrostatic Fields

In all implicit schemes the future positions x_{n+1} depend on the accelerations a_{n+1} due to the electric field E_{n+1}. But this field is not yet known, because it depends on the density ρ_{n+1} of particle positions $\{x_{n+1}\}$. The solution of this large system of nonlinear, coupled particle and field equations is the other major implementation issue.

2.1. Solution of the Implicit Equations

In the first method implemented for this solution, the fields at the new time level are predicted by solving coupled field and fluid equations. After the fields are known, the particles are advanced to the new time-level. This approach, described in detail by Denavit (1981) and Mason (1981a), is outlined later in this article.

It is also practical to predict the future electric field $\mathbf{E}_{n+1}$ quite directly by means of a linearization of the particle-field equations. Langdon, Cohen and Friedman (1983) and Langdon and Barnes (1985), together with references they cite, cover the work published on the direct method up to early 1984.

2.1.1. Outline of the "direct method" The essence of the "direct" method is that we work *directly* with the particle equations of motion and the particle/field coupling equations. These are linearized about an estimate (extrapolation) for their values at the new time level n+1. The future values of $\{\mathbf{x},\mathbf{v}\}$ are divided into two parts:

- increments $\{\delta\mathbf{x}, \delta\mathbf{v}\}$ which depend on the (unknown) fields at the future time level n+1, and
- extrapolations $\{\tilde{\mathbf{x}}_{n+1}, \tilde{\mathbf{v}}_{n+\frac{1}{2}}\}$ which incorporate *all* other contributions to the equation of motion.

The charge density $\tilde{\rho}_{n+1}$, corresponding to positions $\{\tilde{\mathbf{x}}_{n+1}\}$, is collected, as are the coefficients in an expression for the difference $\delta\rho(\{\mathbf{E}\}) = \rho_{n+1} - \tilde{\rho}_{n+1}$ between the densities obtained after integration with $\mathbf{E} = 0$ and with the corrected field $\mathbf{E}_{n+1}$. These comprise the source term in Gauss's law:

$$\nabla\cdot\mathbf{E}_{n+1} = \delta\rho(\{\delta\mathbf{E}\}) + \tilde{\rho}_{n+1} \tag{2.1}$$

This becomes a linear elliptic equation for ϕ_{n+1}, with non-constant coefficients.

2.1.2. A one-dimensional realization The direct implicit method is illustrated in the following one-dimensional, unmagnetized, electrostatic example. The position x_{n+1} of a particle at time level t_{n+1}, as given by an implicit time integration scheme, can be written as

$$x_{n+1} = \beta\Delta t^2 a_{n+1} + \tilde{x}_{n+1}, \tag{2.2}$$

where $\beta > 0$ is a parameter controlling implicitness and is equal to $\frac{1}{2}$ for the D_1 scheme; $\tilde{x}_{n+1}$, the position obtained from the equation of motion with the acceleration a_{n+1} omitted, is known in terms of positions, velocities and accelerations at times t_n and earlier. Eliminating $v_{n+\frac{1}{2}}$ between (1.6) and (1.7), we find

$$\tilde{x}_{n+1} = x_n + v_{n-\frac{1}{2}}\Delta t + \tfrac{1}{2}\bar{a}_{n-1}\Delta t^2 \tag{2.3}$$

The charge density $\tilde{\rho}_{j,n+1}$ is formed as in (1.1) by adding the contribution of the simulation particles at positions $\{\tilde{x}_{i,n+1}\}$.

$$\tilde{\rho}_{j,n+1} = \sum_i q_i S(X_j - \tilde{x}_{i,n+1}) \tag{2.4}$$

If we expand S in (1.1) with respect to position, then

$$\delta\rho_{j,n+1} = -\sum_i q_i \ \delta x_i \ S'(X_j - \tilde{x}_{i,n+1}) \tag{2.5}$$

with $\delta x_i = x_{i,n+1} - \tilde{x}_{i,n+1}$ and $S'(X) = dS/dX$. In terms of E_{n+1}, the particle acceleration is obtained from (1.4) evaluated at $\tilde{x}_{n+1}$.

$$m_i a_{i,n+1} = q_i \Delta x \sum_j E_{j,n+1} \ S(X_j - \tilde{x}_{i,n+1}) \tag{2.6}$$

Using (2.6) to express δx in (2.5) provides a "strict" interpretation of the direct method. The care with which we express the increment $\{\delta\rho\}$ is a compromise between complexity, robust behavior, and strong convergence (Langdon *et al*, 1983; Barnes *et al*, 1983; Langdon and Barnes, 1985). Here we discuss only the simpler implementations.

Writing (2.5) as

$$\delta\rho = -\left[\nabla\cdot\sum q \ \delta\mathbf{x} \ S(\mathbf{x} - \tilde{\mathbf{x}}_{n+1})\right]_{\mathbf{x}=X_j} \tag{2.7}$$

we see that (2.5) is a finite element representations of

$$\delta\rho = -\nabla\cdot[\tilde{\rho}\delta\mathbf{x}], \tag{2.8}$$

With (2.1) we are lead to the elliptic partial difference equation

$$\nabla\cdot[1+\chi]\cdot\nabla\phi_{n+1} = -\tilde{\rho}_{n+1} \tag{2.9}$$

where $\chi(\mathbf{x}) = \beta\tilde{\rho}(\mathbf{x})(q/m)\Delta t^2$ summed over species, i.e. $\chi = \beta(\omega_p\Delta t)^2$. Equation (2.9) is an elliptic field equation whose coefficients depend directly on particle data accumulated on the spatial grid. Because of the similarity of (2.9) to the field equation in dielectric media, we call χ the implicit susceptibility. Where $\omega_p\Delta t$ is large, the regime we wish to access, note that $\chi >> 1$ is dominant in the right-hand-side of (2.9).

With the extrapolated charge density $\tilde{\rho}_{j,n+1}$ and a reasonable finite-difference representation of the linearized implicit contribution $\delta\rho = -\partial(\chi E)/\partial x$, the field equation in one dimension is

$$\tilde{\rho}_{j,n+1} = \left[(1+\chi_{j+\frac{1}{2}})E_{j+\frac{1}{2},n+1} - (1+\chi_{j-\frac{1}{2}})E_{j-\frac{1}{2},n+1}\right]/\Delta x, \tag{2.10}$$

Two representation of $\chi_{j+\frac{1}{2}}$ used here are (Langdon *et al*, 1983, Eqs. (28ab))

$$\chi_{j+\frac{1}{2}} = \tfrac{1}{2}\left[\chi_j + \chi_{j+1}\right] \tag{2.11a}$$

$$\text{or } \chi_{j+\frac{1}{2}} = \max\left(\chi_j, \chi_{j+1}\right), \tag{2.11b}$$

$$\text{where } \chi_j = \Delta t^2 \sum_s \left[\beta\tilde{\rho}_{j,n+1}\frac{q}{m}\right]_s \tag{2.12}$$

is a sum over *species* index s. In both (2.10) and (2.12), $\tilde{\rho}_{j,n+1}$ (for each species differing in q/m) is given by (2.4).

In terms of the field

$$E_{j,n+1} = \tfrac{1}{2}\,[E_{j-\frac{1}{2},n+1} + E_{j+\frac{1}{2},n+1}] \tag{2.13}$$

formed from E at half-integer positions, the particle acceleration is evaluated at $\tilde{x}_{n+1}$ using (2.6). This algorithm is the shortest implicit PIC scheme we have seen, and was the most robust in the expansion test problem reported by Langdon and Barnes (1985).

<u>2.1.3. General electrostatic case</u> We return to the multidimensional case, possibly including a magnetic field imposed by external currents. We have

$$\delta x_{n+1} = \delta v_{n+\frac{1}{2}}\Delta t = T\cdot\frac{q\Delta t^2}{2m}\mathbf{E}_{n+1}(\tilde{x}_{n+1}), \tag{2.14}$$

where $T \equiv \frac{1}{2}[I+R_n(\tilde{x}_{n+1})]$, which follows from (1.8).

The implicit term $\delta\rho = -\nabla\cdot(\rho\delta\tilde{x}_{n+1})$ is seen to be

$$\delta\rho = -\nabla\cdot(\tilde{\rho}_{n+1}\delta x) = -\nabla\cdot(\chi\cdot\mathbf{E}) \tag{2.15}$$

The implicit susceptibility

$$\chi = \left[\sum_s \frac{\tilde{\rho}_{n+1,s}q_s\Delta t^2}{2m_s}\,T_s\right] \tag{2.16}$$

is a *tensor* due to the rotation R induced by **B**. The [...] is a sum over *species* s, not each particle. If only the electrons are implicit, only they appear in (2.16). In this case, the terms [...] require only a knowledge of the electrons' ρ [in addition to the *net* ρ used on the right side of (2.17)]. In general, it is sufficient to accumulate $\tilde{\rho}_{n+1,s}$ separately from species with differing q/m. This requires more storage, but *no more computation* than for an explicit code.

We now have everything needed to write an equation for $\mathbf{E} = -\nabla\phi$. On substituting our expressions for $\tilde{\rho}_{n+1}$ and $\delta\rho$ into the field equation (2.1) we again have the elliptic equation (2.9), except that χ has become a tensor.

If spatial smoothing, denoted by the operator $\hat{S}$, is to be applied to ρ and ϕ on the grid, then $\hat{S}$ must be included in χ if the field solution is to take this into account. In some applications, this has been essential (Sections 2.2.2 and 2.2.3; Barnes *et al*, 1983). Inconsistent smoothing has consequences to linear stability (Section 2.3.4).

The field $\mathbf{E}_{n+1}$ is evaluated at positions $\{\tilde{\mathbf{x}}_{n+1}\}$ in (1.4) in integrating the particles to their final positions $\{\mathbf{x}_{n+1}\}$. The error resulting from this approximation, and from the linearization of $\delta\rho$, introduces a possible limitation on Δt that depends on field and density gradients; see section 2.3.1.

After advancing each particle to position x_{n+1}, one can immediately calculate its $\tilde{x}_{n+2}$ and its contribution to $\tilde{\rho}_{n+2}$. In this way, only one pass through the particle list is required per time step, an advantage when the particles are stored on slower memory devices, such as rotating magnetic disk.

2.1.4. The "Moment" Implicit Method The "moment implicit" method uses velocity moment equations to predict ρ in terms of the future E. The moment equations are closed by accumulating flux and stress tensor moments from the particles at the *known* time level.

$$\tilde{J}_{n+\frac{1}{2}} = J_{n-\frac{1}{2}} - \frac{q\Delta t}{m}\left[\frac{\partial P_n}{\partial x} - \rho_n E_{n+1}\right] \qquad \text{-backward Euler}$$

$$\frac{\partial}{\partial x} E_{n+1} = \rho_{n+1} \cong \rho_n - \Delta t \frac{\partial}{\partial x} \tilde{J}_{n+\frac{1}{2}}$$

where $(\rho, J, P) \equiv n\langle q, qv, mvv\rangle$ are formed as sums over known particle coordinates. Eliminating $\tilde{J}_{n+\frac{1}{2}}$,

$$\frac{\partial}{\partial x} E_{n+1} = \rho_n - \Delta t \frac{\partial}{\partial x}\left[J_{n-\frac{1}{2}} - \frac{q\Delta t}{m}\frac{\partial P_n}{\partial x}\right] - \frac{q}{m}\Delta t^2 \frac{\partial}{\partial x}(\rho_n E_{n+1})$$

This equation is solved for E_{n+1}, with which the particles are integrated.

The field predictor (2.9) has some features in common with the field predictor in the moment method. This is not surprising since both methods attempt an approximate solution of similar equations. The relation of terms arising in the implicit moment and direct implicit viewpoints is discussed by Mason (1985) and Langdon *et al* (1983); here we summarize. The equation for E_{n+1} can be rewritten as

$$\frac{\partial}{\partial x} E_{n+1} = \tilde{\rho}_{n+1} - \frac{\partial}{\partial x}(\chi_n E_{n+1})$$

where we have expressed the moment-method terms using symbols from the direct method. The counterpart to χ in the moment method is approximated from the density ρ_n rather than being formed from $\tilde{\rho}_{n+1}$. The source term $\tilde{\rho}_{n+1}$ is replaced by ρ_n plus the divergence of current and stress tensor terms,

$$\tilde{\rho}_{n+1} \cong \rho_n - \Delta t \frac{\partial}{\partial x}\left[J_{n-\frac{1}{2}} - \frac{q\Delta t}{m}\frac{\partial P_n}{\partial x}\right]$$

If there could be no finite-difference errors, this combination would equal $\tilde{\rho}_{n+1}$. These approximations result in a stability constraint $kv_t\Delta t$

$\lesssim 1$ or $k\bar{v}\Delta t \lesssim 1$ (Section 2.3.1) that does not arise in the direct method, as well as requiring more computation.

Insight gained from the direct method shows how to eliminate the stress tensor, which contributes to the $kv_t\Delta t$ constraint in moment codes (Brackbill and Forslund, 1985; Wallace *et al*, 1984). Experience with moment codes using ad hoc spatial differencing encouraged experimentation with simpler differencing in direct codes. As moment and direct codes are borrowing features from each other, the distinction becomes more one of viewpoint in deriving algorithms and less in the resulting codes themselves. Design of optimal codes requires understanding the fruits of both viewpoints.

2.2. Demonstrative results

2.2.1. Free expansion of a plasma slab Denavit (1981) used the one-dimensional expansion of a plasma slab as one check on his early moment-method code. Featuring sharp gradients initially and a large range of densities, this problem is also used to check and improve variations of direct-method algorithms. Langdon *et al* (1984), as outlined here, sought to isolate those forms most suitable for extension to a two-dimensional and electromagnetic code. Without resort to spatial smoothing, adequate accuracy and robust behavior were obtained. With $n\Delta x = 64$ particles per cell in the slab initially, and values of $\omega_{pe}\Delta t$ as large as 120 (much larger than have been reported previously for unmagnetized plasma), they find that two parameters measure the stress on the algorithm. One parameter is $v_t\Delta t/\Delta x$, the ratio of thermal electron transit distance per cycle to the zone size. With $v_t\Delta t/\Delta x > 1$, energy conservation is degraded in the absence of spatial smoothing.

The more important parameter is $\chi_1 = \beta q^2\Delta t^2/m|\Delta x|$, where q and m are the particle charge and mass, and $|\Delta x|$ the zone volume. This is a worst-case measure of the validity of linearization in the field prediction, as stressed by short-wavelength sampling fluctuations in the charge density. Using the algorithm of Section 2.1.2 with either (2.11a) or (2.11b), they obtain reasonable results with χ_1 well over 100. Most other variants suffer nonlinear numerical instability at the edge of the expansion when $\chi_1 \gtrsim 1$.

With χ_1 written as $\beta(\omega_{pe}\Delta t)^2/N_c$, where N_c is the number of particles per cell ($n\Delta x$ in one dimension), the value of being able to run with $\chi_1 >> 1$ is clarified. Many applications of two-dimensional explicit codes require only $N_c \gtrsim 10$. If we were restricted to $\chi_1 \lesssim 1$, we would need a number of particles per cell exceeding $(\omega_{pe}\Delta t)^2$, which would be a severe limitation.

2.2.2. Ion-acoustic fluctuations of a nonequilibrium plasma Ion-acoustic fluctuations of a uniform, thermal, unmagnetized, two-temperature, one-dimensional plasma are examined by Barnes *et al* (1983). Since the ion-acoustic fluctuations represent an extremely small part of the total fluctuation energy of a thermal plasma and are strongly affected by electron Landau damping .in the parameter range studied, these results represent a severe test of the applicability of the model.

The time step was fixed at $\omega_{pe}\Delta t = 10$, a factor of 50–100 over that possible for an explicit algorithm. The fluctuation spectra are compared for the cases in which the resonant electron response is retained by the direct implicit method, and in which the low-frequency electron response is only adiabatic. The ratio of the former to the latter is approximately T_e/T_i, which is 20 in the example. The fluctuation spectrum in the simulation indicates that the resonant electron response at low frequencies is described accurately by the implicit method. Also, fluctuations at mode frequencies higher than the ion-acoustic range have been suppressed.

2.2.3. Gravitational interchange instability of a magnetized plasma
Using the direct method, and treating both the electrons and ions as zero gyroradius particles, an unstable gravitational interchange is studied for an inhomogeneous two-dimensional magnetized plasma (Barnes *et al*, 1983).

The effect of the polarization motion of the ions is very important for this high-density plasma ($\omega_{pi}/\Omega_i \gg 1$). If the polarization motion is neglected, the growth rate becomes unphysically large for high density. For the simulation parameters considered here, the physical growth rate is 1/7 the growth rate found neglecting polarization. In the simulation, an extremely large time step of $\omega_{pe}\Delta t = 10^3$ was used. The observed growth rate is within 15% of the theoretically predicted value, verifying the correct modeling of ion polarization motion.

2.3. Properties of Implicit Particle Codes

Although we have largely overcome the stability limit on $\omega_{pe}\Delta t$, there remain restrictions that involve Δt. In addition to the aspects discussed under the following subheadings, Sections 2.3.3 and 2.3.4 are also relevant.

2.3.1 Doppler frequency limit When a particle moves in one time step a distance greater than a scale length L for spatial variation of fields, the particle does not sample the field sufficiently closely in space to respond accurately to the field structure. In analysis, this is stated in terms of the Doppler-shifted frequency $\omega - \mathbf{k}\cdot\mathbf{v}$. If the thermal spread of electron velocities $v_t = (T_e/m_e)^{\frac{1}{2}}$, and wavenumber k are such that $kv_t\Delta t \gtrsim 1$, then the Doppler-shifted frequency exceeds Δt^{-1} for a large fraction of the electrons. It is not surprising that the collective response is qualitatively incorrect, showing an excess of shielding, as indicated by the dielectric function becoming

$$\epsilon \cong 1 + \beta\omega_{pe}\Delta t^2 + (\text{ion response})$$

instead of the correct result,

$$\epsilon \cong 1 + (k\lambda_D)^{-2} + (\text{ion response})$$

Violation of the restriction $kv_t\Delta t \lesssim 1$ in direct-method electrostatic codes does not, in general, result in instability. In many applications, the sharp field gradients may arise only during time intervals or in spatial regions such that the inaccuracy does not interfere with the purpose of the simulation. In contrast, strong instability observed with the moment method has been traced to a constraint on $kv_t\Delta t$ (Sakagami *et al*, 1981; Denavit, private communication) and is attributed to lack of convergence (Denavit, 1981; Cohen *et al*, 1982a). Such instability, as opposed to innocuous inaccuracy, is disruptive to the simulation. The ability of a direct-implicit code to remain stable is useful with nonuniform meshes where some cell dimensions may be much smaller, and helps evade failures due to uninteresting transients or small regions.

The Doppler frequency, or transit time, limit is analyzed by Denavit (1981), Cohen *et al* (1982), Langdon *et al* (1983), Langdon and Barnes (1985), and references therein.

<u>2.3.2 Limits due to field and density gradients</u> Linearization of the time-advanced charge density breaks down when field or density gradients are too large. Nonetheless, the codes have been made to function successfully (Section 2.2.1), but convergence is not guaranteed.

Evaluation of $\delta\mathbf{x}$ in terms of $\mathbf{E}_{n+1}$ at $\tilde{\mathbf{x}}_{n+1}$ rather than at $\mathbf{x}_{n+1}$ creates an error $\cong \delta\mathbf{x}\cdot\nabla\mathbf{E}$. Denavit (1981) interprets this term in terms of the frequency of oscillation of a particle trapped in a local potential minimum. The relative error is $\beta(\omega_{trap}\Delta t)^2$. To compare the severity of this and the transit-time limitations, note that

$$\beta(\omega_{trap}\Delta t)^2 = \beta(kv_e\Delta t)^2 \, q\phi/T_e$$

Thus, if we take $kv_e\Delta t \lesssim 1$ as given, the linearization is justified if $q\phi \lesssim T$. Put another way, if $kv\Delta t \lesssim 1$, then $(\omega_{trap}\Delta t)^2 = ka\Delta t^2$ is small if $a\Delta t$, the impulse in one time step, is small compared to v. In grossly nonneutral regions, where the net charge $\Delta\rho$ is comparable to the electron charge ρ_e,

$$\beta(\omega_{trap}\Delta t)^2 \cong (\omega_{pe}\Delta t)^2 \, \Delta\rho/\rho_e$$

will not be small unless $\omega_{pe}\Delta t$ is small in that region.

The expression $\delta\rho = - \nabla\cdot(\rho\delta\mathbf{x})$ is valid when $\delta\mathbf{x}$ is smaller than density gradient scale lengths.

The validity of linearization is treated by Denavit (1981), Langdon *et al* (1983), Cohen *et al* (1984), Langdon and Barnes (1985), and references therein.

2.3.3. Energy Conservation and Artificial Cooling To understand better the origins and control of nonphysical cooling [a manifestation of error in codes using damped equations of motion (Adam *et al*, 1982; Barnes *et al*, 1983], the Lenard-Balescu collision operator corresponding to implicit time integration is examined by Langdon and Barnes (1985), based on the kinetic theory in Birdsall and Langdon (1985), Chapters 9 and 12. They find two spurious terms due to phase errors associated with damping. One is a nonresonant contribution to the polarization drag. That is, the Cerenkov wake of plasmons, excited by the passage of a particle, is nonphysically damped. The other term is a spurious nonresonant contribution to dynamical friction and corresponds to the drag calculated by Cohen *et al* (1982) but with the field given by the thermal fluctuation spectrum.

Quantitative calculations of cooling rates based on the kinetic theory have not yet been carried out. However, cooling rates observed in simulations of thermal plasmas (Barnes, *et al*., 1983) indicate that cooling is likely to be dominated by damping of a broad band of thermal fluctuations. Thus, the reduction in cooling for schemes with third-order damping compared with those with first-order damping is much less than the reduction in the damping of low-frequency plasma modes. Further, the most effective means of reducing cooling is to cut the width of the fluctuation spectrum by cutting off in wavenumber (spatial smoothing) or in frequency (by increasing Δt).

2.3.4. Effects of imperfect field solution on linear stability and dispersion In practice, because simplifications are used in the representation of the $\nabla\cdot\chi\cdot\nabla\phi$ term, and interation is not used in practice, the intended equations, (1.1)-(1.4) and (1.6)-(1.7), are not actually satisfied by the code. Analysis of linear stability and dispersion in uniform plasma, as described by dispersion relations that take into account the actual space and time difference methods, provides guidance to behavior to be expected. In particular, the consequences of various simplifications of the field equation operator have been analyzed by Cohen *et al* (1984), and by Langdon and Barnes (1985).

2.4 Generalizations

Many generalizations are possible. "Gyro-averaged particle simulation" (GAPS) removes the cyclotron frequency while retaining finite-Larmor-radius effects. Combinations of fluid and implicit particle descriptions have been made by Mason (1981, 1985) and Denavit (1983, and this meeting). If a component of the plasma is modeled by fluid equations then those equations are linearized to find $\{\delta\rho, \delta J\}$ (Denavit, 1983).

3. Electromagnetic Direct Implicit Method

Until recently, most plasma simulation including the full electromagnetic field was done with explicit differencing of both the particles and fields. The latter adds a Courant-Levy-Friedrichs time-step limitation $\Delta t < c/\lambda$, where λ is approximately the mesh spacing used for the fields. This limitation has been removed in two ways. One is to alter the field equations so that they no longer support wave propagation. A proven approach here is the Darwin, or magnetoinductive, model. Another is to use implicit differencing of the field equations. Recently, codes have used implicit differencing of both fields and particles.

Darwin codes (Hewett, this meeting) eliminate the Courant restriction $\Delta t < c/\lambda$ by dropping Maxwell's transverse displacement current term. These "pre-Maxwell equations" eliminate electromagnetic wave propagation while retaining electrostatic, magnetostatic, and inductive electric fields. Nielson and Lewis (1976) provide many references for the historical development of these codes. For applications not requiring a kinetic description of the electrons, codes using a hybrid of particle ions and fluid electrons are indicated. With Darwin and quasistatic approximations, long time scales are accessible, as in a fully implicit code, but with less noise (Hewett, 1980, and references therein).

Implicit fields reproduce electromagnetic wave propagation at long wavelengths ($\gg c\Delta t$). At short wavelengths, the electrostatic, magnetostatic, and inductive electric fields are retained, as in a Darwin code. Implicit fields can be used with explicit particles. With implicit particles, Langmuir waves are stabilized at all wavelengths, as in an implicit electrostatic code. The electrostatic fields are accurate for wavelengths longer than the electron transit distance ($v_{te}\Delta t$). These properties make an implicit electromagnetic code attractive, e.g., to modeling of reconnection and of intense particle currents.

Here we outline a fully implicit electromagnetic algorithm that has been implemented by Hewett and Langdon. Barnes and his co-workers have another version of this algorithm. Brackbill and Forslund (1985) and Mason (1985) describe moment method electromagnetic algorithms. Wallace, Brackbill and Forslund (1985) describe an algorithm that incorporates aspects of both viewpoints.

3.1 Implicit Electromagnetic Fields

Here we must select a time-differencing scheme for the fields, and find a method for solving the coupled field and particle equations. The Maxwell equations in rationalized cgs units are time-differenced using the D_1 implicit scheme as

$$c\nabla\times B_{n+\frac{1}{2}} = J_{n+\frac{1}{2}} + \frac{E_{n+1} - E_n}{\Delta t} \tag{3.1a}$$

$$-c\nabla\times\bar{E}_n = \frac{B_{n+\frac{1}{2}} - B_{n-\frac{1}{2}}}{\Delta t} \tag{3.1b}$$

where $\bar{E}_n = \frac{1}{2}[\bar{E}_{n-1} + E_{n+1}]$ (3.1c)

is the result of a recursive low-pass filter with phase error $O(\Delta t^3)$. This phase error is an advance, not a lag as one gets if E_{n+1} is not used, so it provides dissipation of inaccurately calculated short wavelengths, in particular, stability when $c\Delta t >> \Delta x$. In the limit $ck\Delta t >> 1$ the fields are close to those in the Darwin approximation. Long wavelenths are weakly damped, $\mathrm{Im}\,\omega(\mathbf{k})/ck = -\frac{1}{2}(ck\Delta t)^3$.

To form a B_n for use in the particle mover, we use, e.g.

$$B_n = B_{n-\frac{1}{2}} - \tfrac{1}{2}\Delta t\ c\nabla\times E_n \tag{3.2}$$

We use E_n, rather than $\bar{E}_n$ as in (3.1b), to simplify the linearized particle equation.

2.2 The Direct Method for Implicit Particles and Fields

As in Section 2.1.3 for the electrostatic case, the source terms in Maxwell's equations (3.1) are the densities $\{\tilde{\rho}_{n+1}, \mathcal{J}_{n+\frac{1}{2}}\}$ and $\{\delta\rho, \delta J\}$ corresponding to the extrapolations $\{\tilde{x}_{n+1}, \tilde{v}_{n+\frac{1}{2}}\}$, plus the increments $\{\delta x, \delta v\}$ that depend on E_{n+1}.

Evaluation of the extrapolated densities. The extrapolated current and charge densities $\{\tilde{\rho}_{n+1}, \mathcal{J}_{n+\frac{1}{2}}\}$ are evaluated as in explicit codes, such as ZOHAR (Langdon and Lasinski, 1976) and the Los Alamos WAVE code, from $\tilde{x}_{n+1}$, $\tilde{v}_{n+\frac{1}{2}}$ and x_n. At the grid point located at X_j, $\tilde{\rho}_{n+1}$ is given by (2.4),

and $$\mathcal{J}_{n+\frac{1}{2}} = \sum q\tilde{v}_{n+\frac{1}{2}}\ S(X_j - \tfrac{1}{2}[x_n+\tilde{x}_{n+1}]) \tag{3.3a}$$

or $$= \sum q\tilde{v}_{n+\frac{1}{2}}\ \tfrac{1}{2}[S(X_j - x_n) + S(X_j - \tilde{x}_{n+1})] \tag{3.3b}$$

Consistency of $\mathcal{J}_{n+\frac{1}{2}}$ and $\tilde{\rho}_{n+1}$ with the finite-differenced continuity equation is ensured by a correction to the longitudinal component of J or of E_{n+1} (Langdon and Birdsall, 1985, and references therein). Here, we replace $\mathcal{J}_{n+\frac{1}{2}}$ with $J'_{n+\frac{1}{2}} = \mathcal{J}_{n+\frac{1}{2}} + (\nabla\psi)/\Delta t$. Langdon and Barnes (1985) discuss solution for ψ, and an alternate correction to E_{n+1}.

Evaluation of the increments due to future fields. As in deriving the expression for $\delta\rho$, Eq. (2.8), δJ can be shown, by linearization of the expression for $J_{n+\frac{1}{2}} - \mathcal{J}_{n+\frac{1}{2}}$, to be of the form

$$\delta J = \rho\delta v - \tfrac{1}{2}\nabla \times (J \times \delta x) \tag{3.4}$$

for each species, where superscript (~) is understood for ρ and J on the right side. Langdon and Barnes (1985) give a heuristic derivation of (3.4). The second term is needed, e.g., to retain Galilean invariance. With (2.14) and (2.15), we write the implicit terms as $\tilde{\rho}_{n+1}\delta v = \chi\cdot E_{n+1}/\Delta t$ and $\mathcal{J}_{n+\frac{1}{2}} \times \delta x = c\xi\cdot E_{n+1}$, where ξ is a tensor given by a sum over species $\mathcal{J}$.

Construction of the Implicit Field Equation We now have everything needed to write an equation for E_{n+1}. On substituting our expressions for $\mathfrak{J}_{n+\frac{1}{2}}$ and δJ into the field equations (3.1), we have

$$c\Delta t\nabla \times \left[B_{n+\frac{1}{2}} + \tfrac{1}{2}\xi\cdot E_{n+1} \right] = J'_{n+\frac{1}{2}}\Delta t + \chi\cdot E_{n+1} + (E_{n+1} - E_n) \quad (3.5a)$$

$$-\tfrac{1}{2}\, c\Delta t\nabla \times [\bar{E}_{n-1} + E_{n+1}] = B_{n+\frac{1}{2}} - B_{n-\frac{1}{2}} \quad (3.5b)$$

The divergence of the Ampere-Maxwell equation (3.5a) recovers our *electrostatic* implicit field equation, (2.9) with $E_{n+1} = -\nabla\phi_{n+1}$.

3.3 Applications of Implicit Electromagnetic Codes

For my interests, implicit methods will be most profitable in electromagnetic and hybrid codes. Applications have been made at Los Alamos to electron transport (Forslund and Brackbill, 1982; Mason, 1985), Weibel instability (Brackbill and Forslund, 1982), lower hybrid drift instability (Brackbill *et al*, 1984), and space plasma problems (Quest *et al*, 1983; Forslund *et al*, 1984). Preliminary results from direct-method electromagnetic codes have been obtained by D. Barnes and his co-workers at IFS and Nagoya, and by Hewett and Langdon (1985).

ACKNOWLEDGMENTS

This article is mainly a summary of Langdon and Barnes (1985).
This work was performed under the auspices of the U. S. Department of Energy by the Lawrence Livermore National Laboratory under Contract No. W-7405-Eng-48.

REFERENCES

J. C. Adam, A. Gourdin Serveniere, A. B. Langdon, *J. Comp. Phys.* 47 (1982), 229.

D. C. Barnes, T. Kamimura, J.-N. LeBoeuf and T. Tajima, *J. Comp. Phys.* 52, 480 (1983).

C. K. Birdsall and A. B. Langdon, *Plasma Physics Via Computer Simulation*, McGraw-Hill, New York, 1985.

J. U. Brackbill and D. W. Forslund, *J. Comp. Phys.* 46 (1982), 271.

J. U. Brackbill, D. W. Forslund, K. Quest and D. Winske, *Phys. Fluids* 27, 2682 (1984).

J. U. Brackbill and D. W. Forslund, 'Simulation of low frequency, electromagnetic phenomena in plasmas', in the volume *Multiple Time Scales* in the series *Computational Techniques*, Academic, 1985.

O. Buneman, *J. Comp. Phys.* 1 (1967), 517.

B. I. Cohen, A. B. Langdon and A. Friedman, *J. Comp. Phys.* **46** (1982), 15.

B. I. Cohen, A. B. Langdon and A. Friedman, *J. Comput. Phys.* **56** (1984), 51.

J. Denavit, *J. Comp. Phys.* **42** (1981), 337.

J. D. Denavit, 'One-Dimensional Time-Implicit Hybrid Simulations', *Laser Program Annual Report-1982*, Lawrence Livermore National Laboratory, Livermore, CA, UCRL-80021-82 (1983), p. 3-50.

D. W. Forslund and J. U. Brackbill, *Phys. Rev. Lett.* **48** (1982), 1614.

D. W. Forslund, K. Quest, J. U. Brackbill and K. Lee, *J. Geophys. Rev.* **89**, 2142 (1984).

D. W. Hewett, *J. Comput. Phys.* **38**, 378 (1980).

D. W. Hewett and A. B. Langdon, *Laser Program Annual Report-1982*, Lawrence Livermore National Laboratory, Livermore, CA, UCRL-80021-84 (1985).

A. B. Langdon and B. F. Lasinski, 'Electromagnetic and Relativistic Plasma Simulation Models', in *Methods in Computational Physics*, (B. Alder, S. Fernbach, and M. Rotenberg, Eds, Volume Ed. J. Killeen), p. 327, Academic Press, New York, 1976.

A. B. Langdon, B. I. Cohen and A. Friedman, *J. Comp. Phys.* **51** (1983), 107.

A. B. Langdon, D. W. Hewett and A. Friedman, *Laser Program Annual Report-1983*, Lawrence Livermore National Laboratory, Livermore, CA, UCRL-80021-83 (1984).

A. B. Langdon and D. C. Barnes, 'Direct Implicit Plasma Simulation', in the volume *Multiple Time Scales* in the series *Computational Techniques*, Academic, 1985.

R. J. Mason, *J. Comp. Phys.* **41** (1981), 233.

R. J. Mason, *Phys. Rev. Lett.* **47** (1981), 652.

R. J. Mason, 'Hybrid and collisional implicit plasma simulation models', in the volume *Multiple Time Scales* in the series *Computational Techniques*, Academic, 1985.

C. W. Nielson and H. R. Lewis, 'Particle Code Models in the Nonradiative Limit', in *Methods in Computational Physics*, (B. Alder,

S. Fernbach, and M. Rotenberg, Eds, Vol. Ed. J. Killeen), p. 367, Academic Press, N.Y., 1976.

K. Quest, D. W. Forslund, J. U. Brackbill and K. Lee, *Geophys. Research Lett.* **10**, 471 (1983).

H. Sakagami, K. Nishihara and D. Colombant, 'Stability of Time-Filtering Particle Code Simulation', Institute of Laser Engineering report ILE8117P, August 10, 1981.

J. M. Wallace, J. U. Brackbill and D. W. Forslund, 'An Implicit Moment Electromagnetic Plasma Simulation in Cylindrical Coordinates', *J. Comput. Phys.*, to be published.

TIME-IMPLICIT SIMULATION OF PARTICLE-FLUID SYSTEMS

J. Denavit
Lawrence Livermore National Laboratory
P. O. Box 808, L-18
Livermore, California 94550

ABSTRACT. This paper presents one-dimensional particle-fluid hybrid simulations in which the strongly collisional components of the plasma (e.g., ions and thermal electrons with $\nu_{cf}\Delta t > 1$) are treated as fluids and the weakly collisional components (e.g., energetic electrons with $\nu_{cp}\Delta t << 1$) are treated as particles. Here ν_{cf} denotes the fluid ion and electron collision frequencies, ν_{cp} is the energetic particle collision frequency and Δt is the time step. Collisions between particle and fluid components are treated by a Monte-Carlo method and mass transfers between the particle and fluid electron components are governed by collision frequency thresholds. The field is computed implicitly to allow time steps with $\omega_p \Delta t > 1$ (ω_p: plasma frequency).

1. INTRODUCTION

Both particle and fluid simulations have proven to be powerful research tools in plasma physics, but have generally been applied to different problems. In particle simulations, a large number of charged particles, representing the plasma, are followed in their self fields. At each time step, the charge and current densities are computed from the particle positions and velocities, and the electromagnetic fields are evaluated using Maxwell's equations. The particles are then advanced, using these fields, over a time increment, Δt, and the process is repeated. This procedure is characterized as time-explicit because it uses known particle positions and velocities, at a time t_o, to compute the fields which are then used to advance the particles beyond this time, to $t > t_o$. Particle simulations give a very detailed description of the plasma, which is particularly useful in the study on non-equilibrium phenomena, where the distribution function exhibits multiple streaming, supra-thermal tails, or other non-Maxwellian features. However, they are generally limited to short time scales, of the order of the plasma period, ω_p^{-1}, because the explicit nature of the algorithms used gives a violent numerical instability for time step sizes, $\Delta t > \omega_p^{-1}$. Here $\omega_p = (4\pi e^2 n_e/m_e)^{1/2}$

Space Science Reviews **42** (1985) 85–102. 0038–6308/85.15

is the electron plasma frequency, n_e is the electron density, $-e$ is the electron charge, m_e is the electron mass and Gaussian cgs units are used.

In fluid simulations, the plasma is usually represented as a single fluid by MHD equations. Quasi-neutrality is assumed and the currents are computed from a generalization of Ohm's law. The transport coefficients used in the fluid equations assume that the distribution function is close to Maxwellian, and exclude such features as multiple streaming or supra-thermal tails, for example. Fluid simulations of this type are not limited by the condition $\Delta t \ll \omega_p^{-1}$, and give a description of the plasma over much longer time and space scales than particle simulations. They play an important role to model experiments, with realistic geometrical and physical parameters. However, they assume that collisions, or other microscopic effects, maintain the plasma sufficiently near local thermodynamic equilibrium to guarantee their validity.

Recently, particle (and Vlasov) simulation methods, using a time-implicit procedure, have been developed to allow particle simulations, including electron inertia, with large time steps $\Delta t > \omega_p^{-1}$. In this procedure, the fields are predicted at time $t_1 = t_o + \Delta t$, <u>before</u> advancing the particles (or the distribution function) to this time level. The particles (or the distribution function) are then advanced from t_o to t_1, using a weighted average of the fields at t_1, t_o and earlier times. In the first time-implicit methods to be introduced [1, 2, 3], the fields are evaluated at the new time, t_1, using a subsidiary fluid representation of the electrons in terms of the continuity and momentum equations. This method, now known as the "moments" method, has been applied to one-dimensional electrostatic problems such as the two-stream instability, ion acoustic waves and plasma expansion into vacuum. It has also been used in two-dimensional electromagnetic simulations [4]. In another method, which was introduced subsequently, the fields at t_1 are evaluated directly in terms of a susceptibility tensor derived from the equations used to advance particles [5]. This "direct" method has been applied to one-dimensional electrostatic problems and to two-dimensional simulations with a static magnetic field [6]. Its application to electromagnetic problems has also been implemented [7].

Implicit particle simulations, using either the "moments" or the "direct" method allow time steps $\Delta t \gg \omega_p^{-1}$. They can represent accurately low-frequency electron inertia effects, such as trapping or acceleration of energetic electrons, while providing rapid damping of high-frequency plasma oscillations. However, these methods remain limited by the condition $\Delta t \ll (kv_{th})^{-1}$, which specifies that an electron near the thermal velocity, v_{th}, must not move across a large fraction of a characteristic wavelength, $\lambda = 2\pi/k$, during a time step. Taken together, the conditions $\omega_p \Delta t > 1$ and $kv_{th}\Delta t \ll 1$ imply $k\lambda_D \ll 1$, where $\lambda_D = v_{th}/\omega_p$ is the Debye length. Thus, implicit methods yield improved efficiency only in the case of <u>long wavelengths</u> or in <u>dense</u> and <u>cold</u> plasmas. However, dense and cold plasmas also have large collision frequencies ν_c and implicit codes operating in this regime must therefore allow time steps with $\nu_c \Delta t > 1$.

These scaling considerations lead to the concept of a hybrid formulation, in which the sparse and hot component, is represented as particles and satisfies the conditions $\omega_p \Delta t \ll 1$, $kv_{th}\Delta t \ll 1$ and $\nu_c \Delta t \ll 1$, while the dense and cold component is given a fluid representation and is characterized by $\omega_p \Delta t > 1$, $kv_{th}\Delta t \ll 1$ and $\nu_c \Delta t > 1$ [8,9]. The fluid representation considered in these codes does not use MHD equations, but is based on separate electron and ion fluids, and includes electron inertia. An implicit determination of the fields generated by fluid electrons is therefore needed to achieve $\omega_p \Delta t > 1$ for this component.

The computations discussed in this paper are one dimensional. The fluid quantities are represented on a uniform Eulerian grid, which is also used for the particles, and flux corrected transport (FCT) is used to minimize numerical diffusion.

2. FLUID COMPONENTS

2.1 Fluid Equations

Since electron inertia is included, the electrons and ions are represented in terms of similar sets of fluid equations for conservation of mass, momentum and energy. This is a more elementary representation of the plasma than the MHD equations. It includes charge separation effects and, for example, the current density in this model is related to the electric field through the dynamics of the plasma components, rather than being computed from a generalization of Ohm's law.

The one-dimensional equations for a non-magnetized multi-species fluid are

$$\frac{\partial n_s}{\partial t} + \frac{\partial}{\partial x}(n_s u_s) = 0 , \tag{1}$$

$$\frac{\partial P_s}{\partial t} + \frac{\partial}{\partial x}(u_s P_s) = -\frac{1}{m_s}\frac{\partial p_s}{\partial x} + \frac{e_s}{m_s} n_s E + \frac{1}{m_s}\sum_{s'} R_{ss'} \tag{2}$$

$$\frac{\partial U_s}{\partial t} + u_s \frac{\partial U_s}{\partial x} = -\frac{1}{n_s} p_s \frac{\partial u_s}{\partial x} - \frac{1}{n_s}\frac{\partial q_s}{\partial x} + \frac{1}{n_s}\sum_{s'} Q_{ss'} \tag{3}$$

Here, n_s, u_s, and T_s denote respectively the density, the drift velocity and the temperature of species s, $P_s = n_s u_s$ defines the momentum, E is the electric field, m_s and e_s are the mass and charge of the particles of species s. Monoatomic ideal gases are assumed with the pressure, $p_s = k_B n_s T_s$ and specific energy $U_s = (3/2)k_B T_s$, where k_B denotes the Boltzmann constant.

The collisional friction force between components s and s' is given by

$$R_{ss'} = - m_{ss'} C_{ss'} n_s n_{s'} (u_s - u_{s'}) \tag{4}$$

where $m_{ss'} = m_s m_{s'}/(m_s + m_{s'})$, and $C_{ss'}$ is a collision parameter. For Coulomb collisions between charged particles, a generalization of Braginskii's results [10] gives

$$C_{ss'} = \frac{\alpha_o \Lambda_{ss'} e_s^2 e_{s'}^2}{(2\pi)^{3/2} \varepsilon_o \sqrt{m_{ss'}}\, (k_B T_{ss'})^{3/2}} \tag{5}$$

where $\Lambda_{ss'}$ is the Coulomb logarithm, ε_o is the dielectric constant of free space and $T_{ss'} = m_{ss'}[(T_s/m_s) + (T_{s'}/m_{s'})]$. Note that all quantities entering in Eq. (4) are symmetrical with respect to s and s', giving $R_{ss'} + R_{s's} = 0$. The thermal gradient force is not considered here.

The heat transfer between species is due to both dissipation and temperature differences,

$$Q_{ss'} = - \frac{m_{ss'}}{m_s} R_{ss'} (u_s - u_{s'}) - 3 \frac{m_{ss'}}{m_s + m_{ss'}} C_{ss'} n_s n_{s'} k_B (T_s - T_{s'}) \ . \tag{6}$$

Finally, the heat flux is

$$q_s = - \kappa_s \frac{\partial T_s}{\partial x} \ , \tag{7}$$

with the heat conduction reciprocal (heat resistivity) equal to the sum of a collisionless contribution, $1/\kappa_{so}$, which gives the flux limit, and collisional contributions, $1/\kappa_{ss'}$,

$$\frac{1}{\kappa_s} = \frac{1}{\kappa_{so}} + \sum_{s'} \frac{1}{\kappa_{ss'}} \ . \tag{8}$$

The collisionless contribution is

$$\frac{1}{\kappa_{so}} = f_s \frac{m_s^{1/2} \, |\nabla T_s|}{n_s (k_B T_s)^{3/2}} \ ,$$

where f_s is the flux limiter coefficient. For Coulomb collisions between particles, a generalization of Braginskii's results [10] gives

$$\frac{1}{\kappa_{ss'}} = \frac{m_s n_{s'} C_{ss'}}{\gamma_s n_s k_B^2 T_s}$$

The numerical constants α_0, γ_e, and γ_i are given in Ref. [10].

2.2 Transport Algorithm

The mass and momentum conservation equations, Eqs. (1) and (2) are both of the form

$$\frac{\partial \rho}{\partial t} + \frac{\partial}{\partial x} (u\rho) = S \ , \tag{9}$$

where ρ is the transported density, u is the velocity, and S is a source term which is zero in the case of Eq. (1). For the Eulerian representation on a fixed mesh, which has been chosen here, the transport algorithm solving Eq. (9) must ensure numerical stability without introducing excessive diffusion. Flux corrected transport (FCT) algorithms are used for this purpose [11]. The system is divided into J-2 cells of length Δx, with two guard cells, cells 1 and J, laying outside the system as shown in Fig. 1. All quantities are defined at the cell centers and the values at the guard cells are set equal to their values at the adjacent boundary cell, e.g., $\rho_1 = \rho_2$, $\rho_J = \rho_{J-1}$, $u_1 = u_2$, $u_J = u_{J-1}$. This choice allows the transport algorithm to be applied uniformly to all physical cells ($j = 2, . . ., J - 1$) and yields emitting or absorbing boundary conditions depending on the sign of u at the boundary cells. In addition, the mass, (or charge), momentum and energy crossing the boundaries of the system are stored separately to allow conservation checks and to compute the electric field.

Following the FCT algorithm, provisional values, ρ_j^{TD}, of the density are first computed from the old values, ρ_j and u_j,

$$\rho_j^{TD} = \frac{1}{2} Q_+^2 (\rho_{j+1} - \rho_j) - \frac{1}{2} Q_-^2 (\rho_j - \rho_{j-1}) + \rho_j (Q_+ + Q_-) \tag{10}$$

where

$$Q_\pm = \frac{\frac{1}{2} \mp \varepsilon_j}{1 \pm (\varepsilon_{j\pm1} - \varepsilon_j)} \ , \tag{11}$$

and $\varepsilon_j = u_j \Delta t/\Delta x$. The source term, S, is not included here and will be introduced when the momentum equations are considered. Equation (10) gives provisional values of the density, which are always positive for $|\varepsilon_j| < 1/2$. However, these values include a large numerical diffusion and are therefore both transported and diffused (TD). The diffusion is corrected by application of an antidiffusion step, which is limited so as not to cause new maxima or minima of the density. The simplest antidiffusion step, is defined by

$$\Delta^{TD}_{j+\frac{1}{2}} = \rho^{TD}_{j+1} - \rho^{TD}_{j}, \tag{12}$$

$$s^{TD}_{j+\frac{1}{2}} = \text{sign}\,(\Delta^{TD}_{j+\frac{1}{2}}), \tag{13}$$

$$\phi_{j+\frac{1}{2}} = s^{TD}_{j+\frac{1}{2}} \max\left\{0, \min\left[s^{TD}_{j+\frac{1}{2}}\Delta^{TD}_{j-\frac{1}{2}}, \frac{1}{8}\left|\Delta^{TD}_{j+\frac{1}{2}}\right|, s^{TD}_{j+\frac{1}{2}}\Delta^{TD}_{j+\frac{3}{2}}\right]\right\} \tag{14}$$

$$\rho^{new}_{j} = \rho^{TD}_{j} - (\phi_{j+\frac{1}{2}} - \phi_{j-\frac{1}{2}}), \tag{15}$$

where ρ^{new}_j is the new density. This algorithm leaves a significant residual diffusion, even in the stationary case, and more advanced algorithms are given in Ref. [11].

2.3 Momentum Equations

The equation of continuity, Eq. (1), which has a right member equal to zero, may be solved by direct application of one of the FCT algorithms. However, Eq. (2) requires the computation of a source term S. In the implicit time step considered here, this source term is split into two parts. The first part, computed from the old values of the variables is applied before transport,

$$\bar{P}_{sj} = P_{sj} + F_{sj}\frac{\Delta t}{2} - \sum_{s'}(1 - \theta_1)\Delta t\,\frac{m_{ss'}}{m_s}\,C_{ss'j}(n_{s'}P_s - n_sP_{s'}); \tag{16}$$

where

$$F_{sj}\frac{\Delta t}{2} = -\frac{\Delta t k_B}{4m_s\Delta x}[(nT)_{j+1} - (nT)_{j-1}]_s + \frac{e_s\Delta t}{2m_s}(n_sE)_j\,. \tag{17}$$

Here, j denotes the grid point, Δx is the mesh size, and θ_1 is a weight factor which allows the friction force to be either time centered ($\theta_1 = 1/2$) or implicit ($\theta_1 = 1$). The second part of the time step, computed from the new values is applied after transport. Since the new values are unknown, this partial step requires iterations. Let the superscripts q and q+1 denote successive iteration levels, then

$$P_{sj}^{q+1} = FCT(\bar{P}_s)_j + F_{sj}^q \frac{\Delta t}{2} - \sum_{s'} \theta_1 \Delta t \frac{m_{ss'}}{m_s} C_{ss'j}(n_{s'}P_s - n_s P_{s'})_j^{q+1} \tag{18}$$

The friction force, which can be very large in dense fluids cannot be included in the iteration without causing divergence for $\nu_{cf}\Delta t > 1$, where ν_{cf} is the fluid collision frequency. Equations (18) must therefore be solved as a system of linear equations,

$$A_{ss'j} P_{s'j}^{q+1} = B_{sj} \tag{19}$$

where

$$A_{ss'j} = -\theta_1 \Delta t \frac{m_{ss'}}{m_s} C_{ss'j} n_s \tag{20}$$

for $s' \neq s$,

$$A_{ssj} = 1 - \sum_{s \neq s'} A_{ss'j} \, . \tag{21}$$

and

$$B_{sj} = FCT(P_s)_j + F_{sj}^q \frac{\Delta t}{2} \tag{22}$$

For $\omega_p \Delta t > 1$, the electric field which enters in Eq. (17), must also be evaluated implicitly, but this question is examined separately in Sec. 4.

2.4 Temperature Equations

Equation (3) for the temperature has different transport terms than the continuity and momentum equations, Eqs. (1) and (2). However, it can be rearranged to allow the electron and ion temperatures to be advanced using one of the FCT algorithms already used for mass and momentum transport,

$$\frac{3}{2} \left[\frac{\partial T_s}{\partial t} + \frac{\partial}{\partial x} (u_s T_s)\right] = w_s + r_s + d_s \tag{23}$$

where

$$w_s = \frac{1}{2} \frac{\partial u_s}{\partial x} T_s + \sum_{s'} \frac{m_{ss'}^2}{m_s} \frac{C_{ss'}}{k_B} n_{s'} (u_s - u_{s'})^2 \tag{24a}$$

is due to compression and dissipation,

$$r_s = -\sum_{s'} \frac{3m_{ss'}}{m_s+m_{s'}} C_{ss'} n_s (T_s - T_{s'}) \tag{24b}$$

is due to temperature relaxation between components and

$$d_s = \frac{1}{n_s} \frac{\partial}{\partial x} \kappa_s \frac{\partial T_s}{\partial x} \tag{24c}$$

is due to heat conduction.

The temperature is updated through a succession of partial steps:

i) Local temperature changes due to terms w_s and r_s

ii) Heat diffusion

iii) Transport.

3. PARTICLE COMPONENT

The particles are generated from the corresponding fluid when the local collision frequency drops below a threshold ν_1, and are re-absorbed into the fluid when their collision frequency becomes larger than another threshold, ν_2. Thus a given region of space may be occupied either by fluids only, by particles only, or by a mixture of both components, which interact with each other by collisions and through the electric field. Since scattering of particles by the fluid components causes a rotation of their velocity, the particle variables for the one-dimensional, non-magnetized case, must include the position, x, and two velocity components, v_x parallel to the x-axis and v_t tranverse (i.e., perpendicular) to this axis.

Several methods have been used in particle simulations to relate the particles,which are located at discrete points x_i, where i designates a given particle, to the grid quantities, which are defined at the grid points (or cell centers), x_j. Among these methods, nearest grid point (NGP) is the simplest. In NGP, the charge of a particle contributes only to the cell in which it is located and the electric force is due only to the field at this cell center. In another method, area sharing, particles are considered to be sheets of finite thickness, Δx, equal to the cell size, and only the fraction of the particle located in a given cell contributes to the charge in this cell, with the remaining fraction contributing to the adjacent cell. The force in this case is computed by linear interpolation of the field at adjacent cell centers. Both methods conserve momentum but area sharing, which gives smoother fields and generates less noise, is generally favored over NGP. However, in the hybrid simulations considered here, consistency between particle and fluid components is easier to achieve with NGP.

The initial location of the particles must be such that the instantaneous electric field array is unchanged by the mass transfer from fluid to particles. The total electric force on the new particles must also be equal to the force on the fluid element they replace. In the case of NGP, this can be achieved by locating the new particles anywhere within the cell, and in particular, they can be distributed uniformly within the cell. In the case of area sharing, the new particles must all be initialized at the cell centers. Any deviation from this location would spread their charge to the neighboring cells, and cause the electric force on these particles to be affected by neighboring values of the electric field. It follows that the initial distribution of particles is more discrete when area sharing is used than when NGP is used. This tends to cancel the advantages of area sharing over NGP in achieving low noise simulations.

Since strongly collisional particles are eliminated from the simulation by absorption into the fluid electron component, the particles can be assumed to satisfy the conditions $\nu_{cp}\Delta t << 1$. The particles can therefore be advanced using a split time step. Velocities and positions are advanced by a simple leap-frog scheme under the action of the electric field but under collisionsless conditions. This partial step is followed by a collisional step under field-free conditions. In the collisional step the particle velocity is decreased to account for energy loss by collisions with the fluid components and it is rotated to account for random scattering.

4. ELECTRIC FIELD

4.1 Implicit Algorithm

The electric field is evaluated from a finite difference approximation of Poisson's equation

$$G_j \equiv E_{j+1}^{t+\Delta t} - E_j^{t+\Delta t} - \frac{\Delta x}{2\varepsilon_0}\left[\sum_s e_s(n_{j+1} + n_j)_s + \rho_{pj}\right] = 0 \qquad (25)$$

where ρ is the charge density due to particles, and j = 2, ..., J-2. Recall that the system consists of cells 2 thru J-1, see Fig. 1, with cells 1 and J as guard cells, used only to satisfy boundary conditions in the transport algorithm.

The quantity G_j may be considered as a multidimensional "vector" function, $\underline{G}(\underline{E}^{t+\Delta t})$ of the multidimensional "vector", $E_j^{t+\Delta t}$, both $\underline{G}$ and $\underline{E}^{t+\Delta t}$ having one component for each grid point j = 2, ..., J - 1 in the system. Recall that since j = 1 and j = J correspond to guard cells they do not contribute independent values of E_j or n_{sj}. Equation (25) is solved using the well-known Newton iteration scheme. The (q + 1)-th approximation of the field is written as the sum, $E_j^{q+1} = E_j^q + \delta E_j^{q+1}$ of the q-th approximation and a correction δE_j^{q+1}. Taylor expansion of Eq. (25)

$$G_j(\underline{E}^q + \delta\, \underline{E}^{q+1}) \simeq G_j(\underline{E}^q) + \sum_{j'} \left.\frac{\partial G_j}{\partial E_{j'}}\right|^q \delta E_{j'}^{q+1} = 0,$$

or

$$\sum_{j'} \left(\frac{\partial G_j}{\partial E_{j'}}\right)^q \delta E_{j'}^{q+1} = - G_j(\underline{E}^q) \quad . \tag{26}$$

This set of equations, with $j = 2, \ldots, J-1$, can be solved for the corrections δE_j^{q+1} after the matrix $(\partial G_j/\partial E_{j'})^q$ and the vector $G_j(\underline{E}^q)$, which depend only on earlier approximations have been evaluated.

It is important note that the new electric field enters in the definition of G_j in Eq. (25), not only in the explicit terms $E_j^{t+\Delta t}$ and $E_{j+1}^{t+\Delta t}$, but also implicitly through the densities $n_{sj}^{t+\Delta t}$ and $\rho_j^{t+\Delta t}$, since these quantites depend on the new electric field via the algorithms described in Sections 2 and 3. Recognition of this implicit dependence is essential to establish an iterative scheme which is convergent for $\omega_p \Delta t > 1$. Here, only the dependence of G_j on $\underline{E}$ via the fluid densities n_s is considered. The dependence of G_j on $\underline{E}$ via the particle charge is computed as in implicit particle simulations [5].

From the transport algorithm, Eqs. (10) and (11), applied to the electron continuity equation,

$$Q_{\pm} = \frac{1}{2} [1 \mp (\varepsilon_{j\pm1} - \varepsilon_j)] \mp \varepsilon_j + O(\varepsilon^2) \ ,$$

and

$$n_{sj}^{t+\Delta t} = \frac{1}{4} (\varepsilon_{j-1} + \varepsilon_j)\, n_{sj-1} + [1 - \frac{1}{4} (\varepsilon_{j+1} - \varepsilon_{j-1})]\, n_{sj} - \frac{1}{4} (\varepsilon_j + \varepsilon_{j+1})\, n_{sj+1} \ , \tag{27}$$

where $j = 2, \ldots, J-1$. In Eq. (27), the nonlinear limitations to the antidiffusion step, given by max and min functions in Eq. (14) have been neglected.

From the fluid momentum equations, Eqs. (19)

$$P_{sj}^{q+1} = \sum_{s'} (A^{-1})_{ss'j} B_{s'j}$$

or

$$u_{sj}^{q+1} = \sum_{s'} (A^{-1})_{ss'j} \frac{e_{s'} \Delta t}{2m_{s'}} E_j^{q+1} + u_{sj}^{*}$$

where u_{sj}^{*} denotes terms independent of E^{q+1} to first order. Evaluating ε_j from this expression and substituting into Eqs. (27) and (25) gives

$$g_{j1} = \frac{\partial G_j}{\partial E_{j-1}} = - \sum_s e_s (n_{j-1} + n_j)_s\, Q_{sj-1} \ ,$$

$$g_{j2} = \frac{\partial G_j}{\partial E_j} = -1 - \sum_s e_s (n_{j-1} + n_j)_s Q_{sj} ,$$

$$g_{j3} = \frac{\partial G_j}{\partial E_{j+1}} = 1 + \sum_s e_s (n_{j+1} + n_{j+2})_s Q_{sj+1} , \tag{28}$$

$$g_{j4} = \frac{\partial G_j}{\partial E_{j+2}} = \sum_s e_s (n_{j+1} + n_{j+2})_s Q_{sj+2} ,$$

where

$$Q_{sj} = \sum_{s'} \frac{e_{s'} \Delta t^2}{32 \varepsilon_o m_{s'}} (A^{-1})_{ss'j} .$$

4.2 Electric Field Solution

The electric field corrections, δE_j^{q+1}, at grid points within the system, j = 2, ..., J-1, are obtained by solving Eqs. (26). The right members, G_j, are given by Eq. (25) and the matrix $\partial G_j / \partial E_{j'}$ is given by Eqs. (28). For some problems, such as in the double layers, a fixed potential difference ϕ must be maintained across the system. In this case, the last of Eqs. (26), corresponding to j = J-1, must be disgarded and replaced by the equation

$$\sum_{j=1}^{J-1} \delta E_j^{q+1} = 0 \quad , \tag{29}$$

which follows from the requirement that both E_j^{q+1} and E_j^q must satisfy the potential condition, $\sum_{j=2}^{J-1} E_j = -\phi/\Delta x$.

To solve this set of equations conveniently by means of a band matrix solver, Eqs. (26), with j=2, ..., J-2 are written in the form

$$g_{22}\,\delta E_2 + g_{23}\,\delta E_3 + g_{24}\,\delta E_4 = -G_2 ,$$

$$g_{31}\,\delta E_2 + g_{32}\,\delta E_3 + g_{33}\,\delta E_4 + g_{34}\,\delta E_5 = -G_3 ,$$

$$\cdots$$

$$g_{J-3,1}\delta E_{J-4} + g_{J-3,2}\delta E_{J-3} + g_{J-3,3}\delta E_{J-2} = -G_{J-3} - g_{J-3,4}\delta E_{J-1} ,$$

$$g_{J-2,1}\,\delta E_{J-3} + g_{J-2,2}\,\delta E_{J-2} = -G_{J-2} - g_{J-2,3}\,\delta E_{J-1} .$$

This system is solved twice, first with $-G_2$, ..., $-G_{J-2}$ in the right members to obtain a set of partial corrections, $\delta\bar{E}_j$, then with 0, ..., 0, $-g_{J-3,4}$, $-g_{J-2,3}$ in the right members to obtain a set of coefficients C_j. By superposition, the complete corrections are

$$\delta E_j = \delta \bar{E}_j + C_j \, \delta E_{J-1} \tag{30}$$

where j=2, ..., J-2, and δE_{J-1} is unknown. To solve for δE_{J-1}, the δE_j's from Eq. (30) are substituted into the potential condition, Eq. (29), from which

$$\delta E_{J-1} = - \frac{\delta \bar{E}_2 + \ldots + \delta \bar{E}_{J-2}}{C_2 + \ldots + C_{J-1}} \tag{31}$$

after which, δE_2, ... δE_{J-2}, are formed from Eq. (30).

5. EXAMPLES

The expansion of a plasma slab into vacuum is considered as an example. The slab is initially occupying a region of width 2L, located in the center of the simulation region of width 8L, where $L = 10^3 \lambda_{Do}$ is the characteristic length, $\lambda_{Do} = (T_o/4\pi e^2 n_o)^{1/2}$, n_o is the characteristic density and T_o is the characteristic temperature. The slab is initially represented as fluids only, defined in normalized units (see Table I) by setting $n_e = n_i = 1$, $u_e = u_i = 0$, $T_e = 1$, $T_i = 10^{-2}$. A mass ratio $m_i/m_e = 900$, and a charge number $Z_i = 1$ are assumed.

According to a similarity solution of this problem [12], which assumes charge neutrality, isothermal electrons and cold ions, the expansion produces a rarefaction wave, which propagates into the plasma at the ion acoustic speed $c_s = (T_e/m_i)^{1/2}$, equal to 1/30 in normalized units. Behind this wave, the electron and ion densities vary according to $n_e = n_i = n_o \exp[-(1 + x/c_s t)]$, where n_o is the unperturbed density and x is measured from the location of the initial density discontinuity. The electrons and ions are accelerated outward according to the velocity profile $u_e = u_i = c_s + x/t$ and the electric field is uniform with a value $E = T_e/(c_s t)$, equal to 30/t in normalized units.

CASE A: Collisionless, Isothermal Fluids Only

In this case, particle generation is turned off so that the simulation proceeds with fluids only, and a collisionless case is considered by setting $\alpha_o = 0$. In addition, the electron and ion temperatures are maintained constant ($T_e = 1$, $T_i = 10^{-2}$) throughout the computation. The electron density profile for this case is given in Fig. 2 at t = 4, 10 and 30, and the corresponding electric field profiles are shown in Fig. 3.

The propagation of the ion front predicted by the similarity solution is clearly visible on these profiles, and the fronts on each side of the slab reach the center of the slab at $t = 30$, in agreement with the predicted speed, $c_s = 1/30$. At $t = 30$, the electric field on each side of the slab is $E = \pm 1$, in agreement with the expression $E = 30/t$ predicted by the similarity solution. At earlier times ($t = 4$ and 10) the field profiles also forms plateaus behind each rarefaction front, which are in approximate agreement with the expression $E = 30/t$, but large peaks are also observed. These peaks are due to the pure electron gas ahead of the ion front [13,14] and their magnitude is given approximately by $E_f \simeq 0.86\ (L/\lambda_{Do})(n_{ef}n_o)^{1/2}T_e$, where n_{ef} the electron density at the front, which also coincides with the peak of the field. At $t = 4$, $n_{ef} \simeq 4 \times 10^{-4}$, see Fig. 2, yields $E_f = 17.2$, which is in approximate agreement with the simulation, $E_f = 19$, observed in Fig. 3. As time progresses, these peaks decrease, move outward (see Fig. 2 at $t = 10$) and disappear into the absorbing boundaries of the system.

In addition to the peaks, the field profile also exhibits oscillations which are primarily localized at the initial slab boundaries ($x = 3$ and $x = 5$). Since the plasma at these points move outward at the ion sound speed, it follows that the perturbation move relative to the plasma at the same speed and can therefore be interpreted as ion sound oscillations. Note that inward from the points $x = 3$ and $x = 5$, the plasma moves at a velocity which is less than c_s, thus allowing the ion-sound oscillations to propagate upstream.

The computations shown in Figs. 2 and 3 were done with three iterations of the electric field at each time step, and ran with a time step $\Delta t = 2 \times 10^{-3}$, which corresponds to $\omega_p \Delta t = 2$.

CASE B: Collisional Simulation with Fluids Only

The same plasma expansion problem is repeated using a full set of fluid equations, with $\alpha_o = 0.51$, $\gamma_e = 3.16$, $\gamma_i = 3.9$ and $C_{ei} = 1.92$. The results at $t = 30$, shown in Fig. 4, are qualitatively similar to the collisionless isothermal case. Note the scale on the electron temperature curve, T_e, in Fig. 4, showing that the electrons remain almost isothermal, but their temperature drops to $T_e \simeq 0.5$ as the slab expands. The ion temperature has increased an order of magnitude in the center of the slab but remains small in the expansion regions.

The energy of the electron fluid (kinetic plus thermal energies) is initially $W_e = 1.5\ n_e T_e \ell = 3$, where $n_e = 1$, $T_e = 1$ and $\ell = 2$ is the initial slab thickness. The ion fluid energy is initially $W_i = 0.03$ and the initial potential energy, W_E, is initially zero, giving a total energy $W_t = 3.03$. The time evolution of these quantities, given in Fig. 5, shows that (i) The electron energy drops steadily and is transferred to the ions, which eventually acquire an energy approximately equal to the electron energy. This is consistent with the increase in both ion velocity and ion temperature. (ii) The potential energy remains small $\sim 10^{-4}$ to 10^{-6}, in agreement with the quasi-neutral evolution of the system and (iii) The total energy remains approximately constant out to $t \simeq 15$, when a significant amount of electron fluid reaches the boundaries of the system. The energy loss through both boundaries is $-\dot{W}_e = [n_e u_e(u_e^2 + 3T_e)]_{xmax}$. For example, at t = 20, the boundary values are $n_e = 1.5 \times 10^{-3}$, $u_e = 1.8$, and $T_e = 0.62$, from which $-W_e = 7.65 \times 10^{-3}$. This value agrees roughly with the slope of total energy curve in Fig. 5.

CASE C: Collisional Simulation with Particles

The plasma slab is again initialized with fluids only, with the same collisional parameters as in CASE B, but the particle thresholds are now set so that particles are generated when ν_e falls below $\nu_1 = 0.1$ and are absorbed when ν_{pe} is above $\nu_2 = 0.25$. In the interior of the plasma slab, $\nu_e = 1.9$, well above the generation threshold and the slab remains a fluid. However, as the electric field expands, the electron fluid elements ahead of the ion front, where $n_i = 0$, become collisionless. These fluid elements are changed into particles with a velocity distribution corresponding to the local values of u_e and T_e. In the present simulation, the maximum particle weight is $w_{max} = 3 \times 10^{-4}$ and the minimum number of particles emitted is per cell is $N_{min} = 20$. After generation, these particles move self-consistently with the electron and ion fluids and form an energetic electron component which is trapped by the slab as shown in the phase plots of Fig. 6, for t = 0.4 and t = 4. The electric field profile, Fig. 7, is similar to the pure fluid cases, but the peaks are now higher, and the profile is noisier.

As the particles transit across the plasma slab, they lose energy by collisions and a fraction of them is re-absorbed into the fluid components as shown in the particle counts of Table II.

REFERENCES

1. T. L. Crystal, J. Denavit and C. E. Rathmann, Comments Plasma Phys. 5 (1979), 17.
2. J. Denavit and J. M. Walsh, in "Proceedings of the Ninth Conference on Numerical Simulation of Plasmas" Northwestern University, June 1980, paper PC3. Also, J. Denavit, J. Comput. Phys. 42 (1981) p. 337.
3. R. J. Mason, Los Alamos Scientific Laboratory Report LA-UR-80-2171, August 1, 1980. Also R. J. Mason, J. Comput. Phys. 41, (1981) p. 233.
4. J. U. Brackbill and D. W. Forslund, J. Comp. Phys. 46 (1982), 271.
5. A. Friedman, A. B. Langdon and B. I. Cohen, Comments Plasma Phys. 6, (1981) 225; A. B. Langdon, B. I. Cohen and A. Friedman, J. Comp. Phys. 51 (1983).
6. D. C. Barnes, T. Kamimura, J-N. Leboeuf and T. Tajima, J. Comp. Phys. 52 (1983), p. 480.
7. D. W. Hewett and A. B. Langdon, Laser Program Annual Report-1984, Lawrence Livermore National Laboratory, Livermore, CA.
8. R. Mason, J. Comput. Phys. 51, 484 (1983).
9. J. Denavit, Laser Program Annual Report-1982, LLNL, Livermore, CA.
10. S. I. Braginskii, in "Reviews of Plasma Physics" edited by M. A. Leontovich, Vol. 1, Consultant Bureau, N.Y. (1965) pp. 203-311.

11. J. P. Boris and D. L. Book, J. Comp. Phys. 11 (1973), 38 and 20 (1976), 397. D. L. Book, J. P. Boris and K. Hair, J. Comp. Phys. 18 (1975), 248.
12. J. E. Allen and J. G. Andrews, J. Plasma Phys. 4 (1970), 187.
13. J. E. Crow, P. L. Auer and J. E. Allen, J. Plasma Phys. 14 (1975), 65.
14. J. Denavit, Phys. Fluids 22 (1979), 1384.

*Work performed under the auspices of the U.S. Department of Energy by Lawrence Livermore National Laboratory under contract #W-7405-Eng-48.

TABLE I. Units used to define normalized quantities.

Quantity	Unit
Length	L (characteristic length)
Velocity	$v_o = (T_o/m_e)^{1/2}$ (electron thermal velocity corresponding to characteristic temperature T_o)
Time	$t_o = L/v_o$
Density	n_o (characteristic density)
Electric Potential	T_o/e
Electric Field	$T_o/eL = m_e v_o^2/eL$

TABLE II. Particle count for plasma expansion into vacuum, Case C.

t	# Emitted	# Absorbed	# Net
0	0	0	1
0.4	2480	744	1737
0.8	5520	1763	3758
1.2	11700	4006	7695
1.6	21580	7173	14408
2.0	35720	11559	24162

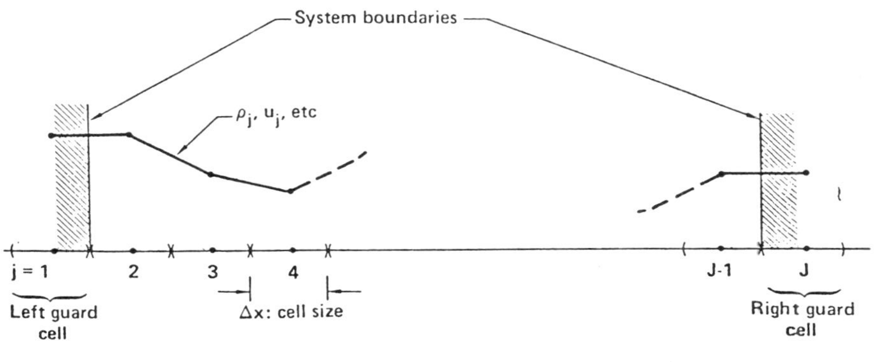

Fig. 1: Uniform Eulerian grid used to represent fluid and particle quantities.

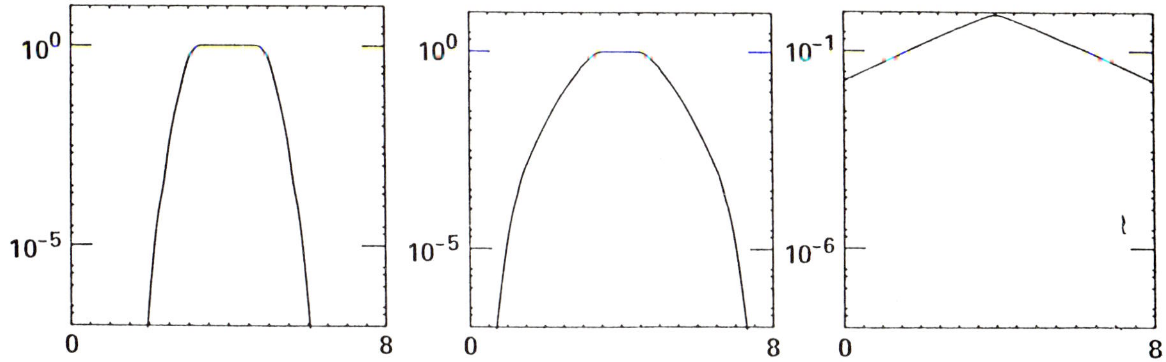

(a) t = 4 (b) t = 10 (c) t = 30

Fig. 2: Ion density for collisionless, isothermal plasma slab expansion into vacuum with fluids only (CASE A).

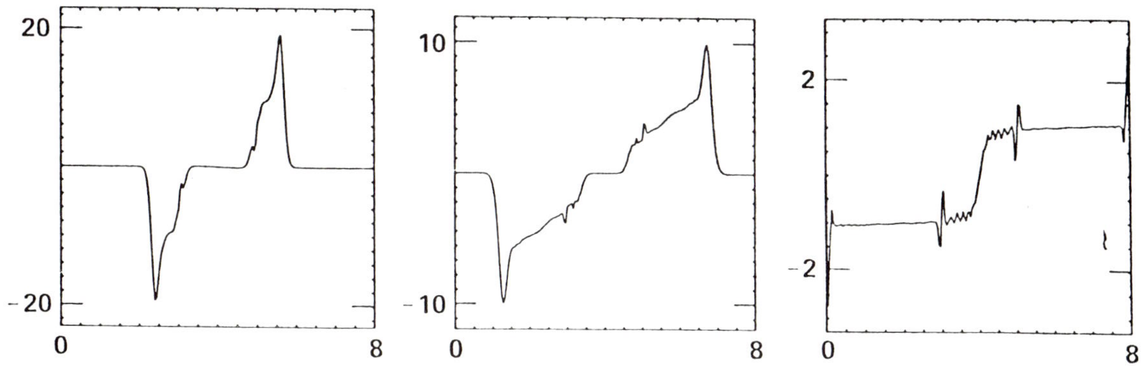

(a) t = 4 (b) t = 10 (c) t = 30

Fig. 3: Electric field for CASE A.

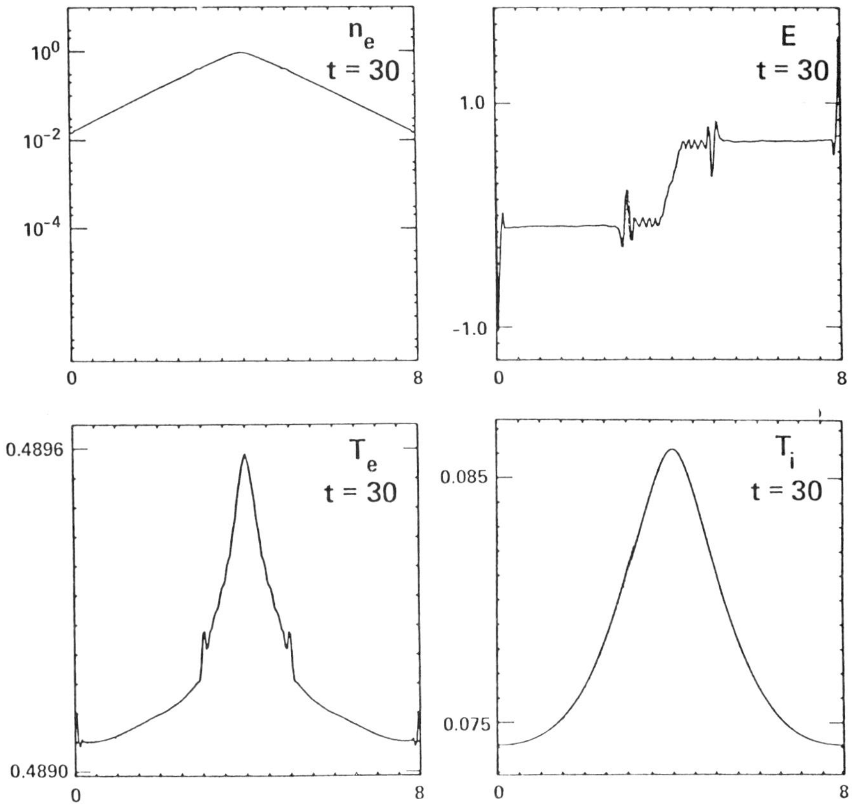

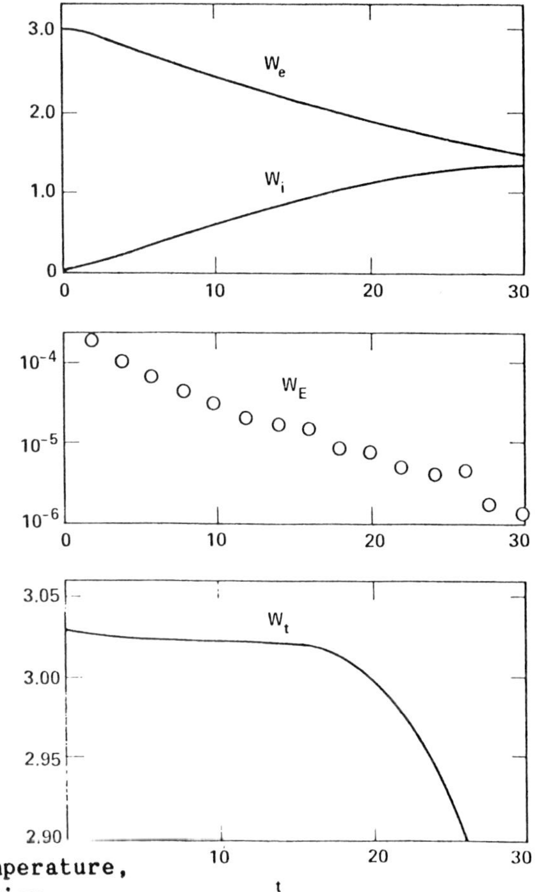

Fig. 4: Electron density, n_e, electric field, E, electron temperature, T_e and ion temperature, T_i, for collisional plasma slab expansion into vacuum with fluids only (CASE B).

Fig. 5: Electron kinetic energy, W_e, ion kinetic energy, W_i, potential energy, W_E and total energy as a function of time for CASE B.

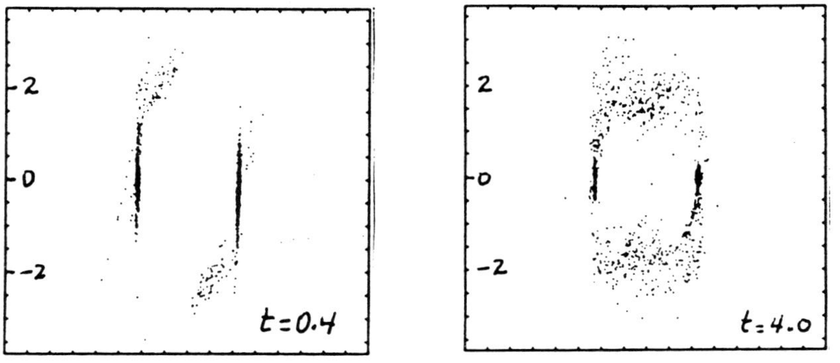

Fig. 6: Electron particle phase plot for plasma slab expansion into vacuum (CASE C).

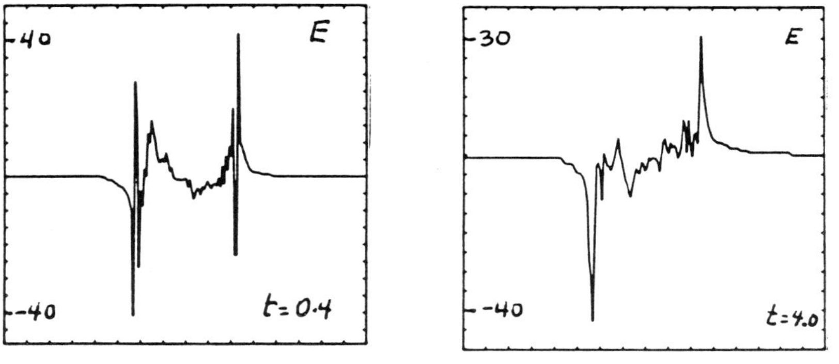

Fig. 7: Electric field for CASE C.

MULTIDIMENSIONAL PARTICLE CODES: THEIR CAPABILITIES AND LIMITATIONS.

Oscar Buneman
Space, Telecommunications and Radio Science Laboratory
Stanford University
Stanford, California 94305

ABSTRACT. Increasing the number of dimensions calls for significant changes in simulation techniques. Demand on computer time and space increases by orders of magnitude and hardware development affects the feasibility. Gridless and Fokker-Planck codes are possible in 1D, but one needs grids and PIC codes in 2D and 3D. This imposes limits on Debye lengths, particle size and spacing, and on resolution. Non-spectral (local) E-M codes also suffer a Courant restriction on δt, in addition to the usual $\omega_p \delta t$ restriction. Spectral methods therefore have an advantage; they also permit convenient filtering, particle shaping and control of resolution. 2D and $2\frac{1}{2}$D codes are well advanced and documented (4,5). 3D codes are in their infancy. Data management, rather than physics or numerical analysis, becomes the major problem (10). Machine-independent 3D codes are too limited in resolution and speed. Parallelism helps greatly but makes the 3D codes machine-dependent. A present-day limit is attempted in a 2*128**3 grid code for CRAYs which process ~ 5 million particles in ~ 2 minutes per time step. Layering is employed to break up the 3D problem into many 2D problems. Fields and particles are packed and buffered in and out of core. Diagnostics are limited by the large volume of information accumulated in a run. Results of runs with 3D codes have tended to show that the third dimension, treated as "ignorable" in 2D simulations, should not have been ignored. With the next generation of highly parallel multiprocessors (which, however, call for the abandonment of spectral and implicit methods) one may hope to do very realistic 3D simulations.

1. THE NEED FOR ADEQUATE SAMPLING.

In this tutorial the successive hurdles will be described which one encounters as one progresses from 1D through $1\frac{1}{2}$D, $1\frac{2}{2}$D, 2D, $2\frac{1}{2}$D to 3D particle simulations. While at first one has to face certain limitations of an analytical nature, ultimately the limits are set by

Space Science Reviews **42** (1985) 103–112. 0038–6308/85.15

data management problems whose resolution depends critically on the available hardware.

A trivial reason for the increasing difficulty of higher dimensional particle simulations is their demand for substantially greater particle numbers. With each added dimension the number of sampling particles has to be multiplied by a certain factor.

This applies also to "half" dimensions. It is customary to denote the inclusion of extra velocity components by referring to them as "half" dimensions: Typically a pure 1D simultaion simulates the plasma as rigid sheet particles, all parallel to the y-z plane, say, and moving in the x-direction. It ignores y- and z- motions of the planes. A $1\frac{1}{2}$D simulation keeps a record of possible y- motions, uniform within each plane. Then the x-ward Lorentz force in the presence of a z-directed magnetic field can be taken into account, as well as the y-wards Lorentz force due to x-motions. In a $1\frac{2}{2}$D simulation, dz/dt would be recorded as well. In a $2\frac{1}{2}$D simulation x,y, dx/dt, dy/dt and dz/dt are tracked (but not z): the particles are rigid straight rods whose motion along their axis is taken into account. $1\frac{1}{2}$D and $1\frac{2}{2}$D simulations have recently found application in space plasma work, namely for simulating the critical ionisation phenomenon (1).

It is reassuring that relatively few samples can often give very good statistics. In many applications the velocity distributions stay close to maxwellian and a modest factor (typically 4) in the sample number may suffice to deal with an added half dimension. One exploits this favourable feature of statistics when initialising thermal velocity distributions: each velocity component is made up as the sum of four random numbers (each uniformily random in a certain interval) The resulting distribution (the "four dice curve", or cubic spline !) is almost indistinguishable from the gaussian.

However, when incrementing by a full dimension, sampling requirements jump dramatically. It is easily checked that the statistical potential energy fluctuations in a granular plasma compete with thermal energies when the particles are spaced of the order of a Debye length apart. Such a plasma would be essentially collision dominated. Mostly one is interested in collective effects, since fluid codes are adequate for collision dominated phenomena. Obviously, one needs several particles per Debye length - again a modest number suffices. Now most of the interesting phenomena to be resolved by simulations are on the scale of many Debye lengths (hundreds, thousands). Therefore, the addition of each full dimension calls for an increase of the number of particles by, typically, two orders of magnitude: a 1D simulation might just be squeezed into a PC, a 2D simulation calls for the campus machine, a 3D simulation for a supercomputer and virtual memory.

2. DISCRETISATION IN TIME AND SPACE

1D, $1\frac{1}{2}$D and $1\frac{2}{2}$D simulations can be done without discretising in space: the electrical interaction of sheets is independent of distance and one only needs to order the sheets to calculate their accelerations. Even if the sheets are of finite thickness or if they are "soft" (i.e., they have a smooth density profile across), only a few operations per sheet are needed to move them one time step. This is an effort of order N where N is the numbers of sheets.

However, in all simulations time must be discretised. By studying the simulation of a simple 1D problem, namely electrostatic oscillations in a cold plasma, and by Fourier transforming one's numerical procedure in time, one finds that while a time step $\delta t = 1/\omega_p$ yields the plasma frequency to 5% accuracy, for $\delta t > 2/\omega_p$ one runs into an instability !

To get over this severe limitation of the speed of simulation in cases where the phenomena of interest are much slower than electrostatic oscillations (typically ion responses), one can either use the implicit methods developed by Langdon (2) and Denavit (3), or one can make one's electrons heavier than real electrons, yet keep them light enough to perform their principal function of rapidly neutralising, and of remaining tied to the magnetic lines. Much has been learned from simulations with ion-electron mass ratios as low as 16:1.

In a $1\frac{1}{2}$D code, one may begin to worry about a possible restriction of $\omega_c \delta t$; this will be discussed briefly below.

The big analytical problems in simulation arise when one advances to two dimensions. Interactions between rods of charge depend on distance, and the many remote rods are as important as the few near ones. For N rods, one has to calculate N^2 interactions and, as pointed out in section 1, N itself might be typically two orders of magnitude larger than for a 1D simulation.

In order to get back to an effort of order N per step in the particle advance, one tabulates the field over a spatial grid and one calculates the self-consistent field from a grid-record of the charge and current-densities to which each particle contributes.

The permissible coarseness of the grid mesh becomes a critical issue and the problem of integrating the finite difference version (now in both space and time) of the field equations is far from trivial. Fortunately, both these subjects have been advanced to a state of relative completion and are exhaustively covered in two texts, "Computer Simulation using Particles" by Hockney and Eastwood (4) and "Plasma Physics via Computer Simulation" by Birdsall and Langdon (5).

Very briefly and broadly, one can state that the grid should be fine enough to resolve Debye lengths, and that smoothing or filtering of high spatial frequencies should be practiced in order to minimise "aliassing". This is the stroboscopic phenomenon of high frequencies parading as low frequencies (long wavelenghts): Many physical instabilities set in preferentially at long wavelengths and can thus be excited numerically through aliassing. Grid effects can often be studied and checked in 1D where grids are optional. Smoothing can be achieved by particle shaping, i.e, spreading point particles into soft balls with a smooth bell-shaped internal density profile. Likewise, splines and finite element techniques help.

Regarding the field update from the charge-current record: fast non-iterative methods for solving Poisson's equation over an L by M mesh have been developed, such as cyclic reduction in rows and columns, or Fourier transforming in one of the two dimensions (say that of M). This is an effort of the order $L\,M\log_2 M$. Two-dimensional simulations go back historically to Hartree who, during WW2, initiated the simulation of the pure electron plasma which circulates in the magnetron. He used Fourier transforms in the ϑ-direction and radial marching - but he kept only very few Fourier harmonies: the FFT has not been invented yet !

Hartree also pioneered the time-centered update of the particles from Lorentz's equation

$$\frac{d\vec{v}}{dt} \pm \frac{e\vec{B}}{m} \times \vec{v} = \pm \frac{e}{m}\vec{E}$$

using $(\vec{v}^{\,new} + \vec{v}^{\,old})/2$ in the Lorentz term and solving the linear equations for $\vec{v}^{\,new}$ explicitly. No limitation of $\omega_c\delta t = (eB/m)\delta t$ arrises from this method, except that for large values of $\omega_c\delta t$ the phases of the gyromotion are misrepresented. For small $\omega_c\delta t$ one gets the same results as with cycloid fitting, i.e., joining solutions of the type

$$\vec{v}^{\,\perp} = \vec{E} \times \vec{B}/B^2 + \text{gyration}$$

for the components of the velocity transverse to $\vec{B}$. As regards this particle update, there is no significant increase in effort when advancing from $1\frac{1}{2}$ D to 2D and 3D.

A further timestep limitation is encountered when one wants to integrate the full electromagnetic equations over the grid. Because Maxwell's equation (not including Poisson's) are hyperbolic, i.e., they contain a natural $\partial/\partial t$ or "update" term, they can be solved in an effort which is of the order of magnitude of the number of grid points, LM in the 2D example above. Essentially, one solves the wave equation. However, this process becomes unstable unless one observes the Courant speed limit $\delta t < \delta x/c$ in 1D, $\delta t < \delta x/c\sqrt{2}$ in 2D, $\delta t < \delta x/c\sqrt{3}$ in 3D for square and cubic meshes of side δx. In many applications, scales chosen from other considerations are such that c is a large number and this restriction of δt results in a severe slow-down.

3. SPECTRAL METHODS AND INTERPOLATION

The Courant condition can be overcome by doing the entire field update in the transform domain. One checks that Maxwell's curl equations combine into one for the complex field vector $\vec{F} = \vec{D} + i\vec{H}/c$. When Fourier transforming, this equation becomes

$$\frac{d\vec{F}}{dt} - c\vec{k} \times \vec{F} = -\vec{j}$$

for the spatial harmonic which goes like exp $(i\vec{k}\cdot\vec{r})$. This field equation is surprisingly similar to that for the particle velocities and has the corresponding solution for the transverse part $\vec{F}^{\perp}$ of $\vec{F}$:

$\vec{F}^{\perp} = \vec{j} \times \vec{k}/ck^2$ (magnetostatic field)

+ circularly polarised wave rotating at angular velocity ck.

The time intervals at which one joins successive solutions of this form are dictated by the rate at which $\vec{j}$ changes, not by the magnitude of ck. (Note: $k_{max} = \pi/\delta x$, $\pi\sqrt{2}/\delta x$ or $\pi\sqrt{3}/\delta x$ according to the number of dimensions.)

Fourier transforming all field-like quantities has many advantages. For instance, the longitudinal part of $\vec{F}$ (which is just $\vec{D}$) can be obtained from Poisson's equation as $\vec{k}\rho/k^2$. Of course, transforming in two dimension rather than only one (as in the fastest Poisson solvers) makes for an effort of the order $LM(\log_2 L + \log_2 M)$. On the other hand, the ready availability of well-programmed FFT's, and the additional benefits of spectral methods make up for this increase in effort.

In the transform domain one can perform the filtering, the particle shaping, an optimzation for the spline fitting process and the truncation of the interaction to be discussed in the following section on boundary conditions. One does not have to use any spatial finite difference calculus for the field equations. However, a grid is still necessary since we have only discrete numerical Fourier transforms between $\vec{r}$-space and $\vec{k}$-space.

This leaves the problem of interpolation in the mesh. By using high order interpolation, one can greatly reduce aliassing and improve accuracy. Quadratic and cubic splines have been used, but this soon becomes expensive: in a 3D, E-M code, cubic splines would require each particle to look up 384 data to interpolate the $\vec{E}$ and the $\vec{B}$ which acts on it !

Linear interpolation is most commonly used. Interpolation is also needed when the particles contribute their charge and current to the ρ, $\vec{j}$ arrays. Linear interpolation is then, in 2D, equivalent to "area weighting". For 3D, we have cut down data look-up (or deposit) effort for linear interpolation by using a tetrahedral mesh: each particle references only the four nearest mesh-point data. The tetrahedra result from drawing the space diagonals into a cubic mesh and introducing cubic center data (6). Interpolation of currents must be done twice in each step of each particle: once at its old position and once at its new position, since the current is that due to the movement between the two.

4. BOUNDARY CONDITIONS

Internal conducting boundaries can readily be introduced into an electrostatic field solver by means of the "capacitance matrix". This was used by the author in an early space-related simulation, namely that of the photo-electron sheath surrounding a satellite (7), using a very primitive computer. Another method is to introduce a species of very light electrons confined to the conducting surface and moving under friction, i.e., simulating the process of surface conductivity.

External rectangular conducting boundaries are readily simulated in a spectral field solver by using only sine- or only cosine-transforms. A major problem in space applications is to simulate free space conditions outside the computer domain. Complex Fourier methods (with $\exp(i\vec{k}\cdot\vec{r})$) imply periodic repeats of the computed domain in all dimensions. If the simulation is to represent typical phenomena in a rather larger plasma, such repeats are acceptable, but for an isolated plasma of limited extent they become unrealistic. The problem of calculating the electrostatic field of an isolated plasma was solved by Hockney (4, section 6-5-4). He keeps a generous empty buffer zone around the domain containing particles and truncates the interaction between charges beyond a certain radius so that the nonphysical repeats introduced by the Fourier method cannot influence the central plasma. This was first applied to gravitational simulations.

The most elusive boundary problem for space plasmas is the radiation condition. It is easy in 1D where it seems to have been first applied by Morse and Nielsen (8). To decide what part of the field in the charge-current-free space outside the plasma is outgoing and what is incoming presents no problem in 1D and the incoming part can be suppressed.

In 2D the decision is more difficult. It requires information not only in the source-free boundary layer at any time but also over its past history. It almost seems as if, in principle, the entire past history is needed for the decision. However, E. Lindman (8) found that a fairly short history (such as 3 past time steps) of the boundary suffices for an algorithm which will suppress all but 1% of the incoming radiation at all but the shallowest angles of incidence. The author is working on a method employing spherical harmonics which should be 100% effective (9).

5. DATA MANAGEMENT

In 2D, or even $2\frac{1}{2}$D, the stage has been reached where codes can be easily built by following the instructions give in the texts (4,5). One may also call a ready-made code, such as Langdon and Byers' "Zohar" from a library. These codes call for a large computer, but not necessarily a supercomputer. They are in Fortran and transportable, and they can be accommodated in core.

Against this, 3D codes are still in their infancy: little progress has been made since the survey paper by the Stanford team in 1980 (10).

No further analytical or numerical analysis work appears to be necessary or helpful in advancing to 3D. The difficulties lie entirely in the data management. For this reason, and also because computational speed and efficiency is at a premium in 3D work, such codes depend heavily on machine architecture. They are not transportable except between like machines.

Using a modest factor of 4 in the total particle number for each velocity dimension and a factor 64 for each spatial dimension, one concludes that 3D simulations ought to employ some 2^{24}, i.e., 16 million particles - each requiring 6 data (x,y,z,u,v,w) to be recorded: 100 million data.

Obviously, there is no hope for doing a plausible 3D simulation without (1) introducing some economies and (2) storing the particles outside core and calling them in only in small batches.

Similarly, the 64^3 linear scale range calls for over a million field data: fields cannot reside in core. "Layering" is therefore practiced in 3D codes. In the TRI-dimensional STANford code "TRISTAN" a grid of $2*128^3$ points (cubic mesh, plus cube centers) is used for recording field data, and particles are kept ordered into 128 layers. At present the number of particles is $4\frac{1}{2}$ million: this could easily be increased to 8 million. Only 2 field layers are in core at any time, and 4 charge current layers (the latter because after moving, particles can have dropped below or risen above their original layers*). Triple buffering is used when the particles of each layer are passed through core for processing, in batches of about 5000. Images of the batches which straddle layer borders are kept in core, for depositing particles that have dropped or risen. Some sorting of the particles is necessary here.

Fields of the next layer above are brought in when a layer of particles has been completed and the lowest of the four charge-current layers is put out to disk. These two moves are accompanied by Fourier transforming within the layer dimensions (x and y, say, when layering is in z, to k_x to k_y). Filtering of high harmonies at this stage helps with I/O economy.

When all layers have been processed, the original field record on the disks has been converted into a charge-current record, indexed in a hybrid manner, namely according to k_x, k_y and z.

6. THE BOTTLENECK: GLOBAL VERSUS LOCAL METHODS

At this point, field solving can begin, but it will require prior Fourier transforming in z. The $\vec{j}$, ρ data have to be read back into core, but in different order, from scattered disk areas. Again, layering must be used but now according to k_y (say). TRISTAN uses 16 sectors at this point. Fourier transforming cannot begin till all

*The code is relativistic and Einstein prevents particles from straying further.

reading of a sector is complete : I/O cannot be overlapped with computation. A similar bottleneck occurs after the Maxwell field update when the FFT's in z have to be completed before the new fields of a sector can be written to disk.

It is difficult to asscess the cost of such waiting for I/O completion - this depends strongly in the operating system. A timestep overall (particle plus field update) in TRISTAN takes about 2 minutes on a CRAY.

The hold-up due to reading and writing data in different order is characteristic of all "global" methods for updating a database, i.e., whenever the advance of each item depends on all the information, not only on close neighbors. Since communication speeds across the computer memory are already limiting factors, such global methods will become increasingly unsuitable in supercomputers. In the multiprocessor devices of the future the preference for local methods (update from a close neighborhood) will be very pronounced.

Abandoning Fourier ("spectral") methods may become necessary ultimately. Local, finite difference methods can readily cope with hyperbolic equations (involving a $\partial/\partial t$ or natural "update" term). Fortunately, it has been checked (11) that the Poisson equation can be eliminated in favor of a more natural retarded potential construction: after all, according to Einstein, distant data ouside our past light-cone can never determine the present state.

The large data volume, and the heavy I/O traffic in a 3D code calls for data-compression. Particles can be comfortably packed into two CRAY words, with x, y, z recorded, 21 bits each, in one word and u, v, w in the other. Field components are conveniently packed 2:1, in a squashed floating point format. Resulting precision is still equivalent to IBM single precision. In view of the importance of exploiting the CRAY's vector capabilities, TRISTAN was written entirely in Assembler. So was its modest CDC 7600 predecessor, "SPLASH", and a now dormant 64^3 IBM code.

7. DIAGNOSTICS, INITIALISATION: PRE- AND POST-PROCESSING

Particle and field records reside entirely outside core at certain instances in the time step cycle. It is therefore possible to initialise the simulation independently of the main program, and to interrupt the program for diagnostic purposes, or even for changing parameters. Initialisers for TRISTAN can set up quiet starts, uniform or non-uniform thermal distributions as well as streaming configurations.

Regarding diagnostics, one has to decide before any run what one wants to observe. To keep all the information on all the time steps would overstain any disk system surrounding the computer: only occasional steps are worth saving in totality. An interruption every so often, and saving of the most recent state, is advisable for the purpose of restarting after system crashes.

However, "continuous" (i.e., step by step) records of certain variables are also wanted in most applications. TRISTAN puts 2K tagged particles into a separate file at every step, for making movies. It also records certain field harmonics continuously. A standard test for any new code is to initialise a uniform thermal plasma and to monitor various spatial harmonics for, typically, 256 time steps. Then one Fourier transforms each spatial harmonic in time and plots its fluctuation spectrum. Theory can predict what this should be. Significantly, the most direct comparison of these spectra is with those received from space by radar backscatter techniques!

Movies of the tagged particles have demonstrated the capabilities of 3D codes most dramatically. The simulation of a z-pinch by SPLASH showed the spontaneous development of helical structures -- again most directly confirmed by experiments and, at the time, not familiar among theorists. The initialisation, and the boundary conditions, were axially uniform but a 2D code would not have reproduced this phenomenon. Similarly, a field reversed mirror configuraiton showed a five-fold flute instability involving all three dimensions.

8. OUTLOOK

TRISTAN has taken 3D codes close to the limits of what is feasible with CRAY hardware at this time -- perhaps too close: it is rather inflexible, allowing for few potential modifications each of which will be difficult to implement. It is only transportable between CRAYS.

The prospect for a more "user friendly" 3D code are not good unless one is willing to accept results from codes with substantially fewer particles and coarser mesh, perhaps by ignoring the limitations of a more analytical nature which were found in the course of developing 2D codes.

The next step in particle simultion is likely to be in the direction of even greater hardware dependence. In view of a universal interest in multiprocessors (array processor) from both the user community and from computer designers, one might devote effort to thinking what architecture would be most appropriate for better 3D particle simulations than those feasible with present supercomputers.

Primarily, the demand is for more particles and finer meshes. As stated before, global methods may have to give way to more local methods: each processor would only have access to a limited region of memory, the latter containing an image of a limited region of physical space. Communiation between processers would be limited by the speed of light, just as it is in real space plasmas!

REFERENCES

1. M. Tanaka and K. Papadopoulos, "Creation of High-Energy Electron Tails by means of the Modified Two-Stream Instability", Phys, Fluids 26(7), p. 1697 (1983).
2. A.B. Langdon, B.I. Cohen, A. Friedman, "Direct Implicit Large Time-Step Particle Simulation of Plasmas", Journ. Comput. Phys. 51(7), p. 107 (1983).
3. J. Denavit, "Time Filtering Particle Simulations with $\omega_{pe}\Delta t >> 1$", Journ. Comput. Phys. 42(8), p. 337 (1981).
4. R.W. Hockney and J. Eastwood, "Computer Simulation using Particles", McGraw Hill, (1981).
5. C.K. Birdsall and A.B. Langdon, "Plasma Physics via Computer Simulation", McGraw Hill, (1985).
6. O. Buneman, "Tetrahedral Finite Elements for Interpolation", SIAM J. SCI. STAT. COMPUT. 1(2), p. 223 (1980).
7. O. Buneman, "Computer Simulation of the Satellite Photosheath", ESRO report, ESRIN, Frascati (1970).
8. E.L. Lindman "Free-Space Boundary Conditions for the Time-Dependent Wave Equation", Journ. Comput. Phys. 18(5), p. 66, (1975).
9. O. Buneman "Radiation Condition for Computer Simulations using Spherical Harmonics", (to be reported elsewhere).
10. O. Buneman, C.W. Barnes, J.C. Green, D.E. Nielsen, "Principles and Capabilities of 3D, E-M Particle Simulations", Journ. Comput. Phys. 38(1), p. 1 (1980).
11. O. Buneman, "Global vs. Local Methods in Particle Simulations", paper od-2, 9th Conf. of Numerical Simulation of Plasma, Northwestern University, Evanston, Ill. June 1980.

WAVE-PARTICLE DIAGNOSTICS FOR PLASMA SIMULATION

Viktor K. Decyk
University of California
Dept. of Physics
405 Hilgard Avenue
Los Angeles, CA 90024 USA

ABSTRACT. One important type of problem for particle simulation of plasmas, is one which is bounded and has external sources and sinks. For example, there are problems with antennas for studying RF heating or current drive in fusion plasmas, or problems in space simulation where particles are injected at one boundary with some specified energy or momentum distribution. In understanding such simulation results, it is useful to know how energy and momentum are flowing inside the plasma. This can be accomplished in electrostatic particle simulations on the basis of some theorems for energy and momentum flow. An important application of these theorems occurs when many waves are involved in producing some effect, e.g., generating a current by RF. One can then extract the contribution of each wave to the effect to identify which are most important. Examples of such wave-particle diagnostics will be given.

Today I will talk about diagnostics for particle simulation. I will begin by giving a general overview of my philosophy of organizing diagnostics and give some examples of generally useful diagnostics. Then I will specialize and give more details about diagnostics for bounded electrostatic plasmas which are driven by external sources of energy and momentum.

Diagnostics, I believe, are where the real physics get done in a simulation code, where you finally get to understand what is happening and can explain it to someone else. Simulation codes generate much more information than one can possibly use--no one wants to know the trajectories of all the particles. Instead, the purpose of the simulation, as of physics, is to explain the behavior of the physical system under study in terms of a small number of simple concepts.

Most of the time we want to represent the simulation data in terms of concepts which are known to be useful, or which appear in some theory. For example, the idea of a particle distribution function or plasma potential function appear in most theoretical descriptions of plasmas and are necessary diagnostics to include in most any simulation. Occasionally, the concepts which best describe the phenomena under study are not even known, and their determination may,

Space Science Reviews **42** (1985) 113–130. 0038–6308/85.15

in fact, be part of the research goal. For example, the best or most simple way to describe strongly turbulent systems probably has not yet been found, so one can be creative and try to contribute to the evolution of new ideas.

The dynamics part of a particle simulation code usually evolves into a fairly standard type that one does not change much.[1-2] But the diagnostics one needs and uses are constantly changing as one tackles new physical problems. Therefore, I have found it useful to do most of my diagnostics in a separate post-processing program, perhaps even on a different computer.

The main simulation code generates a database, or a collection of datafiles, which are very general, and stored on an external device, such as a disk, as the run proceeds. This includes things like the time history of the velocity distribution functions or the time history of a subset of the Fourier modes of the potential. In addition, the main program generates certain key diagnostics whose primary purpose is to point out those things in the database which need to be examined further. This would include things like the time-averaged fluctuation energy of the modes of the potential, from which one can decide which modes have been excited. Thus, the main program can be regarded as producing a database which is a first level reduction of data.

From the database, a further reduction of data occurs, as more detailed diagnostics specialized to the problem at hand are done. This could include things like frequency analysis, cross-correlations, etc. This can be done interactively, which is especially useful in the preliminary runs on a new problem when one is still feeling ones way around. Once one has learned which particular diagnostics are most useful, then this post-processing can be done automatically as a batch job. As an example, when one first starts a problem, one may not know which frequencies are most significant. But eventually one decides what is worth looking at, so that future runs can produce output in the frequency range of interest automatically.

This style of doing diagnostics requires additional off-line storage, but this is not usually a severe constraint, and it has the advantage that as one thinks of new particular diagnostics, they can often be calculated from the database, without having to do the run all over again. (It is rare that I think of the best diagnostics for a new problem the first time around.)

From that general outline, let me now illustrate by specific examples. The model I will be talking about is electrostatic, with two spatial dimensions, and three velocity dimensions. One of the spatial co-ordinates is periodic, and the other is bounded.[3] Generally, there is an electrostatic antenna on one boundary driving the plasma, since this code was used for simulation of rf heating for fusion applications.

For such problems, the main code could generate five different data files. These were:

I. Trajectories of test particles, $\vec{r}(t)$.

II. Velocity distribution functions, either one dimensional, such as

$$F(v_x) \equiv \int_{v_x}^{v_x+\Delta v_x} dv_x' \int_{-\infty}^{\infty} dv_y' \int_{-\infty}^{\infty} dv_z' \; f(v_x', v_y', v_z')$$

or two dimensional, such as

$$F(v_\perp, v_\parallel) \equiv \int_{v_\perp}^{v_\perp+\Delta v_\perp} v_\perp' dv_\perp' \int_{v_\parallel}^{v_\parallel+\Delta v_\parallel} dv_\parallel' \int_{o}^{2\pi} d\theta' \; f(v_\perp', v_\parallel', \theta')$$

These are calculated much like charge distribution is calculated by accumulating number density on a velocity-space grid. Occasionally, the distribution of some other quantity as a function of velocity can also be useful, such as particles' energy gain $\vec{F}\cdot\vec{v}$, which can identify velocity space resonances.

III. Plasma profiles versus the bounded or inhomogeneous co-ordinate x. Usually, the particle density, drift velocity, and kinetic energy versus x are calculated. Sometimes the energy absorption $\vec{F}\cdot\vec{v}$ or energy flux versus x are useful.

IV. Ion density for various Fourier modes in the periodic direction versus x,

$$\rho_m{}^i(x) \equiv \frac{1}{L_y} \int_o^{L_y} \rho^i(x,y) e^{-2\pi imy/L_y} dy$$

This is expecially useful for looking at ion waves. Since the ions do not respond to high frequency waves, this data automatically has the high frequencies filtered out.

V. The potential for various Fourier modes versus x, $\phi_m(x)$.

Most of these quantities are calculated and stored every 5 or 10 time steps.

In addition to generating these datafiles, the main program prints or displays the zeroth, first, and second moments of the velocity distribution functions, to tell me which components are evolving in time and need to be examined further. Occasionally, phase space diagrams of the particles are displayed. At the end of the run, the time histories of the field and kinetic energies and momentum are displayed. The work done and momentum imparted by the antenna is subtracted from the total energy and momentum of the plasma to determine how well the code is conserving these quantities. I will talk in more detail about this later. Finally, the integrated electric field fluctuation spectrum versus Fourier mode number is displayed to indicate which modes have high energy levels.

From this kind of database, let me give some examples of what kinds of things the post-processing programs calculate. The first few examples I will show come from a problem where we were driving lower hybrid waves by an external antenna for fusion applications.[4] The plasma was inhomogeneous with a density profile which increased away from the antenna in the bounded direction. The magnetic field was in the periodic direction.

The program which analyzes the potential datafile, $\phi_m(x,t)$ calculates quantities such as the integrated frequency spectrum per Fourier mode m:

$$P_m(\omega)=\int_0^{L_x} dx \left|\frac{1}{T}\int_0^T \nabla\phi_m\,(x,t)e^{-i\omega t}dt\right|^2$$

The length of integration T depends on how much resolution one needs, but is also limited by how fast the waves are changing in the plasma. The best choice may not be known ahead of time. Figure 1 shows the spectrum for mode m = 1, which corresponds to $k_\parallel \lambda_{DE}$ = .034, and one sees different peaks which correspond to different eigenmodes in the x direction. Since the plasma is inhomogeneous in x, the eigenmodes are not sinusoidal functions of that coordinate. Picking one of the frequencies ω_o which correspond to the peaks, one can obtain the actual eigenfunction by going back to the original data and, using the cross-correlation program, beat $\phi_m(x,t)$ with a reference signal at ω_o to produce the interferogram defined by:

$$C_m(x,\tau) \equiv \frac{1}{T-\tau}\int_0^{T-\tau} \phi_m(x,t+\tau)\sin(\omega_o t)dt.$$

When the eigenfunction is excited only by the thermal fluctuations, as in this example, the length of integration time T must not be longer than the correlation time of the wave. The top graph of Fig. 2 shows the eigenfunction obtained from the simulation and the bottom one shows the WKB solution.

The spectrum in Fig. 1 showed the natural frequencies of oscillation of this inhomogeneous plasma. If the antenna frequency does not match one of these eigenfrequencies, then the plasma responds with constant amplitude driven oscillations, much like a harmonic oscillator which is driven off resonance. I will discuss later what happens if the frequencies do coincide.

In the next example, shown in Fig. 3a, one sees the interferogram of such a driven lower hybrid wave, obtained from

$$E_\perp(x,\tau) \equiv \frac{1}{T-\tau}\int_0^{T-\tau} -\frac{\partial\phi_m}{\partial x}(x,t+\tau)\,\sin(\omega_o t)dt$$

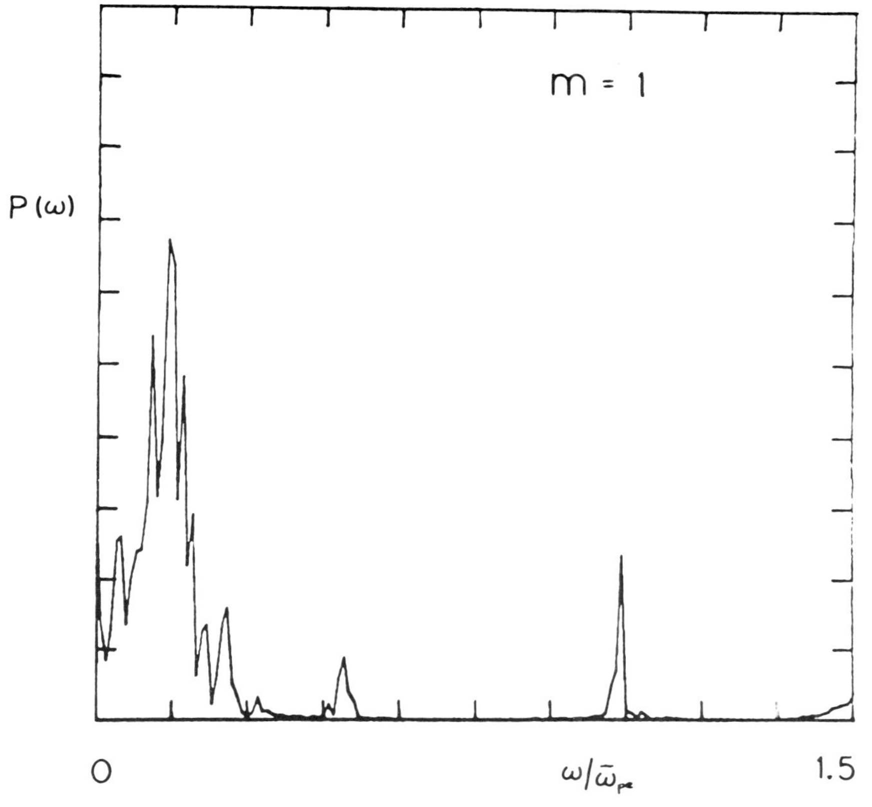

Fig. 1 One fourier component (m=1) of electric field spectrum for plasma bounded in x and periodic in y.

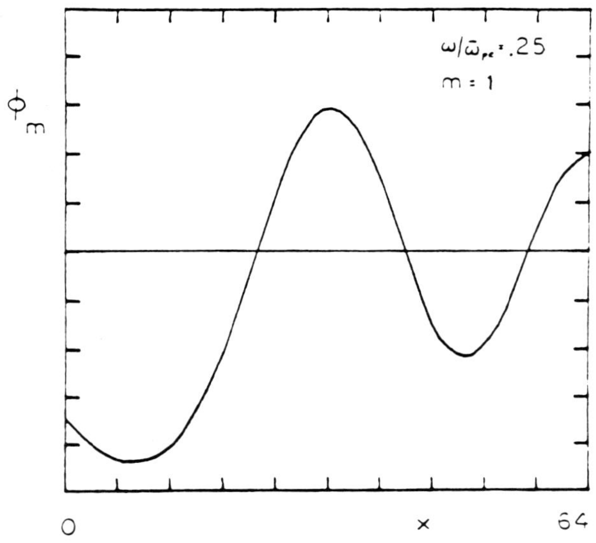

Fig. 2b. Eigenmode of bounded, inhomogeneous plasma, extracted from thermal fluctuations in simulation.

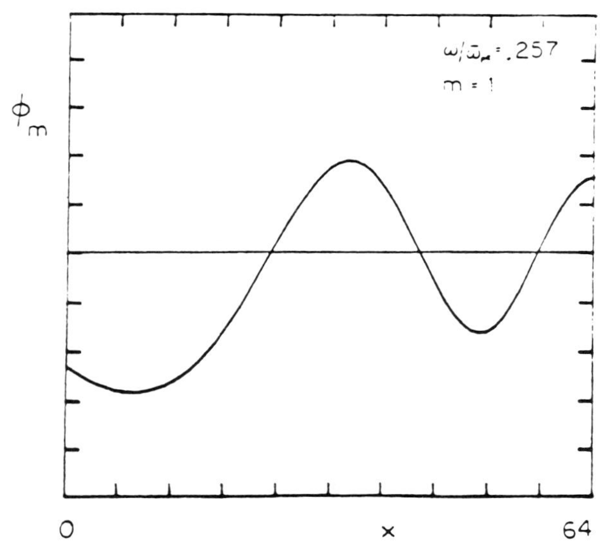

Fig. 2b Eigenmode of bounded, inhomogeneous plasma, from W.K.B. solution of fluid equations.

where ω_0 is the driving frequency and $\partial\phi_m/\partial x$ was obtained from the database by finite differencing. By displaying the waveform at different values of τ, one can determine the direction of propagation of the wave fronts. In this example, the antenna is on the left, so that one can see that we have a backward wave, as one expected for a lower hybrid wave. Once can also see that the wave amplitude becomes very small, indicating that the wave is nearly completely absorbed.

Another quantity one can calculate from this datafile is the local wavenumber. If we have a traveling wave of the form

$$E_\perp(x,\tau) = A(x) \sin[\int k_\perp(x')dx' - \omega_0\tau + \phi]$$

Then the amplitude is given by

$$A(x) = \text{L.M. } [E_\perp(x,\tau)]$$

where L.M. refers to the local maximum as a function of time for each x. The derivative

$$\frac{d}{dx} [\frac{E_\perp(x,\tau)}{A(x)}] = k_\perp(x) \cos[\int k_\perp(x')dx' - \omega_0\tau + \phi]$$

so that

$$k_\perp(x) = \text{L.M. } [\frac{d}{dx} \{ \frac{E_\perp(x,\tau)}{A(x)} \}]$$

The dots shown in Fig. 3b represent these measured values from the simulation and the solid line comes from a local (at each x) solution of the dispersion relation.

These are just a sample of the kinds of diagnostics which are useful from this data file. The datafile of ion density is treated by the same post-processing programs to analyze ion waves.

Figure 3c shows an example of a plot from the datafile which contains the plasma profiles. It shows the energy absorption $\vec{F}\cdot\vec{v}$ versus x, averaged over time to remove the oscillating part. What it illustrates is where in space ion heating occurs. One sees that heating occurs in a rather localized region of space in the interior, as well as in a region just in front of the antenna. This was not seen in a plot of particle kinetic energy versus x, because the amount of heating was small, and because the particles could move away from the heating region, giving a much more uniform distribution of kinetic energy.

Having such data in a file also means that one could superimpose graphs illustrating the change in time of some quantity, which makes for clearer presentations.

The trajectories of test particles in phase space can sometimes be the most revealing diagnostic of all. However, it can be the most difficult to use in practice, since one is usually looking for those few particles which clearly illustrate the effect one is looking for out of a large number.

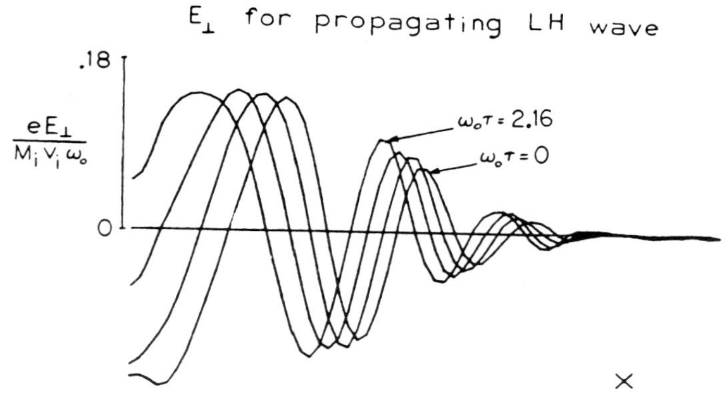

Fig. 3a. Perpendicular electric field of driven lower hybrid wave at different phases.

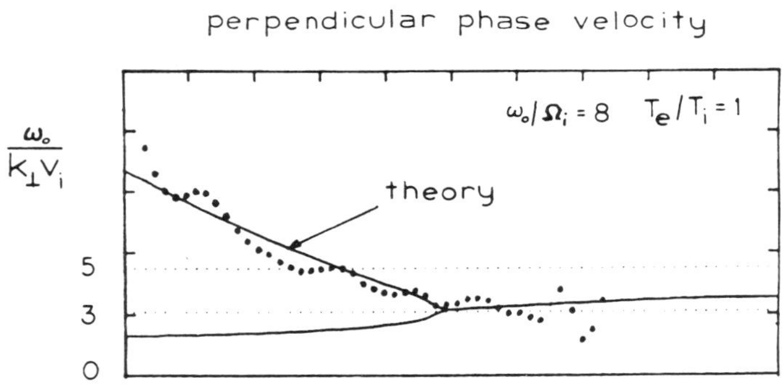

Fig. 3b. Perpendicular phase velocity from fluid theory (solid line) and simulation (dots).

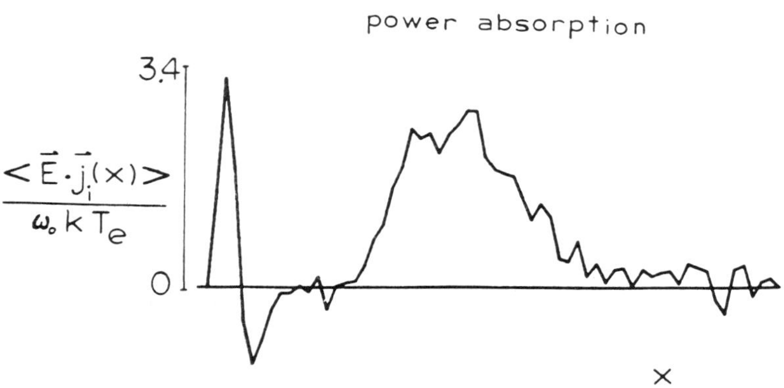

Fig. 3c. Power absorption by ions as a function of space.

One trick is to take advantage of the fact that simulation runs are absolutely reproducible, at least on most computers. In the case of the lower hybrid heating simulations, at the end of the run, we identified which ions had heated the most. The current co-ordinates had been saved in a restart file and were therefore available, and the initial co-ordinates could be reconstructed with the initial random number generator. Once they had been identified, we ran the same run over again, this time saving the trajectories of those test ions. Then we could examine the time histories of these particles to determine how they had gained their energy.

The next graph (Fig. 4) shows one such orbit in perpendicular velocity space. One sees that the ion is mostly going around in its cyclotron orbit, then occasionally makes a quantum jump to another orbit. The reason for this is that if $v_\perp > \omega_0/k_\perp$, then the particle can be traveling along with the wave during part of its gyroperiod and can interact strongly with the wave. If the perturbation electric field is strong enough, the subsequent interaction in the next gyroperiod is decoupled from the previous one, and the particle performs a random walk in velocity space. This is called stochastic heating.[5]

I think that this has given you a flavor of the variety of diagnostics which are useful and possible in a particle simulation. The only advise I would give to someone just starting out is to be flexible and creative--all the information you could possibly want is out there in the data somewhere.

For the last part of my talk, I will discuss energy and momentum flow in electrostatic systems. I was first faced with these issues when someone always asked me after I presented by heating results, how did I know that this heating was not numerical. When a simulation is closed, one can always point out that the code conserved energy and momentum to some acceptable level. But when antennae or other sources are present, the plasma energy and momentum are not constant and I never had a good answer.

When I first started simulation, I was taught that the Poynting vector described how electromagnetic energy was flowing into or out of a system, but that for electrostatic models such as I was using, the Poynting vector did not exist. Eventually I came to the realization that this was not correct.[6]

An electrostatic Poynting vector can be derived very easily. Let

$$\vec{j}_t = \vec{j} - \frac{1}{4\pi} \vec{\nabla} \frac{\partial \phi}{\partial t}$$

This can be regarded as the total plasma current, conduction plus displacement. These are well-defined objects in electrostatic systems. From the equation of charge continuity and Poisson's equation, one readily can show that

$$\nabla\cdot\vec{j}_t = \nabla\cdot\vec{j} - \frac{1}{4\pi}\frac{\partial}{\partial t}\nabla^2\phi = \nabla\cdot\vec{j} + \frac{\partial\rho}{\partial t} = 0$$

Thus, $\vec{j}_t$ is also the transverse plasma current. Since $\nabla\cdot\vec{j}_t = 0$, the divergence of the vector $\vec{R}$ defined by

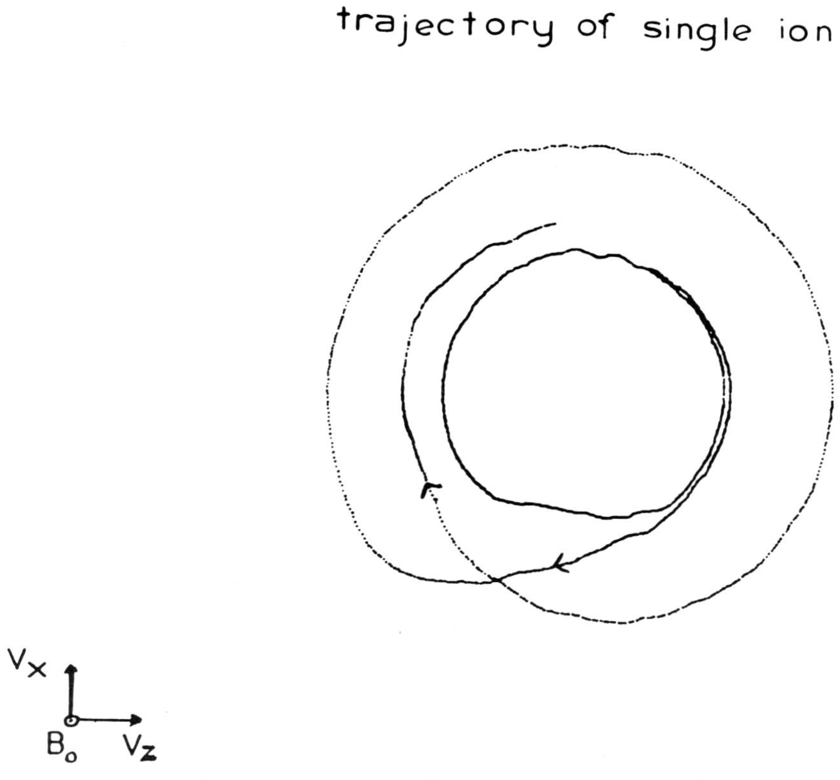

Fig. 4. Ion gyro-orbit in velocity space shows stochastic heating.

$$\vec{R} = \vec{j}_t \Phi$$

can be shown to be:

$$\nabla \cdot \vec{R} = \vec{j}_t \cdot \nabla\Phi = \vec{j} \cdot \nabla\Phi - \frac{1}{4\pi} \nabla\Phi \cdot \frac{\partial}{\partial t} \nabla\Phi$$

$$= - \vec{j} \cdot \vec{E} - \frac{\partial}{\partial t} \left(\frac{E^2}{8\pi}\right)$$

since $\vec{E} = -\nabla\Phi$. Here we have an equation of continuity for electric energy

$$\nabla \cdot \vec{R} + \frac{\partial}{\partial t} \left(\frac{E^2}{8\pi}\right) = - \vec{j} \cdot \vec{E}$$

with the right hand side representing an energy sink or source. Thus, $\vec{R}$ is an energy flow vector for electrostatics, an electrostatic Poynting vector, if you will.

One thing most of us have read somewhere but tend to lose sight of is that the Poynting vector is not unique. Only its divergence has physical significance, and other vectors with equal divergence could do just as well, although we've all rather gotten used to $\vec{E} \times \vec{B}$ for electromagnetics.

To illustrate this, let's derive another energy flow vector. Since $\nabla \cdot \vec{j}_t = 0$, then it means that $\vec{j}_t$ can be expressed as the curl of some vector potential $\vec{V}$,

$$\vec{j}_t = \nabla \times \vec{V}$$

Another vector $\vec{R}'$ with the same divergence as $\vec{R}$ can be constructed merely by adding the curl of some vector to it, since the divergence of a curl always vanishes. Let us then construct the vector

$$\vec{R}' = \vec{R} - \nabla \times (\Phi \vec{V})$$

Multiplying this out

$$\vec{R}' = \vec{j}_t \Phi - \Phi \nabla \times \vec{V} - \nabla\Phi \times \vec{V} = \vec{E} \times \vec{V}$$

since the first two terms on the right hand side cancel by construction. This alternative form of the energy flow vector is in fact exactly the Poynting vector, since the defining equation for $\vec{V}$

$$\nabla \times \vec{V} = \vec{j}_t = \vec{j} + \frac{1}{4\pi} \frac{\partial \vec{E}}{\partial t}$$

is just Ampere's Law without the constants $4\pi/c$! But magnetic fields do not exist in an electrostatic universe, so $\vec{V}$ is merely the vector potential for the transverse current.

Thus if one had calculated the magnetic field from the currents in the electrostatic simulation, ignoring the constants $4\pi/c$, and constructed the Poynting vector one would have gotten the correct energy flow in a purely electrostatic system, even though magnetic fields nowhwere enter into the dynamics. This equally well applies to gravitational systems, even though no one believes that a gravitational analogue to the magnetic field exists.

I have already pointed out that the energy flow vector is not unique. It is equally true that the energy density itself is also not uniquely defined. One can always add the divergence of any vector to the definition of energy and subtract the time derivative of the same amount from the energy flow vector. As an illustration, the set

$$W \equiv \frac{E^2}{8\pi} - \nabla\cdot\left(\frac{\Phi\nabla\Phi}{8\pi}\right)$$

$$\vec{R}'' = \vec{R} + \frac{\partial}{\partial t}\left(\frac{\Phi\nabla\Phi}{8\pi}\right)$$

also satisfy the equation of continuity for energy

$$\nabla\cdot\vec{R}'' + \frac{\partial}{\partial t}(W) = -\vec{j}\cdot\vec{E}$$

If we multiply these out, we find that W is just

$$W = \frac{E^2}{8\pi} - \frac{\nabla\Phi\cdot\nabla\Phi}{8\pi} - \frac{\Phi\nabla^2\Phi}{8\pi} = \frac{1}{2}\rho\Phi,$$

a definition of potential energy we have been using all along. The corresponding energy flow vector can be simplified to

$$\vec{R}'' = \vec{j}\Phi + \frac{1}{8\pi}\left[\frac{\partial\Phi}{\partial t}\nabla\Phi - \Phi\nabla\frac{\partial\Phi}{\partial t}\right]$$

With all these choices, I will chose the following form because it is convenient to calculate in my particle simulation code:

$$\nabla\cdot\vec{R} + \frac{\partial}{\partial t}\left(\frac{E^2}{8\pi}\right) = -\vec{j}\cdot\vec{E}$$

The energy source, represented by the term $-\vec{j}\cdot\vec{E}$ is energy that came from the particles. So to complete the picture, we need to examine the flow of kinetic energy. In a simulation with the finite-size particles, with shape function $S(\vec{r})$, one can define the kinetic energy density and flux as follows:

$$U(\vec{r},t) = \frac{1}{2} \sum_i m_i v_i^2(t)\ S[\vec{r} - \vec{r}_i(t)]$$

$$\vec{K}(\vec{r},t) = \frac{1}{2} \sum_i m_i v_i^2(t) \vec{v}_i(t)\ S[\vec{r} - \vec{r}_i(t)]$$

where the sum is over particles, and I have neglected, as before, the effects of grids and finite time steps. By straightforward differentiation, one can show that these qualities satisfy

$$\nabla \cdot \vec{K} + \frac{\partial U}{\partial t} = \sum_i m_i \vec{v}_i(t) \cdot \frac{d\vec{v}_i(t)}{dt}\ S[\vec{r} - \vec{r}_i(t)]$$

The acceleration of a particle in the simulation is given by the Lorentz force

$$m_i \frac{d\vec{v}_i(t)}{dt} = q_i \ \{ \langle \vec{E}[\vec{r}_i(t)] \rangle + \vec{E}_{ext} + \vec{v}_i(t) \times \vec{B}_{ext}/c \}$$

where

$$\langle \vec{E}[\vec{r}_i(t)] \rangle = - \int S[\vec{r}\,' - \vec{r}_i(t)] \nabla \Phi(r') d^3 r'$$

is the averaged self-consistent field accelerating the particle. If we use the definition

$$\vec{j}(r,t) = \sum_i q_i \vec{v}_i(t)\ S[\vec{r} - \vec{r}_i(t)]$$

Then one can add the equations of continuity for field and energy to give:

$$\nabla \cdot (\vec{K} + \vec{R}) + \frac{\partial}{\partial t} (U + \frac{E^2}{8\pi}) = \vec{j} \cdot \vec{E}_{ext} +$$

$$\sum_i q_i \vec{v}_i(t) \cdot \{ \langle \vec{E}[\vec{r}_i(t)] \rangle - E(\vec{r}) \}\ S[\vec{r} - \vec{r}_i(t)]$$

The second term on the right hand side represents an internal tension force which is constraining the finite size particle to move together even when different parts of the particle feel different forces. When we integrate over the plasma volume that term vanishes, and using the divergence theorem one can write:

$$\oint_S (\vec{K} + \vec{R}) \cdot d\vec{a} + \frac{\partial}{\partial t} [\int_V \frac{E^2}{8\pi} dV + \frac{1}{2} \sum_i m_i v_i^2(t)] = \int_V \vec{j} \cdot \vec{E}_{ext}\ dV$$

This is the result we have been seeking. If $\vec{E}_{ext} = 0$, meaning that only self-consistent electric forces are included in the equation of motion, then by integrating in time, one obtains an expression for global energy conversion:

$$W_E + W_K + \int_0^t \oint_S (\vec{K} + \vec{R}) \cdot d\vec{a} = \text{constant}$$

where $W_E = \int_V \frac{E^2}{8\pi} dV$ and $W_K = \frac{1}{2} \sum_i m_i v_i^2(t)$

Let me show you how I applied this to the case of RF heating simulations[4]. The antenna could be represented by specifying on one boundary $\Phi(x = 0,y,t)$ and setting $\Phi = 0$ at $x = L_x$. In the y direction the system was periodic. Since we used specular reflection for particles in the $\hat{x}$ direction, the surface integral involving $\vec{K}$ vanishes, and only the surface at x = 0 contributes to the surface integral over $\vec{R}$. We then have:

$$W_E + W_K + \frac{1}{4\pi} \int_0^t dt \int_0^{L_y} \Phi \frac{\partial}{\partial t} \left(\frac{\partial \Phi}{\partial x}\right) dy \Big|_{x = 0} = \text{constant}$$

The last term on the left hand side represents the negative of the work done on the plasma by the antenna. Since the system is periodic in y, this expression can be written in terms of the Fourier harmonics:

$$W_E + W_K - \sum_{m=-\infty}^{\infty} W_m = \text{constant}$$

where $W_m = -\frac{L_y}{4\pi} \int_0^t dt \Phi_m^* \frac{\partial}{\partial t} \left(\frac{\partial \Phi_m}{\partial x}\right) \Big|_{x = 0}$ is the work done by each Fourier mode.

The following example was a case where we set the frequency of the antenna equal to one of the eigenfrequencies of the plasma which I showed earlier. In such a case, the plasma responds with secular growth of the mode in time, much like a harmonic oscillator, until some nonlinearity stops the growth. This particular example is a very severe test. The wave phase velocity along B_{ext} was fast, so that the wave amplitude grow large until particle trapping occurred. At peak the wave electric field was 1/3 of the initial plasma kinetic energy. By the end of the run, the plasma kinetic energy increased by more than a factor of five. The calculated work done, $\sum_m W_m$ was able to account for all but 0.2% of that growth.

Figure 5 shows the antenna spectrum versus parallel mode number m, where $k_\parallel = 2m\pi/L_y$. It also shows a plot of W_m versus m, which shows the amount of work done by each component of $k_\parallel$. One can see that mode m = 1 did almost all of the work, which was to be expected since it was one of the eigenmodes with that mode number which was being excited. This is a very important diagnostic in RF heating because it tells us which modes in an antenna are most effective and one which cannot be readily obtained by any other means.

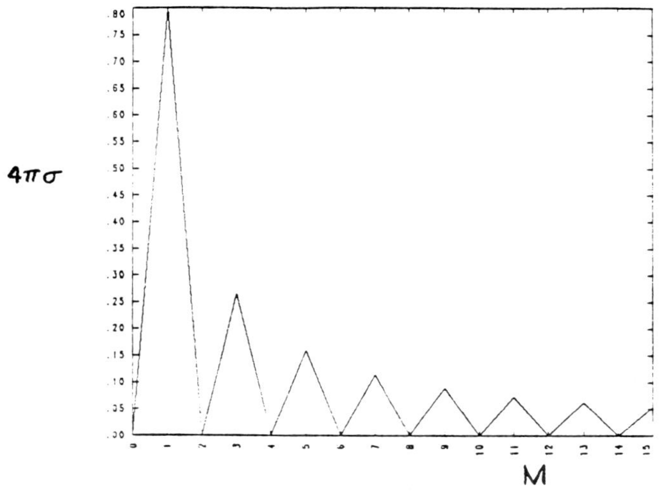

Fig. 5a. Antenna input spectrum for each mode.

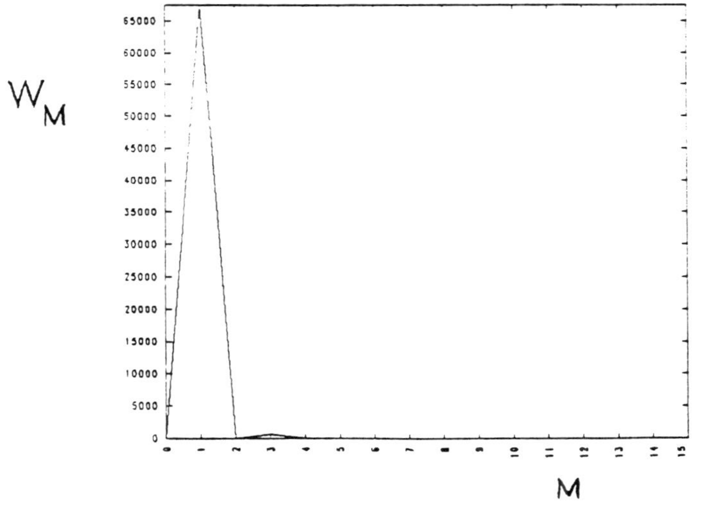

Fig. 5b. Work done by antenna for each mode.

I have shown you how to account for energy flow in electrostatic simulations. The same kind of development can be carried out for momentum. Since the derivation is almost the same, I will not go through the details except to mention the things that are different.

The equation analogous to the equation of continuity for energy $\nabla \cdot \underset{\approx}{T} = \rho \vec{E}$ where $\underset{\approx}{T} = \frac{1}{4\pi} [\underset{\sim\sim}{EE} - \frac{1}{2} (\underset{\sim}{E} \cdot \underset{\sim}{E})\underset{\approx}{I}]$ is the Maxwell stress tensor with the magnetic field missing. This can be shown easily by carrying through the indicated differentiation:

$$\nabla \cdot \underset{\approx}{T} = \frac{1}{4\pi} [(\underset{\sim}{E} \cdot \nabla) \underset{\sim}{E} + \underset{\sim}{E} (\nabla \cdot \underset{\sim}{E}) - \frac{1}{2} \vec{\nabla} (\underset{\sim}{E} \cdot \underset{\sim}{E})] = \rho \vec{E} - \frac{\vec{E} \times (\nabla \times \vec{E})}{4\pi} = \rho \vec{E}$$

since $\nabla \times \vec{E} = 0$. Note that there is no term involving the time derivative of field momentum, as there was for energy. Evidently, electrostatic fields do not carry momentum. One can derive an equation of continuity for particle momentum,

$$\nabla \cdot \underset{\approx}{M} + \frac{\partial \vec{P}}{\partial t} = \sum_i m_i \frac{d\vec{v}_i(t)}{dt} S[\vec{r} - \vec{r}_i(t)]$$

where

$$\vec{P} = \sum_i m_i \vec{v}_i(t) \; S[\vec{r} - \vec{r}_i(t)]$$

$$\underset{\approx}{M} = \sum_i m_i \underset{\sim}{v}_i(t) \underset{\sim}{v}_i(t) \; S[\vec{r} - \vec{r}_i(t)]$$

are the particle momentum density and momentum flux tensor, respectively. Adding the field and particle eqations then gives:

$$\nabla \cdot \underset{\approx}{M} + \frac{\partial \vec{P}}{\partial t} = \nabla \cdot \underset{\approx}{T} + \rho \vec{E}_{ext} + \vec{j} \times \vec{B}_{ext}/c$$

$$+ \sum_i q_i \{\langle \vec{E}[\vec{r}_i(t)] \rangle - E(\vec{r})\} \; S[\vec{r} - \vec{r}_i(t)]$$

Integrating over the plasma volume yields:

$$\oint_S (\underset{\approx}{M} - \underset{\approx}{T}) \cdot d\vec{a} + \frac{\partial \vec{\Pi}}{\partial t} = \int_V [\rho \vec{E}_{ext} + \vec{j} \times \vec{B}_{ext}/c] dV$$

where $\vec{\Pi} = \sum_i m_i \vec{v}_i(t)$ is the total momentum content inside the simulation volume.

In fusion applications one is interested in the momentum along the magnetic field. For that case, the expression for global momentum conservation is expressed in terms of a vector:

$$\Pi_{\parallel} + \int_0^t dt \oint_S (\vec{M}_{\parallel} - \vec{T}_{\parallel})\cdot d\vec{a} = \text{constant}$$

Further simplifying to the same geometry as before, one has:

$$\Pi_{\parallel} + \frac{1}{4\pi} \int_0^t dt \int_0^{L_y} E_x E_y dy \Big|_{x=0} = \text{constant}$$

In terms of the Fourier harmonics, one can write:

$$\Pi_{\parallel} - \sum_m P_m = \text{constant}$$

where

$$P_m = \frac{ik_m L_y}{4\pi} \int_0^t dt\, \Phi_m^* \frac{\partial \Phi_m}{\partial x}\Big|_{x=0}$$

is the momentum input for each Fourier mode.

For the same example I used to illustrate the energy conservation, we did not expect to generate any momentum in the parallel direction, because the antenna elements were phased symmetrically and waves were being launched in both the parallel and antiparallel directions equally. However, the plasma was not symmetric and the plasma momentum in fact did not remain constant. But the term $\sum_m P_m$, representing the net momentum input from the antenna, was able to account for the momentum fluctuation within about a percent. In this case, the momentum input per mode had a large variation, with some modes contributing positive momentum and some negative.

When we changed the wave phasing to launch waves primarily in one direction, then the electrons acquired a net drift of $2.5v_{the}$. The term $\sum_m P_m$ was able to account for all but 0.3% of this momentum increase, which was all coming from mode m = 1.

The last illustration shows an example from some work carried out by Hiratada Abe and myself at Kyoto University.[8] We applied a wave spectrum similar to that in lower hybrid current drive experiments, including all the very low amplitude secondary peaks. We discovered tht the momentum input from the secondary peaks was very significant and in fact crucial to the success of the current drive. When the low amplitude modes were suppressed, the current was drastically reduced.

Acknowledgments

This work has been supported by NSF and USDOE.

References

1. R.W. Hockney and J.W. Eastwood, Computer Simulation Using Particles, (McGraw-Hill, New York, 1981).

2. Charles K. Birdsall and A. Bruce Langdon, Plasma Physics via Computer Simulation, (McGraw-Hill, New York, 1985).

3. V.K. Decyk and J.M. Dawson, J. Computational Phys., 30, 407 (1979).

4. V.K. Decyk, G.J. Morales, and J.M. Dawson, Second Joint Grenoble-Varenna International Symposium on Heating in Toroidal Plasmas, Como, Italy, Sept., 1980 [Commission of the European Communities, Brussels, 1980], Vol. I, p. 365.

5. C.F.F. Karney, Phys. Fluids, 21, 1584 (1979).

6. V.K. Decyk, Phys. Fluids, 25, 1205 (1982).

7. V.K. Decyk, J. Computational Phys., 56, 461 (1984).

8. V.K. Decyk and H. Abe, Bull. Am. Phys. Soc., 29, 1307 (1984).

PARTICLE CODE SIMULATIONS WITH INJECTED PARTICLES

Charles H. Aldrich
Los Alamos National Laboratory
Los Alamos, New Mexico

ABSTRACT. As problems we are interested in become more complex, we often find our simulations stretching the limits of available computer resources. For example, an interesting problem is simulation of dissipation processes in sub-critical collisionless shocks. To simulate this system our simulation box must contain the shock and its upstream and downstream regions over the entire length of a run. If the shock moves with any appreciable speed the box must then be considerably larger than the shock thickness making it hard to resolve the shock front itself with a reasonable number of grid points.

A solution to this problem is to run the simulation in the frame of reference of the shock. Particles are injected upstream of the shock and leave the simulation box downstream. With the shock stationary in the simulation box, we only need to contain enough of the up and downstream regions for the fields, etc., to settle down and separate the shock from the box boundaries.

In this tutorial we consider some basic algorithms used in a practical particle injection code, such as the two dimensional WAVE code used at Los Alamos. We will try to present these ideas in a simple format general enough to be easily included in any particle code. Topics covered are:

- Smoothly Injecting Particles.
- Generating the Distribution Functions.
- Time Dependent Injection Density.
- Boundary Conditions on Fields and Particles.
  (Flux and Charge Conservation)

1. INTRODUCTION

As we have seen in earlier tutorials, particle simulations codes have been an extremely useful tool used to investigate plasma physics. This is particularly true when one wants to understand self consistent shock structures. One problem of interest is how a subcritical shock evolves. The cross-field currents which generate the magnetic field for the shock excite micro-instabilities within the shock front, these instabilities are responsible for the dissipation that allows the shock to exist and strongly effect the basic structure of the shock front including the cross-field

Space Science Reviews **42** (1985) 131–144. 0038–6308/85.15

currents. In order to resolve both of these processes which are occurring perpendicular to each other (requiring a two dimensional code) we often must make as efficient use of the simulation grid as possible.

Early studies often produced shocks with a simple moving piston. Figure 1 shows a shock created by a piston moving into the box from the right side[*Forslund et al. 1984*]. Initially the simulation box is full of particles at rest. As the piston moves into the box from the right wall the particles in front of it are reflected to form a beam moving to the left with twice the piston velocity. The front of this beam eventually forms a shock wave that moves through the box slowly separating away from the piston.

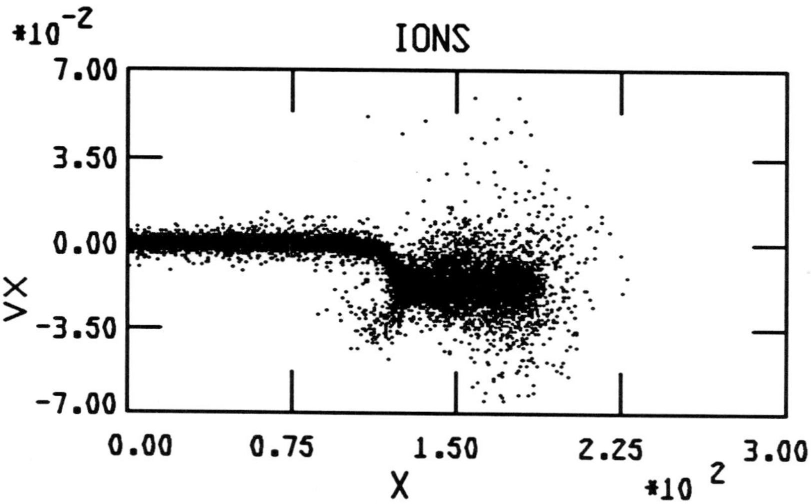

Figure 1. An example of the ion phase space of a shock produced by a piston traveling from the right end wall.

Clearly a way to do this problem making better use of the simulation box would to be running in the frame of reference of the piston. This requires injection. Figure 2 shows a shock where particles are injected on the left hand side of the box. When the problem was initialized the particles in the box were uniformly moving to the right. The particles hit the right hand side of the box and are reflected off a stationary piston and forming a shock downstream. As particles near the left hand side of the box move deeper into the box the space that they vacate is filled with injected particles.

A major problem with the previous two methods is the presence of the piston itself. It takes a long time for the shock front to become distinctly separate from the piston structure. Indeed it may be nearly impossible to separate effects, such as particle heating, occurring at the shock front and the piston from each other. Also, some structures don't lend themselves to formation using such a simple system.

An example of this structure is a magnetic soliton. Figures 3 and 4 show the phase space for electrons and ions for a perpendicular soliton. The velocity, density, and field profiles for the soliton can be easily calculated from a simple

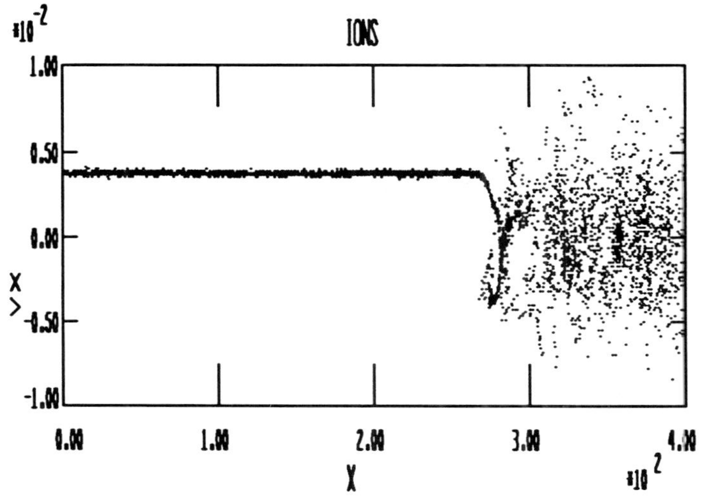

Figure 2. An example of the ion phase space of a shock produced by a piston located at the right wall, with particles injected on the left.

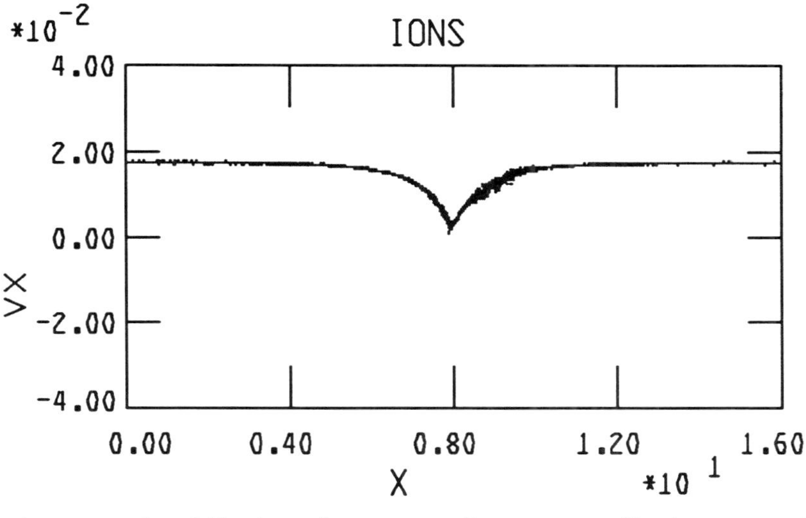

Figure 3. An example of the ion phase space for a perpendicular magnetic soliton held stationary in the simulation box by particle injection.

differential equation [*Tidman and Krall, 1971*]. When the soliton is initialized at the beginning of the run the known values for these quantities are used and the velocity of the injected ions and electrons is chosen so that the soliton is stationary in the center of the simulation box, with particles injected on the left hand side of the box as before. Note that since the drift velocity for the electrons is smaller

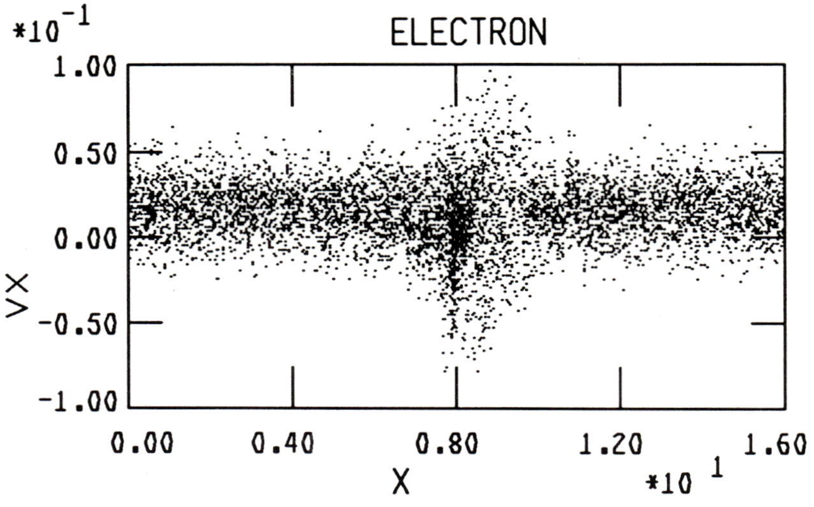

Figure 4. An example of the electron phase space for a perpendicular magnetic soliton held stationary in the simulation box by particle injection.

than the electron thermal velocity a population of electrons will be moving right to left through the soliton structure, requiring an injection of a few electrons from the right hand side of the box with negative velocities (going left). This configuration is particularly suited for this problem, the soliton is completely stable and will continue to sit stationary in the middle of the box if nothing happens. The cross field currents within the soliton will tend to excite instabilities that in turn cause dissipation. This in turn will change the structure of the soliton evolving the structure towards a shock. Having the soliton stationary allows the simulation to run to long enough times for the instabilities to develop and have some visible effect.

Similarly, a stationary shock structure could be initialized in the simulation box. The velocity, density and field jumps across the shock front can be calculated using the Rankine-Hugoniot (R.H.) equations [*Tidman and Krall, 1971*]. The shock speed calculated from the R.H. equations can be used to set the injection speed to make the shock stationary in the simulation box.

In this tutorial we start off in section 2 by describing two ways to produce a system of simulation particles according to some given distribution function. In section 3 we determine how many particles are crossing the simulation boundary and how to put the particles into the simulation box. This includes a discussion of how to choose the velocities of the injected particles randomly to produce the probability distribution we want in the bulk of the box. We then briefly consider injecting a beam with a time and/or spatial variation in section 4. We finish with section 5 with a couple of cautions with respect to the boundary conditions on the particles and electromagnetic fields.

2. THE PROBABILITY DISTRIBUTION FUNCTION.

In the next two subsections we will consider the general problem of throwing random numbers to obtain a set of particles distributed according to some given distribution function. We need this to create the particles as they cross the boundary - as well as when the initial particles are put in the simulation box. The first method maps a uniform dice into a more complex distribution using an analytic formula. The second approach considered uses a monte carlo method where throws of the dice are kept or tossed out according to a rule.

2.1. Analytic.

One important distribution function used in most simulations represents a group of particles in thermal equilibrium with some thermal velocity v_{therm}. In suitable units this can be written (in two dimensional cartesian coordinates),

$$P(v_x, v_y) \sim exp(-(v_x^2 + v_y^2)/v_{therm}^2) dv_x dv_y, \tag{1}$$

or in cylindrical coordinates,

$$P(v, \theta) \sim v \exp(-v^2/v_{therm}^2) dv d\theta, \tag{2}$$

where v is the particle velocity and θ is the angular coordinate.

Most computers have a random number function that will throw a pseudo-random number with uniform probability between the limits 0.0 and 1.0. Lets denote this function by rand(). As an example we will construct an analytic formula to map this uniform distribution into a gaussian distribution.

Consider a variable ς bounded in the range $0 \leq \varsigma \leq 1$. If we choose values for ς with uniform probability within this range, ($\varsigma = \text{rand}()$), then the probability of landing in some range of values $d\varsigma$ wide is simply $d\varsigma$ and the probability of having a value less than ς is ς.

Next, consider another variable v that is bounded in some range $0 \leq v \leq v_{max}$. The two variables are related by a single valued function f such that each point in ς is mapped to one and only one point in v and the order of the points are preserved,

$$v = f(\varsigma). \tag{3}$$

This is shown schematically in figure 5. By drawing several equal width intervals on the ς axis and mapping them onto the v axis using the graph in figure 5, one can see that points equally spaced on the ς axis will be more densely spaced on the v axis near $v = 0$. More quantitatively, the probability of any point being between 0 and v on the v axis, given equation 3, must equal the probability of landing between 0 and ς on the ς axis. Since the ς values are all chosen with equal weight, this probability is ς, or in terms of v, $f^{-1}(v)$, where the -1 denotes the inverse function. The density function $P(v)$ is simply the derivative of this quantity, so the probability of landing a point in some range dv can be written,

$$P(v)\ dv = \frac{\partial f^{-1}(v)}{\partial v}\ dv. \tag{4}$$

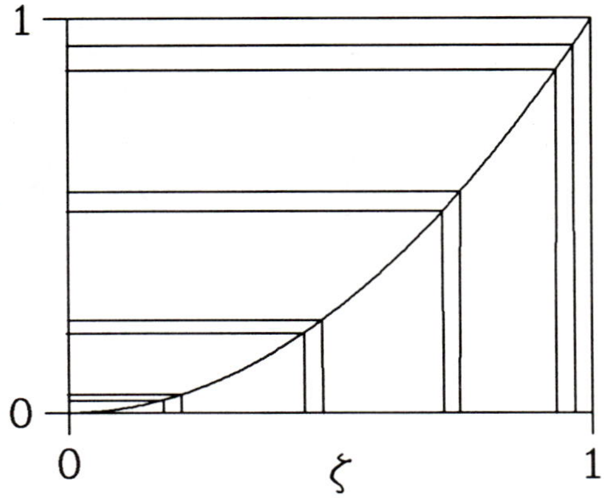

Figure 5. Example of an analytic function used to map a distribution function from one probability space to another. (see section 2.1).

Returning to the gaussian distribution, we can let

$$v = f(\varsigma) = \sqrt{-\log(1-\varsigma)}, \tag{5}$$

or

$$\varsigma = f^{-1}(v) = 1 - \exp(-v^2). \tag{6}$$

The probability of being in some range of values dv is then , by equation 4,

$$P(v)\ dv = 2v \exp(-v^2)\ dv. \tag{7}$$

This equation has the form of equation 2. The velocity vector can point randomly in any direction, so the angular variable is equally probable in the range 0 to 2π. To set up a set of particles according to the gaussian distribution we can set,

$$\begin{aligned} v_x &= v \sin(\theta), \\ v_y &= v \cos(\theta), \end{aligned} \tag{8}$$

where,

$$\begin{aligned} v &= v_{therm} \sqrt{-\log(1 - \text{rand}())}, \\ \theta &= 2\pi\ \text{rand}(). \end{aligned} \tag{9}$$

For an arbitrary distribution function $P(v)$ we can integrate equation 4 to obtain $\varsigma = f^{-1}(v)$ and then invert to obtain the mapping function. As we have just seen, the gaussian distribution turns out to be easy to generate in cylindrical coordinates. If we try to generate a gaussian in three dimensions, for example, using these methods directly, the transformation function f^{-1} will contain error functions and the like which are hard to invert in order to obtain $f(v)$. The solution to this problem is to throw two out of the three coordinates, say x and y with the cylindrical formula, equation 8. We can then use equation 8 again to throw two more coordinates, one of which is thrown away while the other is used for z. If one is simply interested in a one dimensional gaussian, just throw away the coordinate not needed.

2.2. Numeric - Monte Carlo.

Often the distribution function we want is too complex to integrate the equations in section 2.1 analytically. In this case we can resort to a more general method that will work for almost any distribution function. In this method we will be tossing the dice and either keeping the results as our answer or throwing out the toss and redoing it until the toss fits certain rules that we arrange. As an example we can generate the distribution in equation 2. The first step in this process is to represent the function graphically in a box of some finite size. This is done in figure 6, where the high velocity tail of the distribution is being ignored ($v_{therm} = 1.0$).

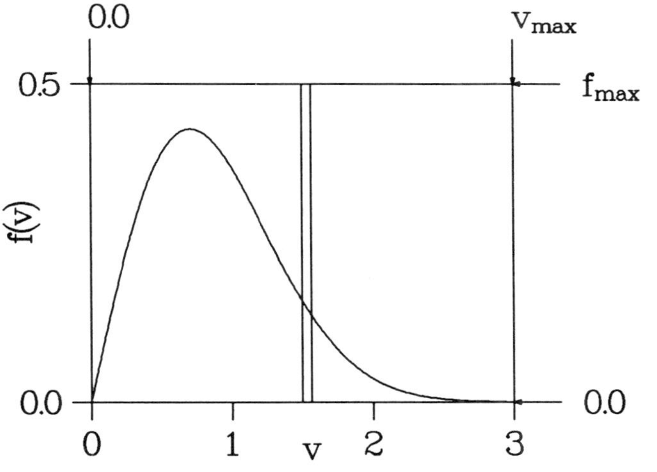

Figure 6. Graph used to produce the distribution function $f(v)$ using numerical monte-carlo method (section 2.2).

We can split the graph up into many vertical bins, one of that is shown in the figure 6. If we randomly choose a value of v within the graph (between 0 and v_{max} the chance of it being in a particular bin is dv/v_{max}. Throwing another number representing the vertical position on the graph (between 0 and f_{max}) the fraction of points that will hit under the curve is $f(v)/f_{max}$. Thus by throwing two random numbers for each point and keeping only points that fall under the curve we fill each bin with an average of

$$N_{bin}(v) = \left(\frac{f(v)dv}{v_{max}f_{max}}\right) N_{thrown} = f(v)dvN_{kept}, \tag{10}$$

where N_{kept} is the number of points that are saved. The probability of being in a bin is then $f(v)$. This means that if we randomly throw points onto the graph keeping the first one that hits under the curve $f(v)$ results in generating the distribution we want. This method has the advantage of being simple to implement and have working on a computer in little time.

3. SMOOTHLY INJECTING THE PARTICLES

In this section we will discuss how to inject a beam of particles into the left hand side of the simulation box as used to create the soliton in figures 3 and 4. This is shown diagrammatically in figure 7 to establish the notation for the rest of the paper. Outside the box the particles are assumed to in thermal equilibrium with some thermal velocity v_{therm} and moving with some drift velocity v_{beam}. The distribution function can be written as a gaussian, offset in velocity,

$$f(v) \sim \exp(-(v_x - v_{beam})^2 - v_y^2 - v_z^2)/v_{therm}^2)). \tag{11}$$

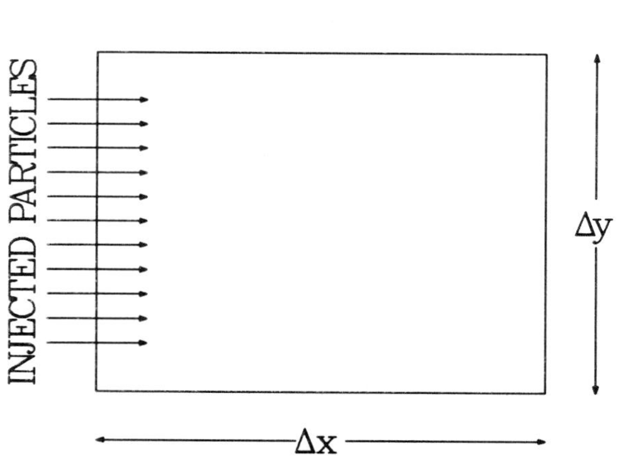

Figure 7. Simulation box with injected particles.

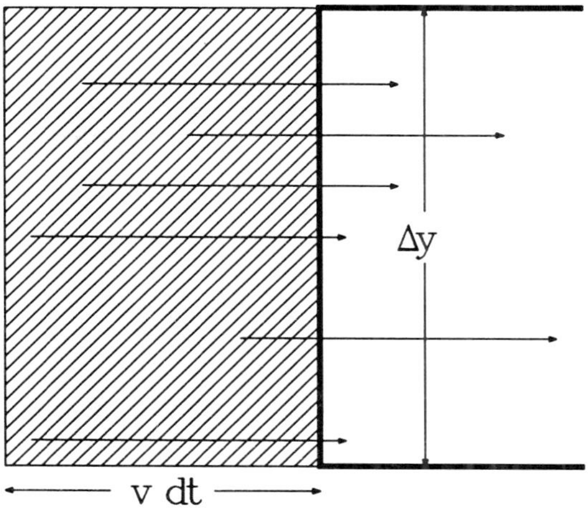

Figure 8. Particles with velocity v_x in the shaded area are swept across the boundary in one time step.

3.1. **How Often?**

We can relate the average time between injections to the drift velocity and injection density. If a particle has velocity v it will travel vdt in one time step. The boundary will have a rectangle of size Δy by vdt full of particles sweep across in one time step(see Figure 8).

Thus the number of particles in this region is the particle density $f(v)\ d^2v$ (assuming a constant injection density along the y direction) times the area $\Delta y\ vdt$, where $f(v)$ is the distribution function for the injected particles, or,

$$\Delta y\ vdt\ f(v)\ d^2v. \tag{12}$$

Integrating over the entire velocity space, the total charge crossing the boundary is then,

$$Q = \rho \Delta y\ dt < v_x > . \tag{13}$$

where,

$$< v_x > = \int v_x\ f(v)\ d^2v. \tag{14}$$

For our injected beam $< v_x >$ can be calculated analytically, giving,

$$< v_x > = \frac{1}{2}\left(\frac{v_{therm}}{\sqrt{\pi}} \exp\left(\frac{-v_{beam}^2}{v_{therm}^2}\right) + v_{beam}\left(1 + \operatorname{erf}\left(\frac{v_{beam}}{v_{therm}}\right)\right)\right). \tag{15}$$

If the charge on a single simulation particle is $q_{species}$ then we find the number of simulation particles crossing the boundary per time step,

$$N = \frac{\rho \Delta y \; dt < v_x >}{q_{species}}, \tag{16}$$

or equivalently, the average time between injecting two particles,

$$dt_{inject} = \frac{q_{species}}{\rho \Delta y < v_x >}. \tag{17}$$

3.2. Where?

As far as the injection algorithm is concerned any particle code can be viewed as a series of black boxes,

- Calculate Fields and Potentials (Forces on particles).
- Push particles with Forces, Calculate Charges.
- Inject Particles.

These steps are repeated over and over again, once for each time step. Each cycle steps a time dt, the nth step beginning at time $t = (n-1)dt$.

To perform particle injection in a smooth manner we wish to decouple the injection process as much as possible from the size of the time step. To do this we inject the particles uniformly in time separated by time dt_{inject}. So that we inject the mth particle at time $t_{inject} = m dt_{inject}$. The time between particle injections may be a fraction of a time step dt or many time steps. Since these two characteristic times are not necessarily multiples of each other the time when a each particle is injected will lie somewhere in the middle of a time step. We can move the particle into the box for this time period, $dt_{move} = t - t_{inject}$. If many particles are injected in one time step this will spread them out evenly in space.

To isolate the simulation problem from effects of the boundary conditions most calculations will be set up so that very little is happening at the boundaries of the box, that is no steep density gradients, large fields, etc. are allowed. The time step is usually small enough so that the particles travel only a small distence into the box. It is usually a good approximation to ignore forces on the particle for this fractional part of a time step and simply move them into the box by setting;

$$\begin{aligned} x_{inject} &= dt_{move} v, \quad and \\ y_{inject} &= \Delta y \text{ rand}(). \end{aligned} \tag{18}$$

This formula could be changed to inject relativistic particles by using the momentum. One could also be completely consistent and move the injected particles with the same particle pusher used in the second step above.

3.3 Distribution Function.

From equation 12 we can see that we have to pick the x velocities according to the probability formula,

$$f(v_x) \sim v_x \exp(-(v_x - v_{beam})^2 / v_{therm}^2). \tag{19}$$

This formula is not easy to generate analytically and the techniques of section 2.2 are used to produce this distribution. For a full three dimensional velocity space the other two directions are a simple non-offset gaussian considered in section 2.1. Thus

$$\begin{aligned} v_y &= v_{par} \sin(\theta), \\ v_z &= v_{par} \cos(\theta), \end{aligned} \tag{20}$$

where,

$$\begin{aligned} \theta &= 2\pi \ \mathrm{rand}(), \\ v_{par} &= \sqrt{-\log(1 - \mathrm{rand}())}. \end{aligned} \tag{21}$$

4. NONUNIFORM DENSITIES.

Generalizing this injection scheme to nonuniform density profiles both in space and time are relatively straightforward. We will quickly consider injecting a shaped beam and a pulsed injection.

4.1. A Shaped Beam.

To produce a uniform simulation across the side of the simulation box the y position was chosen randomly with equal probability of entering anywhere. To change this profile we can choose values for the coordinate in the y direction with one of the methods of section 2. For example if we wanted a beam whose density is zero at the top and bottom edges like,

$$\rho(y) = \rho_{max} \sin(y\pi/\Delta y). \tag{22}$$

Let $v = \Delta y \; y/\pi$ and therefore, from equation 4,

$$\frac{\partial f^{-1}(v)}{\partial v} = \frac{\sin(v)}{2}, \qquad (23),$$

or, integrating,

$$\varsigma = f^{-1}(v) = \frac{Constant - \cos(v)}{2}, \qquad (24),$$

and

$$\varsigma = \frac{1 - \cos(\Delta y\ y/\pi)}{2}, \tag{25}$$

We can then choose y according to,

$$y = \frac{\Delta y}{\pi} \cos^{-1}(1 - 2\mathrm{rand}()), \tag{26}$$

where rand() is the uniform generator mentioned earlier.

The average density ρ used in equation 17 for the time between injections t_{inject} becomes

$$\rho = \frac{2}{\pi}\rho_{max}. \tag{27}$$

4.2. Pulses.

Time variations in the injection parameters can be accomplished in several ways. If the time step is small enough and the variations in the injection velocity or density is slow enough to be resolved one can simply change the parameter as a function of time at each time step and use the new dt_{inject} in equation 17 to inject the next set of particles.

If on the other hand we would like to inject a fast changing density profile, for instance, we can change the time a particle is injected to a nonlinear function of time. If we would like to have one pulse of the shape,

$$\rho(t) = \rho_{max} \sin^2(t\pi/T_0), \tag{28}$$

where T_0 is the length of the pulse, we can let

$$t_{inject} = T(m\ dt_{inject}), \tag{29}$$

where

$$T^{-1}(t) = \frac{t - \frac{T_0}{2\pi}\sin\left(\frac{2\pi}{T_0}t\right)}{2}. \tag{30}$$

In general for arbitrary $T(t)$, the density injected will be

$$\rho(t) = \rho\ \frac{\partial T(t)}{\partial t}. \tag{31}$$

Note for $T(t) = t$ then $\rho(t) = \rho$.

5. BOUNDARY CONDITIONS

Except for the process of injection itself the boundary conditions are not significantly differently from a particle code without injection. There are a couple of points worth noting in the following two subsections.

5.1. Fields

If we are simulating a shock structure and transforming into the frame of reference of the shock we have to be careful of the boundary conditions on the fields. In the frame of reference of the upstream particles we have some magnetic field $B_{upstream}$. Since the particles have zero velocity in this frame, they feel no electromagnetic force. Transforming to the frame of reference of the shock one has to be aware of how the fields transform with changes of reference frame. That is, if the shock is moving with velocity v_{shock} in the upstream frame , an electric field component $\vec{E}_{upstream} = \vec{v}_{shock} \times \vec{B}_{upstream}$ will be present in the simulation frame upstream of the shock and at the boundaries, resulting again in zero net force on the particles. Neglecting this "flux conservation boundary condition will result in an electromagnetic disturbance moving in from the boundary in a fairly obvious way.

5.2. Particles

In figures 3 and 4 we showed the phase space for a subcritical perpendicular magnetic soliton, where both electrons and ions are injected simultaneously. The drift speed of the electrons is less than the electron thermal speed but much greater than the ion thermal speed. The right hand boundary can then be described as

- Let all ions leave the box.
- Let all electrons leave the box.
- Inject a few electrons moving to the left.

The electrons are injected with equation 19. This process is the same discussed earlier except that the sign of v is different and we are injecting the backward going tail.

Statistically the number of electrons crossing the boundary during any particular time step will fluctuate between time steps causing a corresponding fluctuation in the charge imbalance at the boundary. This often can create havoc with the code. To essentially guarantee charge neutrality at the boundary we can move the ions first, note how many ions leave the system, then move the electrons. When we move the electrons, if an electron crosses the boundary we let it leave if and only if there is a positive charge imbalance at the boundary. Since the thermal is much higher for the electrons, many more electrons cross the boundary in each time step. Once enough electrons leave to assure charge neutrality at the boundary the rest are injected according to the distribution, equation 31. This virtually assures neutrality except in the case of large fluctuations in the number density which will be rare.

Acknowledgement

This work was supported by the United States Department of Energy.

References

D. W. Forslund, K. B. Quest, J. U. Brackbill, and K. Lee, *Journal of Geophysical Research*,**89** ,2142,1984.

D. A. Tidman and N. A. Krall, *Shock Waves in Collisionless Plasmas*, John Wiley and Sons,Inc., 1971.

OPTIMIZING COMPUTATIONAL EFFICIENCY AND USER CONVENIENCE IN PLASMA SIMULATION CODES

Christopher W. Barnes
Los Alamos National Laboratory
Los Alamos, New Mexico

ABSTRACT. It is usually important to write plasma simulation codes in such a way that they execute efficiently and are convenient to use. I discuss here practical techniques to achieve this goal. Numerical algorithms must be well formulated and advantage taken of machine architecture in casting the algorithm into a high level language such as Fortran. The advantages of writing critical routines in Assembler are discussed. For large simulation codes, disks must often be used as a temporary store for working data. Efficient methods for doing this are presented. Codes must not only be well organized for ease of implementation and maintenance, but also for ease of use. Ways are suggested for packaging codes such that setup, batch production, restarting and diagnostic postprocessing is facilitated. Particular emphasis is placed on graphics postprocessors, since they must be used in real time with graphics terminals as well as with hardcopy devices.

1. INTRODUCTION

The purpose of this tutorial lecture is to present what experience has shown to be good programming practice in the realization of plasma simulation codes for use on modern computers. I assume that you are well versed in a language such as Fortran, and have used large computer installations in your everyday work. You may not yet have written and used a program that consists of many thousands of lines of source code and that consumes many hours of computer time. The implementation of non-trivial programs such as this requires careful planning and mindfulness of the physical nature of the computer system to be used, if efficiency and convenience are important. These considerations are likely to be important in a large code since the programming manpower needed is only justified by a long production life of the code.

It seems to be true that plasma simulation codes are never written once, and used forever without modification. Instead, there is always new physics to include, new numerical algorithms to try or a change needed in the diagnostics package. You should assume that your new code will never be "finished" and therefore you should make ease of maintenance a primary consideration in its design.

Space Science Reviews **42** (1985) 145–151. 0038–6308/85.15

Can we now identify the principal characteristics of plasma simulation codes that are relevant to optimizing efficiency and convenience? The most prominent features are the need for large amounts of computing time and the large amount of data used. The data base can exceed the storage capacity of the machine's central memory and might therefore need to be stored on disk and brought into memory as needed. The amount of computation done on a unit of data each time it is brought in from disk could be an issue, if it is desired to overlap disk transfers with computation. Particle codes are the worst offenders here. The particle move time can be as little as two microseconds on a CRAY-1 per timestep while the disk read/write time is several times this per particle. Simulation codes calculate the evolution of a plasma over time in a series of small timesteps, each step typically consisting of a particle or fluid element advance followed by an updating of the self consistent electric and magnetic fields. As the calculation proceeds, diagnostic data is produced in vast quantities and is either saved in raw or partly reduced form or is displayed or plotted directly. The diagnostic portions of a code can represent as much as 80 percent of the total number of source code lines in the program. Fortunateley they usually can be made to use only a small part of the total compute time used in a run. They therefore need not be highly optimized.

Since these codes calculate the evolution of a plasma over time, it is desirable to be able to checkpoint the data occasionally so that a problem can be restarted without having to start from scratch. Often, a problem will be executed in several runs because of time constraints or to allow analysis of the problem results part way through. The ability to save the working data at a particular timestep and in a subsequent run restart from it is therefore considered essential in plasma simulation codes.

Section 2 following, dicusses code optimization with a view toward making it cost a little as possible to run the code. Section 3 contains suggestions on ways to make a convenient human interface for running the code and ways to make it easy to maintain.

2. OPTIMIZATION

In this section, I will discuss various ways to ensure that the code runs as efficiently as possible, without an excessive amount of programming effort. We will find that a rough knowledge of the computer architecture being used and some understanding of how the compiler makes use of that architecture will allow us to make substantial increases to execution efficiency.

2.1. A Vector Machine

The CRAY series of computers is now in wide spread use and is a prime example of what is known as a vector or pipeline machine. It has the ability to effectively perform an operation on a vector or array of data almost in parallel. Once the operation is initiated and after a short latency, the results appear one per machine cycle. The real power of this architecture lies in the fact that the overhead for index and address generation for the arithmetic or logical operation to be performed need only be done once, for the whole vector.

All vector machines also have the ability to do scalar operations, since some code is inherently unvectorizable. It is easy to show that vectorization can provide a factor of five speed increase over code executed only in the machine's scalar units. The objective when writing in Fortran for instance, is to convince the compiler to vectorize as much of the code as possible to achieve maximum speed. Even so, it is sometimes worth hand coding small portions of the program.

The maximum vector length allowable on the CRAY series of computers is 64 operands. For vectors longer than this, the compiler must break them into a series of vectors of length 64 or less. It is true that the hardware performs best when the vector length approaches 64. The breakeven point against scalar speed is a vector length of about 5.

A constraint against use of the vector units is that vector operands must be stored in memory with a constant 'stride', or spacing in address value. This property makes writing of particle movers especially difficult, since the cell locations for a vector of particles are random, certainly not of constant 'stride'. It is worth noting that CRAY's upcoming X-MP48 will have the capability of doing 'scatter/gather' vector memory references, ie, non-constant stride. Bank conflicts will still degrade performance, however.

Memory is arranged in banks, or physical units such that as one steps through sequential memory addresses, a different bank is accessed each time until the number of banks has been cycled through, typically 16. Since the memory cycle time is much greater than a machine cycle, rapid hits to the same bank can slow access significantly. One should therefore avoid vector operations with strides which contain factors of 8 or 16 (or some similar power of two, depending on the number of banks in that particular machine).

This brief list is the minimum hardware information needed to allow intelligent vectorization efforts in Fortran codes. The relevant hardware reference manual should be consulted by any serious Fortran programmer.

2.2. **Fortran Optimization**

Optimization of Fortran codes is best done experimentally, since one can not always predict how the compiler will treat a given style of coding. Find a timing routine which reads a microsecond clock, or preferably a machine cycle clock. Then two calls to this routine will measure the time spent in the piece of code bracketed by the two calls. The timing effect of coding changes can easily be seen using this method.

In scalar code, or code to be executed on a non-vector machine, little dramatic improvement can be expected from reprogramming, unless the program formulation was poorly done originally. In optimizing any code, the most fruitful place to begin is with an examination of all DO loops. In any code unit containing nested DO loops, the innermost non-trivial loop will be where the majority of the time is spent. Unnecessary expressions should be kept out of the inner loop (sometimes the compiler will do this for you). Index calculations should be kept as simple as possible with the idea that each time through the loop, there will be as little calulation needed as possible. Sometimes non-changing portions of the index can be calculated outside the inner loop. Multi-dimensional arrays should be avoided. It is usually best to explicitly compute the mapping of a multi-dimensional index into a one-dimensional array. Use can be made of the fact that in Fortran, the first listed index is the fastest varying one, that is, is the contiguous one in storage.

In Fortran, vectorization is achieved not by special vector notation, but by finding suitable DO loops which lend themselves to vectorization. The array indices must not improperly overlap and the flow of the loop must not be interrupted. This means that there can be no IF tests within the loop. For instance, in a particle moving routine, the tests to see when a particle crosses a boundary will apparently not vectorize in Fortran, and should not be put in a loop with other operations which can. Interestingly, the machine instruction which performs the comparison has a vector form. The inclusion of a branch as a part of the IF construct makes it non-vectorizable. There is usually a chapter at the end of Fortran manuals which has a discussion on vectorization. It is well worthwhile to read it.

When there are nested DO loops which are vectorizable, it is sometimes of advantage to reverse their order, although this is not always logically possible. I recently did this in a small FEL code in which the inner loop had a span of around 20 and the outer a span of over 200. The speedup under these conditions was significant although as the span of 20 was increased towards 64, the difference became negligible, as one might expect.

I learned an interesting lesson about libraries in working on this code. At least half the time in the code was spent in calculating sine and cosine pairs (of the same argument). This was done using a call to a vectorized sine and then a vectorized cosine library routine. I took the trouble to look at the CAL source code for these routines. I discovered that the same routine was called for both sine and cosine, and that because the vector arguments could be in any octant, that the routine always calculated both sine and cosine, and threw away the one not needed. I was able to modify the routine so that it returned both sine and cosine with one call, eliminating the need for the second call, which significantly reduced the running time of the code. Someone had thought of this before, of course. I found an entry point in the library to a COSSIN routine that did what I just described. Except it made calls to a scalar routine which would have been far slower. Apparently the person who wrote and maintained the library had time only to make the COSSIN function look vectorizable to the calling routine, but not to actually provide the code needed. The lesson is apparently that library writers are human too.

2.3. Why Hand Code?

Some people are born hackers, and have written in Assembler for as long as they can remember. Others need some persuasion to learn and use assembly language in their codes. The truth is that it is often justified in codes such as particle codes, where most of the time is taken up by a few hundred lines of Fortran, the hand coding of which could achieve a speedup of a factor of two. Vector code is easy to hand code because you don't have to worry about the timing of the scalar part (indexing, counting, etc). In fact, vector code is the only kind I would consider worth the effort of hand coding.

The Cray machines have eight vector registers and a number of functional units each of which performs a speciallized group of operations. They have names like "Floating Point Add Unit", "Floating Point Multiply Unit", "Shift Unit" and "Logical Function Unit". Memory references can be thought of as being performed by a special functional unit. In rough terms, the rules are that you can have as many functional units going at once as you want, so long as the rules of register conflict are observed. It is therefore possible to keep four or five functional units going simultaneously in a hand coded particle mover.

Since it is easier to change working code than to write a new routine from scratch, I suggest a way that novice hand coders can get started painlessly. Take a working Fortran subroutine and produce a CAL listing of it. This can then be assembled to produce the same object as the compiler would have produced itself. Changes can then be made to the CAL version, one by one without much danger of losing control of the program logic. This way, the inner loops of the routine can in effect be hand coded, without having to bother to write the whole routine in Assembler.

3. CONVENIENCE

When a new code is being conceived and written, or when an old code is being impruved, the rush to get the job done often allows us to do a poor job of planning, commenting and documenting the code in spite of our best intentions. The result is a code that is inconvenient to run, particularly by the code's non-authors. Code maintenance may also be nearly impossible, even by the author. We quickly forget what we intended when we wrote the code, and if there are no comments embedded in the code, we may be reduced to analyzing the Fortran to try to figure out what it does. If the code is well structured, this may not be difficult. If it is a mass of GO TO statements, it could be impossible. I wrote a data handling routine once for a particle code that had such convoluted logic that I could never understand the algorithm well enough to follow the logic of the code I had written. At one point, it passed all the tests, so I knew it worked, so I stopped working on it knowing that if it ever broke I wouldn't be able to fix it. By the time I finished the job, I knew a lot about structured programming and documentation.

The following paragraphs set out my recommendations on how to lay out and write a code for ease of use and maintenance and how to organize it so that it can be used conveniently without need for an intimate knowledge of how it works.

3.1. Overall Design

I always design my codes so that the graphical output is generated in a postprocessor. The main code saves diagnostic data periodically in as raw a form as possible. Decisions as to what data is to be taken and at what intervals and with what completeness and precision are made when the physics code is set up and run. The postprocessor reduces and processes the data and then plots it graphically. Many decisions, particularly about graphics options can be made at postprocess time. Since the postprocessor can be rerun cheaply by comparison to the original physics run, lost fiche or a mistyped graphical parameter is less likely to cause grief. The other advantage of keeping the graphics routines in the postprocessor is that they cannot invade and overrun the physics code and thus destroy its legibility. There is no temptation to put graphics calls inside a particle mover, for instance.

The code is designed to be restartable. At predetermined intervals, a copy of all the working data is written to a set of files with suitable names. at the end of the run, these files, along with the diagnostic files are written to permanent mass store. If it is desired to continue this run for say another hour, the data files are made available so that the code will not start from scratch, but resume where it left off. In addition to providing convenience of running the code, this design helps bullet proof the code against machine crashes.

The code is set up to look for data or commands to be entered from the keyboard during operation. Once each timestep, it looks for prearranged signals from the keyboard. It is therefore possible to stop the run, force a data dump, ask for data to be printed at the terminal, etc as the code runs. In principle, there is no limit to the degree of interactivity one could incorporate. It should be noted that the exact same code can be run in a batch subsystem without any need to interact with a terminal.

Parameter initialization is done using Namelist variables. All parameters are initialized to default values inside the code and only those that differ from the default values for a given run need be mentioned in the input deck. The input deck is either a file or it can be typed in at run time from the terminal if need be.

3.2. Postprocessor

The postprocessor is designed to produce graphical output from the data saved in the running of the physics code. The output can then be sent to a plotter of some kind or viewed directly on a graphics terminal. A more sophisticated postprocessor might allow one to design plots on the fly, to better examine some feature in the data for instance.

3.3. Internal Structure of the Code

Internally, the code should be organized into many, logically independent subroutines. No code block should be more than a few hundred lines in length, excluding comments, which should double its length. Liberal use of common blocks makes argument passing almost unnecessary. The idea is to make the code as easy to understand as possible.

In hand coded particle movers, I like to use four or five separate subroutines for the CAL mover, each to perform a logically separate part of the move. With each set of calls, a vector of 64 particles is moved. Within a Fortran bookkeeping subroutine, the sequence might be as follows: A CAL routine is called which unpacks the particles and calculates cell numbers and weights. Another CAL routine calculates the field values for the 64 particles. A third CAL routine advances the particles. A fourth CAL routine recalculates the cell numbers and weights and repacks the particles. Finally, the weights are used to accumulate the charge and current arrays in the last CAL routine. This is the basic cycle. Between these calls to the CAL mover, one can insert calls to Fortran diagnostic routines, particle injectors and routines to handle unusual boundary conditions. Thus it is possible to tightly intermix CAL and Fortran routines cleanly. Even at the level of moving only 64 particles, the overhead of all these subroutine calls is small enough to be ignored.

In the previous section, I mentioned that particles can be packed and unpacked. By that I meant that it is often possible to compact the particle data by a factor of two in order to save space in memory or to reduce the time needed to transfer the data to and from disk. This is a common practice in codes such as this.

3.4. Using Disks as Memory Extension

In particle codes especially, memory requirements may greatly exceed that available to the user. It is then necessary to use disks as temporary storage for data, to be moved in and out of memory often. Particle information is often kept on disk in large particle codes, and moved in and out once each timestep as it is updated. Because the computing time per particle can be comparable to the time to move the data to and from disk, the management of disk i/o can not be left to the system.

One must make use of the fairly primitive i/o operations that do unbuffered, asynchronous reads and writes. The user himself must manage his own buffers and he must retain control after issuing an i/o request. By retaining control, he can continue to do useful computing during i/o transfer with the penalty that he must check for i/o completion himself. With care, one can keep several i/o channels busy simultaneously and move particles at the same time.

The difficulty with this on a time shared machine is that the system will not let you have as much control as you would like over the disks. One often does not have the ability to force contiguous file assignments. One may also have difficulty creating files distributed over the disk units desired. With faster machines, it is even more difficult to overlap i/o with computing when using disks. The SSD or Solid State Disk is now available on some machines. SSD has much higher transfer rates and much smaller latency time than ordinary mechanical disks. Those fortunate to have an SSD available will find that the abovementioned i/o headaches will disappear.

4. CONCLUSIONS

This document gives only a brief introduction to the issues that arise in the proper implementation of a plasma simulation code. To take advantage of the full capabilities that the computer has to offer, one must master a large collection of facts and techniques which have never been well documented or collected. One must also acquire programming skills through actual personal experience. With time, the application of this knowledge becomes almost transparent.

FLUID MODELING OF MAGNETIZED PLASMAS

Jeremiah Brackbill
Los Alamos National Laboratory
Group X-1 MS E531
P. O. Box 1663
Los Alamos, New Mexico 87545

ABSTRACT. The simplified description given by fluid models of magnetized plasmas makes it possible to simulate large scale problems such as the interaction of the solar wind with the earth's magnetic field. However, the accurate numerical solution of the fluid equations is made more difficult by the singular nature of the flow when magnetic reconnection occurs. During substorms, for example, details of small features of the flow in the magnetotail appear to cause changes in the global solutions. New methods for treating singular problems have been developed. Some of these are reviewed, including adaptive meshes and other tricks taken from fluid calculations. Several illustrative problems are described.

1. INTRODUCTION

Fluid equations should only be applied to the dynamics of collisional, fully ionized plasmas in a magnetic field. The solar corona, solar wind, and magnetosphere are not collisional, and sometimes not fully ionized. Nevertheless, observations suggest their applicability [Levy, 1964] and fluid equations are widely used, especially in global modeling of the magnetosphere. [Walker, 1983].

Fluid models can be used to describe the global dynamics of the magnetosphere because something very like collisions occur on much shorter length scales. For example, in the earth's bow shock, current driven instabilities drive turbulence that provides coupling on a much shorter length scale than a mean free path. [Forslund et al., 1984; Quest et al., 1983]. Thus, in applying fluid models, one must include these collision-like processes in a phenomenological way through transport coefficients or other, non-ideal effects.

Exactly how these effects are to be included in modeling space plasmas is still an incompletely answered question. While more detailed measurements of plasma properties exist for space than for laboratory plasmas, the very detail and locality of the measurements makes it difficult to reconstruct what is happening on a global scale. [Gosling, et al., 1984]

Space Science Reviews **42** (1985) 153–167. 0038–6308/85.15

The principal difficulty in making comparisons among various phenomenological models, aside from the effort and cost, is the difficulty in obtaining accurate enough solutions to the global magnetosphere problem to distinguish their effects. The range of time and space scales is so wide that a single numerical computation cannot encompass them all, and one must choose between leaving some of them out or artificially compressing the range. For example, to make the dimensions of the reconnection region more nearly comparable to those of the whole domain, the Lundquist number in most numerical computations is $O(10^2 - 10^3)$ at best compared with $O(10^5 - 10^6)$ in a space plasma. Luckily, in the case of reconnection, the results are insensitive to the magnitude of the resistivity. [Petschek, 1964]

Here, we identify two aspects of typical numerical computations which offer the greatest opportunities for improvement, namely the calculation of convective transport and the resolution of the flow. After first describing the magnetohydrodynamic (MHD) model and its numerical representation, we show, by example and analysis, the impact of numerical error in calculating convective transport on the results obtained in modeling a substorm. We then describe a variational method for adaptive gridding and illustrate its use in global magnetospheric modeling.

2. THE MODEL MAGNETOSPHERE

As in many early studies of the global magnetosphere [Lyon et al.,1980; C.C. Wu et al., 1981, Lyon et al., 1981], we consider the flow in two dimensions of a magnetized plasma over a line dipole. The model has inherent limitations. There are no steady solutions when the interplanetary magnetic field (IMF) is in the plane of the computation, other than the Dungey model (which requires ten times the observed reconnection rate [Levy et al., 1964]) and a symmetry is imposed that does not obtain in the real world. [Gosling et al., 1984]. However, more and better computations can be done at less cost in two dimensions than in three, and the results in two and three dimensions are similar in some respects. [C.C. Wu, 1981]

The earth's magnetosphere has the structure shown in Figure (1). Prominent features of the magnetosphere are the reconnection of the IMF with the earth's field on the day side, and the stretching of the magnetotail on the night side. The transport of flux from day side to night side causes energy storage in the magnetotail; reconnection leads to its sudden release during substorms. [Hones, 1982].

We also remark the dimensions of the magnetosphere. The reconnection region on the day side and in the magnetotail are small features compared with the standoff distance of the bow shock and the length of the magnetotail. The variation in scales is obviously a difficulty in numerical calculations.

We model the flow by magnetohydrodynamic equations which describe the evolution of the dependent variables: mass, momentum, magnetic field intensity and energy. The momentum equation includes the Lorentz force and viscosity,

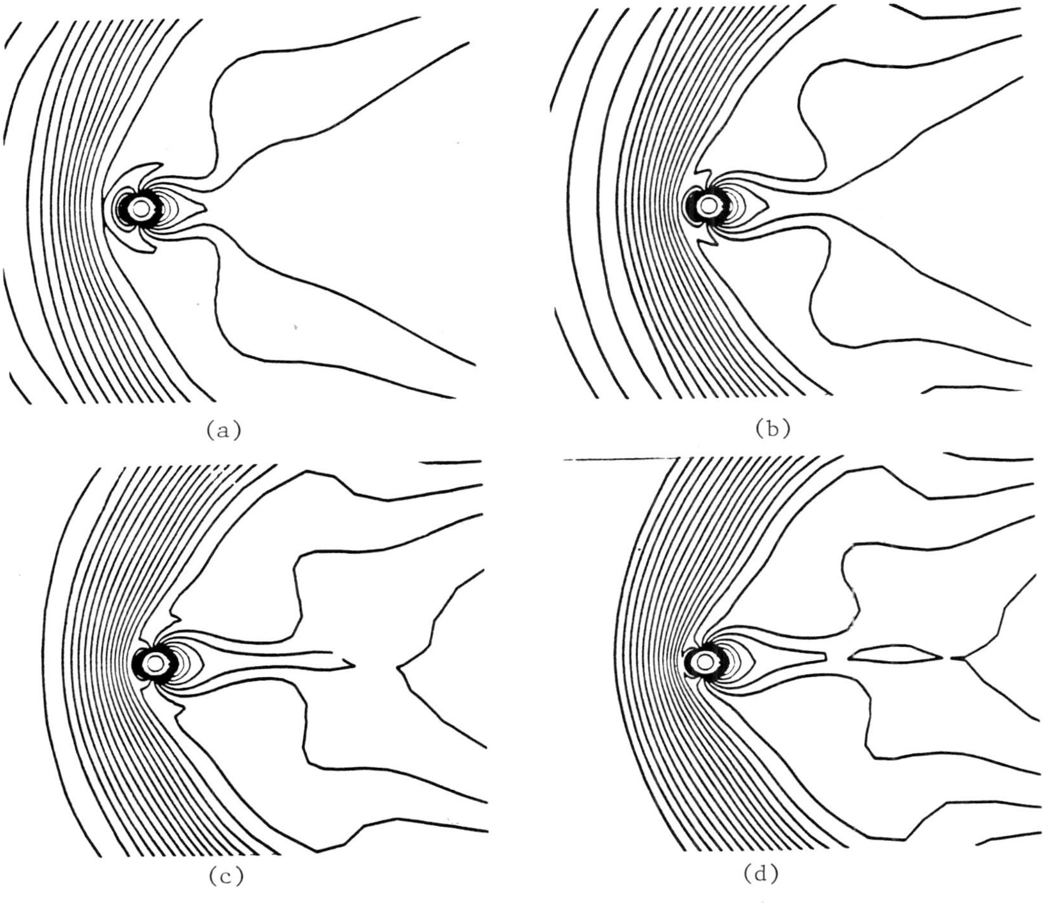

Figure 1 The earth's magnetosphere is depicted in a sequence of magnetic flux plots from a two dimensional calculation. In (a), at t=50, an X-point has formed on the day-side (left), allowing transport of flux to the night-side (right). This stresses the magnetotail, causing it to thin (b), at t=70, and stretch (c), at t=75. Finally, reconnection releases a plasmoid (d), at t=80. The center of the plasmoid is at the magnetopause; X-points lie inside and outside. (Time is measured in solar wind transit times across the earth.)

$$\rho \frac{d\underset{\sim}{u}}{dt} = - \nabla(p + \frac{\underset{\sim}{B}^2}{2}) + \underset{\sim}{B} \cdot \nabla\underset{\sim}{B}$$
$$+ \nabla\mu(\nabla \cdot \underset{\sim}{u}) + \nabla x(\mu + \lambda)(\nabla x\underset{\sim}{u}) , \quad (1)$$

where ρ is the mass density, $\underset{\sim}{u}$ the fluid velocity, p the pressure, $\underset{\sim}{B}$ the magnetic field intensity, and μ and λ the bulk and shear viscosity coefficients. Faraday's law also includes transport,

$$\frac{\partial\underset{\sim}{B}}{\partial t} = - \nabla x\underset{\sim}{E} , \quad (2)$$

where $\underset{\sim}{E}$ is the electric field given by,

$$\underset{\sim}{E} = \underset{\sim}{u} x\underset{\sim}{B} + \eta\underset{\sim}{J} \quad (3)$$

η is the resistivity, and $\underset{\sim}{J}$ is the current, which is calculated from Ampere's law, $\underset{\sim}{J} = \nabla x\underset{\sim}{B}$. The model does not include thermal conduction or thermoelectric effects.

Among the transport coefficients, μ and λ determine the thickness of shock transitions and boundary layers, and the resistivity determines the rate of magnetic field diffusion and reconnection.

To model reconnection phenomenologically, we introduce a nonlinear resistivity [Sato, 1979],

$$\eta = \eta_0[1 + |J|^2/J_0^2]/L, \quad (4)$$

where J_0 is a reference current, $J_0 = B_s/R_e$, (where B_s is the IMF and R_e the radius of the earth), η_0 is a reference value of the resistivity, $\eta_0 = u_s R_e$, (where u_s is the solar wind speed), and L is the specified Lundquist number for the flow.

As in singular problems generally, the larger L, the smaller the reconnection region will be compared with R_e. However, the nonlinearity of η tends to concentrate the effect of resistivity even when L is $O(10^2 - 10^3)$. The nonlinearity increases the diffusion at the neutral point, and generates a perpendicular field component nearby,

$$\nabla x \ \eta\underset{\sim}{J} = - \eta\nabla^2\underset{\sim}{B} + \nabla\eta \ x \ \underset{\sim}{J}. \quad (5)$$

When $\nabla\eta$ is parallel to the neutral field line, as shown in an idealized case in Figure (2) where two, oppositely directed, curved field lines are in contact at a neutral point, the first term in Equation (5) will cut the field lines at the neutral point, and the second, by generating a perpendicular component, will rejoin them. The variation in η tends to decrease the size of the reconnection region along magnetic field lines, but still preserve the usual model. [Petschek <u>et al.</u>, 1964]

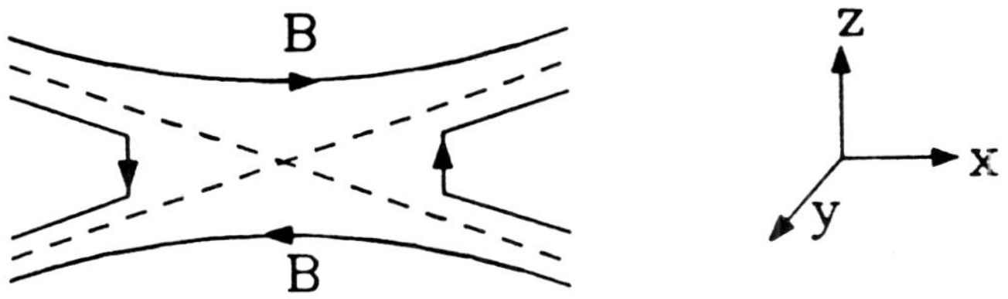

(a)

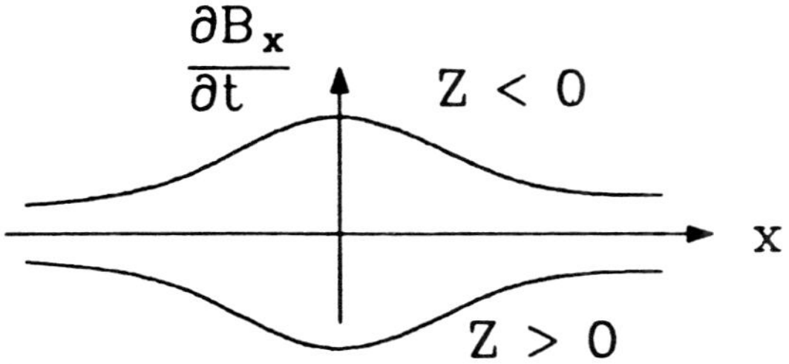

(b)

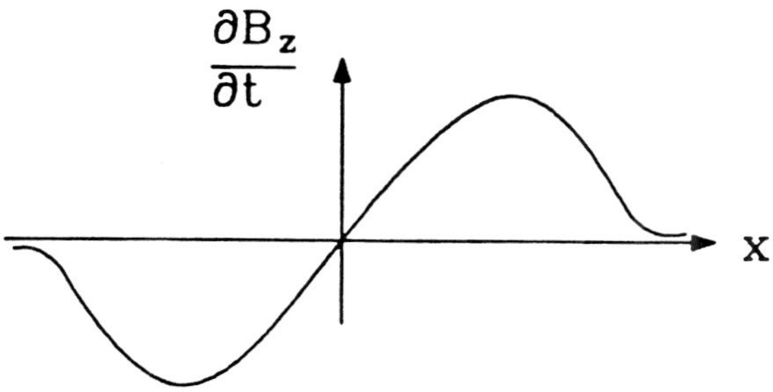

(c)

Figure 2 Two, oppositely directed, curved magnetic field lines are in contact at a neutral point, producing reconnection (a). A nonlinear resistivity enhances the normal tendency for merging at the neutral point (b), and reconnection nearby (c).

The nonlinear dependence of the resistivity on the current has some basis in the results of plasma simulation of the lower hybrid drift instability [Brackbill et al.,1984] and of collisionless shocks, [Forslund et al., 1984]. However, the value of the numerical resistivity is typically much higher than the physical value. This is unavoidable, given the limitations of numericalapproximations. Just as in fluid dynamics, where one makes similar arguments about artificial viscosity, the results of numerical calculations are interesting and useful if the dimensions of the reconnection region are small compared with other lengths of interest.

3. CONVECTIVE TRANSPORT

As in other numerical studies [Walker, 1983], we use finite difference methods in our examples. Our grid is shown in Figure (3), and consists of nested ellipses with one foci at the earth. The zones are very small near the earth, and large near the boundary to remove the specified solar wind flow conditions on the boundary as far as possible from the region of interest at an acceptable cost.

Because flow through the grid contributes the dominant terms to the dynamical equations, numerical diffusion in approximating convective transport is also the dominant source of numerical diffusion. In our calculations, in which we solve the equations of motion in Lagrangean form and then project the solutions back on to the grid, we must solve equations of the form of the equation for conservation of mass,

$$\frac{\partial}{\partial t} \int_V \rho dv = - \oint_S (\hat{n} \cdot \underset{\sim}{u}) \rho ds, \tag{6}$$

for each of the conserved quantities, mass, momentum, energy and magnetic flux.

Differencing Equation (6) requires computing a mass flux ψ, through the surface of each of the cells of the grid shown in Figure (3). In one dimension, an approximation to Equation (6) can be written,

$$\rho_{j+1/2}(t + \Delta t) = \rho_{j+1/2}(t) - \{\psi_{j+1} - \psi_j\} \Delta t/\Delta x \ , \tag{7}$$

where vertices of the mesh are labeled j, and cell centers j+1/2 .

The various methods for solving Equation (7) differ in their evaluation of ψ. Two standard methods, donor cell and Lax-Wendroff (LW) differencing, are written, [Richtmyer and Morton, 1964],

$$\psi_j = \begin{array}{ll} \rho_{j-1/2}\, u_j \ , & u_j > 0 \\ \rho_{j+1/2}\, u_j \ , & u_j < 0 \ , \end{array} \tag{8}$$

for donor cell, and

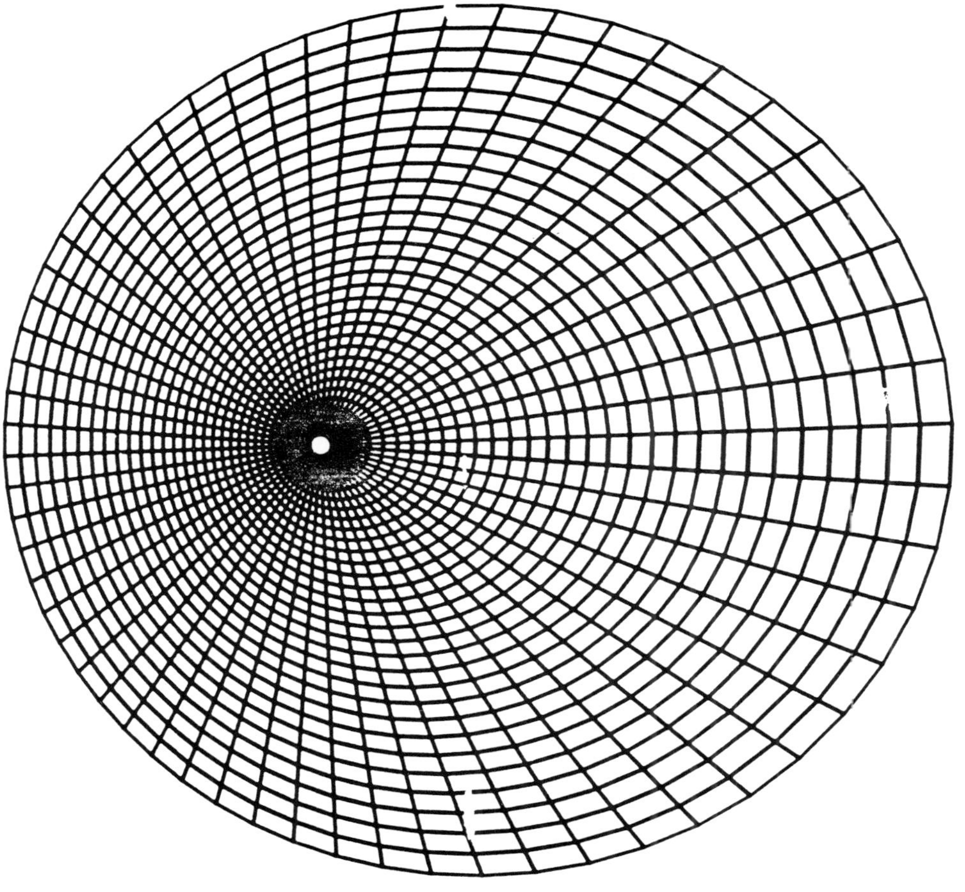

Figure 3 The grid for the numerical computations is formed by nested ellipses with the earth at one focus. The grid extends 30 R_e sunward (left) and 60 R_e anti-sunward. The 30 radial zones increase geometrically from 0.01 R_e at the center. The 60 angular zones are equally spaced in angle.

$$\psi_j = \tilde{\rho}_j u_j \tag{9}$$

for LW differencing. Equation (9) is time-advanced explicitly by substituting spatial for time derivatives in evaluating $\tilde{\rho}$,

$$\frac{\partial \rho}{\partial t} = - u \frac{\partial \rho}{\partial x} ,$$

yielding,

$$\tilde{\rho}_j \equiv 1/2\{\rho_{j-1/2}(1 + u_j \theta \Delta t/\Delta x) + \rho_{j+1/2}(1 - u_j \theta \Delta t/\Delta x)\}. \tag{10}$$

For $1/2 < \theta \leq 1$, Equation (9) is linearly stable.

Using donor cell and LW differencing can give very different results. In Figure (4) are shown flux contours for a calculation using LW differencing with $L = 10^{-3}$, no anomalous resistivity ($J_0 = \infty$), and no IMF. The solar wind is supersonic ($A = 5$). Where time is measured in units of (R_e/u_s), the sequence of contours show a progressive stretching and thinning of the magnetotail, followed by a rapid reconnection and release of a plasmoid to drift in the flow. Also in Figure (4) are shown the corresponding contours with donor cell differencing. No stretching of the magnetotail is evident, nor is there visible reconnection. We can understand the reason for the difference by comparing the velocity plots for the two calculations, shown in Figure (5). With centered differencing, the bow shock is clearly distinguished. With donor cell differencing, the bow shock is spread over so many zones that it is difficult to locate. Evidently, the reason the results are so different is that donor cell differencing is very diffusive.

The diffusivity can be estimated by truncation error analysis [Hirt, 1968] Briefly, one assumes a smooth solution so that each term can be expanded in a Taylor series about (x,t). When coefficients of each derivative are collected, a differential equation results that contains the equation one is solving plus extra terms of $O(\Delta x, \Delta t)$ due to truncation errors. The differential equation corresponding to Equation (7) contains a diffusion term,

$$\frac{\partial \rho}{\partial t} = - u \frac{\partial \rho}{\partial x} + \nu \frac{\partial^2 \rho}{\partial x^2} + O(\Delta x^2, \Delta t^2) . \tag{11}$$

The diffusivity for donor cell differencing is,

$$\nu = \frac{\Delta x}{2} |u| (1 - \frac{|u| \Delta t}{\Delta x} - \frac{\Delta x}{2} \frac{\partial u}{\partial x}) \tag{12}$$

and for LW differencing is,

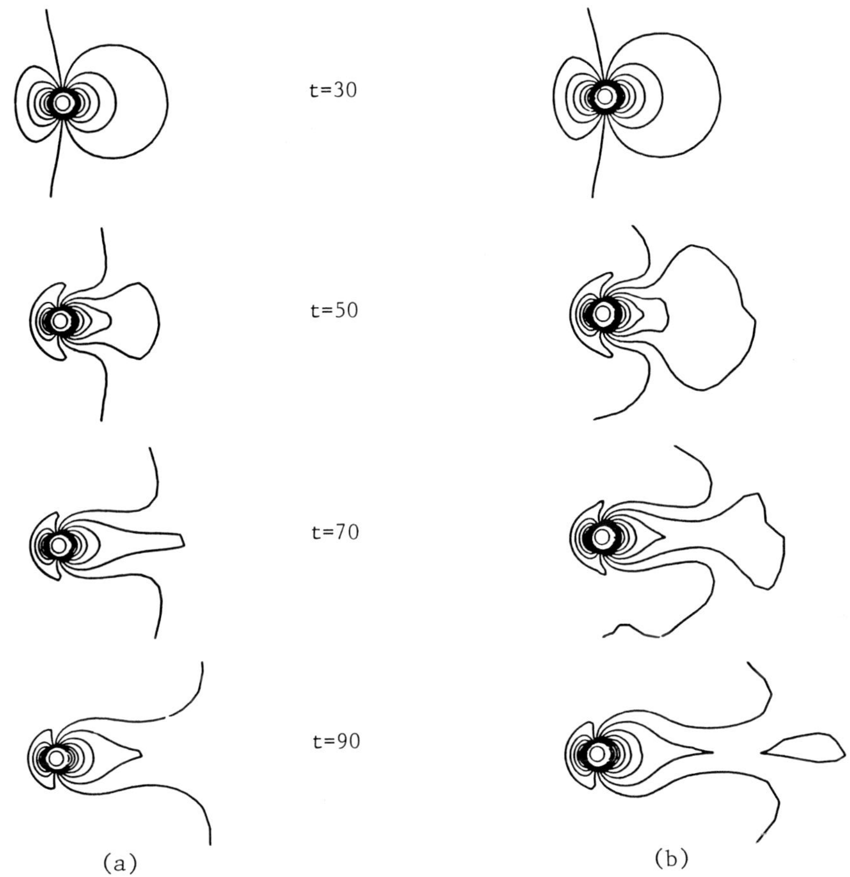

Figure 4 Even in the absence of an IMF, reconnection occurs. However, the results depend on the level of dissipation. With donor cell convection (a), which is diffusive, rapid reconnection causes the magnetotail to shrink as the solar wind exerts stress. With Lax-Wendroff convection (b), reconnection is delayed, and when it occurs the stretched magnetotail rapidly relieves accumulated stress by shedding a plasmoid. Times are in earth transit times.

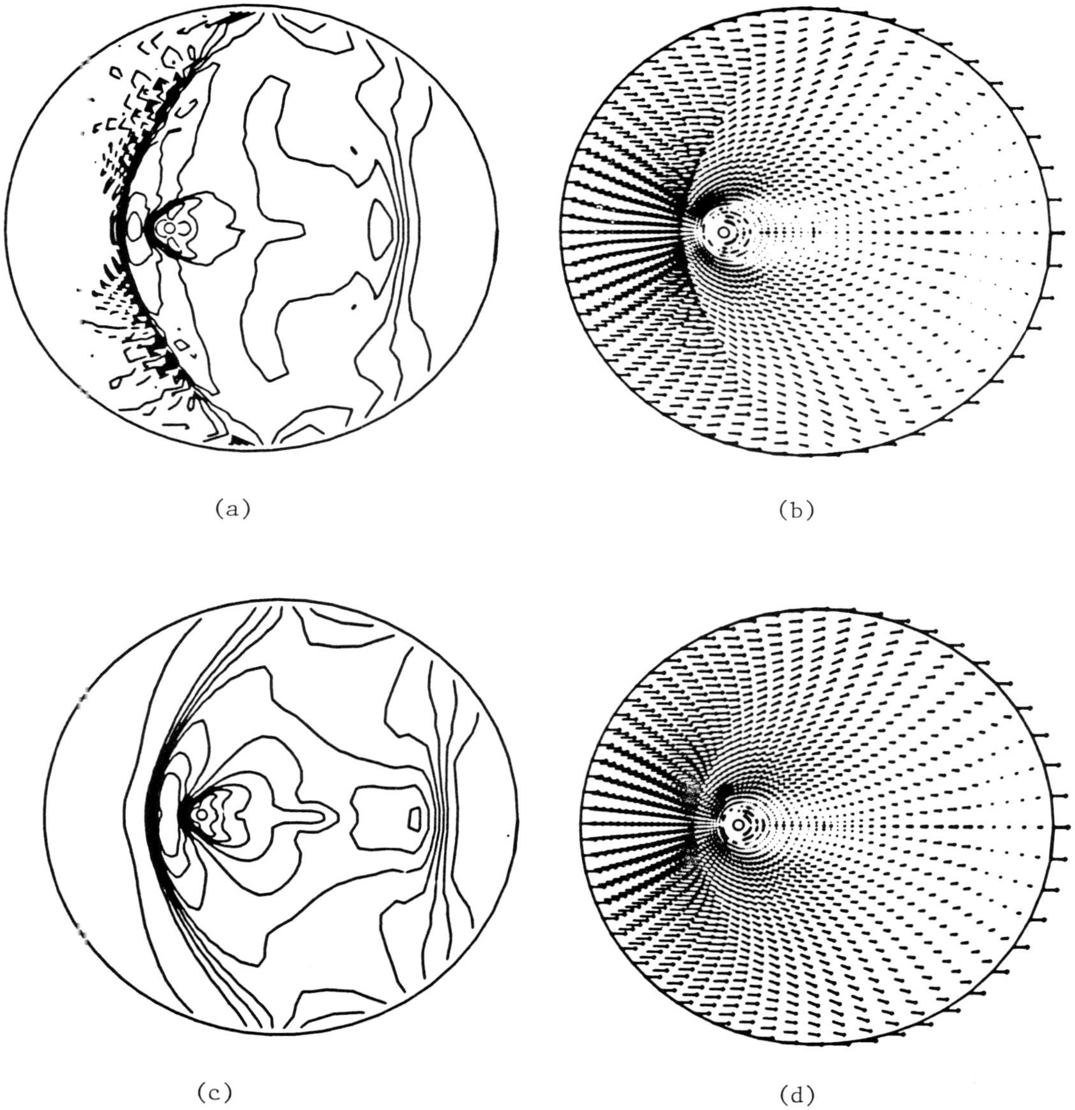

Figure 5 The larger dissipation in donor cell convection is illustrated by a comparison of results with donor cell (c and d) and Lax-Wendroff convection (a and b). Donor cell convection heats the interior of the magnetosphere (a and c where ion temperature contours are shown), and spreads the shear discontinuity at the bow shock over several zones (b and d where velocity vectors are plotted).

$$\nu = \frac{\Delta x}{2}|u| \quad (2\theta - 1) \quad \frac{|u|\Delta t}{\Delta x} - \frac{\Delta x}{2}\frac{\partial u}{\partial x} \tag{13}$$

The two diffusivities are similar. The principal differences are the $O(\Delta x)$ term in donor cell differencing (which accounts for the extra diffusion) and the positive sign of the $O(\Delta t)$ diffusivity in Equation (13), which arises because we have calculated ψ at an advanced time.

For stability, both methods require $|u|\Delta t/\Delta x \lesssim 1$, and both give negative diffusivity when the gradient scale length of the velocity approaches Δx in rarefactions. Because donor cell differencing is more diffusive, it is more robust where there are strong gradients in the dependent variables. (It is also less likely to cause negative densities.) To reduce overall diffusion, yet employ donor cell differencing where it is better, hybrid schemes are used. [Zalesak, 1980]. In the "centered" calculation, Figure (4), we have used a simple hybrid scheme with ψ given by,

$$\psi_j = 1/2[(1 + \alpha_j)\rho_{j-1/2} + (1 - \alpha_j)(\rho_{j+1/2}] \quad , \tag{14}$$

The interpolation coefficient, α depends on u and its gradients. As the length scale of variation of the velocity approaches the mesh spacing, α changes in value from $\theta\ u\Delta t/\Delta x$ to 1 or -1 . α is given by,

$$\alpha_j = (1 - \Delta_j)\theta_j u_j \Delta t/\Delta x + \Delta_j \text{ sign } (1, u_j), \tag{15}$$

where $\Delta_j = \min\ (1, |(u_{j+1} - u_{j-1})/2u_j|)$. This method is easily extended to two-dimensional, non rectilinear meshes.

(Methods that represent shock transitions in one zone are compared in a recent review [Woodward and Collela, 1984]. It has been remarked these methods are especially suited to high speed flow. They have not been applied to MHD flow.)

4. ADAPTIVE GRIDDING

With a given finite difference method, one can improve the accuracy only by increasing the number of zones to resolve the data better. If the zones are uniform in size, increasing the resolution of boundary layers or reconnection regions requires increasing the density of zones everywhere. This is obviously more expensive than using variable zoning, with mesh points clustered in regions of strong gradients and dispersed elsewhere. When boundary layers form or move during the course of calculation, it is difficult to anticipate, *a priori*, where one must cluster points. Instead, one must devise adaptive grid generators to cluster points in response to changes in the data. A calculation using an adaptive grid is illustrated in Fig (6), where the computation mesh and flux plots are shown at a sequence of times. In this example, the IMF is southward and the nonlinear resistivity (Equation (4)) is used with $L = 10^{-3}$. Magnetic flux is introduced on the left by the solar wind flow. It's interaction with

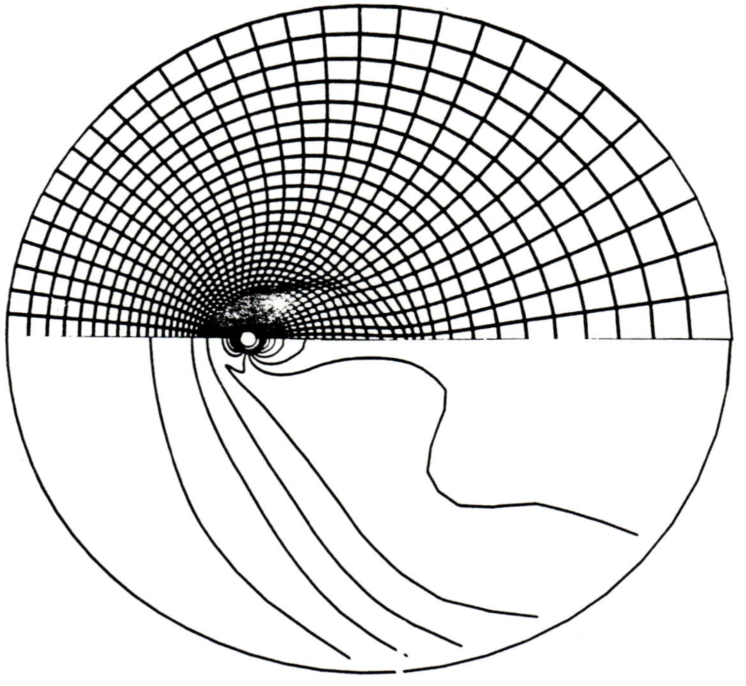

Figure 6 The results of an adaptively zoned calculation are represented by a plot of the grid above, and the magnetic flux contours at the corresponding time below. The mesh points are clustered by adaptive zoning in the dayside X-point, the expansion fan, and the magnetotail where the currents are large.

the earth's magnetic field causes reconnection on the day side and transport of magnetic flux to the magnetotail. Eventually, reconnection in the magnetotail forms a plasmoid, which then moves tailward. In response to changes in the current density, the mesh first clusters at the day side reconnection point, then in the magnetotail in response to changes in the current density.

The adaptive mesh generator we have used is based on a variational formulation described in Brackbill and Saltzman (1982). In deriving the variational principle, we observe that a grid of quadrilaterals is the image of a logical grid of square cells of unit dimension in the space of natural coordinates, (ξ,η). (At vertices, the natural coordinates assume integer values. (i,j)). The connectivity of the mesh is described by the logical grid; the deformation by the metric of the mapping from logical to physical space.

The size of the zones is measured by the Jacobean,

$$J = \det\{\partial(x,y)/\partial(\xi,\eta)\} .$$

The size is controlled by minimizing the integral,

$$I = \iint_V dV\, wJ, \tag{16}$$

where $w(x,y)$ is a weight or control function determined by the data. It can be shown that this integral is minimized when wJ^2 = const. Thus, where w is large, the zones will be small. When w is defined by,

$$w = J^2/B^2, \tag{17}$$

the cells will be smallest where reconnection is most likely to occur (because $\eta \propto J^2$).

In global magnetospheric modeling, we observe that reconnection in the magnetotail is much weaker than on the dayside. The additional resolution in the magnetotail shown in Fig. (6), while modest, is sufficient to allow us to use the same small value of resistivity on day and night sides.

5. PHYSICAL RESULTS

A comparison of the magnetosphere calculation without IMF in Fig. (4), and with IMF in Fig. (1) suggests that a southward pointing field is not essential to produce substorms. In detail, however, the two cases are quite different.

Without an IMF or with a weak one, the magnetotail is stretched and compressed by the solar wind until reconnection releases a plasmoid. The stress is exerted on this magnetotail by a momentum transfer due to viscosity. The dependence of the reconnection on the viscosity can be learned from this case.

With a strong IMF, the structure of the earth's field is altered by reconnection. If we locate the magnetopause by following the junction between a terrestrial and IMF field line from the subsolar x-point to the magnetotail as marked by the kink or shear discontinuity we find the conjugate x-point in the magnetotail is inside the magnetopause. The location appears to coincide with the junction of lobes which have been draped over and under the earth by reconnection. In contrast to the case without IMF, the stress on the magnetotail appears to be exerted by the magnetic field.

REFERENCES

Brackbill, J.U. (1982), "Numerical modeling of magnetospheric reconnection", LAUR 82-483, Los Alamos National Laboratory.

Brackbill, J.U. and J. S. Saltzman, (1982), "Adaptive zoning for singular problems in two dimensions", J. Comp. Phys., 46, 342.

Brackbill, J.U., D.W. Forslund, K.B. Quest, and D. Winske (1984), "Nonlinear evolution of the lower-hybrid drift instability", Phys. Fluids 27, 2682.

Forslund, D.W., K. B. Quest, J. U. Brackbill and K. Lee, (1984), "Collisionless dissipation in quasi-perpendicular shocks", J. Geophys. Resch., 89, 2142.

Gosling, J.T., D. N. Baker and E. W. Hones, (1984), "Journeys of a spacecraft", Los Alamos Science, 10, 33.

Hirt, C.W., (1968), "Heuristic stability", J. Comput. Phys. 2, 339.

Levy, R.H., H.E. Petschek, and G.L. Siscoe, (1964), "Aerodynamic aspects of the magnetospheric flow", AIAA Journal 2, 2065.

Lyon, J.S., S. H. Brecht, J. D. Huba, J.A. Fedder, and P. J. Palmadesso, (1981), "Computer simulation of a geomagnetic substorm", Phys. Rev. Lett., 46, 1038.

Lyon, J.S., S.H. Brecht, J.A. Fedder, and P. Palmadesso (1980), "The effects on the earth's magnetotail from shocks in the solar wind", Geo. Rsch. Lett. 7, 721.

Petschek, H.E. (1964), "Magnetic field annihilation", AAS-NASA Symposium on the Physics of Solar Flares, W.N. Hess ed., NASA SP-50.

Quest, K.B., D. W. Forslund, J. U. Brackbill and K. Lee, (1983), "Collisionless dissipation in quasi-parallel shocks", Geophys. Resch. Lett., 10, 471.

Richtmyer, R.D. and K.W. Morton (1967), "Difference methods for initial-value problems", Interscience, New York.

Sato, T. and T. Hayoski (1979), "Externally driven magnetic reconnection and a powerful magnetic energy converter," Phys. Fluids 22, 1189.

Walker, R. (1983), "Modeling planetary magnetospheres", Rept. No. 2352, UCLA Institute of Geophysics and Planetary Physics.

Wu, C.C., R.J. Walker and J.M. Dawson (1981), "A three-dimensional MHD model of the earth's magnetosphere", Geophys. Rsch. Lett. 8, 523.

Zalesak, S.T. (1979), J. Comput. Phys. 31, 335.

GLOBAL SIMULATIONS USING MHD CODES: A FEW POINTS TO CONSIDER BEFORE YOU TRY ONE

Stephen H. Brecht
Berkeley Research Associates
P.O. Box 241
Berkeley, CA 94701

ABSTRACT: A survey of all global time-dependent MHD simulations is presented. The survey includes a discussion of the physical situations simulated by the respective authors as well as the numerical algorithms employed. Finally a discussion of the advantages enjoyed by certain numerical schemes and the problems that a researcher will very likely encounter if he should undertake construction of such codes is presented.

INTRODUCTION

Beginning in 1978 global MHD simulations have been used to study a variety of large scale problems associated with the dynamics of the earth, venus, comets, and the solar wind. In the future they will be used to study other objects in the solar system. The purpose of these global simulaitons is to obtain information concerning the general topological structure of the object, i.e., the earth's magnetosphere, and to address large scale cause and effect relationships. An example of such a relationship is the correlation between the dynamics of the solar wind and the earth's magnetotail where magnetic substorms represent one of the most dynamic phenomena. These large scale calculations are quite difficult to construct and operate. The difficulty arises because of the dimensionality involved, the fact that the set of MHD equations are hyperbolic in form and that the boundary conditions are very difficult to implement especially around an object like the earth.

It is my intention to present to researchers, who are considering either using the results of global simulations or building and using such a tool themselves, a brief review of such calculations and some of the choices and decisions that will confront them. While presenting this brief review, I will endeavor to present the differences that exist between the various approaches used. This will include the numerical schemes employed as well as the boundary conditions and where possible some of the physical assumptions that were used. Because of the limited amount of time and space, no attempt will be made to provide much detail. Hopefully the references provided will give the interested reader the necessary starting point and information. This

Space Science Reviews **42** (1985) 169–185. 0038–6308/85.15

paper will be limited to the discussion of large scale time dependent MHD simulations. There are many other physical problems to which MHD simulations have been applied most notably the study of tearing modes. However, I will leave this and other very interesting problems to other researchers to discuss.

GLOBAL SIMULATIONS

I will begin the comparison of the various global MHD simulations by noting that all of these simulations have one thing in common. They are time dependent solutions of the MHD equations. Other than boundary conditions, none of these calculations assume a particular structure for the magnetosphere nor do they have any adjustable parameters to affect the structure. The number of ad hoc assumptions is held to those required for the validity of the MHD equations. Frequently, one hears discussions of magnetospheric simulations as MHD "models." In a sense this is correct in that certain physical assumptions were made in the derivation of the MHD equations. Customarily, in a model the macroscopic structure may be manipulated, e.g., the magnetopause boundary might be positioned at will. The simulations to be discussed are not models in this sense, in fact the only thing that can be construed as a model are the boundary conditions. In these simulations, only the microscopic structure, the underlying physics of the MHD equations, is determined. The macroscopic structure and dynamics then either fits what is being simulated or it does not. What the simulations test is whether the physics embodied in the equations solved is adequate to explain the simulated system. To date it has been found that these large scale simulations do accurately represent the dynamics of the systems under study consistent with the dimensionality of the simulations and the scale imposed by the equations.

What is there then to compare between the various simulations? Obviously, the results when similar simulations are performed. In addition, one should compare the numerical schemes being applied to the solution of the MHD equations. Other topics include the dimensionality and the boundary conditions. The numerical scheme is important to the accuracy and speed with which the MHD equations can be solved. The dimensionality is directly related to the fidelity of the simulation to the real system. The boundary conditions are perhaps the most crucial. Since these calculations are initial-value problems, the initial conditions determine the physical system to be simulated. They also can determine the results that are obtained from a simulation. Typically, the boundary conditions are the most difficult aspect of the simulation to implement correctly and they frequently produce unexpected results.

In the following paragraphs I will categorize the global MHD simulations in a variety of ways. First, a categorization by problem area will be presented, then by numerical scheme and dimensionality. Finally where possible a categorization by boundary condition, e.g., IMF orientation or inflow conditions, will be presented.

As mentioned earlier in this paper, global simulations of a variety of objects in our solar system have been performed. Table I shows a breakdown by object of the simulations performed to date. Only the first authors names have been included for brevity but virtually all of this research has involved multiple authors as the reference list shows. Table I indicates that the earth has been the focus of most of the research. Other objects such as comets and the planet Venus have not received very much attention. Venus will receive more in the coming year as two research groups have simulation efforts underway. The solar wind is also the subject of large scale simulation with different researchers applying various numerical techniques. These are MHD treatments but due to space constraints are only mentioned in passing. For convenience the earth simulations are broken into two categories depending on whether or not the ionospheric currents were computed.

The simulations of the earth's magnetosphere were initially undertaken in two dimensions. These simulations were by LeBoeuf et al. (1978) and Lyon et al. (1980, 1981). LeBoeuf et al. (1978) used a fluid particle approach in the MHD modeling and were able to reproduce a Dungey (1961) type magnetosphere. The numerical scheme used however was quite diffusive and yielded few details of the dynamics. Lyon et al. (1980) solved the standard MHD equations using a combination of hybrid schemes of higher accuracy to address the dynamic problem of shock focusing. In this work they studied the effect of solar shocks on the earth's magnetosphere. It was found that solar shocks interacted with the magnetosphere in such a way as to focus the shock in the tail region at -15 to -20 R_E. The focusing of the shocks produced increased temperature and pressures in this region. It was speculated in the paper, as well as in Brecht et al. (1980), that such effects might well trigger magnetic substorms. Following this work Lyon et al. (1981) studied the problem of substormlike events triggered by southward directed IMF. This research was performed with a very high order numerical scheme. The results of these simulations produced behavior in the tail that tended to support the neutral line model for magnetic substorms (Dungey, 1961; Hones, 1973; McPherron et al., 1973).

Following these simulations attention was turned to performing magnetospheric simulations in three dimensions. As can be seen from Table I, a considerable number of 3-D simulaitions have been reported. The first calculation was by LeBoeuf et al. (1981). In this simulation a southward IMF interacted with a current loop. The calculation used periodic boundary conditions and particle fluid approach as had been done in the 2-D calculation. This calculation was reported to reproduce a Dungey (1961) like magnetosphere but due to the diffusive nature of their scheme only a very short tail was created and the ambient earth magnetic field seriously eroded.

Brecht et al. (1981) studied the effects of an east-west component of the IMF being added to a solar wind carrying a -2 nT southward field. This calculation used no planes of symmetry and a hybrid numerical scheme. The results showed that a rotation of the IMF toward an east-west direction rotated or skewed the entire magnetotail and produced an induced B_y in the plasma sheet of the size seen in the solar wind. These results were found to be consistent with experimental data (Hardy et al., 1979; Lui, 1979) and in agreement with an analysis calculation by Cowley (1981).

During this same year Wu et al. (1981) addressed the case where no IMF existed in the solar wind. The calculation was performed on a quadrant of the magnetosphere meaning that both the meridional and equatorial planes were planes of symmetry. The calculation reproduced, apart from the diffusion effects noted by the authors, many of the major features of the magnetosphere.

A time dependent simulation of the magnetosphere was performed by Brecht et al. (1982) to look at the dynamics of the magnetotail when it was undergoing a series of reconnection events. This simulation showed the very discrete nature of the dynamics and the production of multiple islands. These islands had a strong resemblance to the plasmoid picture presented by Hones et al. (1984). In addition it was seen that the flow speed of the plasma during these events was directly correlated to the mass loading in the magnetotail.

At this point the emphasis of the 3-D simulations changed. Instead of considering the magnetotail dynamics or simply proving that such calculations could be performed, the researchers turned their attention to the coupling of the magnetospheric dynamics to the ionospheric current systems. Fedder and Lyon (1984) used an earlier 3-D calculation to map the magnetospheric currents into the ionosphere and obtained surprisingly good agreement with data. The result was surprising because the calculation used had a minimum cell size of 2 R_E on a stretched mesh and had been designed to study the magnetotail with a minimum resolution in the bow shock region and near earth region. The total simulation grid was 29x21x21. This result may imply that the ionospheric current system is controlled by very large scale dynamics. During the same year a series of simulations was reported that addressed the magnetosphere-ionosphere area (Ogino and Walker, 1984; Ogino, 1984; and Ogino et al. 1984). These reports discussed the creation of various region 1 and 2 current systems in the simulations and compared them with data. The IMF was varied and the change in the current patterns as a function of IMF orientation was correlated. Some of the results were compared to the Rice convection model (Wolf et al., 1982) which self-consistently treats the connection of the region 1 and 2 currents but must assume the connection of the current systems to the magnetospheric dynamics. Unfortunately, Ogino 1984 did not spend much time on comparing his calculation to the Rice model to determine the significance of actually calculating the magnetosphere-ionosphere coupling as opposed to the assumptions in the model.

These simulations represent all of the global simulations of the earth's magnetosphere that the author is aware of at this time. A glance at Table I shows that only a few others have been performed on other objects. The first simulation of a comet was by Schmidt and Wegmann (1980). They used a numerical scheme based on the Godunov approach (Godunov, 1959) that did not maintain the divergence of B equal to zero. They studied plasma flow between a hydrodynamic shock at some distance from the comet and an arbitrary fixed contact surface near the comet nucleus. Their basic results appeared reasonable for the inner regions of a comet magnetosphere but the numerical difficulty with the divergence of B casts considerable doubt on their results. Research by Brackbill and Barnes (1980) has demonstrated the difficulties to be

TABLE I

EARTH		COMETS	VENUS	OTHER
MAGNETOSPHERE - IONOSPHERE				
BRECHT (1981)^3	FEDDER (1984B)^3	SCHMIDT (1980)^3	BRECHT (1984)^3	DRYER (1984)^2
BRECHT (1982)^3	OGINO (1984A)^3	FEDDER (1984A)^3		
LE BOEUF (1978)^2	OGINO (1984B)^3			WU (1983)^2
LE BOEUF (1981)^3	OGINO (1984C)^3			
LYON (1980)^2				
LYON (1981)^2				
WU (1981)^3				
WU (1982)^3				

THE NUMBER ABOVE THE DATE OF EACH REFERENCE IS THE DIMENSIONALITY OF THE CALCULATIONS.

TABLE II

IMF

NONE	NORTH	SOUTH	EAST-WEST	TIME-DEPENDENT
WU (1981)^3	OGINO (1982B)^3	BRECHT (1982)^3	BRECHT (1981)^3	BRECHT (1981)^3
LYON (1980)^2	OGINO (1984C)^3	LYON (1981)^2	OGINO (1984B)^3	FEDDER (1984)^3
	OGINO (1984A)^3	LE BOEUF (1978)^2	OGINO (1984C)^3	
	WU (1982)^3	LE BOEUF (1981)^3	FEDDER (1984A)^3	
		FEDDER (1984B)^3		
		WU (1982)^3		

expected if the divergence is not kept to exactly zero in a magnetohydrodynamic calculation. Fedder et al. (1984a) used a code very similar to Brecht et al. (1981, 1982) to study the interaction of the solar wind with a comet. They showed that for a steady solar wind and interplanetary magnetic field the cometary plasma has a distinctive structure; a spheroidal head and a long ribbon-like tail. Rotational discontinuities in the IMF lead to changes in the tail structure with the production of ray-like structures and tail disconnection events.

Apart from the simulations of Venus which are in the early stages of development and those of the solar wind which I have mentioned for completeness, there are no other global simulations.

BOUNDARY CONDITIONS

The previous discussion has touched upon the various areas of study undertaken with the codes. For comparison, these simulations have also been categorized by the boundary conditions used in each. In Table II the various calculations have been broken down by the IMF used in them. The IMF represents the inflow boundary condition in all of these calculations, as well as, different physical systems. Two of the calculations used no IMF. The rest used at least one and a few (Brecht et al. 1981 and Fedder et al. 1984) allowed the IMF to change direction and strength as the simulation proceeded. Readers familiar with the major questions concerning the dynamics of the earth's magnetosphere understand the interest in the orientation of the IMF. I will not address this topic in this paper.

The other aspect of boundary conditions has been summarized in Table III. The no symmetry category means that these simulations used no internal planes of symmetry and that the full three dimensional space of the object was simulated. It turns out that all of the simulations in this category used the same external boundary conditions; outflow on all sides but the sunward one where inflow was used. The one plane symmetry was used in the 2-D simulations of Lyon and the 3-D simulations of Brecht and Smith (1984). The latter is using a bi-spherical coordinate system that basically demands such a symmetry plane. All of the 2-D simulations by Lyon et al. (1980, 1981) used reflecting boundaries on the sides and a outflow boundary on the backside. Care was taken in these simulations to insure that the reflected material did not interfere with the simulation. The two plane symmetry means that these calculations only addressed a quadrant of the magnetosphere. These simulations used an inflow condition on the sunward plane, a mirror boundary at the symmetry planes and a 45 degree outflow boundary on the remaining boundaries. I will have more to say about these boundary conditions in the discussion section of this report.

NUMERICS

The numerical schemes used by the various research groups to perform these global simulations cover the complete spectrum of techniques available for solving the hyperbolic set of MHD equations using explicit techniques. Table IV presents a categorization by scheme and calculation. The equations solved by these codes are the MHD equations as shown

$$\frac{\partial \rho}{\partial t} = -\nabla \cdot \bar{v} + D\,\nabla^2 \rho$$

$$\frac{\partial \bar{v}}{\partial t} = -\bar{v} \cdot \nabla \bar{v} - \frac{1}{\rho}\,(P + Q) + \frac{\bar{J} \times \bar{B}}{c} + \bar{g}$$

$$\frac{\partial P}{\partial t} = -\nabla \cdot P\bar{v} - (\gamma - 1)(P + Q)\,\nabla \cdot \bar{v} - \eta J^2 + D_p\,\nabla^2 P$$

$$\frac{\partial B}{\partial t} = -c\,\nabla \times \bar{E}$$

where $\bar{E} = -(\bar{v} \times \bar{B}) - \eta \bar{J}$ and $\bar{J} = c\,\nabla \times \bar{B}/4\pi - \frac{1}{c}\,\frac{\partial E}{\partial t}$

In these equations I have included the diffusion terms used in the Lax-Wendroff scheme of Ogino et al. (1984a,b,c). For this scheme the diffusion coefficient, D, is included in both the density and pressure equations. The term Q in the momentum and pressure equations is the artificial viscosity used by several of the researchers in the shock region to guarantee the Rankine-Hugoniot relations. Most of the simulations set the resistivity, eta, to zero. It may be nonzero in the work by Ogino et al. (1984a,b,c). Finally, the displacement current is carried in the simulations of Brecht et al. (1981, 1982) and Fedder et al. (1984a,b). The significance of some of these terms will be discussed later in the text. The reader is referred to any number of numerical texts or papers for a complete explanation of artificial viscosity or diffusion terms necessary for the Lax-Wendroff scheme, e.g., Roach (1982) or Woodward and Colella (1984).

The work by LeBoeuf et al. (1978, 1981) used a fluid particle approach where the plasma was treated as discrete particles but the transport was done in the usual fluid manner. As shown in Table IV, they used the Lax scheme (Lax, 1954) which does shocks reasonably well but has poor accuracy and a large amount of numerical diffusion. The diffusion provided a quiet calculation and attainment of a steady state but diffused away many important features. The Rusanov scheme employed by Wu et al. (1981, 1982) is less diffusive than the Lax scheme. However, a diffusion term must be added to the algorithm for stability of the scheme. This term adds a numerical viscosity as well as numerical conductiviaty and mass diffusion throughout the calculation. In this regard the Rusanov scheme is very similar to the Lax scheme, see Roach (1982). The Godunov scheme (see Roach, 1982; Woodward and Colella, 1984; and Godunov, 1959) is a first order scheme of a complicated form because it does not require the assumption that the solution of the equations be smooth. This approach allows the calculation to produce accurate and very well behaved treatment of shock discontinuities without unphysical oscillations. However it is slow and it has poor convergence properties. The Lax-Wendroff scheme is a very popular scheme

TABLE III

BOUNDARY CONDITIONS

NO SYMMETRY	1 PLANE SYMMETRY	2 PLANE SYMMETRY	PERIODIC
BRECHT (1981)[2]	LYON (1980)[2]	WU (1981)[3]	LE BOEUF (1978)[2]
BRECHT (1982)[3]	LYON (1981)[2]	OGINO (1984A)[3]	LE BOEUF (1982)[3]
FEDDER (1984A)[3]	BRECHT (1984)[3]	OGINO (1984B)[3]	
FEDDER (1984B)[3]		OGINO (1984C)[3]	

TABLE IV

EXPLICIT NUMERICAL SCHEMES

LAX	RUSANOV	LAX-WENDROFF	GODUNOV	HYBRID (PDM-FCT)	HIGH-ORDER
LE BOEUF (1980)	WU (1981)	OGINO (1984A)	SCHMIDT (1980)	LYON (1980)	LYON (1981)
LE BOEUF (1981)	WU (1982)	OGINO (1984B)		BRECHT (1981)	
		OGINO (1984C)		BRECHT (1982)	
				BRECHT (1984)	
				FEDDER (1984A)	
				FEDDER (1984B)	

and has been used in its two step formulation by Ogino and Walker (1984a), Ogino (1984b) and Ogino et al. (1984b). This scheme has a low numerical diffusion. Their formulation made use of artificial viscosity to obtain the proper description of the shock. However, as shown in the equations numerical diffusion is also added because while the scheme has a small amount of diffusion it will shed waves. In order to remove these waves diffusion is added. The result is frequently a shock structure that may require 5-10 cells to resolve.

This brings us to the final two categories listed on Table IV. In my opinion they represent the future of global simulations such as have been discussed in this paper. These are the hybrid schemes and the high order schemes. As alluded to in this section, one of the problems of calculating a discontinuity like a shock is that these structures cause numerical oscillations which are at the very least a nuisance. The traditional way of removing these oscillations is the addition of diffusion into the numerical scheme. This works but causes the whole calculation to be diffusive when one really only wants the shock area to be smoothed. One way to hybridize a numerical scheme is to use a non-diffusive high order and then go back and add diffusion locally to guarantee monotonicity. A scheme called FCT, flux corrected transport, developed by Boris and Book (1973) and generalized by Zalezak (1979) is such a scheme. A hybrid scheme can also be formulated which uses a low order algorithm that guarantees monotonicity and then applies a high order algorithm that produces the sharp features expected from, for example, a shock. The partial donor cell method, PDM, scheme developed by Hain (1978) is such a method. Both of these schemes can resolve a shock in about 3 cells. In the work by Lyon et al. (1980) both the FCT and the PDM schemes were used. The next 2-D simulation report by Lyon et al. (1981) used a leap-frog time integration scheme and a 20th order finite difference approximation for the spatial derivatives along with the FCT scheme. This produced the minimum possible numerical dissipation consistent with a solution showing no extraneous extrema. All of the simulations by Brecht et al. (1981, 1982, and 1984), Fedder et al. (1984a) and Fedder and Lyon (1984b) used the PDM algorithm. These calculations used an artificial viscosity to insure that the Rankine-Hugoniot relations were satisfied throughout the shock region. The artificial viscosity and any diffusion necessary for a solution that meets the monotonicity requirements were activated in the region of the simulation where there was compressive flow, divergence of velocity less than zero, leaving the rest of the calculation with minimum numerical diffusion and no artificial viscosity.

This concludes the discussion of the numerical methods used by the various research groups in their global time dependent simulations. The discussion is meant to give the reader just the flavor of the considerations and limitations of the numerical algorithms. In the next section I will list some of the considerations that should be included when one is selecting a scheme or evaluating the results of such a calculation.

DISCUSSION: POINTS TO CONSIDER

In this section I would like to address several points that I think are critical in the development of a global simulation capability. These are the main areas where decisions must be made. They are: the numerical scheme, the boundary conditions and the model and physics of the problem. I cannot emphasize enough the importance of the numerical scheme. Many people undertaking such simulations use schemes with which they are familiar and assume that as long as it works it is fine. In my opinion there are several important aspects of the numerical scheme to be considered: the accuracy, the number of cells necessary to resolve important features and in the case of global simulations the speed. These categories are related. The accuracy topic covers several areas. The Lax scheme in my opinion will never be accurate enough in a time dependent simulation to be useful due to the diffusion inherent in it. Any scheme used must be able to resolve the features of interest to the researcher doing the simulations. Certainly the other schemes will do so if enough cells are provided. However, it has been shown (Woodward and Colella 1984) that many of the more common schemes do not increase accuracy of representation linearly with the increase in the number of cells in the structure in nonlinear situations. However, the hybrid schemes do approach linearity. The crucial point here is that a Lax type scheme may take 10 or more cells to resolve a shock and the Lax-Wendroff algorithm may use 5-10 cells but the hybrid schemes need only three and some of the high order schemes need only one. Consider the comments of the earlier section about the various schemes used and then look at Figure 1. In this figure a cone of vector potential is simply convected around a cartesian grid. As one can plainly see the Lax scheme and to a similar degree the Rusanov scheme simply diffuse away all details. If one wants a steady state this algorithm will surely provide it. In the third panel one sees the results of the Lax-Wendroff algorithm. In this case the convection only made it half the way around the grid before the wave amplitude of the waves being created by this scheme made any further convection meaningless. One can see that to reduce this scheme to a very quiet one will require the extra diffusion terms displayed in equations 1 and 3. Finally a very high order scheme was employed, one sees that the shape of the potential has been modified very little. A similar test was performed by Zalezak (1979) for his hybrid formulation which produced little diffusion of the object being transported and no rippling behavior.

This brings me to a very important point concerning numerics and large scale global simulations, speed. This is the most crucial parameter to performing good global simulations because the code must be used repeatedly if it is to provide the information for which a global simulation is best suited, cause and effect relationships. Speed is very important due to the difficulty of doing these calculations and handling the boundary conditions correctly. One can anticipate doing a large number of these calculations before a simulation or collection of simulations are fit to publish. If the code takes enormous computational resources the computer time necessary for a proper simulation will not be available. Therefore, if an algorithm can resolve a given feature in

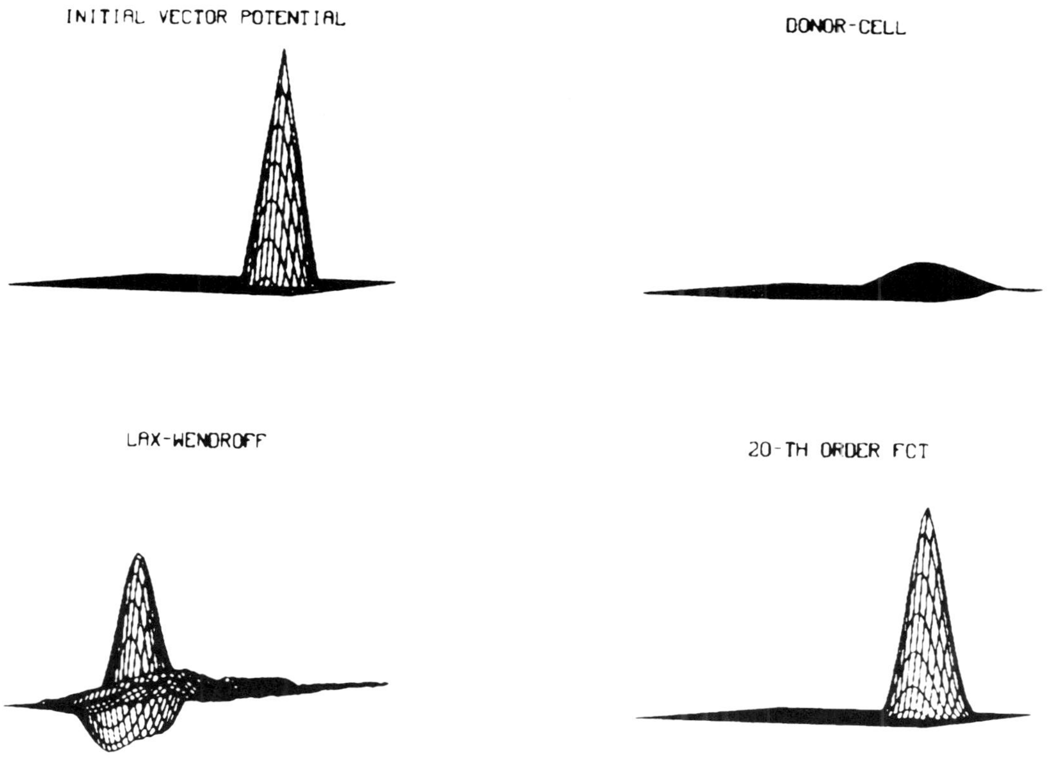

Figure 1. This displays the results of convecting a conical vector potential, (a), around a cartesian grid. Three numerical schemes are shown: donor cell, (b), Lax-Wendroff, (c), and a 20th order FCT, (d). This figure was generated and furnished by John Lyon of NRL.

in half as many cells per direction but takes twice as long per cell to execute one will see a factor of four increase in speed in a three dimensional calculation. The difference between a 10 cell resolution of the shock and a three or one cell resolution of the same feature is enormous and might well justify the builder of such a code investing the extra time initially to realize such savings when the code is used.

In the next few paragraphs I would like to discuss some aspects of simulations that are frequently relegated to a few words in a paper but usually require the most effort, intuition and knowledge. These are the boundary conditions and the physics that can and in some cases must be applied to the problem. Let us begin with boundary conditions. I have already mentioned what the simulations to date have used. Some are much harder than others to implement. For instance the reflecting boundary condition, while not necessarily simple, can be implemented much faster and with better confidence than the outflow conditions. The main consideration with the reflecting boundary conditions is whether or not the reflected material and magnetic field interfere with the dynamics understudy. The outflow conditions, particularly on the sides where shock and magnetic structures must leave at a variety of angles, require one to be very careful that the boundary does not reflect unwanted waves. This turns out to be non-trivial since a structure created by such wave reflection may take thousands of time steps to appear. Debugging such an event can be expensive from a computational and human standpoint but doing so on a rapid code is clearly an advantage.

The next consideration is whether or not to use symmetry boundary conditions. It is frequently argued that using such conditions allows greater accuracy since one can use more cells to cover only a quadrant. However, one must ask why the code is being built in the first place. A large scale 3-D code provides a unique capability to study transient effects on an object like the earth's magnetosphere. If a two symmetry plane calculation is used, then asymmetric changes in the IMF are excluded as well as the three dimensional response of the magnetosphere to ostensibily symmetric wind conditions interacting with the tipped dipole of the earth. I suggest such codes should be built with the minimum number of symmetry planes and if a high resolution calculation needs to be done switch to the symmetry boundary condition. One will even be able to tell if the symmetry boundaries are producing spurious effects. Further, one of the best numerical diagnostics available to a researcher developing a three dimensional code is the preservation of symmetry during the course of a calculation that should be symmetric. This tool has proven invaluable in diagnosing boundary condition difficulties and the more mundane indexing problem and is only available if one does not assume a symmetric calculation beforehand.

Without a doubt the most difficult aspect of simulating an object in our solar system is handling the internal boundary that is a magnetized planet. This boundary condition consumed at least 50% of our research time when developing both the two dimensional code (Lyon et al. 1980, 1981) and the three dimensional code (Brecht et al. 1981, 1982). Why is it difficult? First, if one is doing a large scale calculation that may extend as far as 30 R_E sunward and 100 R_E in the tail and 40 R_E

on the sides, cell sizes on order of the earth's radius must be used. Unfortunately, near the earth the dipole field enhances the Alfven speed and the cells are usually the smallest. This means that the region where the least resolution of detail is occurring is controlling the timestep of the calculation. Further, with R_E cell sizes near the earth one may have cells in which the earth's field realistically should turn a good fraction of 90 degrees. Such lack of resolution will produce tremendous numerical current systems. All researchers that have published research on the global dynamics of the earth have had to address this problem and have solved it in some fashion or other. I suspect that they did what I did and resorted to a trial and error method. It should be obvious to the reader that a fast code is a tremendous advantage in testing out different models or conditions to be used in the very near earth region.

Exact details of what the various research groups did to handle this boundary are sketchy and in some cases fall into the category of trade secrets. However, all are constrained to obtain results that compare well with data. Wu et al. (1981) attacked the problem by using an adaptive grid and having many small cells in the very near earth region. This cut the timestep of the calculation tremendously. His solution to this was to advance the small cell region many timesteps before advancing the outer magnetospheric region. This two timestep process saved enormous amounts of time. The work by Ogino et al. (1984a,b,c) took a slightly different approach. First, the current was calculated using the difference between the total magnetic field and the dipole field of the earth. This reduces some of the problems due to the dipole field curvature. Resistivity in the calculation also helps here. In addition, a mass density model was included to mimic the ionosphere. This serves two purposes: one it reduces the local Alfven speed in the code and two it mass loads this region so that local velocities are greatly reduced. Both of these effects are physically realistic and help increase the timestep of the simulation. One of the difficulties of adding mass near the earth is, if for example one wants to use 2 R_E resolution one only has 6-7 cells from the dipole to the shock, the numerical scheme will tend to diffuse mass to remove the gradient and fill the region out to the shock with mass or artificially modify the lobe and tail region dynamics. We made use of several methods and tried a variety of others. One of the more useful physical models that we used was the Alfven correction. This is the recognition that the displacement current acts as a retarding force and one can then limit the maximum speed of propagation of the Alfven wave (Boris, 1970). A very brief description of how it is used is given in appendix A. One sets the speed of light to the maximum wave speed that is desired in the calculation and makes the modification to the momentum equation shown in appendix A. This allows use of the usual parameters in the magnetospheric region but limits the Alfven speed near the earth where the B field is increasing along with the density. One then does not have to have an exact ionospheric mass loading to keep the near earth region from controlling the simulation. Some mass loading may still be warranted to keep the near earth velocities under control. The diffusion problem can be avoided by realizing that in the near earth region one does not have to transport the mass; simply reduce the acceleration by the effective mass. These statements are of course more applicable if the very near earth region is to be poorly resolved so that other regions of the magnetosphere can be simulated. We have found that these and many other physical models can be used in the very near earth region. They have the effect of eliminating spurious effects caused by poor resolution of the dipole field and speeding up the simulation so that we could do full simulations in an hour's time on a machine much slower than the CRAY 1.

The discussion presented has been intended to give the reader some idea of the models and numerical adjustments that have or could be used in these simulations. Anyone building such codes should consider these points and use them if they fit into their scheme. It should be remembered that the data is the final arbitrater as to the validity of such short cuts and the researcher should have a very good understanding of the assumptions made.

CONCLUSION

It has been the intention of this paper to familiarize the reader with the evolution of large scale time dependent global simulations of objects in the solar system. The emphasis of the research to date has been on the earth's magnetosphere. The simulation capability has evolved to the point where 3-D simulations have been performed that agree well with data. This tool has not been used to its fullest extent. There is significant room for improvement of the numerics and the physics that the equations being solved contain. An obvious improvement of global simulations would be the inclusion of an anistropic pressure which would allow more electromagnetic codes to evolve in the calculation as they probably do in space. All of these calculations are difficult to perform but they do produce a unique global view of the full dynamics of large systems such as the earth's magnetosphere. With care and effort they will become both more accurate and useful to the study of these systems.

REFERENCES

Boris, J.P., and D.L. Book, 'Flux-corrected transport, 1, SHASTA, a transport algorithm that works,' J. Comp. Phys., 11, 38, 1973.

Boris, J.P., 'Physically motivated solution of the Alfven problem,' Naval Res. Lab. Memo. Rept. No. 2167, 1970.

Brackbill, J.U. and D.C. Barnes, 'The effect of non-zero $\nabla \cdot B$ on the numerical solutions of the magnetohydrodynamic equations,' J. Comp. Phys., 35, 426, 1980.

Brecht, S.H., J.G. Lyon, J.A. Fedder, and P.J. Palmadesso, 'Comments on plasma dynamics in the earth's magnetotail,' Comments Plasma Phys. and Controlled Fusion, 6, 59, 1980.

Brecht, S.H., J.G. Lyon, J.A. Fedder, and K. Hain, 'A simulation study of east-west IMF effects on the magnetosphere,' Geophys. Res.

Lett., 8, 397, 1981.
Brecht, S.H., J.G. Lyon, J.A. Fedder, and K. Hain, 'A time dependent three dimensional simulation of the earth's magnetosphere: Reconnection events,' J. Geophys. Res., 87, 6098, 1982.
Brecht, S.H. and D.F. Smith, 'Three-dimensional simulations of the Venusian magnetosphere,' EOS Trans., AGU, 65, 1041, 1984.
Cowley, S.W.H., 'Magnetospheric asymmetries associated with the y-component of the IMF,' Planet. Space Sci., 29, 79, 1981.
Dryer, M., S.T. Wu, G. Gislason, S.M. Han, Z.K. Smith, J.F. Wang, D.F. Smart and M.A. Shea, 'Magnetohydrodynamic modelling of interplanetary disturbances between the sun and earth,' Astrophys. and Space Sci., 105, 187, 1984.
Dungey, J.W., 'Interplanetary magnetic field and the auroral zone,' Phys. Rev. Lett., 6, 47, 1961.
Fedder, J.A., S.H. Brecht, and J.G. Lyon, 'MHD simulation of a comet magnetosphere,' (submitted to Icarus, 1984) NRL Memo. Rpt. No. 5306, 1984a.
Fedder, J.A. and J.G. Lyon, 'Region 1 Birkland currents from a global simulation of the magnetosphere,' (in press JGR) NRL Memo. Rpt. No. 5397, 1984b.
Godunov, S.K., 'Finite difference method for numerical computation of discontinuous solutions of the equations of fluid dynamics,' Matematichekii Sbornik, 47, 271, 1959.
Hain, K., 'The partial donor cell method,' NRL Memo. Rpt. No. 3713, Naval Res. Lab., Wash., D.C., 1978.
Hardy, D.A., H.K. Hills and J.W. Freeman, 'Occurrence of the lobe plasma at lunar distances,' J. Geophys. Res., 84, 72, 1979.
Hones, E.W., Jr., 'Plasma flow in the plasma sheet and its relation to substorms,' Radio Sci., 8, 979, 1973.
Hones, E.W., Jr., D.N. Baker, S.J. Bame, W.C. Feldman, J.T. Gosling, D.J. McComas, R.D. Zwickl, J.A. Slavin, E.J. Smith, and B.T. Tsurutani, 'Structure of the magnetotail at 220 R_E and its response to geomagnetic activity,' Geophys. Res. Lett., 11, 5, 1984.
Lax, P.D., "Weak solution of nonlinear hyperbolic equations and their numerical computation,' Commun. Pure Appl. Math., 17, 159, 1954.
LeBoeuf, J.N., T. Tajima, C.F. Kennel and J.M. Dawson, 'Global simulations of the time-dependent magnetosphere,' Geophys. Res. Lett., 5, 609, 1978.
LeBoeuf, J.N., T. Tajima, C.F. Kennel and J.M. Dawson, 'Global simulations of the three-dimensional magnetosphere,' Geophys. Res. Lett., 8, 257, 1981.
Lui, A.T.Y., 'Observations on plasma sheet dynamics during magnetospheric substorms,' in Dynamics of the Magnetosphere, pp. 563-597, D. Reidel, Hingham, Mass., 1979.
Lyon, J., S.H. Brecht, J.A. Fedder, and P.J. Palmadesso, 'The effect on the earth's magnetotail from shocks in the solar wind,' Geophys. Res. Lett., 7, 712, 1980.
Lyon, J., S.H. Brecht, J.D. Huba, J.A. Fedder, and P.J. Palmadesso, 'Computer simulation of a geomagnetic substrom,' Phys. Rev. Lett., 46, 1038, 1981.

McPherron, R.L., C.T. Russell and M.P. Aubry, 'Phenomenological model for substorms,' J. Geophys. Res., 78, 3131, 1973.

Ogino, T. and R.J. Walker, 'An MHD simulation of the bifurcation of the tail lobes during intervals with northward interplanetary magnetic field,' UCLA Rpt. PPG-782, 1984a.

Ogino, T. 'A three-dimensional MHD simulation of the interaction of the solar wind with the earth's magnetosphere: The generation of field aligned currents,' UCLA Rpt. PPG-796, 1984b.

Ogino, T., R.J. Walker, M. Ashour-Abdalla and J.M. Dawson, 'An MHD simulation of B_y-dependent magnetospheric convection and field aligned currents during northward IMF,' UCLA Rpt. PPG-805, 1984c.

Roach, P.J., Computational Fluid Dynamics, 5th Ed. Hermosa Publishers, P.O. Box 8172, Albuquerque, NM 87108, 1982.

Schmidt, H.U. and R. Wegman, 'MHD-calculations for cometary plasmas,' Comp. Phys. Comm., 19, 309, 1980.

Wolf, R.A., M. Harel, R.W. Spiro, G.-H. Voigt, P.H. Reiff, and C.-K. Chen, 'Computer simulation of inner magnetospheric dynamics for the magnetic storm of July 29, 1977,' J. Geophys. Res., 87, 5949, 1982.

Woodward, P. and P. Colella, 'The numerical simulation of two-dimensional fluid flow with strong shocks,' J. Comp. Phys., 54, 115, 1984.

Wu, C.C., R.J. Walker, and J.M. Dawson, 'A three-dimensional MHD model of the earth's magnetosphere,' Geophys. Res. Lett., 8, 525, 1981.

Wu, S.T., M. Dryer, and S.M. Han, 'Non-planar MHD model for solar flare-generated disturbances in the heliospheric equatorial plane,' Solar Physics, 84, 395, 1983.

Zalezak, S.T., 'Fully multidimensional flux-corrected transport algorithms for fluids,' J. Comp. Phys., 31, 335, 1979.

APPENDIX A

As discussed in the text one can limit the propagation speed of the waves in a calculation by using an Alfven correction. In effect this is like adding a mass, see Boris (1970) for many of the details. How-even, just changing c is not sufficient. One must also change the perpendicular momentum equations as well. In this appendix one can see how this is done.

ALFVEN CORRECTION

Use

$$\bar{v} \times \bar{B} = \frac{1}{c}\frac{\partial \bar{E}}{\partial t} + \frac{4\pi}{c}\bar{J}$$

$$\frac{\partial B}{\partial t} = -c \quad \bar{\nabla} \times \bar{E}$$

$$\bar{E} = -\frac{\bar{v} \times \bar{B}}{c}$$

Apply the equation to the momentum equation

$$\frac{\partial \bar{v}}{\partial t} = \frac{\bar{J} \times \bar{B}}{c\,\rho} - \frac{\bar{\nabla} P}{\rho}$$

and consider only the perpendicular momentum

$$\bar{B} \times \frac{\partial \bar{v}}{\partial t} = \frac{B \times \bar{J} \times B}{c\,\rho} - \frac{\bar{B} \times \overline{\nabla P}}{\rho}$$

The result of considerable vector algebra is

$$\bar{B} \times \frac{\partial \bar{v}}{\partial t} = \frac{1}{4\pi\rho} B \times \frac{\nabla B^2}{2} - \frac{\bar{B} \times \overline{\nabla P}}{\rho} - \frac{B^2}{4\pi\rho c^2} \bar{B} \times \frac{\partial \bar{v}}{\partial t}$$

or

$$\frac{\partial \bar{v}_\perp}{\partial t} = \frac{-\frac{1}{4\pi\rho}\frac{\nabla B^2}{2} - \frac{\nabla P}{\rho}}{1 + v_A^2/c^2}$$

In one sets c to the maximum propagation velocity one wants to allow in the code then Alfven wave speed is limited to this speed. If one has flow exceeding this speed the results are dubious.

THE FUTURE OF SPACE PLASMA SIMULATION

John M. Dawson
University of California, Los Angeles
Department of Physics
405 Hilgard Avenue
Los Angeles, California 90024

ABSTRACT. Space plasma simulation is a subject which is in its infancy, but which is already having an important impact on space science. Its growth is being spurred onward by the continuing increase in capacity (speed and memory) of computers and by advances in the sophistication of numerical models. These advances are making it possible to simulate more realistic situations using more complex models. Already significant three-dimensional MHD calculations of the magnetosphere and its interaction with the solar wind have been carried out. In addition multi-dimensional particle simulations are illuminating many of the microscopic physics processes which go on (instabilities, saturation levels and wave nonlinearity, shock structure, etc.). Notwithstanding these advances, the surface has only been scratched; many challenges and opportunities are provided by simulation both for the space physicist and the model builders (also for computer designers). In MHD models more physics need to be included (Hall effect, gyroviscosity, accurate models of boundaries, how do we put microscopic physics effects into macroscopic codes, etc.). For model builders correct treatments of systems containing a large range of important space and time scales, magnetic field strengths, Alfven wave velocity, etc. present real challenges. What are the best ways to diagnose these complex models and obtain meaningful information? What quantities should be looked at? How should they be displayed? A discussion of the promises, the prospects, and the challenges of the above topics will be given with examples taken from recent work.

I was asked to talk on the future of space plasma simulation. Any projection of the future is always a risky business; our view can at best penetrate a short distance down this twisting and turning path; it is obscured by what may at present seem like dominant features and is highly influenced by our own limited experience; predictions are almost always wrong. Certainly, if a few years ago I had attempted to predict the scope and rapid development of space plasma simulation as shown by the contents of this workshop I would have fallen far short of the mark. Nevertheless I will try to discuss some developments which I

Space Science Reviews **42** (1985) 187–208. 0038–6308/85.15

think will come; I will also discuss some areas which I believe need development and I hope some of you will take them up.

One guide to the future is to look at the developments of the past; I will therefore start by discussing some small amount of the history of plasma simulation. I will stick mainly to things I have been involved with or close to as I am most familiar with them.

Plasma simulation started about 25 years ago when high speed computers began to become generally available. Oscar Buneman might argue with this; he often quotes earlier calculations by Hartree done on sheet models with desk calculators and we heard of the orbit calculations done by Störmer and his 17 graduate students (involving some degree of parallel processing I am sure although I have not checked the degree to which his students overlapped). In any case, I think 25 years ago is quite a reasonable place to start. It started in efforts to model fusion plasmas and also the operation of klystrons and other microwave tubes. I think the first calculations which I would class as simulation would be the early 1D electrostatic and magnetostatic particle models. By simulation I mean that one makes a numerical model of the system of interest and carries out numerical experiments on it so as to gain insight into nonlinear or other complex behavior which the system displays. This was clearly one of the things Von Neumann had in mind when he developed the computer. There were many large calculations done earlier which were used to evaluate complex calculations that arose in theoretical treatments; however these were more as an adjunct to the theory rather than what I would classify as simulation. Of course, there is a fuzzy boundary here but in my view simulation should aim to show up new processes or seek enhanced physical insights. It is most valuable when its results can be condensed into a simple physical model which will quickly yield the most important results. Its purpose is not to generate mountains of computer output, it is to generate understanding, insight and an intuition.

Over the past 25 years the power of computers in both speed and memory has increased on the average by a factor of 2 every 1.5 years. The cost of the largest computers has remained roughly constant at about $\$10^7$ (probably this is what people are willing to pay). At present there is no sign of saturation in this trend.

We can now carry out particle calculations using fully electromagnetic models which are relativistic in 2 1/2 dimensions (x,y,v_x,v_y,v_z) which involve 10^6 to 10^7 particles and 10^5 grid points. Some fully three dimensional calculations are beginning to be done and I believe that soon many more such calculations will be seen. We are already carrying out quite significant three dimensional MHD fluid calculations using 5×10^5 grid points.

Smaller more compact integrated circuits continue to be developed. These keep increasing the power of machines and are reducing their costs. Machines with multiple CPU's and larger memories are becoming possible. Parallel processing is being developed rapidly. The major problems here are the efficient organization of such machines. A large part of this problem rests with the system software developers.

The development of cheaper more versatile chips are also making possible mini computers with the power of super computers of only a few years ago. In fact, it is now possible to buy a PC which has more power than the computer I first used for simulation. Many powerful mini computers is one form of parallelism. If you have a computer to yourself or for just a few people you can run very large problems by letting it grind away for a long time.

For the biggest problems the biggest most powerful super computers are still needed. For three dimensional problems an increase in power of a factor of 10^3 would find immediate application.

Although the advances in computers have probably been the most spectacular aspect of plasma simulation there have also been important advances in computational techniques. To name just a few: (1) efficient large particle codes have been written, (2) efficient large MHD codes have been developed, (3) hybrid particle-fluid codes have been constructed, (4) long time step long space scale techniques have been developed and employed in codes to solve a number of problems. Although the art of model building has come a long way from the early one dimensional models there is still a great deal to do here. In fact, this is an open area since there is no God given method to model a system; there are only approximations some of which are better than others and some apply to one type of problems and not to others. I believe there are problems here to challenge the best minds we have; they provide opportunity for a good deal of creativity and ingenuity.

Let me now turn to some examples of the progress made in space plasma simulation. The first problem I want to consider is the problem of the solar wind magnetospheric interaction. The situation is shown in Fig. 1. The solar wind blows in from the left; a bow shock forms in front of the earth and complex flow and magnetic field patterns form around the earth. To the best of my knowledge, the first dynamic calculation of this interaction was made by Le Boeuf[1] et al. using a particle MHD model. This calculation was made in the late 70's. The flow patterns found are shown in Fig. 2. This figure shows the flow patterns found in three planes; the top one is a plane through the equator, the middle one is a north south plane and the bottom one is a plane perpendicular to the undisturbed solar wind flow and located some distance behind the earth. The most interesting figure is the top one which clearly shows the bow shock; it also shows vortexes forming behind the earth; there is some indication that there is a wake shock formed some distance down stream. I believe these were the first MHD calculations showing the existance of vortex flow.

Recently a much more refined MHD fluid model has been used by Ogino[2] et al., to ivestigate the flow around the earth when the interplanetary magnetic field is northward. They find that the auroral zone becomes bifurcated as occures in Θ aurora due to the vortex flow. Some samples of their results are shown in Figure 3. Here the various quantities, pressure P, parallel vorticity, $\Omega_{\parallel}$, field aligned current, $J_{\parallel}$, and field aligned flow, $V_{\parallel}$, are projected along the field lines onto the polar region of the earth. The complex structure caused by the intricate vortex flow is evident. The actual vortex flow patters are shown in Figure 4. These results show strikingly more detail than

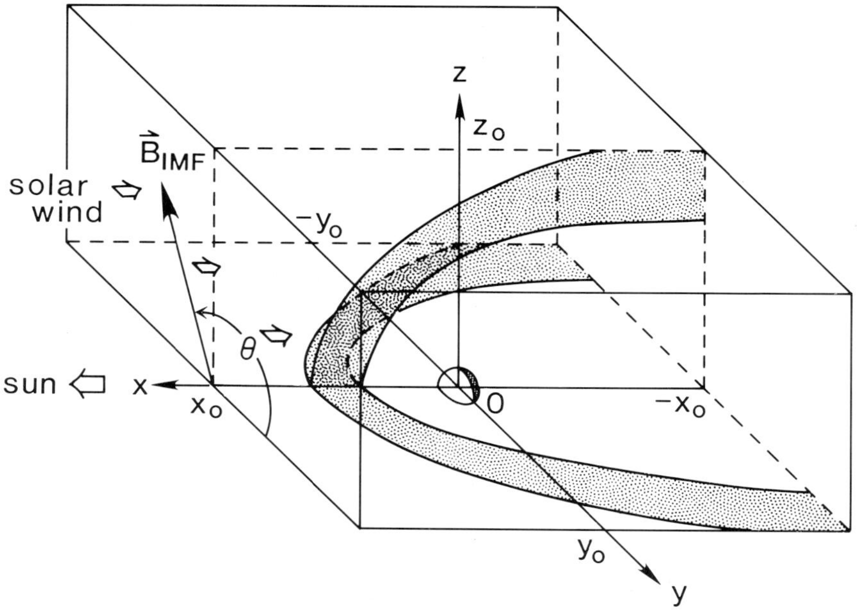

Figure 1. Model of the magnetosphere and solar wind

those done just a few years before. Moreover, they seem to be making real contact with observation in connection with the observations of the Θ aurora. They are also beginning to show the complexity of the flow which is possible.

Although these calculations have come a long way there is still much to be done. One of the critical problems is a proper treatment of the earth and the magnetospheric ionospheric coupling. As shown in Figure 4, the earth occupies one or perhaps a few zones in these calculations. No real detail of the magnetosphere-ionosphere coupling can be put in. But there is a lot of physics going on in this region which at some stage will be important to the observations and certainly plays an important role in what is going on. Some examples of important phenomena are the degree of slip of the B lines through the ionosphere and flux limits on the amount of current that can be drawn (such things will give rise to double layers and reduced coupling of the magnetosphere and ionosphere).

One problem that is encountered in this connection is the high Alfven velocity near the earth. Combined with the desire to use finer zones there, this leads to real problems. The Courant stability condition limits one to very short time steps. If the whole calculation is done at this limiting time step then the calculation is very slow. A method has been developed by C.C. Wu[3] whereby a variable size mesh can be used so that a fine mesh can be used near the earth (and also near the shock) where it is needed. A coarser mesh is used further away from the earth. Such a mesh is shown in Figure 5. Near

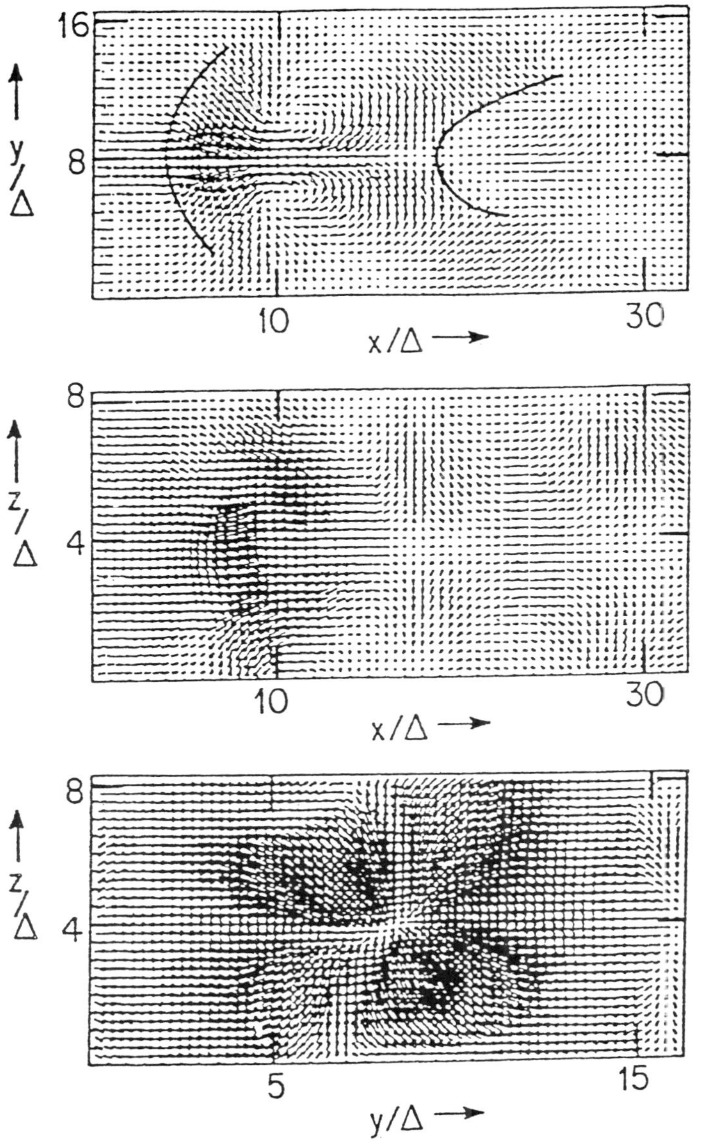

Figure 2. Velocity field plots. The local fluid velocity minus the solar wind velocity is displayed. Top inset: equatorial plane; middle inset: North-South plane; bottom inset: cross-sectional plane at $\chi = 20\Delta$. The bow shock and wake shock are drawn in to guide the eye.

the earth (the shaded region in Fig. 5) a short time step is used. Far from the earth a longer time step is employed. The solutions in the two regions are made to match by matching boundary conditions where

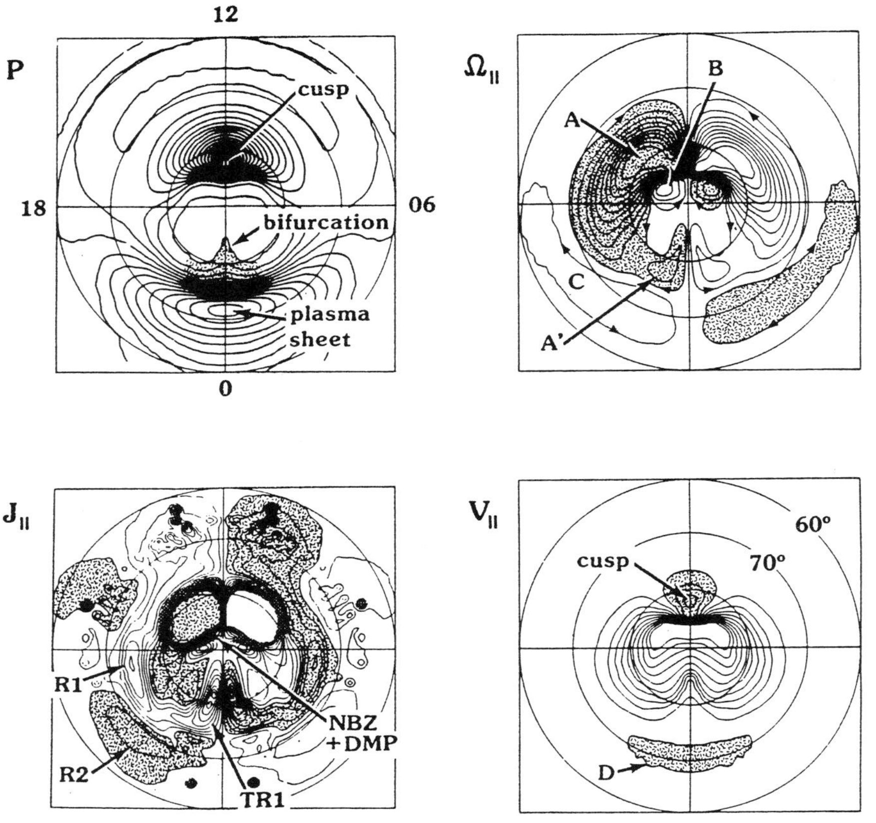

Figure 3. Projection of various quantities along B lines onto the polar region - taken from MHD simulation of the Magnetosphere[2].

they meet. This method speeds up the calculation by a factor of 100 over the use of a single time step over the whole mesh. Even though there has been a lot of progress here we are still a long way from being able to give a completely satisfactory treatment of the ionosphere coupled to the magnetosphere.

Another area which needs to be worked on is that of improving the physics in the fluid models of the magnetosphere and the solar wind. Simple MHD, while being a good start, does not contain all the physics of the real world. Of course one would have to go to the particle level to get at all the microscopic physics and then it is impossible to model the large scale phenomenon; some kind of hybrid mixture of particle models and fluid models might ultimately be satisfactory, but we are a very long way from any such thing. All this notwithstanding, it is possible to put a great deal more physics in fluid models than they now contain. One example of this is the inclusion of the Hall term; this term allows for slippage of the ions relative to the

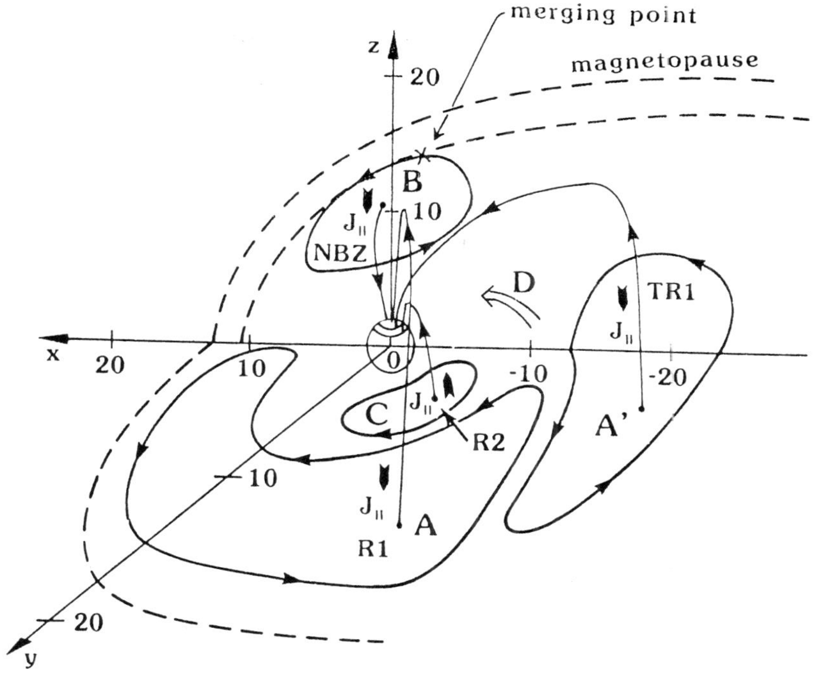

Figure 4. Vortex flow pattern from MHD model of Ogino[2].

electrons, and its introduction splits the Alfven waves into two branches, one goes to ion cyclotron waves while the other goes to the whistler branch.

For simple one fluid MHD we use the equations

MHD

$$\frac{\partial \rho}{\partial t} + \underset{\sim}{\nabla} \cdot \rho \underset{\sim}{v} = 0 \tag{1}$$

$$\rho\left(\frac{\partial \underset{\sim}{v}}{\partial t} + \underset{\sim}{v} \cdot \underset{\sim}{\nabla} \underset{\sim}{v}\right) = \frac{\underset{\sim}{j} \times \underset{\sim}{B}}{c} - \underset{\sim}{\nabla} P \tag{2}$$

$$E + \frac{\underset{\sim}{v} \times \underset{\sim}{B}}{c} = \eta \underset{\sim}{j} \tag{3}$$

$$\underset{\sim}{\nabla} \times \underset{\sim}{E} = -\frac{1}{c} \frac{\partial \underset{\sim}{B}}{\partial t} \tag{4}$$

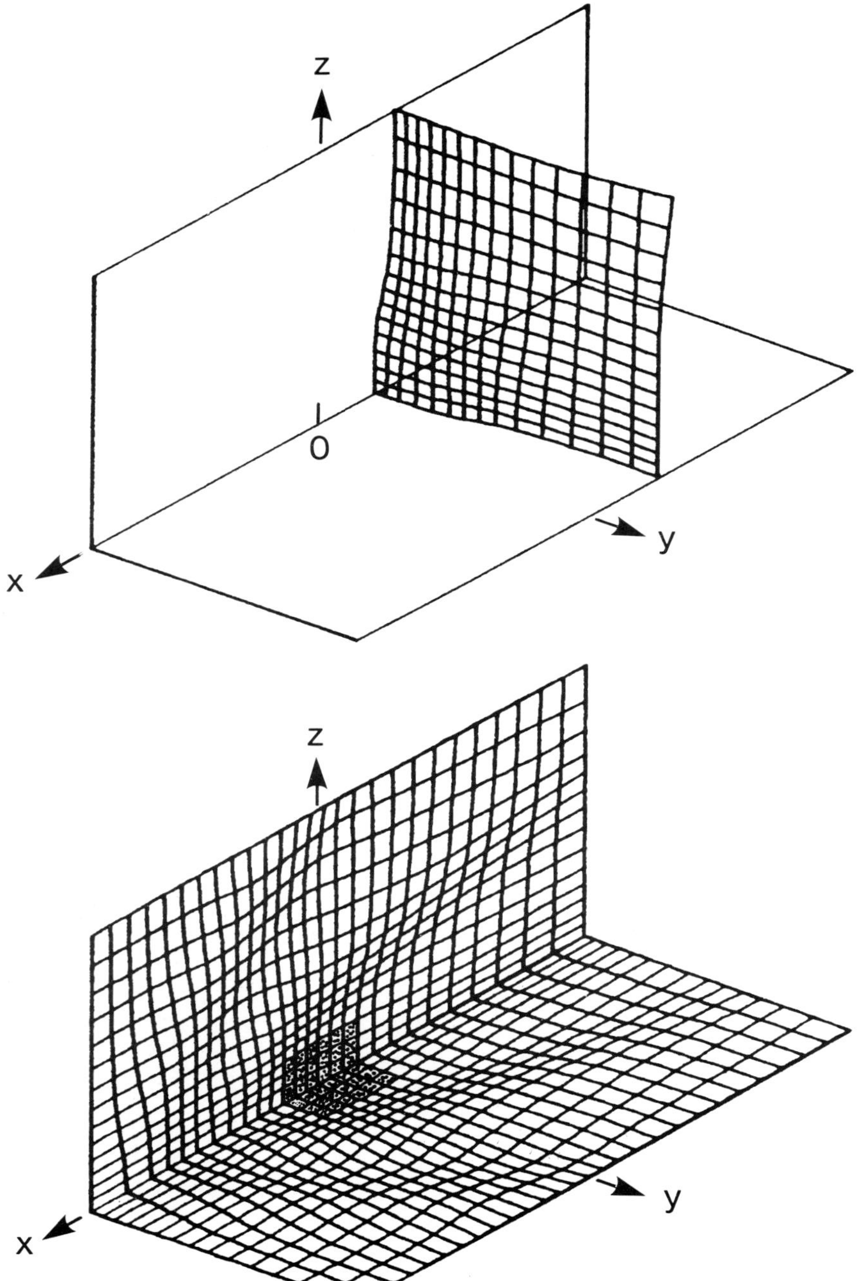

Figure 5. Schematic of variable mesh of Wu3.

$$\underset{\sim}{\nabla} \times \underset{\sim}{B} = \frac{4\pi \underset{\sim}{j}}{c} \tag{5}$$

$$d(P/\rho^{\gamma})/dt = 0 \tag{6}$$

Equations 3, 4, and 5 lead to the equations for B

$$\frac{\partial B}{\partial t} = \underset{\sim}{\nabla} x (\underset{\sim}{v} x \underset{\sim}{B}) + c^2 \eta \nabla^2 B \tag{7}$$

For MHD with the Hall term equation (3) changes to

$$\underset{\sim}{E} + \frac{\underset{\sim}{v} \times \underset{\sim}{B}}{c} - \frac{\underset{\sim}{j} \times \underset{\sim}{B}}{nec} = \eta \underset{\sim}{j} \tag{8}$$

where $\underset{\sim}{v} - \frac{\underset{\sim}{j}}{ne} = \underset{\sim}{v_e}$ = electron velocity.

$$\frac{\partial \underset{\sim}{B}}{\partial t} = \underset{\sim}{\nabla} \times (\underset{\sim}{v_e} \times \underset{\sim}{B}) + c^2 \eta \nabla^2 \underset{\sim}{B} \tag{9}$$

In simple MHD if η is put equal to zero the $\underset{\sim}{B}$ field is frozen in the fluid while with the Hall term it would be frozen in the electrons.

Such a Hall Term Code was constructed a number of years ago by Leboeuf[4] et al. It was a particle MHD code. They tested wave propagation in this system and showed that it correctly gave the splitting of the Alfven branch. Their results for wave propagation parallel to B are shown in Figure 6.

Now I would like to turn to another area which I believe will be of critical importance to the success of modeling, that of diagnostics. Three dimensional models are capable of exhibiting very complex behavior. Indeed, if they are to successfully model space plasma phenomenon their results should be comparable in complexity to those of observation. That this will be the case is already showing up in the magnetospheric results shown in Figure 3. Another example of this, also taken from Reference 2, is the spider diagram shown in Figure 7; this figure shows the magnetic field lines around the earth. The complexity of the situation is quite evident.

The aim of computer modeling is to enhance our understanding, to improve our intuition, to enable us to pick out the crucial processes in a complex phenomenon and condense them into a physical picture or simpler theory which can be more easily handled. To do this we need powerful diagnostic methods which will allow us to look at what is happening in a number of different ways so that we can see in detail what is important. One problem that is faced here is presentation of three-dimensional data in the two-dimensional format of the journals. One possibility is stereopic pictures. This approach has already been introduced by Gekelman and Stenzel[5] for displaying experimental data.

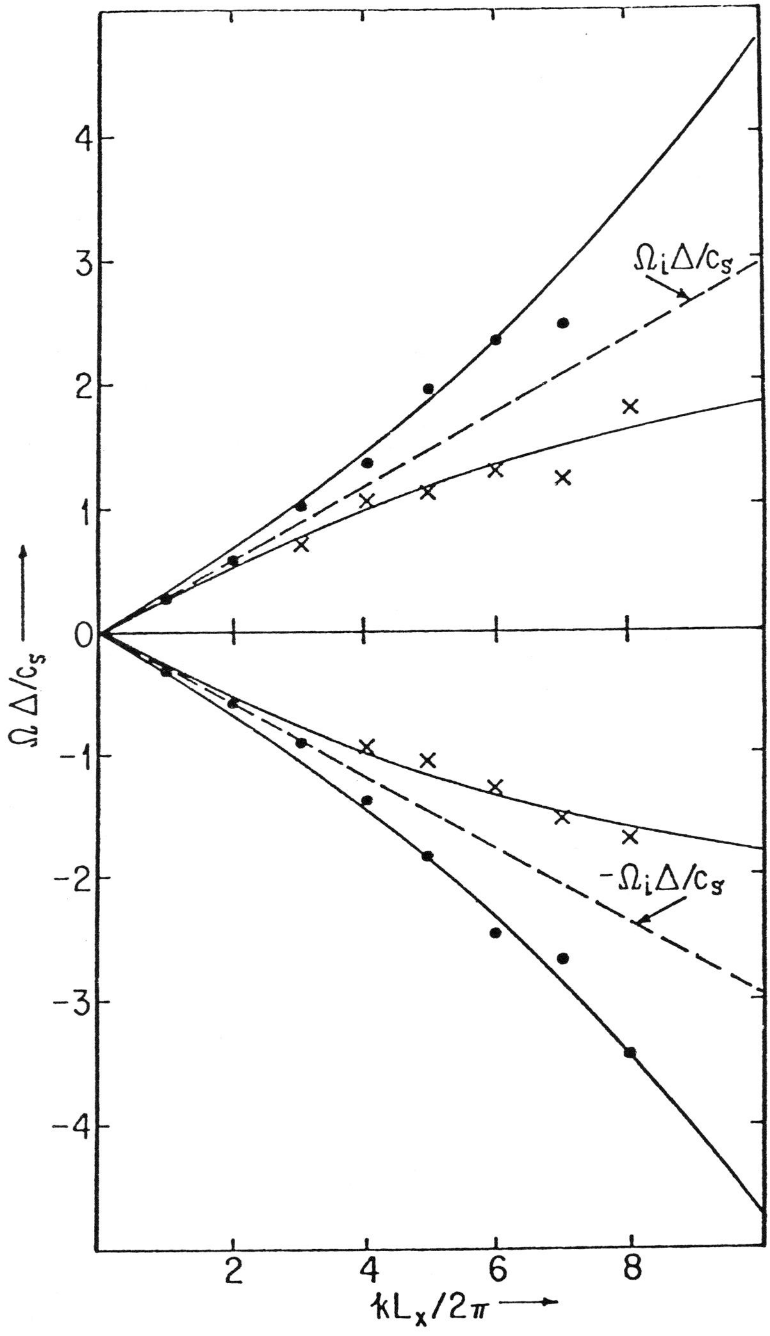

Figure 6. Splitting of Alfven wave into an ion cyclotron wave and a whistler wave by inclusion of the Hall term in the MHD equations.

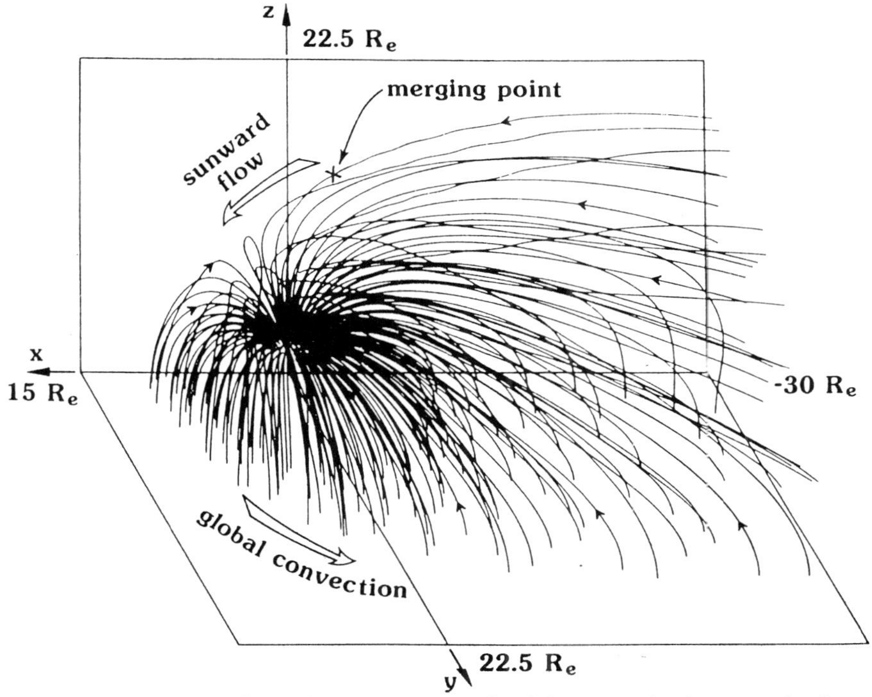

Figure 7. Spider diagram for the magnetic field around the earth from Ogino's MHD calculation[3].

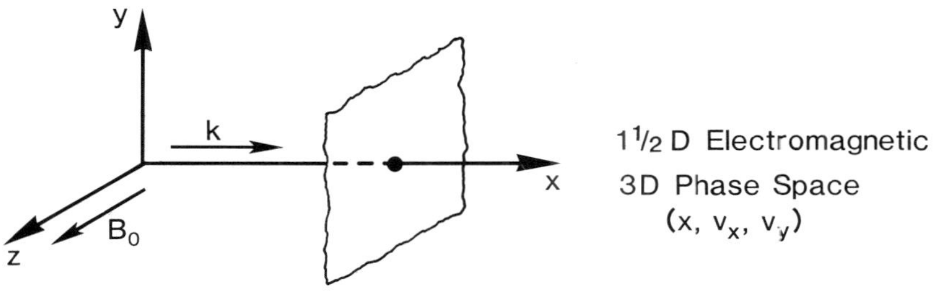

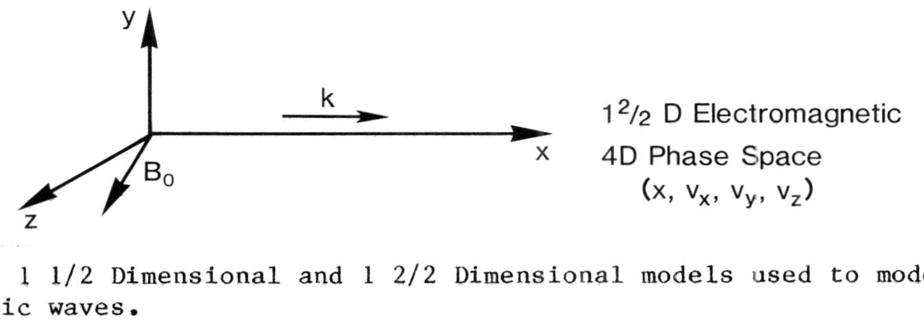

Figure 8. 1 1/2 Dimensional and 1 2/2 Dimensional models used to model magnetosonic waves.

While helping to some extent, this approach does not really solve the problem; the actual data is much too rich to be conveyed in a three-dimensional picture from one viewing angle. One possible solution might be for authors to provide computer disks of selected samples of their results at a nominal cost to interested readers. The figures could then be displayed on local computers and rotated about so that the reader could see as much detail as he desired. He could even analyze the data in ways that the author might not have thought of. If this is to happen, standardization of data format would have to be agreed upon.

In connection with the above, it should be mentioned that very powerful graphic displays are being developed for many other applications (industrial computer graphics, computer generated movies, etc.), and I believe that the plasma simulation community should make use of these developments.

One step we have taken in this direction at UCLA is given by the display of some data we needed to analyze magnetosonic waves propagating perpendicular to a magnetic field; the model used is illustrated in Fig. 8. It is what we call a 1-1/2 dimensional model. It is made up of charged sheets perpendicular to x and allowed to have both an x and y velocity; however, all quantities vary only in x. x is a full dimension but in y, since only V_y is considered, we have what we call a half dimension. The y velocity takes into account the transverse motion associated with the B field which is in the Z direction. This model has a three-dimensional phase space (X, V_x, V_y). A generalization of the model is shown at the bottom of Fig. 8. Here B is allowed to lie in the x,z plane and the charged sheets now acquire transverse z velocities as well as y velocities; all quantities still vary only in x. We call this a 1-2/2 dimensional model with x being a full dimension and y and z each being half a dimension through V_y and V_z; this model has a four-dimensional phase space.

Figure 9 shows a display of the position of a sample of the particles in phase space produced by an intense magnetosonic wave. In the middle of a three-dimensional plot (each point being the position of a particle) while around it are shown projection on the p_x, x plane ($p = mv$), the py,x plane and the p_x, p_z plane. This display was made by Bertrand Lembege using some routines developed by Walter Gekelman.

In the end, the proof of the value of computer modeling will come from the comparison of its results with experiments or observations. Lembege also looked at the case of magnetosonic waves propagating at an angle to B using the lower model in Fig. 8. In some discussions with R. Bingham of Rutherford Appleton Laboratory, I was told that they had a theory for the acceleration of electrons in the magnetosphere and that they had measured such waves and the electron distribution function that occurred there. I gave him some of Lembege's data and he compared it with their observations; the result is shown in Fig. 10.

Another thing that computer modeling can be used for is to find the normal modes of a plasma. We have done this with some of our bounded plasma models (see Decyk[6]) and it should be possible to do similar things with the magnetosphere. The results of such studies could be compared with observations.

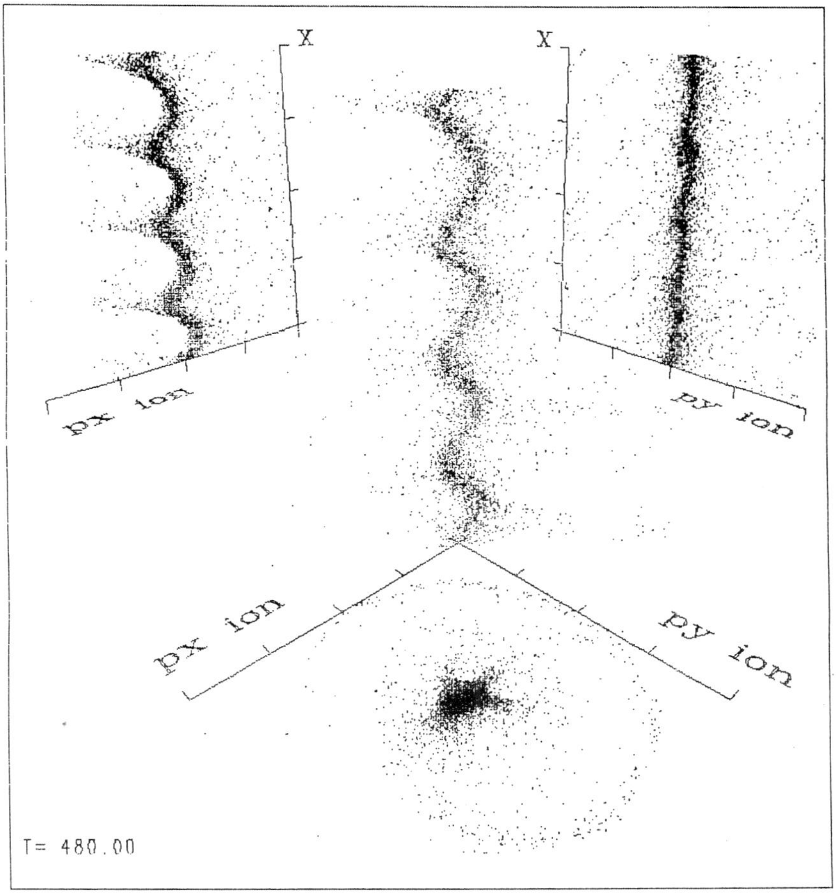

Figure 9. Phase space plots for intense magnetosonic waves propagating perpendicular to B.

The method can be applied to a system in which the modes are thermally excited (excited by the random motion of the particles) or systems which receive some external disturbance. One records some quantity of interest as a function of position and time, some examples are $\phi(\underline{\Omega},t)$, $\underline{B}(\underline{\Omega},t)$, $\underline{E}(\underline{\Omega},t)$, $n(\underline{\Omega},t)$ . . . One time spectral analyses the quantity of interest; let us use ϕ for an example so that we would compute

$$\phi(\underline{r},\omega) \text{ and } |\phi(\underline{r},\omega)|^2 = I(\underline{r},\omega) \tag{10}$$

In general, $I(r,\omega)$ will have the form shown in Fig. 11. There will be sharp spikes in the spectrum at say ω_1, ω_2, ω_3, ω_4, and in general a smeared out broad continuous spectrum. To find the eigenfunction

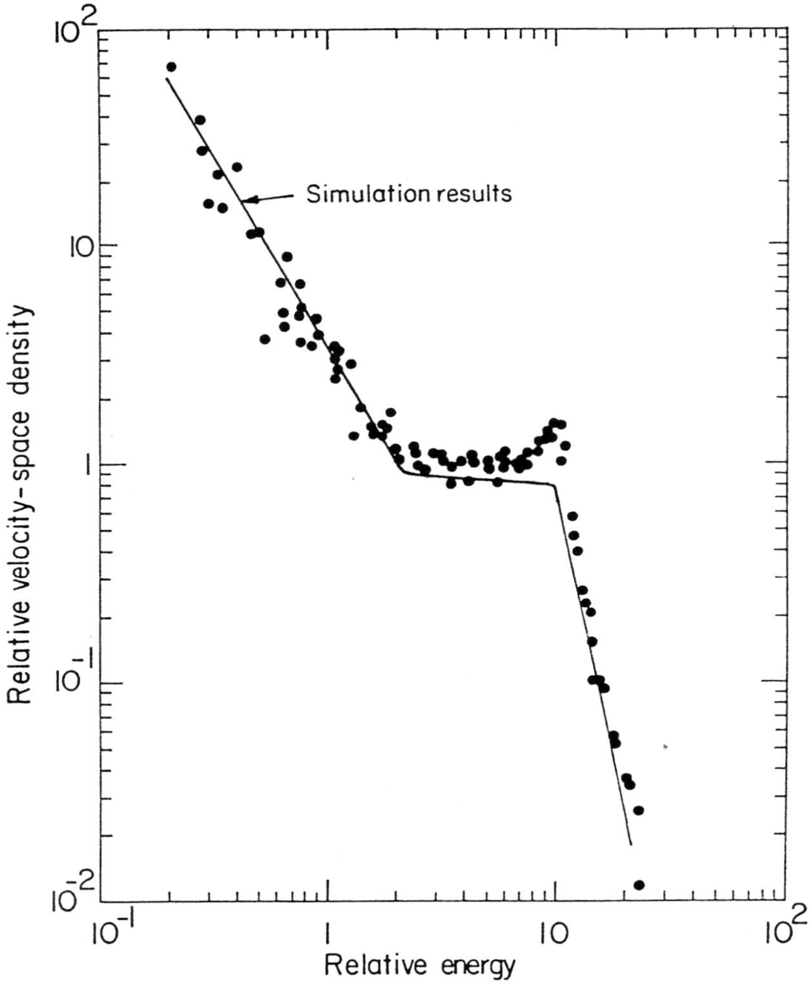

Figure 10. Auroral arc electron distribution function. The dots represent experimental results (R. Bingham, private communication) and solid line is from particle simulation from magnetosonic waves at an angle to B.

associated with say ω_1, one splits out the part of ϕ which has an ω_1 dependence; that is, we write

$$\phi(\underline{r},t) = \phi_s((\underline{r},\omega) \sin \omega_1 t + \phi_c(\underline{r},\omega) \cos \omega_1 t + \tilde{\phi}(\underline{r},t) \tag{11}$$

One can solve for ϕ_s and ϕ_c by computing the integrals

$$\frac{1}{T} \int_0^T \phi(\underline{r},t) \sin \omega_1 t dt = \frac{\phi s(\underline{r},\omega 1)}{z} = I_s \tag{12}$$

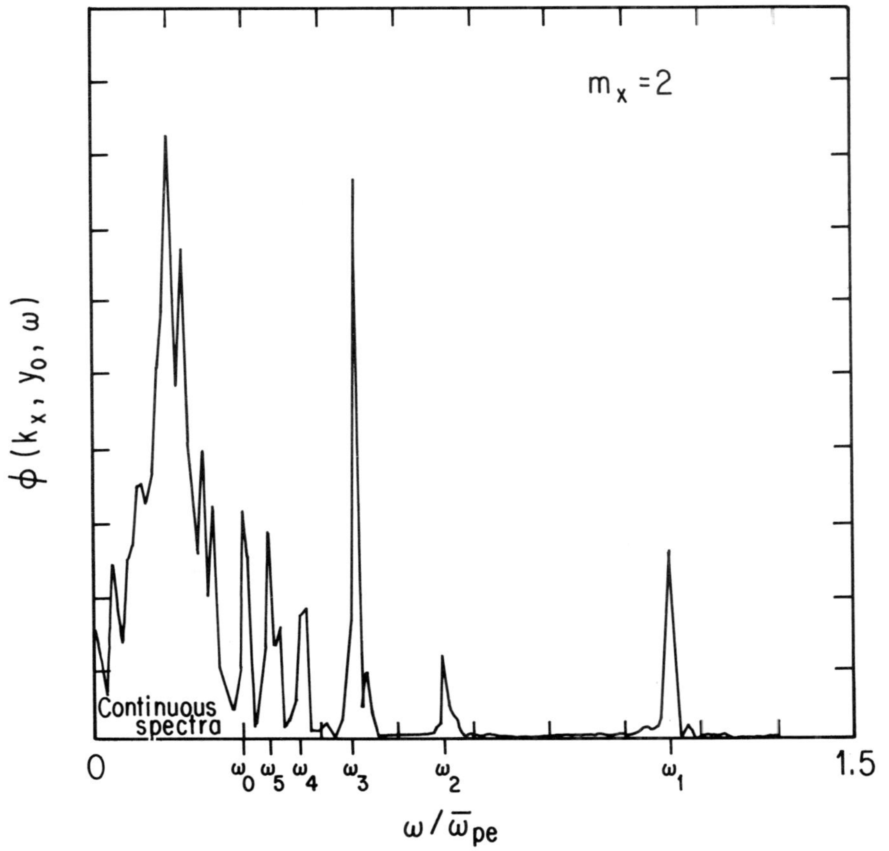

Figure 11. Discrete and continuous spectra for plasma oscillations of a non-uniform plasma.

and

$$\frac{1}{T} \int_0^T \phi(\underline{r},t)\cos(\omega_1 t) = \frac{1}{2} \phi_c \ (r,\omega_1) = S_c \tag{13}$$

If we write

$$\phi(\underline{r},\omega_1) = \phi_{\omega 1}(r)\sin(\omega_1 t + \theta_1(r)) \tag{14}$$

$$\theta_1(r) = \tan^{-1}(S_c/S_s) \tag{15}$$

The same technique can be employed to find the nature of the waves in the continuous spectrum. There one picks a frequency in the continuous range and proceeds as above. One then usually finds some localized regions of the plasma oscillating at this frequency which

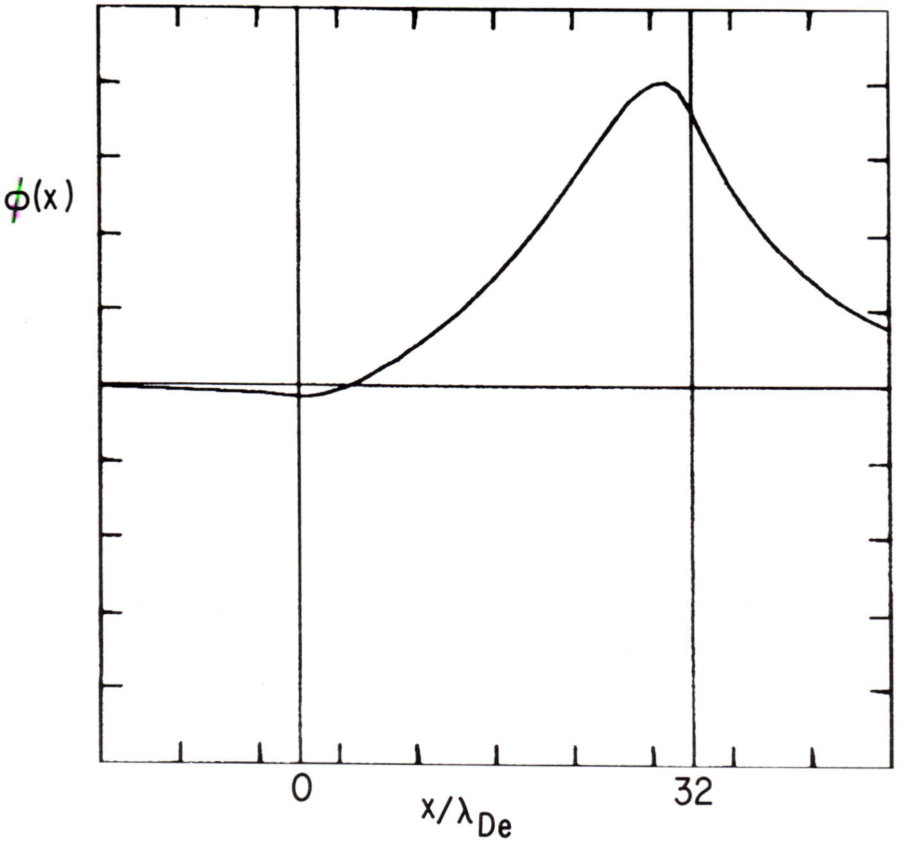

Figure 12. Normal mode Eigenfunction for right hand electron surface wave ($\vec{B}_o = B_z\hat{z}$, $\Omega_e/\omega_{pe} = 1/2$, $\omega/\omega_{pe} = 1.084$, $k_y\lambda_{De} = 0.098$, $m_e/M_i = 1/9$, $L_x/\lambda_{De} = 32$).

generally characterize the local properties of the plasma (the local plasma freqency for example).

Viktor Decyk has applied this technique to look at the normal modes of a uniform density magnitized plasma slab and also a slab of non-uniform density; these are shown in Fig. 11.

Some results for the uniform density slab case are shown in Figs. 12, 13, and 14. Figure 12 shows the waves function vs. x for a mode with a given k_y (since the plasma is uniform in the y direction, we can Fourier analyze in this direction). The mode shown is a surface mode. Figure 13 shows the potential contour for ϕ for this case. Figure 14 shows a comparison of the frequencies of various modes vs. k_y and their comparison with a fluid theory. The different branches are symmetric and antisymmetric body and surface waves. The agreement is reasonably good except for an additional branch which shows up in the simulation.

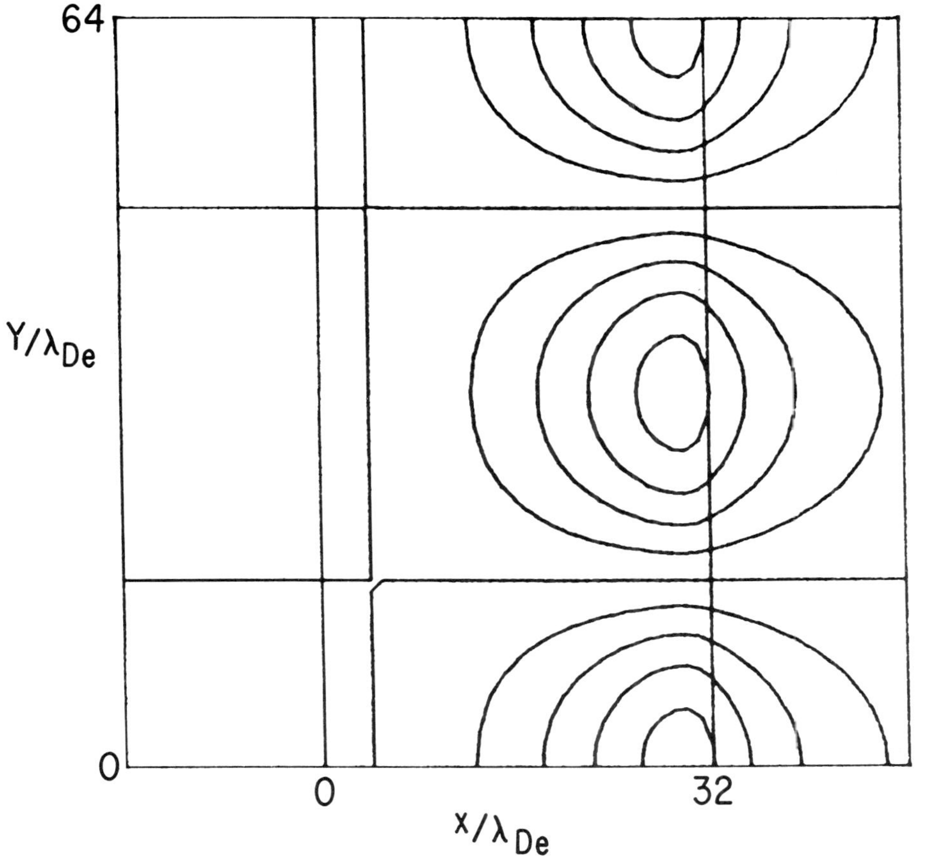

Figure 13. Equipotential plot of normal mode Eigenfunction for right hand electron surface wave ($\vec{B}_o = B_z\hat{z}$, $\Omega_e/\omega_{pe} = 1/2$, $\omega/\omega_{pe} = 1.084$, $k_y\lambda_{De} = 0.098$, $m_e/M_i = 1/9$, $L_x/\lambda_{De} = 32$).

The simulation shows that this is a surface mode; we believe it is a kind of surface Bernstein mode.

Another example is shown in Fig. 15 for a slab of nonuniform density in the x direction. Here again the system was Fourier analyzed in y and the wave function vs. x is shown. The upper curve shows the results from the simulation which the lower curve shows the wave form predicted from WKB theory. The agreement is quite good. This shows the potential of the method for finding normal modes of complex systems where the analysis is difficult.

Figure 16 sketches the form of the wave function for a frequency in the continuous spectrum for the plasma slab of non-uniform density sketched at the top. In this case, there is a very localized region of the plasma in which there is a strong oscillating E field. The upper

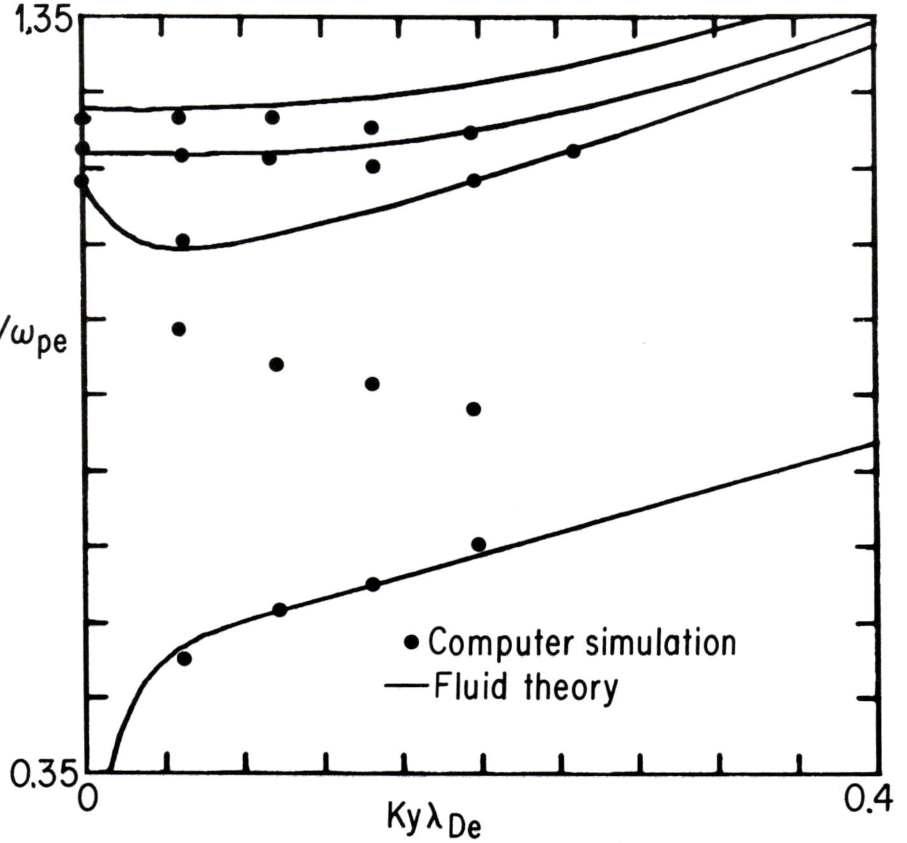

Figure 14. Dispersion relation for electron waves ($\vec{B}_o = B_z\hat{z}$, $\Omega_e/\omega_{pe} = 1/2$, $m_e/M_i = 0$, $L_x/\lambda_{De} = 32$, 32 × 128 grid, 28 × 128 particles, $a_x = a_y = 1$, $\lambda_{De} = 1$.

hybrid frequency of the plasma matches the choosen frequency. In this case, the potential jumps across the region of strong oscillation.

It should be possible to apply this technique to the magnetosphere as illustrated in Fig. 17. One would introduce a fluctuating solar wind that would excite the modes of the magnetosphere. Their wave functions and frequencies could then be found and the results could be compared with observations (either from the earth or from satelites). One would thus get some tests of the model and if there is agreement, one should be able to obtain information about the magnetosphere. The magnetosphere should show both discrete global modes, as well as a continuous spectra; the latter would correspond to Alfven waves or other types of waves propagating along relatively localized flux tubes.

I have given you a thumbnail sketch of some things which I think will happen in space plasma modeling and some which I think are important and I hope will happen. I know there are many developments

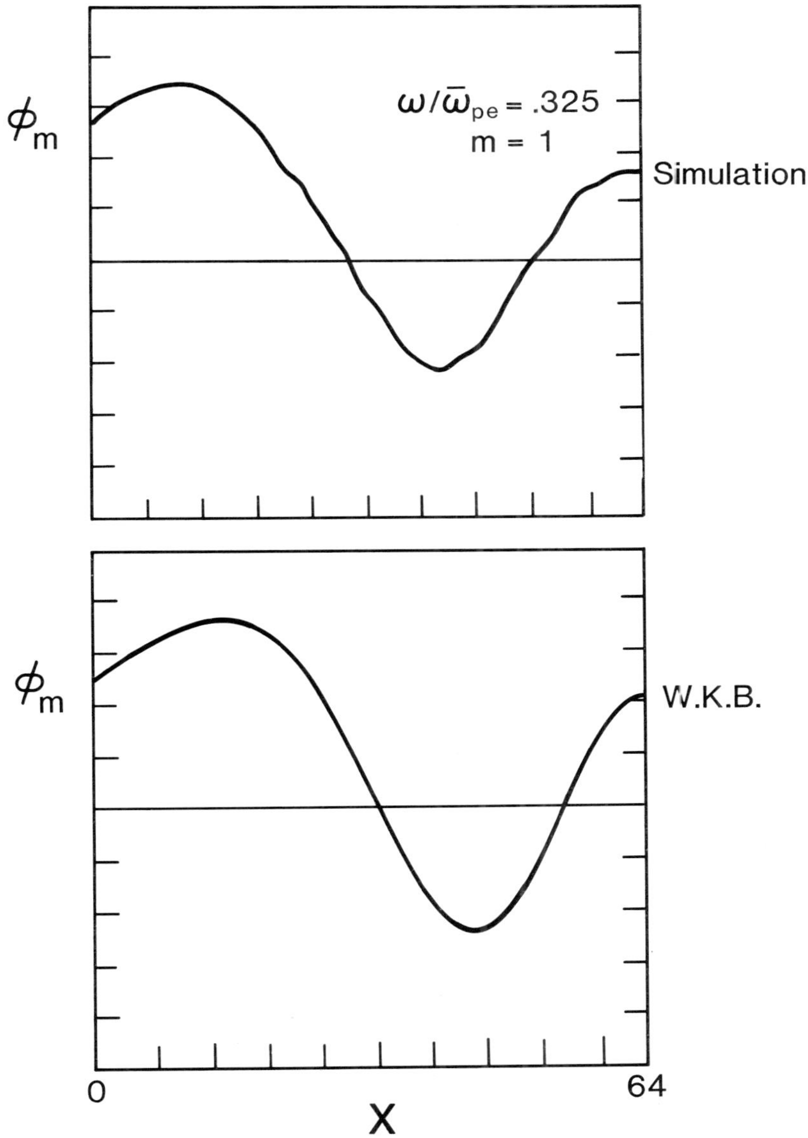

Figure 15. Comparison of wave functions for a nonuniform plasma obtained from simulation and W.K.B. Theory.

which will take place which I have not covered or am not aware of. However, I think you can see this is a subject in its infancy and there will be many exciting developments to come.

I might close by briefly summarizing the major points I have made.

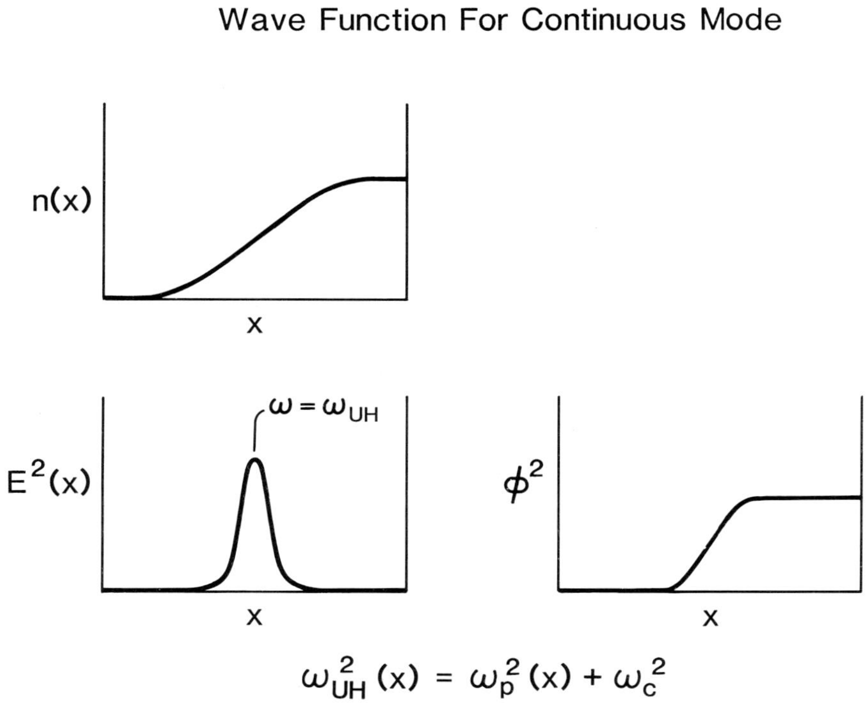

Figure 16. Nonuniform plasma density and the wave function form for the continuous spectral region.

SUMMARY

1. Computers will continue to grow in speed and memory at roughly the same rate (a factor of 2 every ~1.5 years). Larger, more realistic models will be possible.

2. Computers will become much more available (both supercomputers and powerful minicomputers). This will open modeling to a much larger community.

3. The power of computational techniques and the physics sophistication of the models will continue to grow for as long as one can see. Techniques will change with computer architecture. The simulation community should probably think about influencing computer architecture.

4. Diagnostics and display of results are relatively neglected areas. To obtain meaningful and useful results, particularly from 3-D codes, depends critically on developments here.

5. Comparison of simulation with observation and experiment are the acid test. Success here is what will determine the future of models and methods in an even larger arena, that of astrophysical plasma simulation.

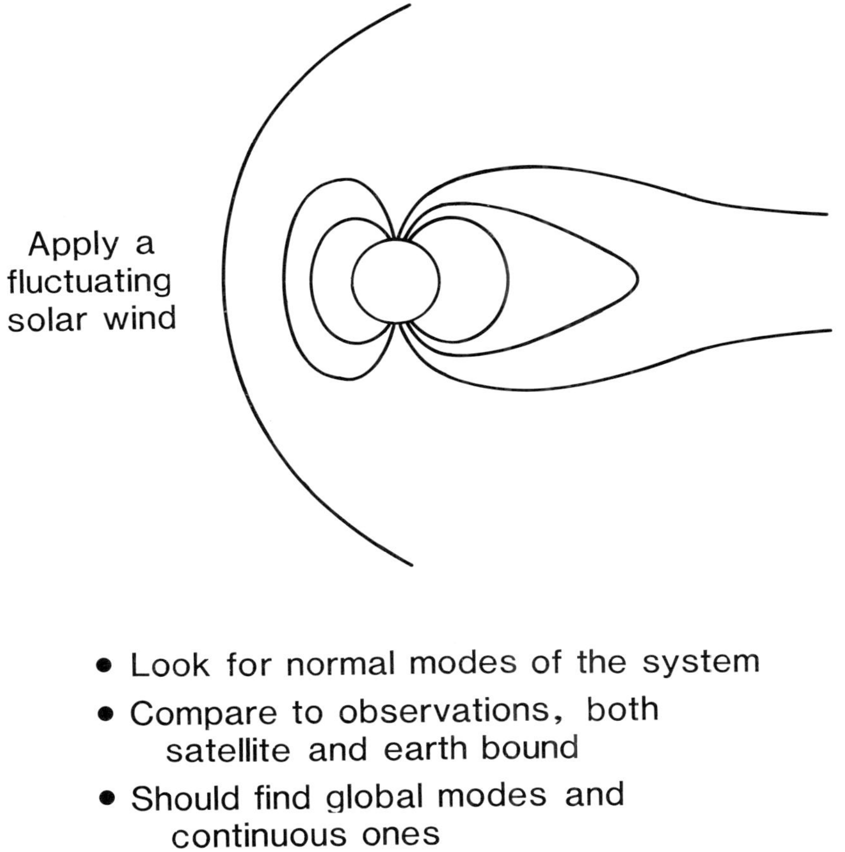

Figure 17. Magnetosphere with potential for global modes and continuous modes.

ACKNOWLEDGEMENTS

The author wishes to acknowledge help received from Drs. M. Ashour-Abdalla, W. Gekelman, T. Ogino, R. Stenzel and C.C. Wu. Work was supported by NSF Contract No. ATM 82-18746.

REFERENCES

1. Leboeuf, J.N., Tajima, T., Kennel, C.F., and Dawson, J.M., Geophysical Research Letters 8, 257 (1981).

2. Ogino, T., Walker, R.J., Ashour-Abdalla, M., and Dawson, J.M., 'A MHD Simulation of the Bidependent Magnetospheric Convections and Field Aligned Currents During Northward IMS', UCLA Preprint Report No. PPG-805 (1984).

3. Wu, C.C., 'MHD Modeling of the Earth's Magnetosphere', in Computer Simulations of Space Plasmas, H. Matsumoto and T. Sato (Eds.), D. Reidel Publishing Co. (1985).

4. Tajima, T., Leboeuf, J.N., and Dawson, J.M., 'A Magnetohydrodynamic Particle Code with Force Free Electrons for Fluid Simulations', J. of Computational Physics 38, 2 (1980).

5. Gekelman, W. and Stenzel, R.L., 'Magnetic Field Line Reconnection Experiments: 6. Magnetic Turbulence', J. of Geophysical Research 89, A5 (1984).

6. Decyk, V.K., Ph.D. Thesis, Physics, UCLA (1977). Decyk, V.K. and Dawson, J.M., J. of Computational Physics 30, 407 (1979).

PART 2 - OBSERVATIONS

PLASMAS IN THE EARTH'S MAGNETOTAIL

L. A. Frank
Department of Physics and Astronomy
The University of Iowa
Iowa City, Iowa 52242
U.S.A.

ABSTRACT. An overview of the general characteristics of plasmas within the Earth's magnetotail and its environs is presented. Present knowledge of the plasmas within these regions as gained via in situ measurements provides the general theme, although observations of magnetic fields, energetic particles and plasma waves are included in the discussion. Primary plasma regimes in the magnetotail are the plasma sheet, its boundary layer, the magnetotail lobes, the boundary layer at the magnetopause and the distant magnetotail. Although great progress in our understanding of these regions is evident in the literature of the past several years, many of their features remain as exciting enigmas to be resolved by future observational and theoretical investigation.

1. INTRODUCTION

Since the solar wind and magnetosheath provide the boundary conditions along the outer skin, or magnetopause, of the magnetotail our first two sections summarize the gross properties of these two regions. The boundary layer at the surface of the magnetotail, or magnetopause boundary layer, is comprised primarily of antisunward flowing plasmas of solar wind origins, that have entered the magnetotail. The nature of this boundary layer at polar and equatorial latitudes is given in Section 4. The largest plasma reservoir in the magnetotail is the plasma sheet, an extensive body of plasma centered on the position of the neutral sheet where the magnetic field exhibits a dramatic reversal of direction. Discussion of the primary features of this plasma sheet and its boundary layer is provided in Section 5. High-speed ion flows, field-aligned currents and broadband electrostatic noise are found in the boundary layer of the relatively docile plasma sheet. This plasma sheet boundary layer is an important topological and energy transport region of the magnetotail. A general paucity of plasma is one of primary observational features of the magnetotail lobes. However recent sightings of ionospheric ion beams and the increasing awareness of polar cap auroras are responsible for stimulating renewed interest. The available information concerning the plasmas in these regions where

Space Science Reviews **42** (1985) 211–240. 0038–6308/85.15

large magnetic energy is stored is offered in Section 6. At great distances in the magnetotail, i.e., beyond the lunar orbit, direct measurements are only available for the occasional passage of an interplanetary spacecraft through these regions until the pioneering survey with ISEE 3. Major results of this important investigation of the distant magnetotail and their relevance to the extensive findings inside the lunar orbit are given in the last section.

Clearly the scope of the present overview must be limited to major features of the magnetotail and its environs. The intent is to provide our colleagues in related research areas with a tutorial monologue with sufficient references to allow pursuit in detail of any of the numerous topics cited in the presentation.

2. SOLAR WIND

The solar wind is the primary source of energy for the sustenance of the magnetotail and one of the two significant sources of plasmas. The other primal source of plasmas for the magnetotail is the Earth's ionosphere. In the spirit of our presentation typical solar wind ion densities and temperatures are 1 to 10 cm^{-3} and 2×10^4 to 2×10^5 °K, respectively (Wolfe, 1972; Feldman et al., 1978). Electron temperatures are generally in the range of 6×10^4 to 3×10^5 °K (Feldman et al., 1975; Montgomery, 1972). Flow speeds at Earth's orbit are typically 200 to 600 km/sec. The solar wind exhibits large-scale plasma features such as high-speed streams associated with coronal holes and low-speed streams near interplanetary sector boundaries (Hundhausen, 1972; Burlaga, 1979). The magnetotail is naturally expected to respond to fluctuations in the solar wind (cf. Burlaga, 1975). For example, the hydrostatic pressure affects the overall geometry of the magnetotail and thus also the associated magnetic energy (cf. Coroniti and Kennel, 1972). Minor ions in the solar wind are important in determining the relative importance of the solar wind and ionospheric sources of ions throughout the magnetotail. The dominant minor ion is He^{++} (Bame et al., 1968; Bame, 1983). Other minor ions, e.g., $^3He^{++}$ and O^{6+}, are expected to increase in utility as tracers for plasma entry and motion in the magnetotail as the measurement capability increases. The magnitude of the interplanetary magnetic field is in the range of 2 to 15 nT and grossly organized into sectors of sunward and antisunward fields in an Archimedean spiral pattern as predicted by Parker (Klein and Burlaga, 1980; King, 1979; Slavin and Smith, 1983). The high degree of responsiveness of the plasma convection and magnetic topology of the magnetotail to the direction of the interplanetary fields is well established (cf. Russell and McPherron, 1973).

3. MAGNETOSHEATH

Ion densities and temperatures within the magnetosheath are in the ranges of ~ 2 to 50 cm^{-3} and $5{\times}10^5$ to $5{\times}10^6$ °K, respectively. The ion bulk flow speed is ~ 200 to 500 km/sec (Bame, 1968; Montgomery et al., 1970; Formisano et al., 1983). Electron temperatures are in the range of ~ 10^5 to 10^6 °K (references above; Feldman et al., 1983). Although such measurements are available, there are no published surveys of ions within the magnetosheath along the flanks of the magnetotail. Often, in order to obtain representative plasma parameters, it is necessary to refer to such hydromagnetic treatments as that of Spreiter et al. (1966). A hot plasma component with ion energies in the range of 10 keV is also reported for the magnetosheath (Asbridge et al., 1978). The source is located either at the bow shock or within the magnetosphere, i.e., leakage through the magnetopause. Sanders (1981) finds an ion component with temperatures similar to those within the solar wind at distances ~ 60 R_E downstream in the magnetosheath. The existence of such an ion distribution may indicate the direct penetration of solar wind plasmas into the downstream magnetosheath at large distances from the subsolar bow shock. A property of ion velocity distributions within the magnetosheath that is similar to those found in many regions of the magnetotail is a high-energy (speed) tail (Formisano, 1973; Peterson et al., 1979; Sanders, 1981). Such ion velocity distributions depart from a Maxwellian form within several characteristic thermal energies. When ion temperature anisotropies are observed, $T_\perp > T_\parallel$. Magnetic fields in the magnetosheath are more extensively investigated than the plasmas (Fairfield, 1976). Along the flanks of the magnetotail the magnitudes of the magnetic fields are similar to instantaneous values for the interplanetary field, and the direction of these magnetosheath magnetic fields more or less exhibits the same orientation as that of the interplanetary field. Departures from such positive correlations of interplanetary and magnetosheath magnetic fields are observed, in particular within the subsolar magnetosheath (Crooker et al., 1984).

4. PLASMA BOUNDARY LAYER INSIDE THE MAGNETOPAUSE

The magnetopause is usually identified as the position at which there is a clear transition from magnetic fields associated with the magnetosheath to those expected for a distorted geomagnetic cavity (Paschmann et al., 1978, 1979; Aubry et al., 1971). The thickness of the magnetopause is generally of the order of 5 to 10 ion gyroradii at 1 keV, or ~ 400 to 800 km (Berchem and Russell, 1982). Near the subsolar region the identification of the magnetopause is typically unambiguous in the magnetic field signature. Along the flanks of the magnetotail the magnetopause location is not as clearly resolved with the magnetic field measurements alone, and plasma observations must be utilized to identify this boundary (cf. Eastman et al., 1985). Near the subsolar magnetopause evidence of large-scale quasi-stationary reconnection is occasionally found (Paschmann et al., 1979). The importance of this mechanism for significant plasma entry into the magnetosphere is questioned by Eastman and Frank (1982) on the basis of the infrequent observations of accelerated ion bulk flows and of inconsistent evidence for open geomagnetic fields in these regions. The recent detection of 'flux transfer events' may in fact resolve these ambiguities with the introduction of the concept of sporadic reconnection of solar wind and geomagnetic fields on a significantly smaller spatial scale over the surface of the subsolar magnetopause (Russell and Elphic, 1979; Russell, 1984). The relatively good efficiency by which magnetosheath plasmas penetrate the magnetopause is demonstrated by the existence of a thick zone of magnetosheath-like plasmas that is present behind the magnetopause locations not only at subsolar positions but along the flanks of the magnetotail at polar and equatorial latitudes. This plasma region is known as the magnetopause boundary layer. The thickness of the magnetopause boundary layer of generally antisunward flowing plasmas is ~ 2,000 to 10,000 km. The mechanism for entry of these plasmas into this relatively thick region behind the magnetopause is not identified, e.g., small-scale reconnection or diffusive transport (Eastman et al., 1985). Quantitative evaluations of the plasma transport in the magnetopause boundary layer are not currently available for these various mechanisms.

The magnetopause boundary layer at polar latitudes includes the entry layer, the distant polar cusp and the plasma mantle as shown in Figure 1. The entry layer is a region of relative magnetic and plasma turbulence that is identified as the location of plasma entry into the polar magnetosphere and the source of the downstream magnetopause boundary layer (Paschmann et al., 1976). The polar cusp is associated with magnetic field lines that provide direct entry of magnetosheath

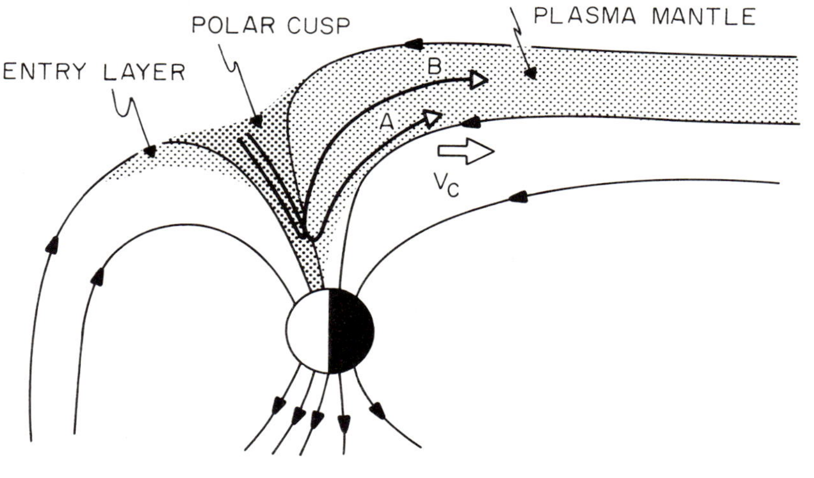

Figure 1. Plasma regions in the polar magnetosphere (noon-midnight meridional plane).

plasmas into the polar ionosphere (Frank, 1971), and the plasma region inside the downstream polar magnetopause with plasmas of direct magnetosheath origins is the plasma mantle (Rosenbauer et al., 1975). The temperatures of ions in these three regions are $\sim 5\times10^5$ to 8×10^6 °K and electron temperatures are $\sim 10^5$ to 10^6 °K. The range of ion temperatures appears to be equal or somewhat higher than that observed within the magnetosheath, and electron temperatures are similar. Densities are in the range of ~ 0.5 to 50 cm^{-3}. Ion bulk flow speeds in the polar magnetospheric boundary layer are typically 100 to 300 km/sec. The ion bulk velocities are approximately field-aligned in the polar cusp and plasma mantle. In the polar cusp field-aligned electron current sheets are observed. At low altitudes above the ionosphere in the polar cusp the densities of the magnetosheath plasma are generally lesser (Heikkila and Winningham, 1971). Within the plasma mantle, density and ion temperature gradients are observed such that the density and ion temperatures decrease with increasing distance from the magnetopause. At least part of these gradients can be accounted for by the velocity filter effect due to antisunward convection and penetration of plasma into the polar cusp (see Figure 1). Slower ion A will arrive at a greater distance from the magnetopause than faster ion B at a given distance downstream from the Earth. The thickness of the plasma mantle is $\sim 3{,}000$ to 20,000 km. Entry of magnetosheath plasmas along the magnetopause that bounds the plasma mantle is neither demonstrated or excluded by in situ measurements. The major source of these plasmas is currently thought to be the magnetosheath via the entry layer.

Plasmas of direct magnetosheath origins are found inside the magnetopause at equatorial latitudes at all local times, i.e., in the subsolar regions and extending along the flanks of the magnetotail (Eastman et al., 1976, 1985; Paschmann et al., 1978; Sckopke et al., 1981). This region is approximately 2,000 to 10,000 km in thickness near the subsolar point and appears to thicken to 5,000 to 20,000 km at distances ~ 20 to 40 R_E (R_E, Earth radii) downstream from the Earth. Ion densities are in the range of ~ 1 to 50 cm^{-3} with values similar to those within the magnetosheath near the magnetopause and decreasing by an order of magnitude at the earthward edge of the magnetopause boundary layer. The ion temperatures in the boundary layer are similar or greater than magnetosheath ion temperatures in the vicinity of the magnetopause and increase with increasing distance from the magnetopause. The ion temperatures in the magnetopause boundary layer at low latitudes range from $\sim 2\times10^6$ °K to 2×10^7 °K. The ion composition is dominantly H^+ and He^{++}, thus further establishing the direct magnetosheath origins of these plasmas (Peterson et al., 1982). Electron temperatures are usually $\sim 10^5$ to 2×10^6 °K. These magnetosheath-like plasmas coexist with hotter ion and electron plasmas of the outer magnetosphere. The ion bulk flows are generally directed antisunward and the characteristic flow speeds are in the range of 100 to 300 km/sec. These speeds are often similar to those observed in the adjacent region of the magnetosheath. The flow speed usually exhibits no or a small variation with increasing distance from the magnetopause. When such a variation is observed, the flow speed decreases with increasing depth relative to the magnetopause position. In contrast to the field-aligned plasma flows in the magnetopause boundary layer at polar latitudes as noted previously, the ion bulk velocity at equatorial latitudes is directed at a large angle to the local magnetic field (cf. Eastman et al., 1985). The polarization electric fields associated with the ion bulk velocities observed within the boundary layer are expected to generate field-aligned currents with closure in the ionosphere. Such currents arising from field-aligned velocity distributions of electrons are observed; the corresponding current densities are $\sim 10^{-8}$ to 10^{-7} A/m^2. At low altitudes these field-aligned currents should correspond to the polar system, or Region 1, of field-aligned currents (Potemra, 1979; Bythrow et al., 1981). The plasma bulk velocity in the magnetopause boundary layer exhibits a component of $\sim$ 10 km/sec in the direction perpendicular to the magnetopause (Eastman et al., 1985). The existence of such a normal flow component over a substantial fraction of the magnetopause indicates that magnetospheric plasma entry into the magnetosphere may not be restricted to subsolar positions. The observation of this large-scale plasma entry can be used to predict the increased dimensions of the magnetospheric boundary layer at the lunar orbit, $\sim 60\ R_E$, downstream in the magnetotail. Such an increase in thickness to 15,000 to 30,000 km is perhaps indicated by plasma surveys at the lunar orbit (Hardy et al., 1979a). Electron angular distributions closer to the Earth at $\lesssim 20\ R_E$ radial distances often indicate that the magnetic field lines threading the magnetopause boundary layer are closed, i.e., geomagnetically trapped electrons. The magnetic field is characterized by a northward direction in most of this region and by large fluctuations of spatial and/or temporal origins. Conservative estimates for the global power input into the magnetopause boundary layer are 10^{11} to 10^{12} watts, values adequate to sustain magnetospheric convection, ionospheric losses and magnetotail outflow (Eastman et al., 1985). Thus a theoretical understanding of the mechanism for formation of the magnetopause boundary layer is expected to be a substantial contribution to the development of global models of solar wind momentum and mass transport into the magnetosphere.

5. THE PLASMA SHEET AND ITS BOUNDARY LAYER

The first direct finding of the existence of an extended magnetotail in the antisunward direction from Earth is reported from magnetic field observations in these regions (Ness, 1965). This magnetotail comprises two large flux tubes with magnetic fields directed in the solar and antisolar directions in the northern and southern flux tubes, respectively. At the common midplane between these two flux tubes and generally parallel to the ecliptic plane, the magnetic field exhibits a dramatic reversal between solar and antisolar directions. This plane is referred to as the neutral sheet. Typical magnitudes of the magnetic field in the flux tubes are ~ 20 to 50 nT at ~ 20 R_E (Fairfield, 1979). A thick zone of lesser, more disordered magnetic fields at the magnetotail midplane is the signature of a sheet of plasma. Plasmas associated with the current sheet that must support this extended magnetotail are first identified by Bame and coworkers with measurements at ~ 17 R_E geocentric radial distance (Bame et al., 1967; Bame, 1968). This plasma sheet extends across the midplane of the magnetotail with typical thicknesses of ~ 4 to 6 R_E. Typical ion and electron temperature ranges are 5×10^6 to 5×10^7 °K and 2×10^6 to 2×10^7 °K, respectively, with large fluctuations during substorms. As within the magnetosheath ion temperatures are greater than electron temperatures by factors of ~ 3 to 5. The reason for this nonequilibration of ion and electron temperatures in the plasma sheet is believed to be associated with the primary magnetotail acceleration mechanism and remains unresolved at this time. Plasma densities within the plasma sheet are ~ 0.1 to 1 cm^{-3}. Ion bulk speeds within the plasma sheet, including its boundary layers that are characterized by high-speed ion streams, vary greatly within the range of 10 to 1000 km/sec (Frank and Ackerson, 1979). Components of the bulk velocity are often of similar magnitude along the Earth-Sun line (X) and cross-magnetotail (Y) directions. A gross average of a large number of plasma flow observations yields an earthward component of ~ 10 km/sec. At close distances to the Earth the plasmas in the plasma sheet are adiabatically heated by earthward convection and increasing magnetic fields to form the terrestrial ring current (Frank, 1967, 1971b).

Direct detection of the currents associated with the current sheet supporting the magnetotail is difficult due to the small ion and electron drift speeds relative to the thermal velocities. Such recent measurements show that the ion and electron motions in the vicinity of the neutral sheet are nonadiabatic. This situation produces currents from the ion and electron drifts that are additive and in the correct sense for the current sheet of the magnetotail (Frank et al., 1983). Magnetic fields and the electric fields perpendicular to these magnetic fields are weak and $\lesssim$ 5 nT and $\lesssim$ 0.5 mV/m, respectively. The ion and electron drift speeds are in the ranges of ~ 100 to 200 km/sec and ~ 500 to 800 km/sec, respectively. These drifts yield a current density ~ 1 to 2 × 10^{-8} A/m^2 in a sheet with thickness ~ 1 to 3 R_E, which is directed generally from dawn to dusk across the magnetotail. Since the motion of ions and electrons is nonadiabatic and thus energy-dependent (cf. Lyons and Speiser, 1982) self-consistent models of the magnetotail current sheet are expected to be realistic only if these motions are computed in detail. Such a well-founded model of the magnetotail current sheet is not available in the literature and is most appropriately yielded by numerical simulation techniques.

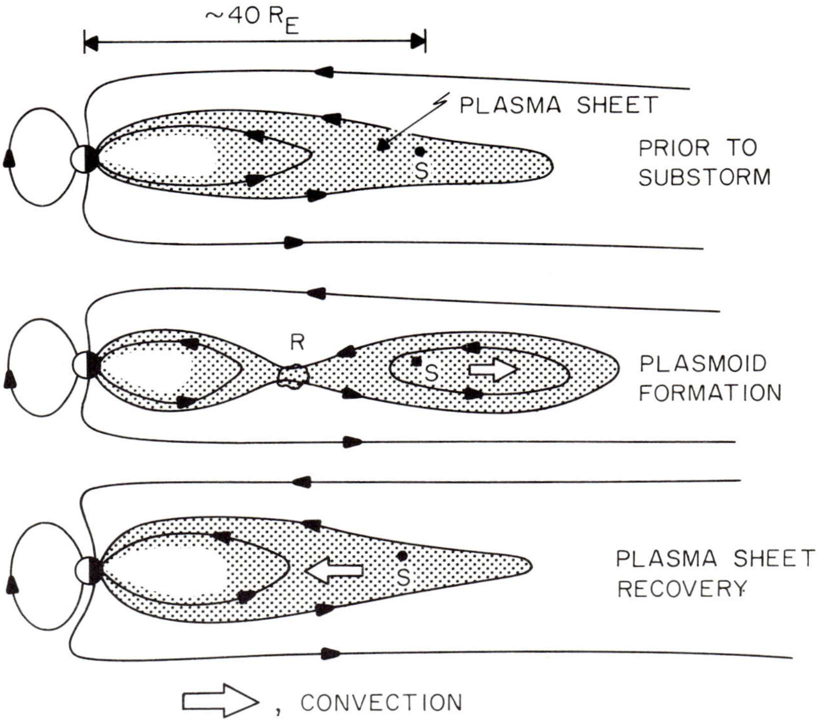

Figure 2. Plasmoid model for magnetotail substorms.

The dynamical behavior of the plasma sheet during magnetic substorms receives the greatest attention of all magnetotail research topics (Hones, 1979, 1984; Hones et al., 1973; Nishida and Hones, 1974; Russell and McPherron, 1973; Nishida and Russell, 1978). The motivation for this intensive investigation arises from the observed dramatic variations of plasmas in the magnetotail during these magnetotail substorms and their close association with auroral substorms in the ionosphere (Akasofu, 1977). The plasma sheet is observed to thin during substorms. Estimates of the rate of this plasma sheet thinning normal to the magnetotail current sheet are ~ 50 to 100 km/sec (Parks et al., 1979). Sometimes during substorms a strong correlation of the magnetic field component perpendicular to the neutral sheet (Z) and the direction of ion flow is found (Hones, 1979). Initially during the substorm antisolar flow and southward magnetic fields are observed. During the substorm recovery earthward flows with northward magnetic fields are present. This correlation is used to deduce that a plasmoid is formed in the plasma sheet during magnetic substorms. The corresponding relationship of the plasmoid relative to the spacecraft position S is sketched in Figure 2. The evolution of the magnetic field topology to a more dipolar geometry at the onset of a substorm and the return to an extended tail configuration with the recovery is discussed by Fairfield (1973). The earthward jetting of plasmas during substorm recovery as seen at the spacecraft position is interpreted in this model as indicative of the motion of the entire plasma sheet.

The magnetotail is a topologically complex and dimensionally immense region to deduce its behavior and geometry from single-point observations. For example, is the earthward motion of plasmas during substorm recovery the signature of earthward convective motion of the entire plasma sheet or of the passage of a more spatially limited zone of streaming plasmas past the spacecraft position? This observational dilemma is resolved with measurements with two spacecraft (DeCoster and Frank, 1979). There is a boundary layer of generally field-aligned ions on the surface of the plasma sheet. The ion bulk speeds in this boundary range from ~ 100 to 1500 km/sec and are directed earthward. In contrast the ion bulk speeds in the plasma sheet are ~ tens of km/sec. The densities in this plasma sheet boundary layer are ~ 0.1 to 1 cm^{-3}. The ranges of ion and electron temperatures are typically 5×10^6 to 5×10^7 °K and 2×10^6 to 2×10^7 °K, respectively, and similar to those observed within the plasma sheet (cf. Frank et al., 1981). The thickness of this boundary is usually 0.5 to 2 R_E. This plasma sheet boundary layer is a durable feature of the plasma sheet during a wide range of magnetic activity (Lui et al., 1983; Eastman et al., 1984). Thus the downstream acceleration region associated with these plasma flows must be continually active.

Field-aligned currents are observed in the boundary layer of the plasma sheet (Frank et al., 1981). The magnitudes of these field-aligned currents are considerably greater than upper limits for such currents within the plasma sheet proper. The electron temperatures are as noted above and the electron drift speeds, V_D, are in the range of 200 to 1000 km/sec, and thus considerably less than the characteristic thermal velocity, V_T. The current densities are ~ 10^{-8} A/m^2. Several such sheets can exist in the boundary layer with oppositely directed current densities, i.e., into and out of the ionosphere. The thicknesses of these current sheets are ~ 0.1 to 1 R_E. These current sheets are mapped into the Region 1 current system over the auroral ionosphere (Potemra, 1979). The ion and electron velocity distributions thus differ significantly from those slowly convecting, quasi-isotropic distributions of the plasma sheet. The character of these velocity distributions is sketched in Figure 3. The electron velocity distributions are characterized by a thermal isotropy $T_{\parallel}/T_{\perp}$ ~ 1.5 with the exception of the superposition of lower energy electrons from the ionosphere to give currents directed into the ionosphere (dashed line, northern flux tube). The ion velocity distribution is greatly distorted from that of a convecting Maxwellian (DeCoster and Frank, 1979). This deformation cannot be accounted for by an adiabatic deformation of an initially Maxwellian velocity distribution. On the other hand, the isodensity contours can be generally accounted for by a field-aligned electrostatic potential (DeCoster and Frank, 1979) or perhaps by nonadiabatic acceleration near the neutral sheet (Lyons and Speiser, 1982). It is noted here that the plasma sheet boundary layer

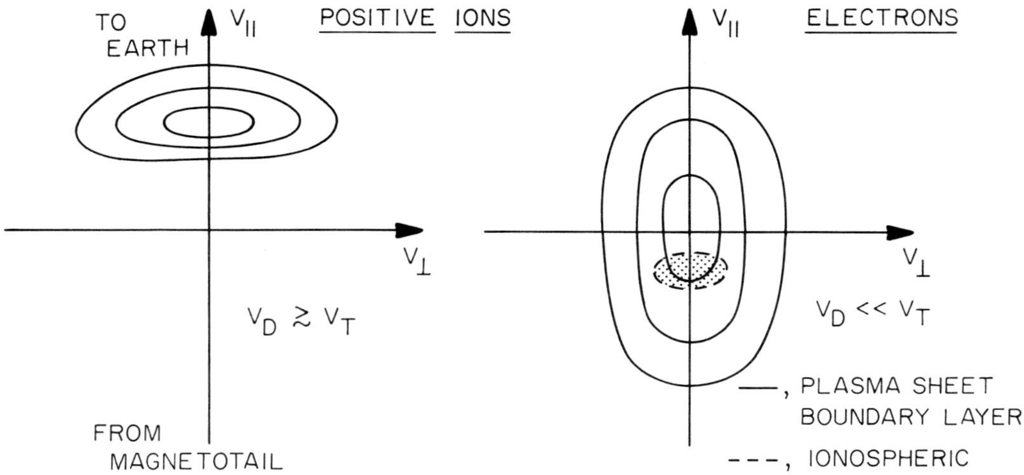

Figure 3. Ion and electron velocity distributions in the plasma sheet boundary layer.

is associated with the discrete auroral emissions in the poleward sector of the auroral oval whereas the equatorward diffuse emissions are identified with the plasma sheet and its extension into the inner magnetosphere as the ring current. Examination of the velocity distributions of ions and electrons in Figure 3 shows that a charge separation earthward along the magnetic field lines will result from the increasing magnetic field. The electric fields arising from the charge separation are presumably modified by the presence of ionospheric plasmas. The complex coupling between these boundary layer and ionospheric plasmas is expected to be usefully studied with numerical simulation. Field-aligned acceleration of electrons at a discontinuity of the convection electric fields is reported for one finding in the plasma sheet boundary layer at ~ 10 R_E (Huang et al., 1984). At low altitudes, ~ 1000 km, such acceleration is seen frequently (Frank and Ackerson, 1972; Evans, 1974; Reiff, 1983). Recently field-aligned electron acceleration is observed in regions extending to altitudes ~ 3.5 R_E over the ionosphere (Lin et al., 1985). Values of plasma parameters given above provide the high-altitude boundary conditions for modeling this as yet unresolved acceleration mechanism; of course, the low-altitude boundary is the ionosphere.

The detection of ionospheric thermal electrons in the magnetotail is rendered difficult by the effects of spacecraft charging and the presence of large fluxes of photoelectrons and secondary electrons. Sojka and coworkers (1985) report several measurements of cold electrons in the vicinity of the plasma sheet boundary layer. The cold ionospheric electron densities are ~ 0.1 to 1 cm^{-3} with a temperature ~ 10^4 to 5×10^4 °K. Observations of cold electrons in the magnetotail deserve further such efforts in order to establish or discount their roles in various possible plasma instabilities in the magnetotail.

Broadband electrostatic noise is present in the boundary layer of the plasma sheet (Gurnett et al., 1976). This electrostatic noise is also observed along these field lines to low altitudes over the discrete aurora (Gurnett and Frank, 1977). On the other hand, these plasma waves are relatively absent in the plasma sheet, including the current sheet centered on the neutral sheet (Anderson, 1983). Various mechanisms are proposed to account for the presence of broadband electrostatic noise (Huba et al., 1978; Grabbe and Eastman, 1984), none of which are completely successful. Since similar broadband electrostatic noise is observed at lower altitudes in the absence of high-speed ion beams, an electron-current driven mechanism may be most likely. The present observations of ion and electron velocity distributions, along with the wave spectral power densities, offer a well-defined challenge for the technique of numerical simulation.

Measurements of the ion composition of the plasma sheet and its boundary layer reveal that both the ionosphere and the solar wind contribute importantly (Shelley et al., 1982; Sharp et al., 1981; Peterson et al., 1981; Eastman et al., 1984). The primary ions used for tracers of the source are O^+ for the ionosphere and He^{++} for the solar wind. Protons are abundant from both sources and are usually the dominant ion in these regions. A variable, large fraction, 0.1 to 0.5, of the hot plasma in the plasma sheet is of ionospheric origin. Thus ionospheric plasmas are present in the magnetotail acceleration region. These ionospheric ions arrive at the downstream acceleration region as field-aligned beams from the ionosphere, that are propagating along magnetic field lines in the region including the plasma sheet boundary layer and extending into the magnetotail lobe and into the adjacent regions of the plasma sheet. These ion distributions are generally field-aligned and flowing in the antisolar direction with speeds in the range of 50 to 200 km/sec. The dominant ions are usually H^+ and O^+. The ion temperature is $\sim 10^5$ to 10^6 °K. Densities are ~ 0.5 to 5 cm^{-3}. The velocity distributions of these ions exhibit the effects of substantial heating perpendicular to the magnetic field at lower altitudes (Frank et al., 1977). Further the different ion species are often, but not always characterized by the same bulk flow speed parallel to the magnetic field. The mechanism for imparting this bulk motion to these ionospheric plasmas is yet to be identified and remains as an important theoretical goal, along with the wave-particle interaction for the ion heating noted above. Note that a field-aligned potential will provide parallel drifts with the same energy per unit charge for each species. Definitive measurements of the electron temperatures are often difficult within the ionospheric ion beams because the velocity distributions are obscured by those of higher temperature electrons associated directly with the plasma sheet boundary layer; otherwise the electron temperatures are $\sim 5{\times}10^5$ to $2{\times}10^6$ °K. The composition of the earthward, or return ion beams in the plasma sheet boundary layer shows that both ionospheric ions and ions within the magnetopause boundary layer are mixed and heated in the downstream acceleration region (Eastman et al., 1984).

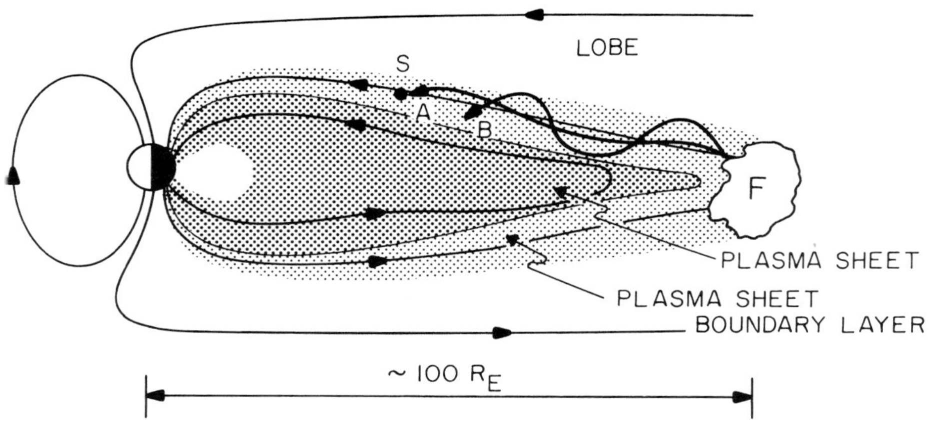

Figure 4. Determination of the location of the magnetotail acceleration region via time-of-flight of energetic ions.

Even though the plasmas within the Earth's magnetotail are characterized by thermal energies in the range of 1 keV to 10 keV, there is firm observational evidence that energetic ions and electrons are efficiently accelerated to energies exceeding 1 MeV (Sarris et al., 1978; Krimigis and Sarris, 1980; Ipavich and Scholar, 1983; Williams et al., 1985). Within the plasma sheet the velocity distribution of energetic ions is the continuous extension of the thermal velocity distribution as a high energy tail beginning at several thermal energies. Examples of such singular ion velocity distribution over the energy range 100 eV to 3 MeV are given by Sarris et al. (1981). This significant enhancement of higher energy charged particle intensities relative to Maxwellian expectations is noted for many of the plasmas in the magnetotail and environs, including the magnetopause boundary layer and the magnetosheath. On the other hand, field-aligned streaming of high-energy ions and electrons which are not directly identifiable as the higher energy component of a thermal plasma is also frequently observed. Such velocity distributions are commonly found in the vicinity of the plasma sheet boundary layer, magnetopause boundary layer and magnetosheath, but are relatively absent within the plasma sheet. These streaming, energetic charged particles are interpreted in terms of either a selective acceleration mechanism for high energy particles, e.g., an inductive electric field from temporal fluctuations of magnetic fields, or the signature of velocity dispersion from a common acceleration region for plasmas and higher energy particles at a remote position in the magnetotail. Williams (1981) provides an analysis of the temporal and pitch-angle fluctuations of energetic ion intensities in the plasma sheet boundary layer, that determines a location for the acceleration region. Consider two particles, A and B, with the same energy and propagating from an impulsive acceleration event at location F in Figure 4. The spacecraft position is S. Particle A is injected at a lesser pitch angle than B and hence will arrive at an earlier time at position S. Observation of the temporal evolution of the pitch angle distributions at S allows determination of the downstream distance to F. The position can be confirmed by repeating a similar analysis for particles (ions) with different energy. The results of this important analysis position the acceleration region at distances typically ~ 100 R_E downstream from Earth. Thus magnetic field lines threading the plasma sheet boundary layer extend downstream into a region of energetic charged particle acceleration. In overview it is a remarkable feature that such charged particles can be accelerated to these high energies in the weak magnetic fields, tens of nT, and relatively cool, flowing plasmas with speeds in the range of hundreds of km/sec. The application of an in situ observation of the acceleration region is of conspicuous importance to energetic charged particle acceleration in the plasmas surrounding other astronomical objects as well.

As plasma convects earthward in the plasma sheet under the influence of the cross-tail electric field, the ions and electrons are heated adiabatically in the increasing geomagnetic field. This region lies earthward of the current sheet of nonadiabatic motions discussed above. The earthward extension of the plasma sheet is the terrestrial ring current (Frank, 1971b). As the gradient drift velocities increase the ions and electrons drift westward and eastward, respectively. The earthward edge of the electron spatial distributions exhibits well-defined gradients in temperature and pressure as shown in Figure 5 (Vasyliunas, 1968; Schield and Frank, 1970). The ring current lies principally inside this region. Typical electron 'temperatures' as derived from average kinetic energies and the pressures at the inner and outer edges of the electron boundary are 10^6 °K and 10^{-9} erg/cm^3, and 5×10^6 °K and 10^{-8} erg/cm^3, respectively. The gross nature of the spatial distributions of ions and electrons can be accounted for in terms of electric and gradient drifts (cf. Kavanagh et al., 1968). During substorm activity, earthward convection of magnetotail plasmas is observed and the electron boundary of the plasma sheet similarly moves toward Earth (Konradi et al., 1975; Moore et al., 1981). The earthward edge of this body of new magnetotail plasma that is injected into the inner magnetosphere is referred to as an injection boundary or front. The speed of this plasma convection is sometimes ~ 100 km/sec or similar to that of motion of the boundaries of the more distant plasma sheet at the onset of substorms. In the simplest consideration the charge separation forced by the gradient drift of ions and electrons produces an electric field (Schield et al., 1969). This electric field should generate field-aligned currents that close in the ionosphere. Relatively large current densities are anticipated at the earthward edge of the electron spatial distributions, and field-aligned ions and electrons are observed in the vicinity of this region. However, to this date there is not a definitive observation of the three-dimensional ion and electron velocity distributions, and thus of the current densities. The lower latitude current system in the auroral oval, Region 2 (Potemra, 1979) is presumed to be associated at least in part with this near-Earth plasma regime. Since the motions of the ion and electron plasmas due to convection and gradient drifts are complex, and a strong coupling to the ionosphere via field-aligned currents is most likely present, numerical simulation is a particularly strong candidate for yielding a satisfactory interpretation of the observed complex plasma behavior and a prediction for the spatial distribution and strength of the field-aligned current system.

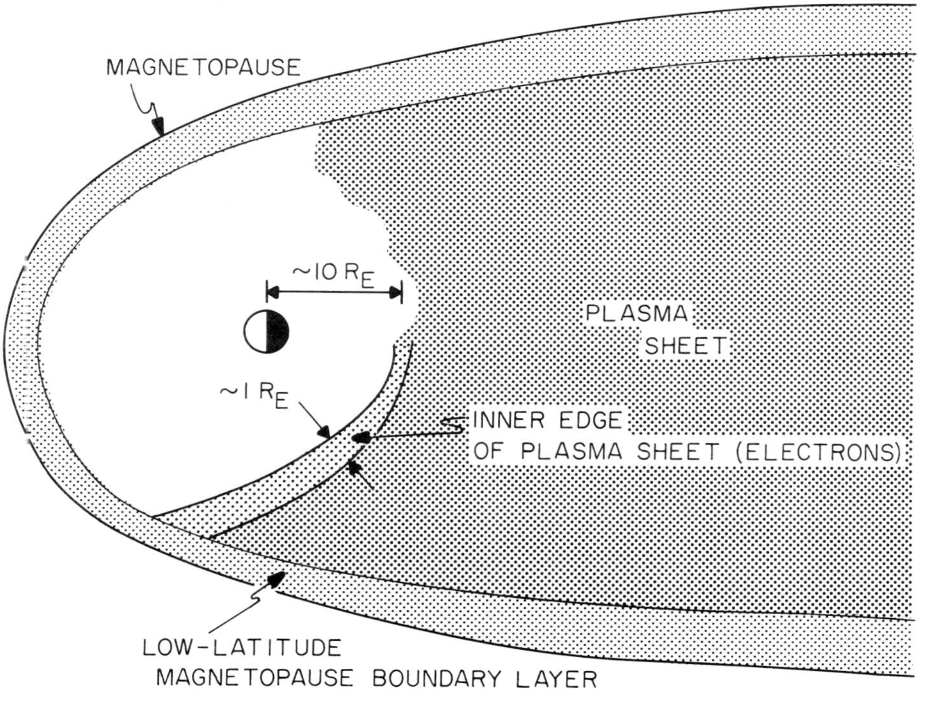

Figure 5. Inner edge of the spatial distribution of plasma sheet electrons.

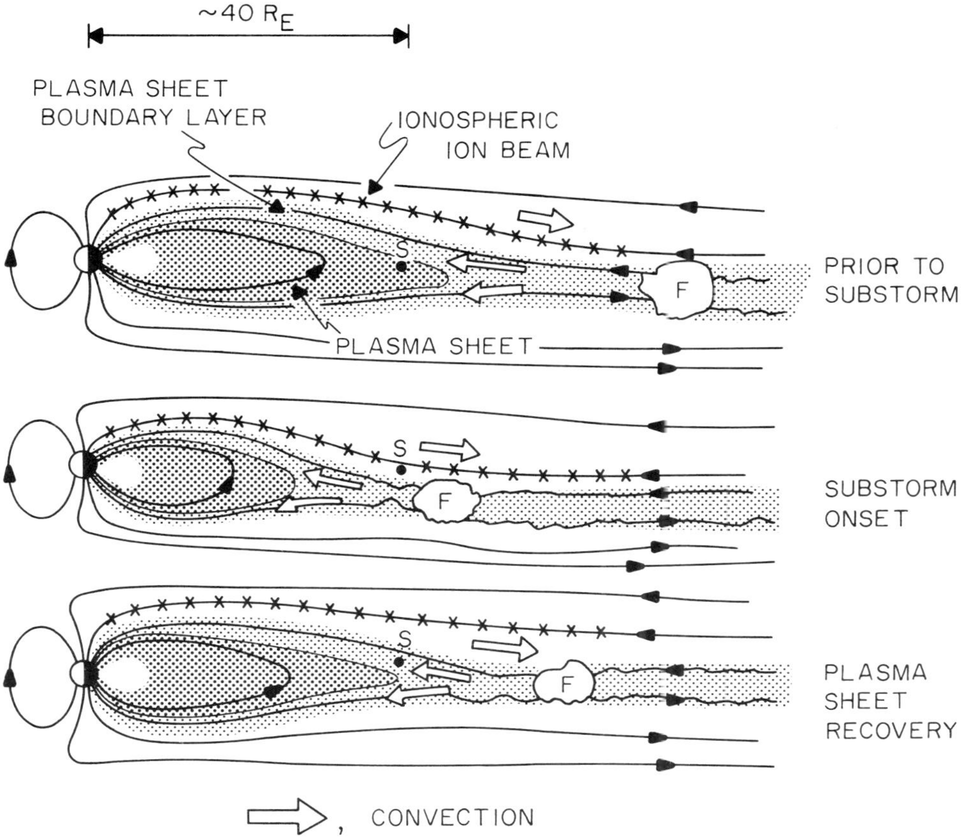

Figure 6. Boundary layer, or fireball (F) model for magnetotail substorms.

An alternative model of the magnetotail is currently being suggested to replace the plasmoid picture that is based on the occasionally observed, striking correlation of the earthward and tailward plasma flows with the sign of the Z-component of the magnetic field during substorms. Recall that the plasmoids are formed by reconnection in the hot plasmas of the plasma sheet and that the plasma sheet boundary layer is not identified as an important dynamical feature in the plasmoid model. The alternative model employs this boundary layer as a primary energy transport region in the magnetotail. Thus it is sometimes referred to as the boundary layer model. The character of ion and electron velocity distributions, the positions of field-aligned currents and ion beams, the location of the magnetotail acceleration region, and the other features of magnetotail plasmas summarized in this writing are accounted for in the boundary layer model. A sketch of this model is shown in Figure 6. A single ion beam from the ionosphere in the magnetotail lobe is shown for simplicity. These ionospheric beams may also be found within the plasma sheet boundary layer and at locations in the plasma sheet just inside this boundary layer. The plasma sheet and its boundary layer are also identified. During relative magnetic quiescence, hot plasmas are convected into the plasma sheet at speeds ~ 10 km/sec. The boundary layer is the principal source of plasmas within the plasma sheet. The expansion of the plasma sheet may also proceed at similar rates. The ionospheric beams, the earthward high-speed ion streams in the boundary layer and the isotropic, slowly convecting plasmas of the plasma sheet are all detected at spacecraft location S during expansions and contractions of the plasma sheet, or of magnetotail flapping motions. The acceleration region F is located at typical distances ~ 100 R_E downstream from Earth and on the boundary of the plasma sheet. This acceleration region is sometimes referred to as a fireball and is not located within the plasma sheet (Frank et al., 1976). Plasmas entering this acceleration region are those of the magnetopause boundary layer and of the ionospheric ion beams.

Acceleration in region F may occur via magnetic reconnection or another mechanism. Magnetic field lines associated with these plasmas are shown to be variously open and closed. The closed magnetic field lines at such distances are sufficiently stretched as to be pinched off by reconnection, if that is the appropriate mechanism. The plasma sheet does not extend beyond the region F, but boundary layers with field-aligned currents and antisunward-directed ion beams are expected to extend considerably tailward. At the onset of a substorm, the plasma sheet is convected rapidly earthward and the acceleration region F follows its outer boundary toward Earth. Thus the plasma sheet thins at the position S. Note that reconnection does not occur in the plasma sheet to form a plasmoid. Currents associated with the acceleration region give large perturbations of the magnetic field near the spacecraft location, at times yielding southward deflections of the magnetic field. As the plasma sheet collapses, ionospheric beams of H^+ and O^+ pass over the spacecraft position. Occasionally these tailward-directed beams are encountered with the above mentioned southward deflection of the magnetic fields associated with the current systems driven by region F.

During the substorm considerable enhancement of tailward plasma flows, tailward-directed intensities of energetic particles, and field-aligned low-energy electrons associated with the magnetotail current system are to be anticipated downstream from region F. During substorm recovery the plasma sheet expands, the magnetic field displays a northward component, and earthward-directed field-aligned ion beams in the plasma sheet boundary layer, as well as occasional tailward ion beams from the ionosphere, are detected. Eventually the plasma sheet expands sufficiently to place the spacecraft S in the hot, slowly convecting plasmas of the plasma sheet. The success of this alternative model, that accounts for many of the observed topological and dynamical features of the magnetotail, stresses the difficulty in interpreting single, or even several-point measurements in terms of unique identifications of the location and details of the acceleration mechanism. A direct encounter with the acceleration region is yet to be convincingly reported. Resolution of many magnetotail issues awaits such an encounter.

6. THE MAGNETOTAIL LOBES

The magnetotail lobes are the two relatively plasma-barren regions in the magnetotail that are bounded at equatorial latitudes by the plasma sheet boundary layers and elsewhere by the magnetopause boundary layers. A cross-section of the magnetotail as seen in a plane perpendicular to the Earth-Sun line and at 40 R_E downstream from Earth is shown in Figure 7. These two regions are magnetically mapped into the northern and southern polar regions of the ionosphere, respectively. The plasma densities are $\lesssim 0.1$ cm^{-3} and the electron and ion temperatures are estimated to be $\lesssim 10^6$ °K and $\lesssim 10^7$ °K, respectively (Bame, 1968). Peterson and coworkers (1984) find typical densities of 10^{-3} to 10^{-2} cm^{-3} with an ion composition comprising H^+, He^{++} and O^+. Thus these diffuse plasmas are of both ionospheric and solar wind origins. Their presence in the magnetotail lobes is unaccounted for; however the composition indicates a source related to the plasma sheet and/or its boundary layer. The magnetic flux in the lobes increases with the southward turning of the interplanetary field; thus the total magnetic energy of the magnetotail similarly increases (Fairfield et al., 1981; Makita et al., 1985; Frank et al., 1985). This magnetic energy is released during a substorm. Whether this release of energy is spontaneous or stimulated by a fluctuation of the interplanetary medium is an issue that is unresolved.

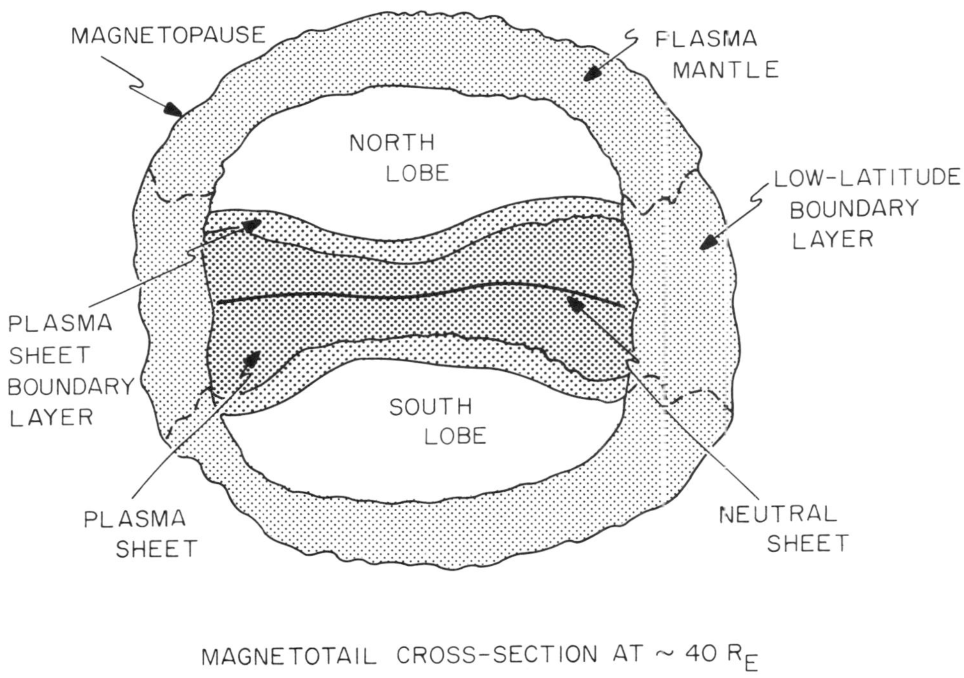

Figure 7. Plasma regions in the magnetotail.

The magnetic field lines threading the magnetotail lobes are connected to field lines in the solar wind, i.e., open field lines. This magnetic field topology is supported by the asymmetric access of low-energy electrons, hundreds of eV, to the magnetotail lobes (Yeager and Frank, 1976; Mizera and Fennell, 1978). The asymmetry is such that the lobe with magnetic field directed opposite to that of the interplanetary field is favored. Fairfield and Scudder (1985) show that these low-energy electrons are to be identified with the severely field-aligned electron intensities in the energy range of several hundreds of eV that comprise the high-speed component of the solar wind electron velocity distribution. These electrons are observed to enter into the magnetotail lobe with little or no distortions of their angular distributions, with the exception of that expected from the increasing magnetic field with their motion into the lobe and toward the polar cap. Thus the magnetotail lobes are often, if not always open to the interplanetary medium.

Interest in the magnetotail lobes is currently revived in part due to the findings of tailward-streaming ion beams from the ionosphere (Sharp et al., 1981) and the remarkably active and puzzling behavior of auroras in the polar cap ionosphere during periods of northward interplanetary fields (Burke et al., 1979; Gussenhoven, 1982; Frank et al., 1985). This auroral activity occurs in the polar cap with relatively little fluctuations in intensity and geometry of the auroral oval. The appearance of a transpolar arc across the polar cap to form a theta aurora is currently inspiring several models for the associated topology and global convection system within the magnetotail. The theta aurora is currently believed to be the signature of a torque exerted on the magnetosphere along the Earth-Sun line by asymmetric access of plasmas at the magnetopause, i.e., correlated with the dawn-dusk component of interplanetary magnetic field (cf. Cowley, 1981). Since the plane of the plasma sheet is not observed to rotate significantly in response to the interplanetary field (Hardy et al., 1979b) a 90° reorientation of the plasma sheet cannot account for the transpolar arc. The available diagnostics indicate that the magnetic field lines associated with the transpolar arc are closed whereas the remainder of the polar cap exhibits an open topology (Frank et al., 1985). Ion composition within the transpolar arc plasmas is similar to that of the plasma sheet and its boundary layer (Peterson et al., 1984). Whereas the plasma convection over the polar cap is typically directed antisunward, the convection within the transpolar arc plasmas is sunward. This convection is indicative of a two-cell pattern within the polar cap. Recent examination of in situ measurements within the magnetotail lobes finds the existence of sheets or tubes of plasmas similar to those of the plasma sheet boundary layer (Huang et al., 1985). These plasmas are probably associated with the polar auroras. Thus the magnetotail lobes are no longer to be considered as relatively featureless and passive domains within the magnetotail.

In general, modeling of the energy and mass transport within the Earth's magnetotail, including its lobes, requires in turn a global model for convection electric fields. A realistic description, excluding empirical fits and gross oversimplifications, is most likely achieved by numerical simulation. Even so, this important task is difficult: (1) the outer boundary condition is sensitive to the solar wind, in particular the orientation of the interplanetary field (cf. Cowley, 1982), (2) the inner boundary is the ionosphere, and (3) the electric field configuration and dimensions of the magnetotail acceleration region are unknown. If we are to understand the coupling of the solar wind with the magnetosphere such modeling must be performed. Two of the first severe tests of any model are the more or less permanent presence of a plasma sheet boundary layer and the appearance of a magnetotail convection pattern corresponding to that associated with a theta aurora.

7. THE DISTANT MAGNETOTAIL

Two Pioneer interplanetary spacecraft provide three brief encounters with the Earth's magnetotail at geocentric radial distances of ~ 500, 1,000 and 3,100 R_E. The character of the plasmas differs considerably from the features of the magnetotail inside the lunar orbit (Wolfe et al., 1967; Siscoe et al., 1970; Intriligator et al., 1979). The general appearance of the plasmas in this region is that expected of a wake, i.e., the sporadic presence at the spacecraft of distorted solar wind ion velocity distributions and, at times, large decreases in plasma densities. Such features as the plasma sheet and its boundary layer are not identified. Correlation of these plasma measurements with those of magnetic fields is generally inconclusive as to whether the magnetotail's structure is lobar as nearer the Earth or filamentary (Ness et al., 1967; Walker et al., 1975). The first systematic exploration of the distant magnetotail is provided by ISEE 3 for the radial distance range of ~ 60 to 240 R_E. These long-awaited observations are briefly summarized here with special attention to their relationship with known features of the magnetotail inside the lunar orbit.

The magnetotail retains its two-lobed structure at the radial distances ~ 60 to 240 R_E surveyed with ISEE 3 (Tsurutani et al., 1984b). A persistent region of weaker, variable magnetic fields relative to those of the lobes is easily identifiable with a sheet of plasma at the midplane of the magnetotail (Tsurutani et al., 1984c). The character of these magnetic fields is grossly similar to those found in the plasma sheet and its boundary layer at lesser distances. The strength of the lobe magnetic fields decreases as ~ $R^{-0.53}$ over radial distances R of 20 to 130 R_E. Beyond 130 R_E the lobe magnetic fields are relatively constant, ~ 9 nT (Slavin et al., 1983, 1985). The magnetotail diameter across its midplane between lobes is ~ 60 R_E

in the distant tail and can be compared to a diameter of ~ 50 R_E in the vicinity of the lunar orbit (Howe and Binsack, 1972; Behannon, 1968; Hardy et al., 1979a; Mihalov et al., 1970). Thus a flaring model is appropriate for the magnetotail. Flaring should terminate at the radial distance at which the lobe field reaches its asymptote, i.e., equilibration of lobe magnetic pressure with the hydrostatic pressure of the solar wind (cf. Coroniti and Kennel, 1972). Such a model agrees reasonably with the observed asymptote, ~ 130 R_E. The average north-south component of the magnetic field is of considerable interest for identification of a signature of reconnection, or alternatively a large-scale current system differing from that of the cross-tail current sheet maintaining the lobe magnetic fields. Slavin and coworkers (1985) show that there is a region centered in the magnetotail midplane at ~ 100 to 180 R_E and ~ 10 R_E in width where the mean magnetic field is directed southward. This region widens to ~ 30 R_E at radial distances ~ 180 to 225 R_E. At distances ~ 210 to 240 R_E the mean magnitude of the southward component is 0.22 nT. The Y-component of the interplanetary field is observed to penetrate the magnetotail lobes and plasma sheet in an asymmetric manner (Tsurutani et al., 1984a; Sibeck et al., 1985). With reference to a cross-section of the magnetotail as divided into north-dawn, south-dawn, etc., quadrants the interplanetary field penetrates more effectively into the north-dawn and south-dusk quadrants when the Y-component is positive. The reverse correlation is also observed. Thus the survey of the magnetotail with ISEE 3 provides an excellent map of the topological evolution of magnetic fields in the magnetotail as a function of radial distance. Only the final intermingling of the magnetotail with the solar wind at greater distances is left unresolved.

The thermal electron plasmas are used to identify various plasma regimes of the magnetotail, e.g., lobes, magnetopause boundary layers, and plasma sheet (Bame et al., 1983). Typical electron densities in the magnetopause boundary layer, i.e., mantle and low-latitude boundary layer, are ~ 1 cm^{-3}. The temperatures are ~ 3×10^5 to 8×10^5 °K with tailward flow velocities ~ 200 km/sec. The densities and temperatures in the lobes are ~ 0.04 cm^{-3} and 8×10^5 °K, respectively. Within the region identified as the plasma sheet these plasma parameters are ~ 0.6 cm^{-3}, 9×10^5 °K and 500 km/sec (tailward). Note that the electron temperatures and bulk speeds are similar in most of these regions. The magnitude of the flow speed in the region identified as the plasma sheet is about an order of magnitude greater than that observed in this region inside the lunar orbit. On the other hand, the flow speed, densities and temperatures are similar to those often observed in the plasma sheet boundary layer at the lesser distances. However these electron bulk velocities are associated with field-aligned currents inside the lunar orbit. Since components of the electron bulk velocity $\vec{V}(V_\perp, V_\parallel)$ are not separated with ISEE-3 measurements, it is not possible to determine whether the observed electron bulk velocity is associated with a field-aligned current ($V_{\parallel}$) or convective motion of a magnetic structure ($V_{\perp}$). Since the electron thermal speeds are large at ~ 10^6 °K, ~ 5,000 km/sec, relative to expected convection velocities in the plasma sheet, ~ 10 to 100 km/sec, it is not possible to determine the important component $V_{\perp}$ with the electron measurement. Inside the lunar orbit the convective velocity $V_{\perp}$ is determined from ion velocity measurements. No thermal ion measurements are available with ISEE 3 with the exception of ion composition. The plasma sheet in the deep tail exhibits significant densities of H^+ and He^{++} with only traces of the ionospheric ion O^+ (Ogilvie and Coplan, 1984). The composition of the plasma sheet inside the lunar orbit usually includes a large fraction of O^+. In consideration of these various features of the plasma sheet in the deep magnetotail it is more likely that this plasma regime is to be associated with the plasma sheet boundary layer tailward of the magnetotail acceleration region, not the plasma sheet as identified at distances inside the lunar orbit.

Gosling and his colleagues (1984) identify clearly the presence of the magnetopause boundary layer in the distant magnetotail. The electron density gradients, temperatures and flow speeds are similar to those observed near the Earth in the high-latitude magnetopause boundary layer, or plasma mantle.

Energetic ions in the energy range of several keV to tens of keV are observed in the distant magnetotail and identified as the high-energy tail of the thermal ion distribution in the plasma sheet (Cowley et al., 1984; Gloeckler et al., 1984; Scholer et al., 1984). These ions are flowing generally tailward with speeds ~ 200 to 1000 km/sec. Again the field-aligned and convective components of the flow velocity are not separated. The ion composition is suggestive of a substantial if not dominant contribution from the solar wind. The temperatures of He^{++} are greater than those for H^+ by factors ~ 3 to 4. The thermal ion temperature is estimated from a straightforward argument for pressure balance to be ~ 5×10^6 °K. A survey of the direction of the anisotropy of energetic ions as a function of radial distance finds equal probability of tailward and earthward flows at ~ 70 R_E and dominant tailward flow at ~ 130 R_E and 210 R_E. This locates the acceleration region at a similar distance from the Earth, ~ 100 R_E, as that determined from time-of-flight measurements (Williams, 1981).

Some of the transient fluctuations of electron plasmas, energetic ions and magnetic fields associated with substorms are identified as plasmoids moving tailward across the spacecraft position (Hones et al., 1984; Gloeckler et al., 1984; Slavin et al., 1984). There is little doubt that the plasmas are moving tailward from an acceleration region inside the spacecraft location during these periods. However, there is reasonable doubt as to whether or not these plasmas convect from the near-Earth plasma sheet at 10 to 30 R_E via the formation of a plasmoid by reconnection at these distances. Observations of the absence of O^+, the relatively low electron temperatures and the high-speed, field-aligned electron flows more typically found in the plasma sheet boundary layer are several hurdles for this interpretation.

There is plausible evidence of the presence of slow mode shocks associated with reconnection in the distant magnetotail (Feldman et al., 1984; Smith et al., 1984). However the convection electric fields associated with the flow across the shock are not determined, and thus a decisive identification of the shock associated with the merging region is not available. The position of this possible slow-mode shock is located at the boundary between the lobe and plasma sheet. Two essential features to observe are the convection ($\vec{V}_\perp$) of plasmas in the lobe toward the midplane of the magnetotail and post-shock convection tailward or earthward near the midplane. It is anticipated that future missions are to include direct measurements of electric fields and of the convection velocity $\vec{V}_\perp$ of the thermal ion plasma. For the plasma sheet in the distant magnetotail Scarf and coworkers (1984) find broadband electrostatic noise similar to that observed in the plasma sheet boundary layer inside the lunar orbit. This electrostatic noise is not found in the more dormant plasma sheet at the lesser distances. The ion acoustic anomalous resistivity associated with these broadband electrostatic noise amplitudes in the distant magnetotail is shown to be insufficient to account for the power dissipation expected in the shock (Scarf et al., 1984). However it is possible that sufficient anomalous resistivity is provided by lower hybrid drift waves.

ACKNOWLEDGMENT

This research was supported in part by the National Aeronautics and Space Administration under grant NGL-16-001-002 and contract NAS5-28700 and the Office of Naval Research under grant N00014-76-C-0016.

REFERENCES

Akasofu, S.-I., Physics of Magnetospheric Substorms, Reidel, Dordrecht-Holland, pp. 71-110, 1977.

Anderson, R. R., Plasma waves in planetary magnetospheres, Rev. Geophys. Space Phys., 21, 474, 1983.

Asbridge, J. R., S. J. Bame, J. T. Gosling, G. Paschmann, and N. Sckopke, Energetic plasma ions within the Earth's magnetosheath, Geophys. Res. Lett., 5, 953, 1978.

Aubrey, M. P., M. G. Kivelson, and C. T. Russell, Motion and structure of the magnetopause, J. Geophys. Res., 76, 1673, 1971.

Bame, S. J., Plasma sheet and adjacent regions, Earth's Particles and Fields, ed. by B. M. McCormac, Reinhold, New York, p. 373, 1968.
Bame, S. J., Solar wind minor ions - recent observations, Solar Wind Five, ed. by M. Neugebauer, NASA Conference Publication 2280, p. 573, 1983.
Bame, S. J., R. C. Anderson, J. R. Asbridge, D. N. Baker, W. C. Feldman, J. T. Gosling, E. W. Hones, Jr., D. J. McComas, and R. D. Zwickl, Plasma regimes in the deep geomagnetic tail: ISEE 3, Geophys. Res. Lett., 10, 912, 1983.
Bame, S. J., J. R. Asbridge, H. E. Felthauser, E. W. Hones, Jr., and I. B. Strong, Characteristics of the plasma sheet in the Earth's magnetotail, J. Geophys. Res., 72, 113, 1967.
Bame, S. J., A. J. Hundhausen, J. R. Asbridge, and I. B. Strong, Solar wind ion composition, Phys. Rev. Lett., 20, 393, 1968.
Behannon, K. W., Mapping the Earth's bow shock and magnetic tail by Explorer 33, J. Geophys. Res., 73, 907, 1968.
Berchem, J. and C. T. Russell, The thickness of the magnetopause current layer: ISEE-1 and -2 observations, J. Geophys. Res., 87, 2108, 1982.
Burke, W. J., M. C. Kelley, R. C. Sagalyn, M. Smiddy, and S. T. Lai, Polar cap electric field structures with a northward interplanetary magnetic field, Geophys. Res. Lett., 6, 21, 1979.
Burlaga, L. F., Interplanetary streams and their interaction with the Earth, Space Sci. Rev., 17, 327, 1975.
Burlaga, L. F., Magnetic fields, plasmas, and coronal holes, Space Sci. Rev., 23, 201, 1979.
Bythrow, P. F., R. A. Heelis, W. B. Hanson, R. A. Power, and R. A. Hoffman, Observational evidence for a boundary layer source of dayside Region 1 field-aligned currents, J. Geophys. Res., 86, 5577, 1981.
Coroniti, F. V., and C. F. Kennel, Changes in magnetospheric configuration during the substorm growth phase, J. Geophys. Res., 77, 3361, 1972.
Cowley, S. W. H., Magnetospheric asymmetries associated with the Y component of the IMF, Planet. Space Sci., 29, 79, 1981.
Cowley, S. W. H., The causes of convection in the Earth's magnetosphere: a review of developments during the IMS, Rev. Geophys. Space Phys., 20, 531, 1982.
Cowley, S. W. H., R. J. Hynds, I. G. Richardson, P. W. Daly, T. R. Sanderson, K.-P. Wenzel, J. A. Slavin, and B. T. Tsurutani, Energetic ion regimes in the deep geomagnetic tail: ISEE-3, Geophys. Res. Lett., 11, 275, 1984.
Crooker, N. U., G. L. Siscoe, T. E. Eastman, L. A. Frank, and R. D. Zwickl, Large-scale flow in the dayside magnetosheath, J. Geophys. Res., 89, 9711, 1984.
DeCoster, R. J., and L. A. Frank, Observations pertaining to the dynamics of the plasma sheet, J. Geophys. Res., 84, 5099, 1979.
Eastman, T. E., and L. A. Frank, Observations of high-speed plasma flow near the Earth's magnetopause: evidence for reconnection?, J. Geophys. Res., 87, 2187, 1982.

Eastman, T. E., L. A. Frank, W. K. Peterson, and W. Lennartsson, The plasma sheet boundary layer, J. Geophys. Res., 89, 1553, 1984.

Eastman, T. E., E. W. Hones, Jr., S. J. Bame, and J .R. Asbridge, The magnetospheric boundary layer: site of plasma, momentum and energy transfer from the magnetosheath into the magnetosphere, Geophys. Res. Lett., 3, 685, 1976.

Eastman, T. E., B. Popielawska, and L. A. Frank, Three-dimensional plasma observations near the outer magnetospheric boundary, (submitted for publication), J. Geophys. Res., 1985.

Evans, D. S., Precipitating electron fluxes formed by a magnetic field aligned potential difference, J. Geophys. Res., 79, 2853, 1974.

Fairfield, D. H., Magnetic field signatures of substorms on high-latitude field lines in the nighttime magnetosphere, J. Geophys. Res., 78, 1553, 1973.

Fairfield, D. H., Magnetic fields of the magnetosheath, Rev. Geophys. Space Phys., 14, 117, 1976.

Fairfield, D. H., On the average configuration of the geomagnetic tail, J. Geophys. Res., 84, 1950, 1979.

Fairfield, D. H., R. P. Lepping, E. W. Hones, Jr., S. J. Bame, and J. R. Asbridge, Simultaneous measurements of magnetotail dynamics by IMP spacecraft, J. Geophys. Res., 86, 1396, 1981.

Fairfield, D. H., and J. D. Scudder, Polar rain: solar coronal electrons in the Earth's magnetosphere, (submitted for publication), J. Geophys. Res., 1985.

Feldman, W. C., R. C. Anderson, S. J. Bame, S. P. Gary, J. T. Gosling, D. J. McComas, and M. F. Thomson, Electron velocity distributions near the Earth's bow shock, J. Geophys. Res., 88, 96, 1983.

Feldman, W. C., J. R. Asbridge, S. J. Bame, and J. T. Gosling, Long-term variations of selected solar wind properties: IMP 6, 7, and 8 results, J. Geophys. Res., 83, 2177, 1978.

Feldman, W. C., J. R. Asbridge, S. J. Bame, M. D. Montgomery, and S. P. Gary, Solar wind electrons, J. Geophys. Res., 80, 4181, 1975.

Feldman, W. C., S. J. Schwartz, S. J. Bame, D. N. Baker, J. Birn, J. T. Gosling, E. W. Hones, Jr., D. J. McComas, J. A. Slavin, E. J. Smith, and R. D. Zwickl, Evidence for slow mode shocks in the distant geomagnetic tail, Geophys. Res. Lett., 11, 599, 1984.

Formisano, V., G. Moreno, F. Palmiotto, and P. C. Hedgecock, Solar wind interaction with the Earth's magnetic field, 1. Magnetosheath, J. Geophys. Res., 78, 3714, 1973.

Frank, L. A., On the extraterrestrial ring current during geomagnetic storms, J. Geophys. Res., 72, 3753, 1967.

Frank, L. A., Plasma in the Earth's polar magnetosphere, J. Geophys. Res., 76, 5202, 1971a.

Frank, L. A., Relationship of the plasma sheet, ring current, trapping boundary, and plasmapause near the magnetic equator and local midnight, J. Geophys. Res., 76, 2265, 1971b.

Frank, L. A., Hot plasmas in the Earth's magnetosphere, Physics of Solar Planetary Environments, Vol. II, ed. by D. J. Williams, AGU, p. 685, 1976.

Frank, L. A., and K. L. Ackerson, Local time survey of plasma at low altitudes over the auroral zones, J. Geophys. Res., 77, 4116, 1972.

Frank, L. A., and K. L. Ackerson, Examples of plasma flows within the Earth's magnetosphere, Magnetospheric Particles and Fields, ed. by B. M. McCormac, Reidel, Dordrecht-Holland, p. 29, 1976.

Frank, L. A., and K. L. Ackerson, Several recent findings concerning the dynamics of the Earth's magnetotail, Space Sci. Rev., 23, 375, 1979.

Frank, L. A., K. L. Ackerson, and R. P. Lepping, On hot tenuous plasmas, fireballs, and boundary layers in the Earth's magnetotail, J. Geophys. Res., 81, 5859, 1976.

Frank, L. A., K. L. Ackerson, and D. M. Yeager, Observations of atomic oxygen in the Earth's magnetotail, J. Geophys. Res., 82, 129, 1977.

Frank, L. A., J. D. Craven, D. A. Gurnett, S. D. Shawhan, D. R. Weimer, J. L. Burch, J. D. Winningham, C. R. Chappell, J. H. Waite, R. A. Heelis, N. C. Maynard, M. Sugiura, W. K. Peterson, and E. G. Shelley, The theta aurora, (submitted for publication), J. Geophys. Res., 1985.

Frank, L. A., J. D. Craven, C. T. Russell, and E. J. Smith, On the relationship of auroral polar cap area with the interplanetary magnetic field, (to be submitted), J. Geophys. Res., 1985.

Frank, L. A., C. Y. Huang, and T. E. Eastman, Currents in the Earth's magnetosphere, Magnetospheric Currents, ed. by T. A. Potemra, AGU Monograph 28, Washington, D.C., p. 147, 1983.

Frank, L. A., R. L. McPherron, R. J. DeCoster, B. G. Burek, K. L. Ackerson, and C. T. Russell, Field-aligned currents in the Earth's magnetotail, J. Geophys. Res., 86, 687, 1981.

Gloeckler, G., F. M. Ipavich, D. Hovestadt, M. Scholer, A. B. Galvin, and B. Klecker, Characteristics of suprathermal H^+ and He^{++} in plasmoids in the distant magnetotail, Geophys. Res. Lett., 11, 1030, 1984.

Gloeckler, G., M. Scholer, F. M. Ipavich, D. Hovestadt, B. Klecker, and A. G. Galvin, Abundances and spectra of suprathermal H^+, He^{++} and heavy ions in a fast moving plasma structure (plasmoid) in the distant geotail, Geophys. Res. Lett., 11, 603, 1984.

Gosling, J. T., D. N. Baker, S. J. Bame, E. W. Hones, Jr., D. J. McComas, R. D. Zwickl, J. A. Slavin, E. J. Smith, and B. T. Tsurutani, Plasma entry into the distant tail lobes: ISEE-3, Geophys. Res. Lett., 11, 1078, 1984.

Grabbe, C. L., and T. E. Eastman, Generation of broadband electrostatic noise by ion beam instabilities in the magnetotail, J. Geophys., Res., 89, 3865, 1984.

Gurnett, D. A., and L. A. Frank, A region of intense plasma wave turbulence on auroral field lines, J. Geophys. Res., 82, 1031, 1977.

Gurnett, D. A., L. A. Frank, and R. P. Lepping, Plasma waves in the distant magnetotail, J. Geophys. Res., 81, 6059, 1976.

Gussenhoven, M. S., Extremely high latitude auroras, J. Geophys. Res., 87, 2401, 1982.

Hardy, D. A., H. K. Hills, and J. W. Freeman, Occurrence of the lobe plasma at lunar distance, J. Geophys. Res., 84, 72, 1979a.

Hardy, D. A., P. H. Reiff, and W. J. Burke, Response of magnetotail plasma at lunar distance to changes in the interplanetary magnetic field, J. Geophys. Res., 84, 1382, 1979b.

Heikkila, W. J., and J. D. Winningham, Penetration of magnetosheath plasma to low altitudes through the day side magnetospheric cusps, J. Geophys. Res., 76, 883, 1971.

Hones, E. W., Jr., Transient phonomena in the magnetotail and their relation to substorms, Space Sci. Rev., 23, 393, 1979.

Hones, E. W., Jr., Plasma sheet behavior during substorms, Magnetic Reconnection in Space and Laboratory Plasmas, ed. by E. W. Hones, Jr., AGU Geophys. Monograph No. 30, p. 178, 1984.

Hones, E. W., Jr., J. R. Asbridge, S. J. Bame, and S. Singer, Substorm variations of the magnetotail plasma sheet from $X_{SM} \simeq -6\ R_E$ to $X_{SM} \simeq -60\ R_E$, J. Geophys. Res., 78, 109, 1973.

Hones, E. W., Jr., D. N. Baker, S. J. Bame, W. C. Feldman, J. T. Gosling, D. J. McComas, R. D. Zwickl, J. A. Slavin, E. J. Smith, and B. T. Tsurutani, Structure of the magnetotail at 220 R_E and its response to geomagnetic activity, J. Geophys. Res., 11, 5, 1984.

Howe, H. C., Jr., and J. H. Binsack, Explorer 33 and 35 plasma observations of magnetosheath flow, J. Geophys. Res., 77, 3334, 1972.

Huang, C. Y., L. A. Frank, and T. E. Eastman, High-altitude observations of an inverted-V event, J. Geophys. Res., 89, 7423, 1984.

Huang, C. Y., L. A. Frank, W. K. Peterson, D. J. Williams, W. Lennartson, D. G. Mitchell, R. C. Elphic, and C. T. Russell, Filamentary structures in the magnetotail lobes, (submitted for publication), J. Geophys. Res., 1985.

Huba, J., N. Gladd, and K. Papadopoulos, Lower-hybrid-drift wave turbulence in the distant magnetotail, J. Geophys. Res., 83, 5217, 1978.

Hundhausen, A. J., Coronal Expansion and Solar Wind, Springer-Verlag, New York, 1972.

Intriligator, D. S., H. R. Collard, J. D. Mihalov, O. L. Vaisberg, and J. H. Wolfe, Evidence for Earth magnetospheric tail associated phenomena at 3100 R_E, Geophys. Res. Lett., 6, 585, 1979.

Ipavich, F. M., and M. Scholer, Thermal and suprathermal protons and alpha particles in the plasma sheet, J. Geophys. Res., 88, 150, 1983.

Kavanagh, L. D, Jr., J. W. Freeman, Jr., and A. J. Chen, Plasma flow in the magnetosphere, J. Geophys. Res., 73, 5511, 1968.

King, J. H., Solar cycle variations in IMF intensity, J. Geophys. Res., 84, 5938, 1979.

Klein, L., and L. F. Burlaga, Interplanetary sector boundaries 1971-1973, J. Geophys. Res., 85, 2269, 1980.

Konradi, A., C. L. Semar, and T. A. Fritz, Substorm-injected protons and electrons and the injection boundary model, J. Geophys. Res., 80, 543, 1975.

Krimigis, S. M., and E. T. Sarris, Energetic particle bursts in the Earth's magnetotail, Dynamics of the Magnetosphere, edited by S.-I. Akasofu, Reidel, Hingham, p. 599, 1980.

Lin, C. S., J. N. Barfield, J. L. Burch, and J. D. Winningham, Near-conjugate observations of inverted-V electron precipitation using DE 1 and DE 2, J. Geophys. Res., 90, 1669, 1985.

Lui, A. T. Y., D. J. Williams, T. E. Eastman, and L. A. Frank, Observations of ion streaming during substorms, J. Geophys. Res., 88, 7753, 1983.

Lyons, L. R., and T. W. Speiser, Evidence for current sheet acceleration in the geomagnetic tail, J. Geophys. Res., 87, 2276, 1982.

Makita, K., C.-I. Meng, and S.-I. Akasofu, Temporal and spatial variations of the polar cap dimension inferred from the precipitation boundaries, J. Geophys. Res., 90, 2744, 1985.

McPherron, R. L., Magnetospheric substorms, Rev. Geophys. Space Phys., 17, 657, 1979.

Mihalov, J. D., D. S. Colburn, and C. P. Sonnett, Observations of magnetopause geometry at lunar distance, Planet. Space Sci., 18, 239, 1970.

Mizera, P. F., and J. F. Fennell, Satellite observations of polar, magnetotail lobe and interplanetary electrons at low energies, Rev. Geophys. Space Sci., 16, 147, 1978.

Montgomery, M. D., Average characteristics of solar wind electrons, NASA SP-308, Solar Wind, ed. by C. P. Sonett, P. J. Coleman, Jr., and J. M. Wilcox, p. 208, 1972.

Montgomery, M. D., J. R. Asbridge, and S. J. Bame, Vela 4 plasma observations near the Earth's bow shock, J. Geophys. Res., 75, 1217, 1970.

Moore, T. E., R. L. Arnoldy, J. Feynman, and D. A. Hardy, Propagating substorm injection fronts, J. Geophys. Res., 86, 6713, 1981.

Nishida, A., and E. W. Hones, Jr., Association of plasma sheet thinning with neutral line formation in the magnetotail, J. Geophys. Res., 79, 535, 1974.

Nishida, A., and C. T. Russell, On the expected signatures of reconnection in the magnetotail, J. Geophys. Res., 83, 3890, 1978.

Ness, N. F., The Earth's magnetic tail, J. Geophys. Res., 70, 2989, 1965.

Ness, N. F., C. S. Scearce, and S. C. Cantarano, Probable observations of the geomagnetic tail at 10^3 earth radii by Pioneer 7, J. Geophys. Res., 72, 3769, 1967.

Ogilvie, K. W., and M. A. Coplan, Composition measurements in the magnetotail between 60 R_E and 240 R_E, Geophys. Res. Lett., 11, 347, 1984.

Parks, G. K., C. S. Lin, K. A. Anderson, R. P. Lin, and H. Reme, ISEE 1 and 2 particle observations of the outer plasma sheet boundary, J. Geophys. Res., 84, 6471, 1979.

Paschmann, G., G. Haerendel, N. Sckopke, H. Rosenbauer, and P. C. Hedgecock, Plasma and magnetic field characteristics of the distant polar cusp near local noon: the entry layer, J. Geophys. Res., 81, 2883, 1976.

Paschmann, G., N. Sckopke, G. Haerendel, I. Papamastorakis, S. J. Bame, J. R. Asbridge, J. T. Gosling, E. W. Hones, Jr., and E. R. Tech, ISEE plasma observations near the subsolar magnetopause, Space Sci. Rev., 22, 717, 1978.
Paschmann, G., B. U. O. Sonnerup, I. Papamastorakis, N. Sckopke, G. Haerendel, S. J. Bame, J. R. Asbridge, J. T. Gosling, C. T. Russell, and R. C. Elphic, Plasma acceleration at the Earth's magnetopause: evidence for reconnection, Nature, 282, 243, 1979.
Peterson, W. K., and E. G. Shelley, Origin of the plasma in a cross-polar cap auroral feature (theta aurora), J. Geophys. Res., 89, 6729, 1984.
Peterson, W. K., R. D. Sharp, E. G. Shelley, R. G. Johnson, and H. Balsiger, Energetic ion composition of the plasma sheet, J. Geophys. Res., 86, 761, 1981.
Peterson, W. K., E. G. Shelley, G. Haerendel, and G. Paschmann, Energetic ion composition in the subsolar magnetopause and boundary layer, J. Geophys. Res., 87, 2139, 1982.
Peterson, W. K., E. G. Shelley, R. D. Sharp, R. G. Johnson, J. Geiss, and H. Rosenbauer, H^+ and He^{++} in the dawnside magnetosheath, Geophys. Res. Lett., 6, 667, 1979.
Potemra, T. A., Current systems in the Earth's magnetosphere, Rev. Geophys. Space Phys., 17, 640, 1979.
Reiff, P. H., Polar and auroral phenomena: a review of U.S. progress during 1979-1982, Rev. Geophys. Space Phys., 21, 418, 1983.
Rosenbauer, H., H. Grunwaldt, M. D. Montgomery, G. Paschmann, and N. Sckopke, Heos 2 plasma observations in the distant polar magnetosphere: the plasma mantle, J. Geophys. Res., 80, 2723, 1975.
Russell, C. T., Reconnection at the Earth's magnetopause: magnetic field observations and flux transfer events, Magnetic Reconnection in Space and Laboratory Plasmas, ed. by E. W. Hones, Jr., AGU Geophys. Monograph No. 30, p. 124, 1984.
Russell, C. T., and R. C. Elphic, ISEE observations of flux transfer events at the dayside magnetopause, Geophys. Res. Lett., 6, 33, 1979.
Russell, C. T., and R. L. McPherron, The magnetotail and substorms, Space Sci. Rev., 15, 205, 1973.
Sanders, G. D., J. W. Freeman, and L. J. Maher, A two-temperature plasma distribution in the magnetosheath at lunar distances, J. Geophys. Res., 86, 2475, 1981.
Sarris, E. T., S. M. Krimigis, A. T. Y. Lui, K. L. Ackerson, L. A. Frank, and D. J. Williams, Relationship between energetic particles and plasmas in the distant plasma sheet, Geophys. Res. Lett., 8, 349, 1981.
Sarris, E. T., D. J. Williams, and S. M. Krimigis, Observations of counterstreaming between plasma and energetic particles in the magnetotail, J. Geophys. Res., 83, 5655, 1978.
Scarf, F. L., F. V. Coroniti, C. F. Kennel, E. J. Smith, J. A. Slavin, B. T. Tsurutani, S. J. Bame, and W. C. Feldman, Plasma wave spectra near slow mode shocks in the distant magnetotail, Geophys. Res. Lett., 11, 1050, 1984.

Schield, M. A., and L. A. Frank, Electron observations between the inner edge of the plasma sheet and the plasmasphere, J. Geophys. Res., 75, 5401, 1970.

Schield, M. A., J. W. Freeman, and A. J. Dessler, A source for field-aligned currents at auroral latitudes, J. Geophys. Res., 74, 247, 1969.

Scholer, M., D. Hovestadt, B. Klecker, G. Gloeckler, and F. M. Ipavich, Average flow between ~ 70 R_E and ~ 220 R_E in the geomagnetic tail, Geophys. Res. Lett., 11, 343, 1984.

Sckopke, N., G. Paschmann, G. Haerendel, B. U. O. Sonnerup, S. J. Bame, T. G. Forbes, E. W. Hones, Jr., and C. T. Russell, Structure of the low-latitude boundary layer, J. Geophys. Res., 86, 2099, 1981.

Sharp, R. D., D. L. Carr, W. K. Peterson, and E. G. Shelley, Ion streams in the magnetotail, J. Geophys. Res., 86, 4639, 1981.

Shelley, E. G., W. K. Peterson, A. G. Ghielmetti, and J. Geiss, The polar ionosphere as a source of energetic magnetospheric plasma, Geophys. Res. Lett., 9, 941, 1982.

Sibeck, D. G., G. L. Siscoe, J. A. Slavin, E. J. Smith, B. T. Tsurutani, and R. P. Lepping, The distant magnetotail's response to a strong IMF By: twisting, flattening and field line bending, (to be published), J. Geophys. Res., 1985.

Siscoe, G. L., F. L. Scarf, D. S. Intriligator, J. H. Binsack, H. S. Bridge, and V. M. Vasyliunas, Evidence for a geomagnetic wake at 500 earth radii, J. Geophys. Res., 75, 5319, 1970.

Slavin, J. A., and E. J. Smith, Solar cycle variations in the interplanetary magnetic field, Solar Wind Five, ed. by M. Neugebauer, NASA Conference Publication 2280, 323, 1983.

Slavin, J. A., E. J. Smith, D. G. Sibeck, D. N. Baker, R. D. Zwickl, and S.-I. Akasofu, An ISEE-3 study of average and substorm conditions in the distant magnetotail, (submitted for publication), J. Geophys. Res., 1985.

Slavin, J. A., E. J. Smith, B. T. Tsurutani, D. G. Sibeck, H. G. Singer, D. N. Baker, J. T. Gosling, E. W. Hones, Jr., and F. L. Scarf, Substorm associated traveling compression regions in the distant tail: ISEE-3 geotail observations, Geophys. Res. Lett., 11, 657, 1984.

Slavin, J. A., B. T. Tsurutani, E. J. Smith, D. E. Jones, and D. G. Sibeck, Average configuration of the distant (< 220 R_E) magnetotail: initial ISEE-3 magnetic field results, Geophys. Res. Lett., 10, 973, 1983.

Smith, E. J., J. A. Slavin, B. T. Tsurutani, W. C. Feldman, and S. J. Bame, Slow mode shocks in the Earth's magnetotail: ISEE-3, Geophys. Res. Lett., 11, 1054, 1984.

Sojka, J. J., L. A. Frank, and T. E. Eastman, Cold ionospheric electrons in the Earth's outer magnetosphere, (to be submitted), J. Geophys. Res., 1985.

Spreiter, J. R., A. L. Summers, and A. Y. Alksne, Hydromagnetic flow around the magnetosphere, Planet. Space Sci., 14, 223, 1966.

Tsurutani, B. T., D. E. Jones, R. P. Lepping, E. J. Smith, and D. G. Sibeck, The relationship between IMF By and the distant tail (150-238 R_E) lobe and plasma sheet By fields, Geophys. Res. Lett., 11, 1082, 1984a.

Tsurutani, B. T., D. E. Jones, and D. G. Sibeck, The two-lobed structure of the distant ($X > 200\ R_E$) magnetotail, Geophys. Res. Lett., 11, 1066, 1984b.

Tsurutani, B. T., D. E. Jones, J. A. Slavin, D. G. Sibeck, and E. J. Smith, Plasma sheet magnetic fields in the distant tail, Geophys. Res. Lett., 11, 1062, 1984c.

Vasyliunas, V. M., Observations of low-energy electrons in the evening sector of the magnetosphere with Ogo 1 and Ogo 3, J. Geophys. Res., 73, 2839, 1968.

Walker, R. C., U. Villante, and A. J. Lazarus, Pioneer 7 observations of plasma flow and field reversal regions in the distant geomagnetic tail, J. Geophys. Res., 80, 1238, 1975.

Williams, D. J., Energetic ion beams at the edge of the plasma sheet: ISEE 1 observations plus a simple explanatory model, J. Geophys. Res., 86, 5507, 1981.

Williams, D. J., D. G. Mitchell, T. E. Eastman, and L. A. Frank, Energetic particle observations in the low latitude boundary layer, (accepted for publication), J. Geophys. Res., 1985.

Wolfe, J. H., The large-scale structure of the solar wind, Solar Wind, ed. by C. P. Sonett, P. J. Coleman, Jr., and J. M. Wilcox, NASA SP-308, p. 170, 1972.

Wolfe, J. H., R. W. Silva, D. D. McKibbin, and R. H. Mason, Preliminary observations of a geomagnetospheric wake at 1000 earth radii, J. Geophys. Res., 72, 4577, 1967.

Yeager, D. M., and L. A. Frank, Low-energy electron intensities at large distances over the Earth's polar cap, J. Geophys. Res., 81, 3966, 1976.

WAVE OBSERVATIONS IN OUTER PLANET MAGNETOSPHERES

Frederick L. Scarf
TRW Inc.
One Space Park
Redondo Beach, CA 90278 U.S.A.

ABSTRACT. The first measurements of plasma waves and wave-particle interactions in the magnetospheres of the outer planets were provided by instruments on Voyager 1 and 2. At Jupiter, the observations yielded new information on upstream electrons and ions, bow shock dissipation processes, trapped radio waves in the magnetospheres and extended Jovian magnetotail, pitch angle diffusion mechanisms and whistlers from atmospheric lightning. Many of these same emissions were detected at Saturn. In addition, the Voyager plasma wave instruments detected dust particles associated with the tenuous outer rings of Saturn as they impacted the spacecraft. Most of the plasma wave activity at Jupiter and Saturn is in the audio range, and recordings of the wave observations have been useful for analysis.

INTRODUCTION

The study of plasma instabilities and wave-particle interactions at Jupiter and Saturn is of particular interest because: (a) the planetary magnetic fields are strong while the outer solar system wind pressure is low, so that the magnetospheres, including the extended tails, are truly enormous--in fact, the magnetic tail of Jupiter is known to reach beyond the orbit of Saturn; (b) the rapid spin rates of Jupiter and Saturn coupled with the strong planetary magnetic fields lead to very high corotation speeds for charged particles, and this introduces certain novel plasma interaction and coupling phenomena; (c) the icy satellites and the ring systems are within the Jupiter and Saturn magnetospheres and these objects act as sources and sinks of plasma as well as particulate matter (dust)--the dust produces plasma upon impact with the spacecraft, and dilute rings of dust also perturb the ambient plasma distribution functions; (d) both Jupiter and Saturn have large magnetospheric satellites (Io and Titan) that generate significant regions of enhancement in the neutral gas density--in the case of Jupiter the satellite gas production is clearly associated with a localized plasma torus of such high density that many inner magnetosphere processes are controlled by torus interaction phenomena.

All of the outer planets plasma wave measurements were provided by

Space Science Reviews **42** (1985) 241–255. 0038–6308/85.15

identical instruments flown on Voyager 1 and 2. The sensor for the Voyager plasma wave investigation is a balanced single-axis electric dipole with effective length of seven meters, and the wave measurements utilize: (a) a 16-channel spectrum analyzer covering the range 10 Hz to 56 kHz with four filters per decade (i.e., 10, 17.8, 31.1 and 56 Hz, . . .) and (b) a broadband channel that records the waveform over the range 50 Hz to 14.4 kHz (28,800 E-field measurements per second). Most important characteristics of the instrument are described in papers by Scarf and Gurnett (1977) and Scarf et al. (1981b), but the real assessment of the measurement capability requires study of the overall science return. Accordingly, the reference section for this chapter contains a complete listing of all completed publications by the Voyager plasma wave investigation groups at TRW and at the University of Iowa; the remaining parts of this chapter will serve as a summary and guide to the Voyager plasma wave literature.

THE JUPITER ENCOUNTERS

During the initial cruise period out to 5 AU, the Voyager plasma wave investigators studied typical interplanetary phenomena such as shocks, Type III solar radio bursts and ion acoustic waves (Gurnett et al., 1978; Gurnett, 1981; Kurth et al., 1979a; Burlaga et al., 1980; Scarf et al., 1981a; 1982a). However as Voyager 1 approached Jupiter in February 1979, the plasma wave instrument began to detect strong radio emissions, as well as increasing ion acoustic wave activity that was interpreted in terms of entrance into the planetary foreshock region (Scarf et al., 1979a). It is of interest to note that many of these upstream noise bursts were subsequently identified as waves generated by escaping sulphur and oxygen ions that emanated from the volcanic regions of the moon Io (Scarf et al., 1979a; 1982c). The surprise associated with this discovery was not an isolated event; in terms of plasma physics, the outer planet magnetospheres were frequently found to behave in novel and unexpected ways.

The Voyager bow shock crossings provided new physical insights along with additional surprises. The initial discussions (Scarf et al., 1979a; 1979b; 1971b; Gurnett et al., 1979c; Gurnett and Scarf, 1983) focused attention on a comparison of bow shock plasma wave spectra in the inner and outer solar system; the high Mach numbers encountered at 5 AU lead to supercritical shocks and the wave spectra have characteristics that differ from those at Earth and Venus. Using the Voyager plasma broadband data, it was also possible to construct detailed fourier transforms that yielded very high resolution in both frequency and time. The bottom panel in Figure 1 contains a plasma wave frequency-time diagram for one Voyager 1 bow shock crossing, and this shows that intense low frequency ($f < 3$ kHz) wave turbulence developed after 1226:45. In the initial discussions of this event, Scarf et al. (1981b) and Gurnett and Scarf (1983) identified the 6 kHz waves detected before 1226:42 as electron plasma oscillations (see Gurnett et al., 1981a) and they associated the shock crossing with the onset of the low frequency turbulence. However, Moses et al. (1985) recently put together the wave data and the B-field profile shown in the upper panel of Figure 1, and

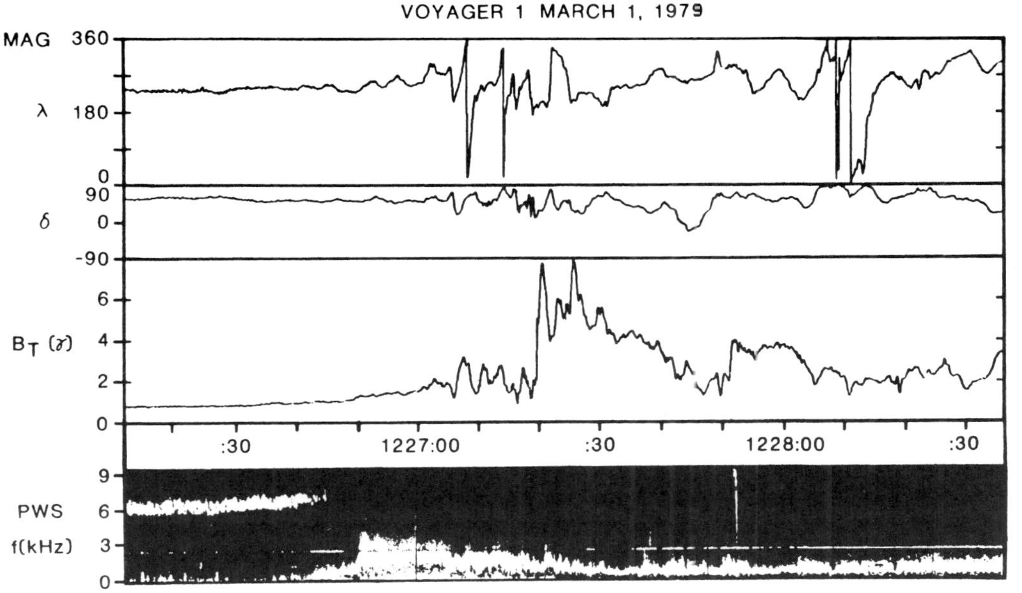

Figure 1. Magnetic field and plasma wave observations at a crossing of the Jupiter bow shock. Here one B-field vector and one E-field wave spectrum are measured every 60 milliseconds; the gray scale in the lower panel is arranged so that the most intense waves are white.

this complete plot shows that the actual bow shock crossing occurred at 1227:20, so that the intense wave activity actually developed in the precursor foot region. Detailed analysis of high frequency waves detected in the upstream region on Voyager 2 revealed other very novel phenomena; Moses et al. (1984) found that strong electron heat conduction from the shock produced broadband electrostatic waves having frequencies well below the electron plasma frequency.

These results by Moses et al. are of considerable physical significance because they reveal a complex sequence of developments in which plasma waves heat electrons upstream from the current ramp, and they also show that simple arguments involving Doppler shifts are not adequate to explain certain prominent Voyager wave observations. More generally, these results illustrate the relatively immature nature of the data analysis program for Voyager, since few high time-resolution multi-instrument correlation studies were carried out in the first few years after encounter. Although very many excellent scientific publications have emerged (see the large number of plasma wave references, for example), much remains to be done.

Within the Jovian magnetosphere, the Voyager instruments detected a great many forms of wave activity. The measurements of continuum radiation (see Figure 2) yielded absolute electron density profiles throughout much of the low latitude magnetosphere (Gurnett et al., 1980; 1971b) and other techniques provided information on proton

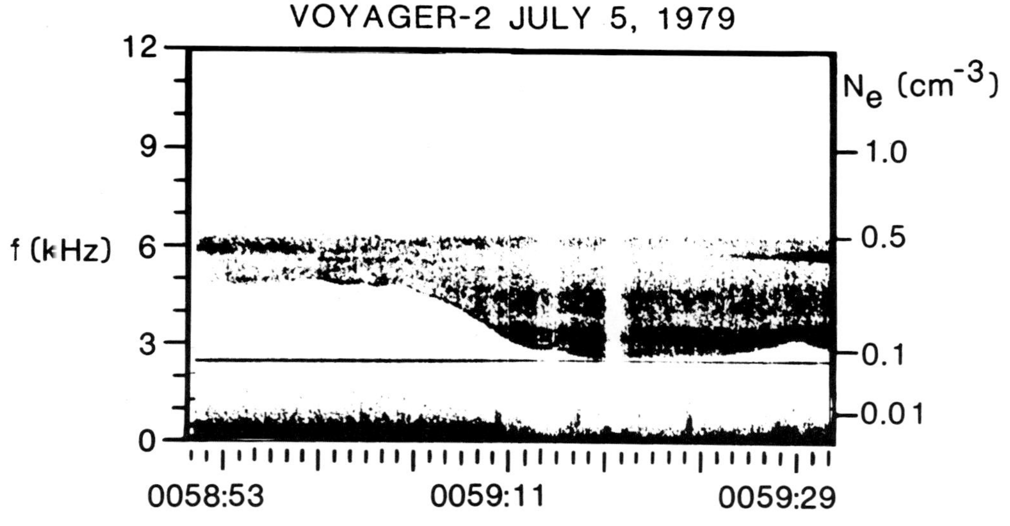

Figure 2. Detection of electromagnetic continuum radiation trapped in the outer magnetosphere of Jupiter. Here (and in Figure 3) the most intense waves have dark gray shading.

densities (Barbosa et al., 1979; Tokar et al., 1982a; 1982b). Studies of lightning from Jupiter's atmosphere (Gurnett et al., 1979b; Menietti and Gurnett, 1980; Scarf et al., 1981c; Borucki et al., 1982; Kurth et al., 1985) and analysis of "auroral hiss" analogs (Gurnett et al., 1979a) provided additional non-local information on density profiles; studies of kilometric radio emission from Jupiter (Kurth et al., 1979b; 1980; Green and Gurnett, 1980; Gurnett and Goertz, 1981) also gave very significant information related to plasma density models.

Many of the important Jupiter observations involved waves of the type associated with plasma instabilities and strong wave-particle interactions. The pre-Voyager expectations were generally focused on the search for electron whistler mode turbulence, as well as ion cyclotron waves. Accordingly, when intense low frequency ($f < f_c/2$) waves appeared in the 16-channel data as Voyager traversed the high density plasma of the Io torus, it was natural to identify this as whistler mode chorus and hiss (Scarf et al., 1979a); pitch-angle scattering calculations then suggested that the hiss and chorus together could produce enough scattering to account for the observed auroral precipitation (Scarf et al., 1979c). However, while analysis of the waveform data (see Figure 3) conclusively established that those waves with f just below $f_c/2$ do represent chorus that effectively precipitates soft electrons (Coroniti et al., 1980; see also Coroniti, 1984; Inan et al., 1983; Reinleitner et al., 1984), other possibilities for wave-particle scattering of ions and energetic electrons have been proposed (Barbosa and Kurth, 1980).

Kurth et al. (1980a) and Barbosa et al. (1981) discussed the detection of relatively weak broadband electrostatic waves and field-

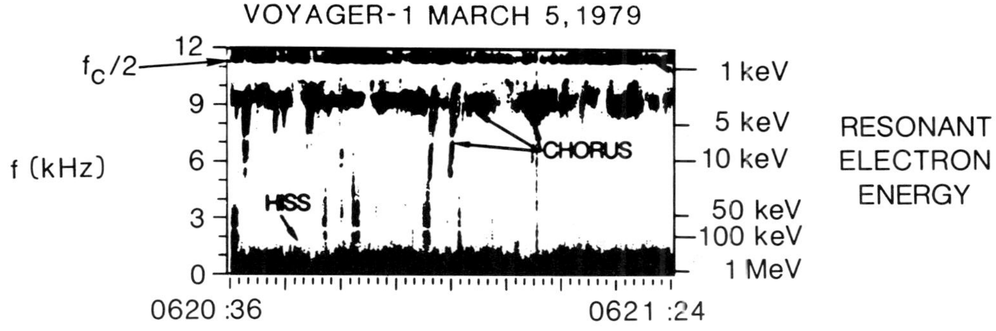

Figure 3. Detection of chorus and $f_c/2$ emissions in the inner magnetosphere of Jupiter. The electron energies for cyclotron resonance with the whistler signals are shown on the right.

aligned current systems near the outer boundary of the equatorial plasma sheet, and Scarf et al. (1981b) noted that within the torus, very intense broadband waves were detected in regions where there were also significant changes in ion phase space density; they conjectured that these strong signals represented Doppler shifted ion acoustic waves, with the shift arising because of the very high corotation speed of the Jovian plasma sheet.

Another area of recent speculation involves waves with frequencies in the range of the lower hybrid resonance and ion cyclotron frequencies. These analyses are complicated by the presence of several spacecraft interference effects that produce spurious signals in the channels below 100 Hz; aside from the normal physical questions of data interpretation, the investigators have had to face significant uncertainties related to the separation of ambient and spacecraft signals. Nevertheless, since the problems are quite important, a number of reports have addressed the analysis of the Voyager low frequency plasma wave measurements at Jupiter. Gurnett and Goertz (1983) retained the interpretation that the low frequency waves are generated in the electron resonant whistler mode, but they introduced the concept of conversion to ion cyclotron waves by a polarization reversal process at the local crossover frequency. This model would give rise to strong ion pitch angle diffusion, and it was suggested that this would provide significant auroral emission. Thorne and Scarf (1984) presented a different interpretation of the observed enhancement in wave energy for $f < f_c^+$; they argued that this implied direct excitation of ion cyclotron waves, presumably associated with strong pitch angle scattering of ions into the loss cone. In a somewhat higher frequency range, Barbosa et al. (1985) identified lower hybrid resonance emissions, and they noted that these waves could provide an important mechanism for heating the plasma. However, as noted above, the Voyager low frequency wave measurements have serious limitations; it is hoped that in the future the Galileo plasma wave investigation will be able to resolve many of the uncertainties that remain in the analysis of the low frequency Voyager data from Jupiter.

THE EXTENDED JUPITER TAIL

Scarf (1979) noted that Voyager 2 would travel from Jupiter to Saturn more or less along the Jupiter-Sun line. He conjectured that the spacecraft might encounter Jupiter's magnetic tail, if the tail extended to several hundred subsolar magnetopause boundary lengths (as might be expected from theoretical considerations, and from scaling of the terrestrial tail), and he also speculated that Saturn might be within Jupiter's magnetic tail during the Voyager 2 flyby. The first evidence of trapped continuum radiation in Jupiter's distant tail was obtained in February 1981, when Voyager 2 was more than 6000 Rj (or 2.8 Astronomical Units) behind Jupiter. Figure 4, adapted from a paper by Scarf et al. (1981e), shows the weak wave levels that suggest a dip in density down to about 0.001 electrons per cubic centimeter. Figure 4 also demonstrates a modulation consistent with the Jupiter rotation period, and this reinforces the conclusion that these waves have a Jovian origin.

Several additional studies (Kurth et al., 1981a; 1982b; Lepping et al., 1983) provided confirmation of the extended tail interpretation. Moreover, some aspects of the Voyager 2 Saturn encounter data immediately suggested (Scarf et al., 1982b) that Saturn was indeed within the Jupiter tail during a part of the magnetosphere traversal, and evidence for subsequent tail encounters was also presented (Scarf et al., 1983c). Thus, it seems assured that Jupiter does have an enormous magnetic tail (length greater than 4 AU) and that plasma wave observations of continuum radiation provide an excellent way to identify the tail posi-

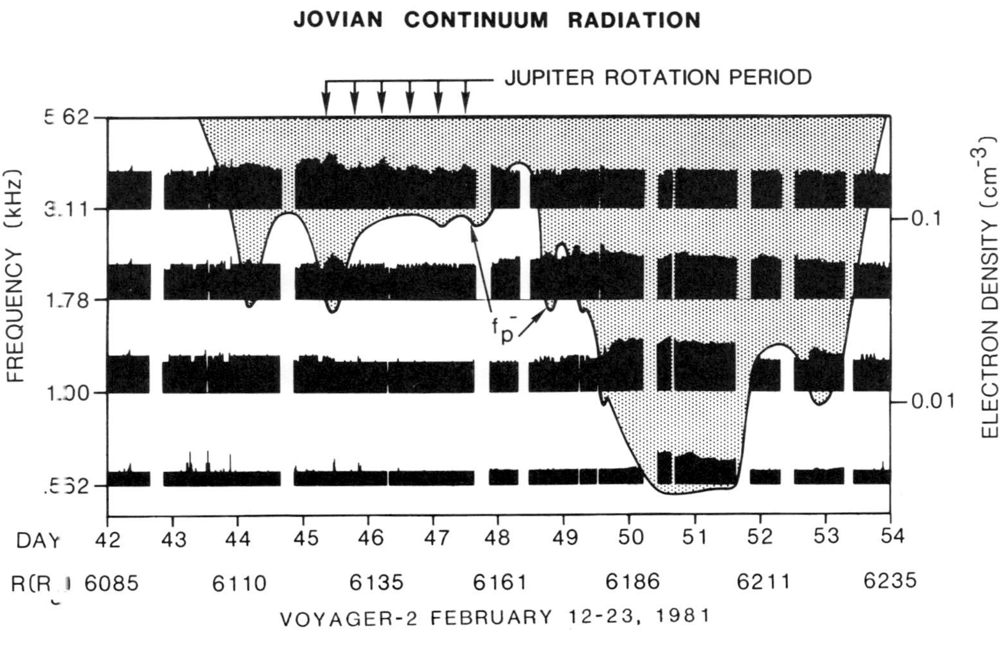

Figure 4. Detection of continuum radiation in the distant tail of Jupiter.

tions. However, there are many unsolved questions related to these investigations. In particular, from the point of view of this review, it is worth noting that no theories have been developed to account for the intensity of the continuum radiation within the distant tail.

THE SATURN ENCOUNTERS

At Saturn there were relatively few opportunities to acquire wideband plasma wave data, and during the bow shock traversals, only spectrum analyzer measurements were available (Gurnett et al., 1981c; Scarf et al., 1982b; 1983c; see also Scarf, 1984). In a paper by Scarf et al. (1981d), these bandpass measurements (see Figure 5) were used to suggest a significant change on the bow shock turbulence spectrum as the heliocentric distance increases, but the new interpretation of the high resolution Voyager 1 shock data (see Figure 1 and associated discussion) raises the possibility that these spectral variations could actually develop in the foot region rather than in the shock layer; thus despite the appearance, the variations in Figure 5 might not be directly related to outer solar system changes in the physics of collisionless shocks.

The Voyager 1 trajectory included a very close Titan flyby designed to provide information about the interaction of the Titan atmosphere

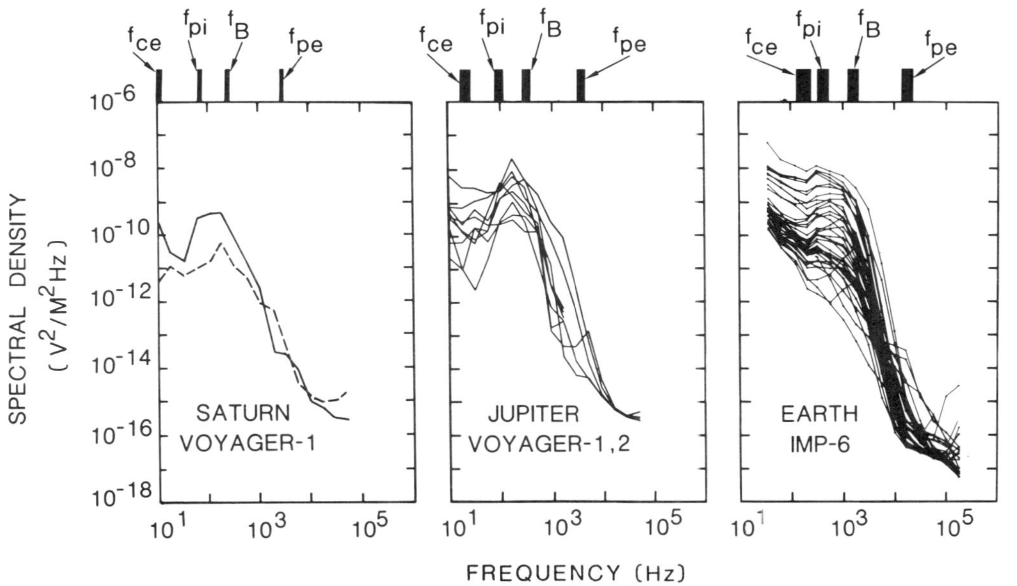

Figure 5. Typical bow shock wave spectra at Earth, Jupiter and Saturn. The characteristic frequencies are the electron and ion plasma frequencies (f_{pe}, f_{pi}), the electron cyclotron frequency (f_{ce}) and the Buneman mode frequency (f_B).

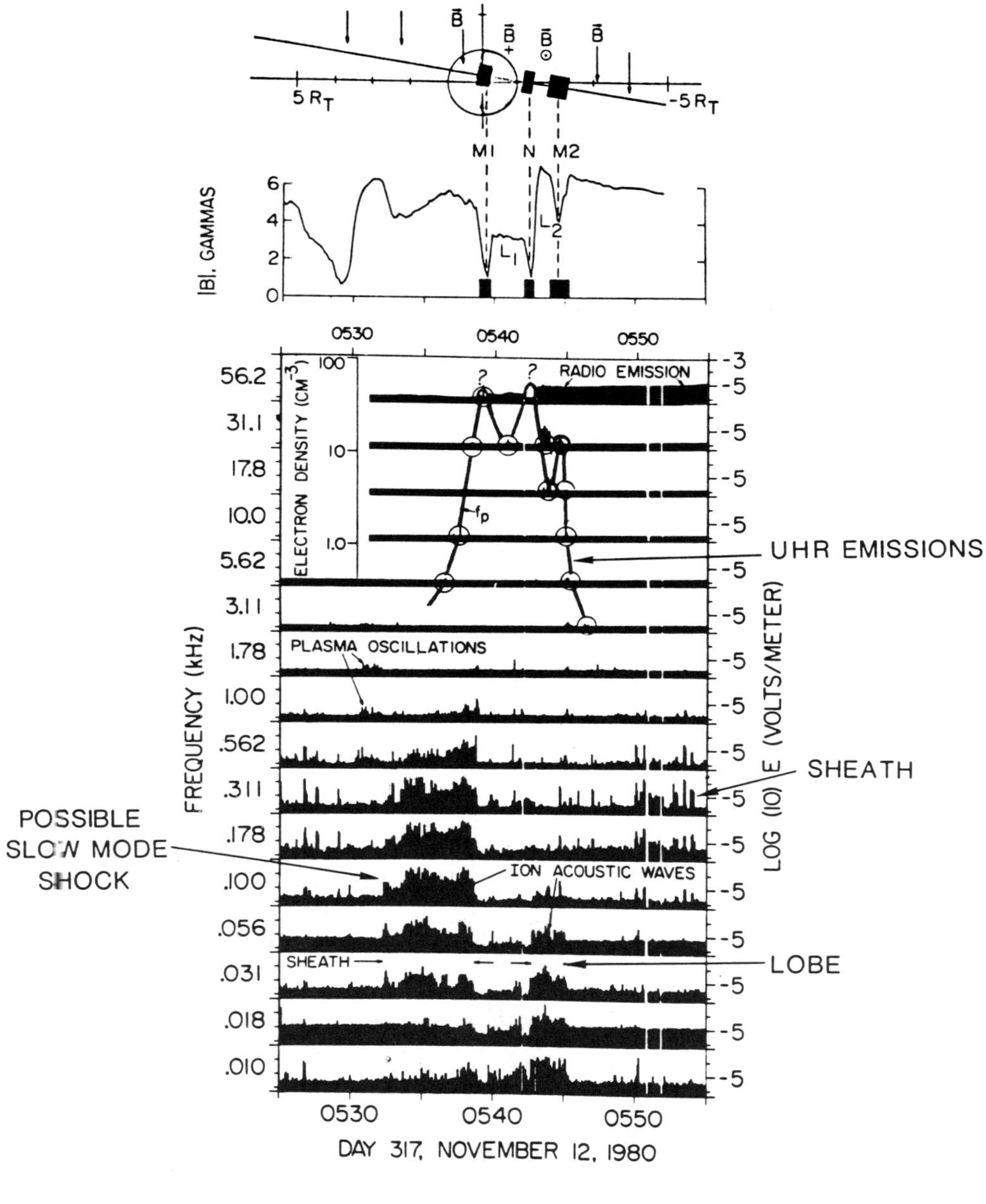

Figure 6. Magnetic field profile and plasma wave amplitude variations as Voyager 1 traversed the Titan wake.

and ionosphere with the ambient plasma. On November 12, 1980, Titan was within Saturn's magnetosphere and the interaction involved the corotating plasma of Saturn's outer magnetosphere. Figure 6 shows the bandpass channel observations of strong turbulence associated with the crossing of Titan's wake. The wave instruments provided crucial information on the density of the cold electrons in this wake region (see Gurnett et al., 1981c; 1982; Neubauer et al., 1984). It is possible that some features of the comet-solar wind interaction are well represented here since the cometary plasma is also likely to have a very low temperature.

As the Voyager spacecraft moved in towards closest approach, the plasma wave and radio astronomy instruments both detected very baffling phenomena associated with the Saturn kilometric radiation. Pioneer 11 had already demonstrated that the spin axis of Saturn is essentially parallel to its magnetic dipole, but still Voyager found that the radio emission amplitude was strongly modulated at the planetary spin rate (see the summary by Kaiser et al., 1984). On Voyager 1 it was also demonstrated that the lowest frequency ($f < 100$ kHz) emissions were affected by the orbital position of Dione, a satellite with strange linear markings on parts of its surface (Kurth et al., 1981b), but during the Voyager 2 encounter the Dione modulation effect was far less evident (perhaps because Saturn was within the Jovian tail for part of the time). In terms of electromagnetic radiation from Saturn, another topic of interest concerns the detection of lightning signals; although no whistler mode impulses were detected during the encounters (Gurnett et al., 1981c; Scarf et al., 1982b), relatively high frequency electromagnetic impulses from nearly equatorial lightning bursts were detected by the radio astronomy instrument (Kaiser et al., 1984; see also Kurth et al., 1983a).

Figures 7 and 8 show the bandpass channel measurements near closest approach. The local wave measurements in the inner magnetosphere have been described in several survey papers (Gurnett et al., 1981c; Scarf et al., 1982b, 1983c, 1984; Kurth et al., 1983b) and a number of specialized reports have also appeared. Conventional magnetospheric plasma wave modes such as chorus, electron gyrofrequency harmonics, upper hybrid resonance emissions and broadband electrostatic waves were clearly detectable during the Saturn traversals, and the f_c, f_p curves in Figure 7,8 can be used to identify these observations. However, even for these "expected" measurements, some unexpected characteristics were found. For instance, in this planet with a low density radiation belt, it is difficult to account for the chorus observations since they are generally thought to be associated with stable trapping limit phenomena. A number of problems that were initially classified as mysteries were resolved after detailed analysis. For instance, the smoothly varying waves with well-defined lower cutoff frequency detected on Voyager 1 after 2000 UT (Figure 7) were first identified as electromagnetic continuum radiation, but a wideband frame recorded at 2252 on November 12 showed a sequence of many narrowband emissions, rather than the expected continuum (Gurnett et al., 1981d). This result was especially surprising because at the time, it was concluded that no such emissions had appeared at Earth or Jupiter; however, Gurnett et al. (1983b) later showed that such narrowband features were present in many Jupiter wideband spectra, and they interpreted them in terms of mode conversion involving upper hybrid resonance emissions and electron cyclotron harmonics. Later, Kurth et al. (1982a) identified a much weaker signal as the actual continuum radiation.

Figure 8 shows that Voyager 2 detected very intense broadband wave activity in a very narrow region near the Saturn ring plane. Scarf et al. (1982b) used high resolution wideband measurements (one data point per 35 microseconds) to show that the noise impulses represented dust impacts, and not plasma waves. Gurnett et al. (1983) later performed a detailed analysis of these observations; they showed how the pulses are generated by impact on the spacecraft, and they deduced very significant information about Saturn's G-ring. Scarf et al. (1983c) noted that similar, but more isolated, impacts had been detected on Voyager 1 near the ring plane and even at high latitudes. These exercises demonstrate the need for great caution in the analysis of wave data from an exotic new planetary system; it is not always possible to use observations from the Earth's magnetosphere to interpret the new measurements.

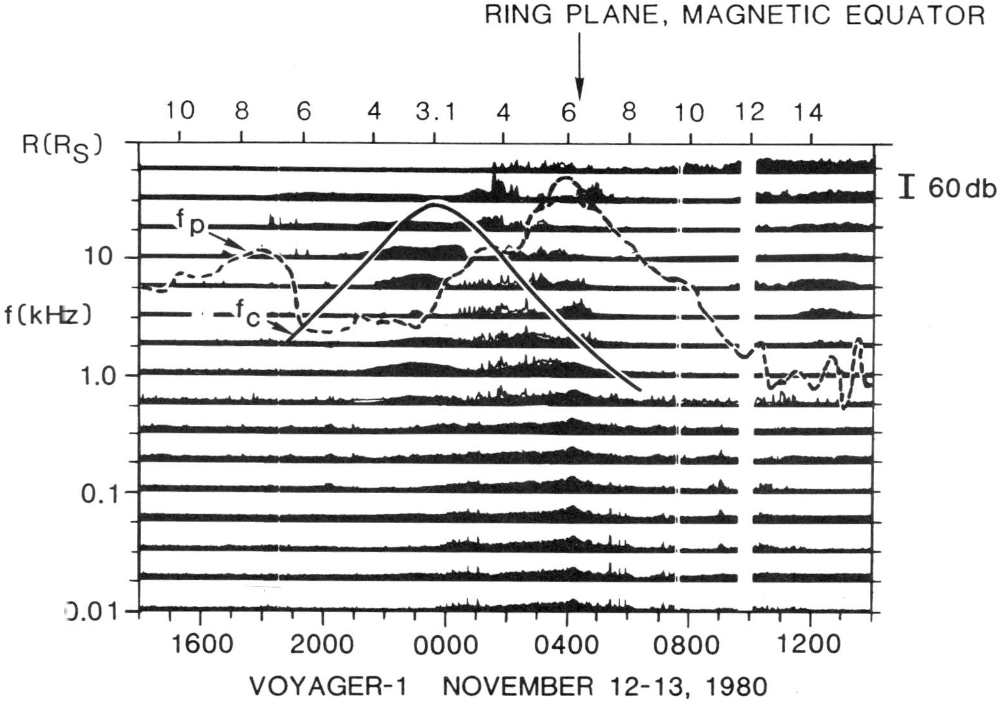

Figure 7. Plasma wave amplitude variations during the Voyager 1 traversal of Saturn's inner magnetosphere. The f_c curve comes from the magnetometer data and the f_p curve is derived from the plasma probe observations.

BEYOND JUPITER AND SATURN

The Voyager 2 post-Saturn trajectory was designed to include encounters with Uranus and Neptune, while Voyager 1 moved well out of the ecliptic plane, and far from all planets, after the Saturn swingby. In 1983 and 1984 the wave instruments on both of these outer solar system spacecraft detected weak radio emissions that could be associated with a fascinating and unexpected source--the boundary between the solar wind and the interstellar medium, frequently called the heliopause.

The observations and interpretations were presented in two reports by Kurth et al. (1984a; 1984b). A weak signal with frequency near 3 kHz and variable amplitude was detected first on Voyager 1 and later on Voyager 2. The top panel in Figure 9 shows the 3 kHz bandpass channel data from Voyager 1 in 1983 and 1984. By comparing data from the two spacecraft, Kurth et al. were able to rule out Jupiter, Saturn and Uranus as sources, and this suggested a source beyond the solar system. The bottom panel in Figure 9 illustrates the explanation offered by the Voyager investigators. It was suggested that a terminal shock forms at the heliopause, and that the radio waves are generated in this region. The specific mechanism indicated in the figure involves the generation, by suprathermal shock-accelerated electrons, of electrostatic waves at f_p and $2f_p$, and the subsequent coupling to electromagnetic waves at $2f_p$. This mechanism operates in the Earth's foreshock,

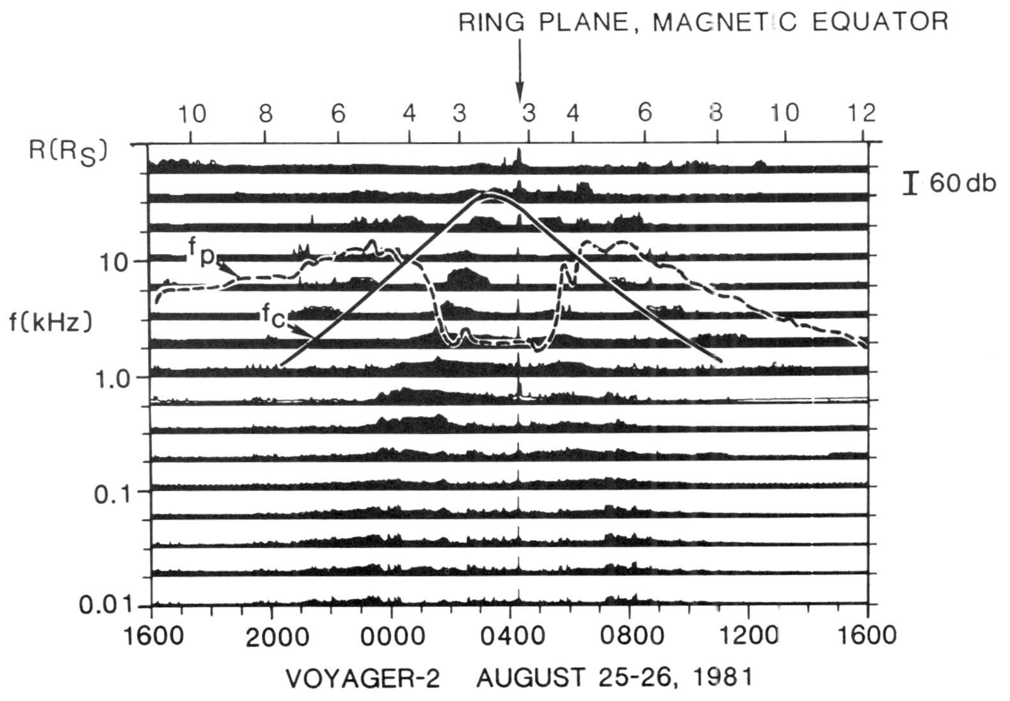

Figure 8. Voyager 2 observations in the format of Figure 7.

and it is commonly invoked to explain Type III solar radio bursts. If this explanation is correct, the Voyager measurements imply that in 1983-84, the heliopause was at a radial distance of about 45 AU. Voyager 1 will reach this distance in 1991-92, and it will provide a long-range opportunity to test this theory.

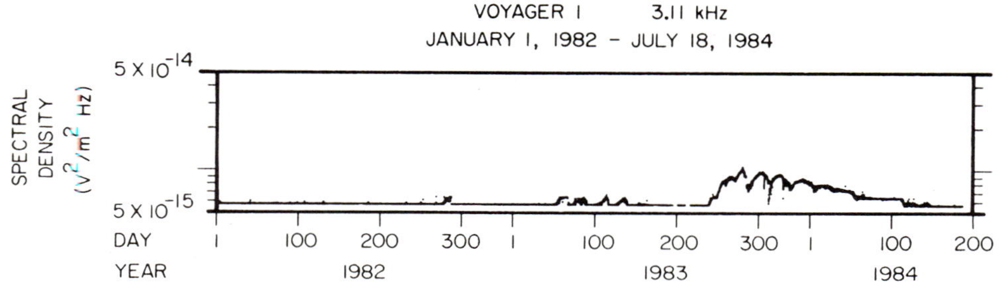

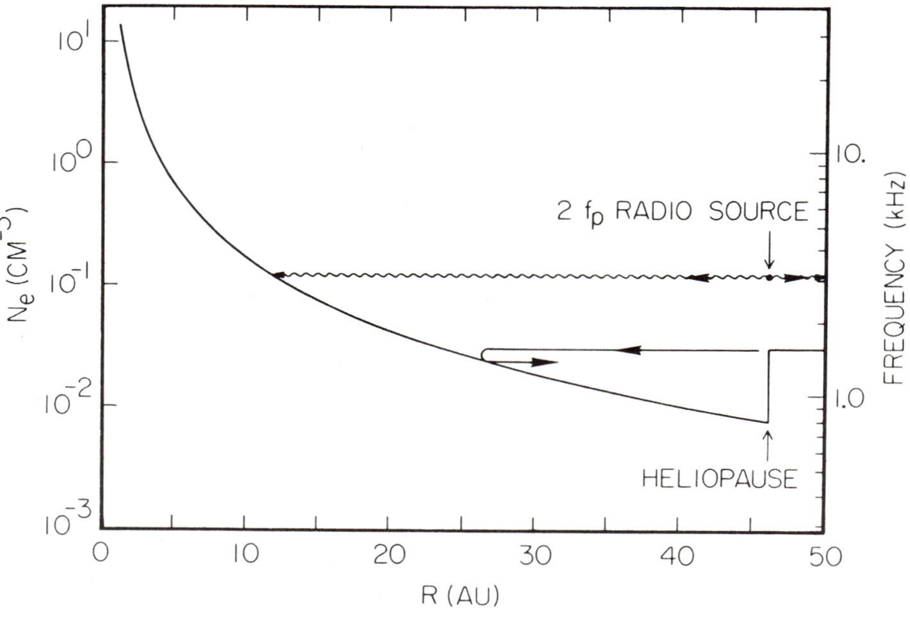

Figure 9. Top: Voyager 1 3 kHz cruise observations from 1982-84. Bottom: An interpretation of the 3 kHz data in terms of a heliopause source.

Acknowledgments. The research was supported by NASA through Contract 954012 with the Jet Propulsion Laboratory.

REFERENCES

Barbosa, D. D., Gurnett, D. A., Kurth, W. S., and Scarf, F. L.: 1979, *Geophys. Res. Lett.*, 6, p. 785.
Barbosa, D. D., and Kurth, W. S.: 1980, *J. Geophys. Res.*, 85, p. 6729.
Barbosa, D. D., Scarf, F. L., Kurth, W. S., and Gurnett, D. A.: 1981, *J. Geophys. Res.*, 86, p. 8357.
Barbosa, D. D., Coroniti, F. V., Kurth, W. S., and Scarf, F. L.: 1985, 'Observations of lower hybrid noise in the Io plasma torus and anomalous plasma heating rates', *Astrophysical Journal* (in press).
Borucki, W. J., Bar-Nun, A., Scarf, F. L., Cook II, A. F., and Hunt, G. E.: 1982, *Icarus*, 52, pp. 492-502.
Burlaga, L., Lepping, R., Weber, R., Armstrong, T., Goodrich C., Sullivan, J., Gurnett, D., Kellogg, P., Keppler, E., Mariani, F., Neubauer, F., Rosenbauer, H., and Schwenn, R.: 1980, *J. Geophys. Res.*, 85, p. 2227.
Coroniti, F. V., Scarf, F. L., Kennel, C. F., Kurth, W. S., and Gurnett, D. A.: 1980, *Geophys. Res. Lett.*, 7, p. 45.
Coroniti, F. V., Scarf, F. L., Kennel, C. F., and Kurth, W. S.: 1984, *J. Geophys. Res.*, 89, p. 3801.
Green, J. L., and Gurnett, D. A.: 1980, *Geophys. Res. Lett.*, 7, p. 65.
Gurnett, D. A., Anderson, R. R., Scarf, F. L., and Kurth, W. S.: 1978, *J. Geophys. Res.*, 83, p. 4147.
Gurnett, D. A., Kurth, W. S., and Scarf, F. L.: 1979a, *Nature*, 280, p. 767.
Gurnett, D. A., Shaw, R. R., Anderson, R. R., Kurth, W. S., and Scarf, F. L.: 1979b, *Geophys. Res. Lett.*, 6, p. 511.
Gurnett, D. A., Kurth, W. S., and Scarf, F. L.: 1979c, *Science*, 206, p. 987.
Gurnett, D. A., Kurth, W. S., and Scarf, F. L.: 1980, *Geophys. Res. Lett.*, 7, p. 53.
Gurnett, D. A.: 1981, 'Plasma waves in the solar wind: A review of observations', *Solar Wind 4*, H. Rosenbauer, ed., Springer, New York-Heidelberg-Berlin, p. 286.
Gurnett, D. A., and Goertz, C. K.: 1981, *J. Geophys. Res.*, 86, p. 717.
Gurnett, D. A., Maggs, J. E., Gallagher, D. L., Kurth, W. S., and Scarf, F. L.: 1981a, *J. Geophys. Res.*, 86, p. 8833.
Gurnett, D. A., Scarf, F. L., Kurth, W. S., Shaw, R. R., and Poynter, R. L.: 1981b, *J. Geophys. Res.*, 86, p. 8199.
Gurnett, D. A., Kurth, W. S., and Scarf, F. L.: 1981c, *Science*, 212, p. 235.
Gurnett, D. A., Kurth, W. S., and Scarf, F. L.: 1981d, *Nature*, 292, p. 733.
Gurnett, D. A., Scarf, F. L., and Kurth, W. S.: 1982, *J. Geophys. Res.*, 87, p. 1395.
Gurnett, D. A., and Scarf, F. L.: 1983, 'Plasma waves in the Jovian magnetosphere', *Physics of the Jovian Magnetosphere*, A. J. Dessler, ed., Cambridge University Press, p. 285.
Gurnett, D. A., Grün, E., Gallagher, D., Kurth, W. S., and Scarf, F. L.: 1983, *Icarus*, 53, p. 236.
Gurnett, D. A., Kurth, W. S., and Scarf, F. L.: 1983b, *Nature*, 302,

p. 385.
Gurnett, D. A., and Goertz, C. K.: 1983, *Geophys. Res. Lett.*, 10, p. 587.
Inan, U. S., Helliwell, R. A., and Kurth, W. S.: 1983, *J. Geophys. Res.*, 88, p. 6171.
Kaiser, M. L., Desch, M. D., Kurth, W. S., Lacacheux, A., Genova, F., Pedersen, B. M., and Evans, D. R.: 1984, 'Saturn as a radio source', *Saturn*, T. Gehrels and M. Matthews, ed., University of Arizona Press, Tucson, pp. 378-415.
Kurth, W. S., Gurnett, D. A., and Scarf, F. L.: 1979a, *J. Geophys. Res.*, 84, p. 3413.
Kurth, W. S., Barbosa, D. D., Scarf, F. L., Gurnett, D. A., and Poynter, R. L.: 1979, *Geophys. Res. Lett.*, 6, p. 747.
Kurth, W. S., Barbosa, D. D., Gurnett, D. A., and Scarf, F. L.: 1980a, *Geophys. Res. Lett.*, 7, p. 57.
Kurth, W. S., Gurnett, D. A., and Scarf, F. L.: 1980b, *Geophys. Res. Lett.*, 7, p. 61.
Kurth, W. S., Gurnett, D. A., Scarf, F. L., Poynter, R. L., and Sullivan, J. D.: 1981a, *J. Geophys. Res.*, 86, p. 8402.
Kurth, W. S., Gurnett, D. A., and Scarf, F. L.: 1981b, *Nature*, 292, p. 742.
Kurth, W. S., Scarf, F. L., Sullivan, J. D., and Gurnett, D. A.: 1982a, *Geophys. Res. Lett.*, 9, p. 889.
Kurth, W. S., Sullivan, J. D., Gurnett, D. A., Scarf, F. L., Bridge, H. S., and Sittler, E. C.: 1982b, *J. Geophys. Res.*, 87, p. 10,373.
Kurth, W. S., Gurnett, D. A., and Scarf, F. L.: 1983a, *Icarus*, 53, p. 255.
Kurth, W. S., Scarf, F. L., and Gurnett, D. A.: 1983b, *J. Geophys. Res.*, 88, p. 8959.
Kurth, W. S., Gurnett, D. A., Scarf, F. L., and Poynter, R. L., 1984a, 'Detection of a radio emission at 3 kHz in the outer heliosphere', *Nature*, 312, p. 27.
Kurth, W. S., Strayer, B. D., Gurnett, D. A., and Scarf, F. L.: 1985 'A summary of whistlers observed by Voyager 1 at Jupiter', *Icarus* (in press).
Kurth, W. S., Gurnett, D. A., Scarf, F. L., and Poynter, R. L.: 1984b, 'A new radio emission at 3 kHz in the outer heliosphere', Proceedings of the Varenna Conference on Plasma Astrophysics, p. 285.
Lepping, R. P., Desch, M. D., Klein, L. W., Sittler, Jr., E. C., Sullivan, J. D., Kurth, W. S., and Behannon, K. W.: 1983, *J. Geophys. Res.*, 88, p. 8801.
Menietti, J. D., and Gurnett, D. A.: 1980, *Geophys. Res. Lett.*, 7, p. 49.
Moses, S. L., Coroniti, F. V., Kennel, C. F., and Scarf, F. L.: 1984, *Geophys. Res. Lett.*, 11, p. 869.
Moses, S. L., Coroniti, F. V., Kennel, C. F., Scarf, F. L., Greenstadt, E. W., Kurth, W. S., and Lepping, R. P.: 1985, *Geophys. Res. Lett.*, (in press).
Neubauer, F. M., Gurnett, D. A., Scudder, J. D., and Hartle, R. E.: 1984, 'Titan's magnetospheric interaction', *Saturn*, T. Gehrels and M. Matthews, ed., University of Arizona Press, Tucson, pp. 760-787.
Reinleitner, L. A., Kurth, W. S., and Gurnett, D. A.: 1984, *J. Geophys.*

Res., 89, p. 75.
Scarf, F. L., and Gurnett, D. A.: 1977, *Space Science Reviews*, 21, p. 289.
Scarf, F. L., Gurnett, D. A., and Kurth, W. S.: 1979a, *Science*, 204, p. 991.
Scarf, F. L., Gurnett, D. A., Kurth, W. S., and Poynter, R. L.: 1979b, *Nature*, 280, p. 796.
Scarf, F. L., Coroniti, F. V., Gurnett, D. A., and Kurth, W. S.: 1979c, *Geophys. Res. Lett.*, 6, p. 653.
Scarf, F. L.: 1979, *J. Geophys. Res.*, 84, p. 4422.
Scarf, F. L., Gurnett, D. A., and Kurth, W. S.: 1981a, 'The first year of Voyager plasma wave observations in the solar wind', *Solar Wind 4*, H. Rosenbauer, ed., Max-Planck-Institut Report Nc. MPAE-W-100-81-31, Lindau, Germany, p. 305.
Scarf, F. L., Gurnett, D. A., and Kurth, W. S.: 1981b, *J. Geophys. Res.*, 86, p. 8181.
Scarf, F. L., Gurnett, D. A., Kurth, W. S., Anderson, R. R., and Shaw, R. R.: 1981c, *Science*, 213, p. 684.
Scarf, F. L., Gurnett, D. A., and Kurth, W. S.: 1981d, *Nature*, 292, p. 747.
Scarf, F. L., Kurth, W. S., Gurnett, D. A., Bridge, H. S., and Sullivan, J. D.: 1981e, *Nature*, 292, p. 585.
Scarf, F. L., Gurnett, D. A., Kurth, W. S., and Shaw, R. R.: 1982a, 'Voyager 1, 2 plasma wave observations for the September 1977 storm period', *UAG Report No. 83*, p. 152.
Scarf, F. L., Gurnett, D. A., Kurth, W. S., and Poynter, R. L.: 1982b, *Science*, 215, p. 587.
Scarf, F. L., Coroniti, F. V., Kennel, C. F., and Gurnett, D. A.: 1982c, *Vistas in Astronomy*, 25, p. 263.
Scarf, F. L., Gurnett, D. A., Kurth, W. S., and Poynter, R. L.: 1983, *J. Geophys. Res.*, 88, p. 8971.
Scarf, F. L., Frank, L. A., Lanzerotti, L. J., Lepping, R. P., McNutt, Jr., R. L., and Gurnett, D. A.: 1984, 'Plasma distributions and wave-particle interactions in Saturn's inner magnetosphere', *Saturn*, T. Gehrels and M. Matthews, ed., University of Arizona Press, Tucson, pp. 318-353.
Scarf, F. L.: 1984, 'Plasma waves in space', in *Radiation in Plasmas*, vol. 1, B. McNamara, ed., World Scientific, pp. 103-133.
Thorne, R. M., and Scarf, F. L.: 1984, *Geophys. Res. Lett.*, 11, p. 263.
Tokar, R. L., Gurnett, D. A., Bagenal, F., and Shaw, R. R.: 1982a, *J. Geophys. Res.*, 87, 2241.
Tokar, R. L., Gurnett, D. A., and Bagenal, F.: 1982b, *J. Geophys. Res.*, 87, p. 10,395.

THE MENAGERIE OF GEOSPACE PLASMA WAVES

Stanley D. Shawhan
Office of Space Science and Applications
NASA Headquarters Code EES
Washington DC 20546 USA

ABSTRACT. Sounding rockets and satellites have discovered a large variety of plasma waves within the Earth's magnetosphere--geospace. These waves are found over a frequency range of millihertz to megahertz. The frequency ranges are generally associated with characteristic frequencies such as the plasma frequency and gyrofrequency. Most waves are generated by hot or streaming magnetospheric plasma; some waves are due to lightning discharges, to intentional man-made transmitters or to incidental radiation from power transmission systems. Propagation of waves from the observation region back to a probable source region can be modelled using ray tracing techniques in a model magnetosphere where the electron number density, ion composition and magnetic field vector is specified. Information in addition to the common amplitude-frequency-time spectrograms can be obtained from the received waves using multiple antennas and receivers. Cross-correlation of the wave electric and magnetic components can provide information on the wave polarization and direction of propagation and on the wave distribution function.

1. INTRODUCTION

Geospace is considered the region surrounding the Earth out beyond the bow shock at 20 R_E on the dayside to more than 250 R_E on the nightside of the Earth. This region contains plasma above about 80 km altitude and plasma is the predominant state of matter above 250 km. Plasma waves have been found to occur throughout geospace through numerous sounding rocket and satellite measurements. The term plasma wave is meant to specify all waves which are generated in a plasma or interact with the plasma medium as they propagate through; space plasma waves are found in geospace, the interplanetary medium, at other planets and in the solar corona. See for example Scarf (1985) and Kimura (1985) in this volume.

Within geospace there is a menagerie of plasma waves types. Figure 1 identifies many of these types by commonly used names. In a given spatial region a number of types can occur simultanenously in

Space Science Reviews **42** (1985) 257–274. 0038–6308/85.15

several different frequency ranges. The total range observed covers millihertz to megahertz. Specific frequencies are dependent on frequencies characteristic of the plasma which generally decrease at greater distance from the Earth since they depend on plasma density which varies by six orders of magnitude and on magnetic field which varies by five orders of magnitude. Both electromagnetic and electrostatic waves can exist; both types can generally be described mathematically through plasma dispersion relations (Stix, 1962). Almost all of the plasma waves types can propagate from the source region to other remote regions, but cannot escape the plasma medium into free space. One type, the Auroral Kilometric Radiation (AKR), does escape to make the Earth a cosmic radio source at a power level of 1 billion watts.

Geospace waves are named for a variety of reasons: for the region in which they occur--such as auroral kilometric radiation and plasmaspheric hiss, for the way they sound when played through a speaker--such as whistlers and lion roars, or for the association with a characteristic frequency--such as the upper hybrid resonance frequency--UHR noise, for example. Naturally occurring plasma waves are found to be the result of electromagnetic energy from lightning discharges coupled into the magnetosphere or to energy transfer from hot or streaming plasma. Man-induced waves are also readily detected. These waves may be due to intentional transmitters such as very low frequency (VLF) code transmitters or to incidental radiation from high voltage power transmission lines. A compendium of wave types, observed frequency range, geospace location and observed characteristics is given in Table 1. Even more detailed reviews about specific wave types are given in works by Helliwell (1965), Al'pert (1974), Gendrin (1975), Shawhan (1979a) and Anderson (1983), for example. This tutorial paper draws heavily on these papers and a few other published works which are cited as appropriate. Each of these works contains an extensive bibliography which should be consulted for additional information.

The purpose of this tutorial paper is to provide some insight as to why geospace plasma waves occur in frequency and space as observed, to identify the types of sources for these waves, to describe a typical plasma wave receiving system and provide illustrations of wave types detected and displayed on spectrograms, to establish a basis for understanding the ray tracing procedure with sample results, to describe the type of information that can be obtained from analysis of multiple wave components including polarization, and up/down propagation and the wave distribution function, and to suggest areas needing further research.

2. CHARACTERISTIC FREQUENCIES AND SOURCE MECHANISMS

By assuming a plane wave solution to the wave equation for an infinite, homogeneous plasma with cold electrons and ions and a permeating magnetic field, a dispersion relation for electromagnetic plasma waves can be derived (Stix, 1962):

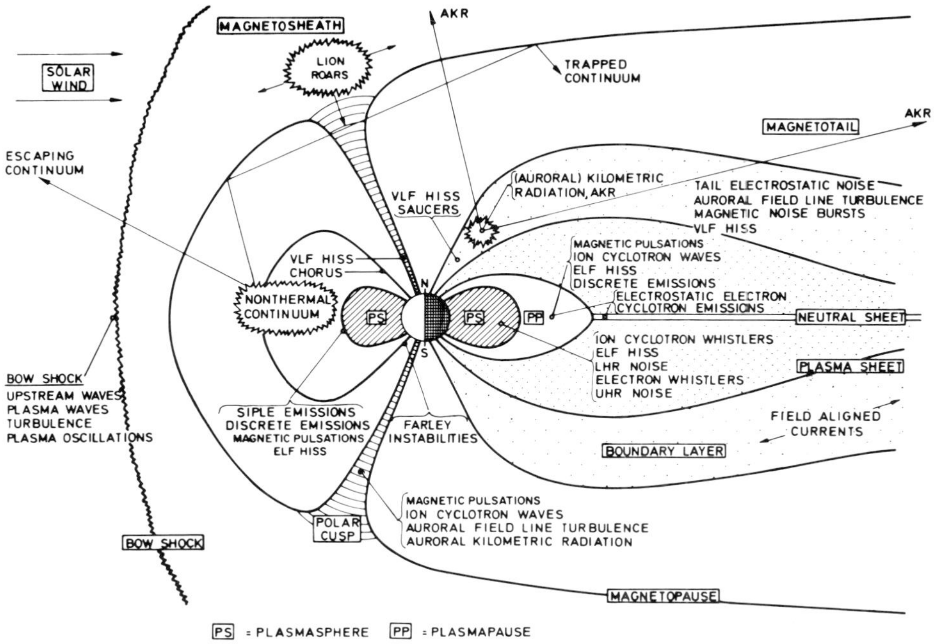

Figure 1. The menagerie of geospace plasma wave types located on a noon-midnight cross-section of the Earth's magnetosophere (Shawhan, 1979a).

$$n = c\underline{k}/2\pi f \qquad (1)$$

where $\underline{k}$ is the wave vector, f is the wave frequency and n is the index of refraction for the medium. Because of the magnetic field, the plasma is anisotropic so that the value of n depends on the angle between $\underline{k}$ and the magnetic field $\underline{B}$ which is called the wave normal angle ψ as well as on the magnitude of B, the plasma density N and the effective ion mass Mi. Figure 2 illustrates the solution to the dispersion relation for a particular set of plasma parameters B, N and Mi, and for two wave normal directions $\psi = 0^o$ and $\psi = 90^o$. For some frequency ranges there can be two solutions which implies that several modes can cc-exist with different characteristics. Waves with $\psi \sim 0^o$ have a polarization which is nearly circular--left-hand (L) or right-hand (R); waves with $\psi \sim 90^o$ tend to have a linear polarization (O-ordinary or X-extraordinary) and to have the power concentrated in the electric field or to be purely electrostatic. Associated with this diagram are many of the wave types identified in Figure 1 and Table I.

The index of refraction n depends on spatial location because the plasma parameters vary with spatial location. Figure 3 provides a model variation characteristic of the polar ionosphere and magnetosphere. According to Stix (1962), n can be expressed in terms of the wave frequency f, the electron plasma frequency f_p, the electron gyrofrequency f_g, equivalent ion frequencies involving the ion mass and the lower and upper hybrid frequencies LHR and UHR. These frequencies

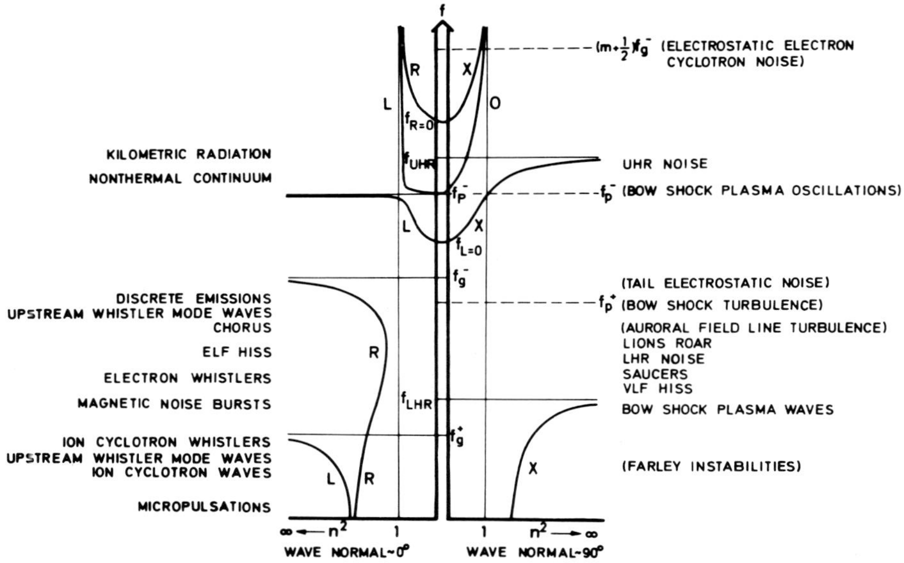

Figure 2. Association of plasma wave types with the characteristic frequencies of a plasma for wave normal directions nearly along the geomagnetic field direction (0°) and nearly transverse (90°). The curves represent the index of refraction, n, for the left-and right-hand (L and R) and the extraordinary and ordinary (X and O) wave modes. Waves in parentheses () are electrostatic. (Shawhan, 1979b).

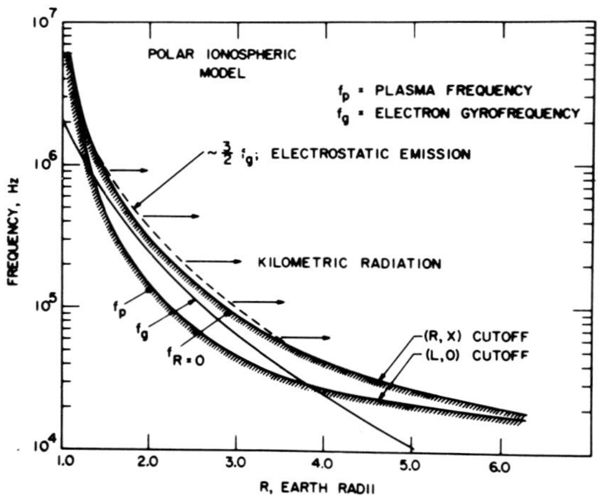

Figure 3. A model of the polar magnetosphere showing the variation of three characteristic plasmas frequencies with altitude--the electron gyrofrequency f_g, the electron plasma frequency f_p, and the right-hand mode cut-off F (R=0). (Gurnett, 1974)

are given as follows:

$$
\begin{aligned}
f_p &= 9\times10^3\ N^{1/2} \\
f_g &= 2.8\times10^6\ B \\
f(LHR) &= (M_i/M_e)(1/f_p^2 + 1/f_g^2)^{-1/2} \\
f(UHR) &= (f_p^2 + f_g^2)^{1/2}
\end{aligned}
\qquad (2)
$$

where N is electrons/cm^3, B is in Gauss, and f in Hertz.

Current plasma wave research concentrates on understanding the various source mechanisms for the menagerie of plasma waves through theory, modelling and simulations. Both man-made and lightning discharge sources are remote from the magnetosphere and are reasonably well understood. The energy for plasma waves generated internal to the magnetosphere comes from the solar wind flow past the magnetosphere through electric fields and particle kinetic energy. Through plasma instabilities, a fraction of the free energy can be converted to a number of wave modes. Energy conversion mechanisms fall into two broad categories--gyroresonance and streaming instabilities. Gyroresonance is involved in the generation of chorus and the precipitation of trapped electrons by power line radiation. Streaming instabilities have a variety of forms--current driven, beam, two stream and drift wave, for example--which are responsible for most of the natural plasma waves. Some review papers on these mechanisms include Gendrin (1975), Anderson (1983) and papers in this volume--Scarf (1985), Ronnmark (1985), Matsumoto (1985), Gendrin (1985) and Fraser (1985).

3. WAVE RECEIVER SYSTEMS AND AMPLITUDE-FREQUENCY-TIME SPECTROGRAMS

The Dynamics Explorer-1 (DE) Plasma Wave Instrument is chosen to be representative of the many plasma wave antenna and receiver systems that have been flown throughout geospace, and the solar system. A detailed description of this system is given by Shawhan, et al (1981); more general discussions of system capabilities are given elsewhere (for example, Shawhan, 1970; S-300 Experimenters, 1978; and Shawhan, 1983). The DE-1 Spacecraft is depicted in the foreground of Figure 4. It exhibits two 6 meter booms; one with a magnetometer and the other with a short electric dipole antenna (ES), a searchcoil magnetic wave field antenna (H) and a 1 m^2 magnetic loop antenna (B). Two electric dipole antennas extend from the body itself--a 200 meter tip-tip wire electric antenna (EX) and a 9 meter tip-tip tubular electric antenna (EZ).

These five antenna systems are identified in the block diagram in Figure 5 along the five receiving systems that process the received electric and magnetic signals. Both the EX and the ES electric antennas are used for sensing the quasi-static electric field from DC-10Hz. The Step Frequency Correlation provides amplitude-frequency-time spectrograms of two wave components simultaneously and a cross-correlation spectrogram, every 32 seconds for 128 frequency steps between 100Hz and 400kHz. The Low Frequency Correlator extends the frequency range downward to 1.78Hz in 8 frequency steps. Examples of the composite

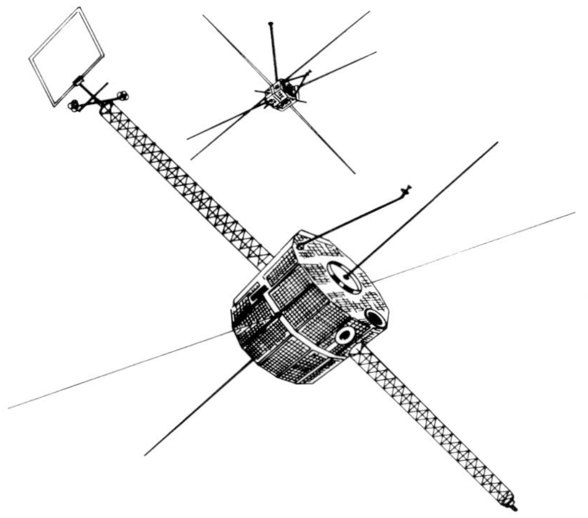

Figure 4. Sketch of Dynamics Explorer Spacecraft with DE-1 in foreground. (Shawhan, et al, 1981)

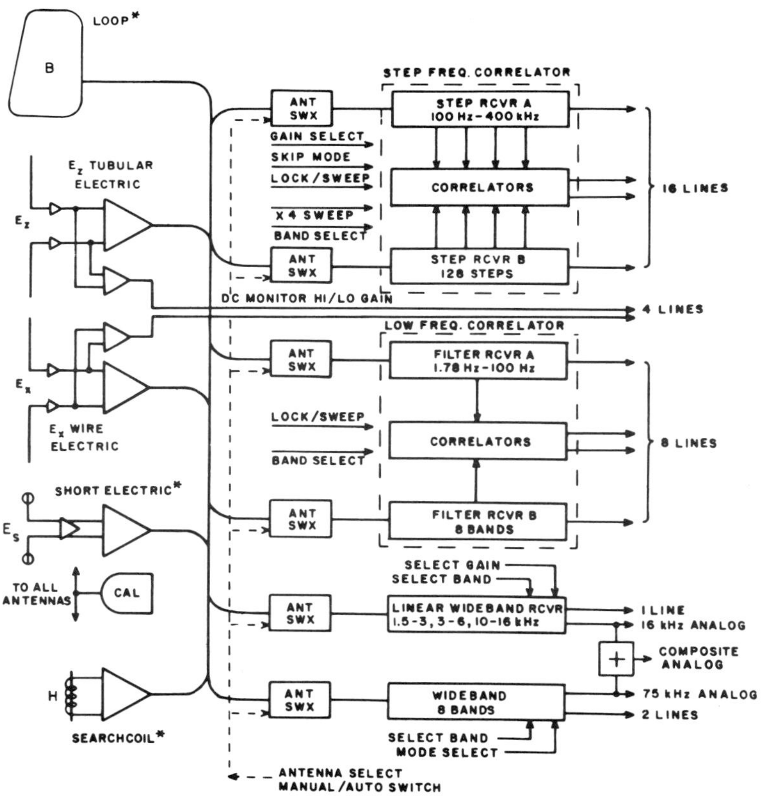

Figure 5. Block diagram of DE-1 Plasma Wave Instrument. (Shawhan, et al, 1981).

receiver electric field amplitude spectrogram outputs are displayed in Figures 6 and 7 from DE, and Figure 8 from ISEE-1. These receiver systems cover a wide frequency range but with frequency and/or time resolution compromised. Both the Linear Wideband Receiver and the Wideband Receiver in Figure 5 provide very high frequency and time resolution through ground processing but for a restricted frequency range. An example of the ISEE-1 Wideband Receiver output for auroral kilometric radiation is displayed in Figure 9.

In the DE-1 amplitude spectrograms of Figures 6 and 7, five orders of magnitude in frequency are displayed for a two hour time period with a 100dB amplitude range compressed into the gray-scale. This sort of display provides the basis for identifying the different species of waves in the menagerie and their dependence on the spatial position of the satellite. Note for example that auroral hiss has a characteristic "funnel shape" appearance, that AKR occurs only above the gyrofrequency, that Z-mode radiation occurs only when the plasma frequency f_p is below the gyrofrequency and that VLF transmitter signals cannot propagate for frequencies above the gyrofrequency. In the equatorial region of Figure 8, there are electrostatic emissions between harmonics of the gyrofrequency. With the poor time resolution of Figures 6 and 7, the AKR appears to be a continuous emission in frequency and time; in Figure 9, however, AKR is seen to be a super-position of narrow band, short-lived, rising and falling tones.

4. RAY TRACING ANALYSIS

Amplitude spectrograms provide information on the existence and the amplitude of plasma waves at the observing point. Waves may exist at that point either because they are generated in that region or because they are propagating through that region. It is essential to know the source region location in order to carry out measurements of the source plasma and to calculate growth rates for particular instability mechanisms. Ray tracing is a technique (as in optics) whereby the path of energy flow can be uniquely determined between the source and observer. In the magnetosphere the medium is inhomogeneous, anisotropic and supportive of more than one mode. In this volume, Kimura (1985) presents the basis for the technique and results for the Earth and other planets.

Ray tracing equations have been formulated by Haselgrove (1985). For two dimensions

$$\frac{dr}{dt} = \frac{1}{n^2}\left(P_r - n\frac{\partial n}{\partial P_r}\right)$$

$$\frac{d\theta}{dt} = \frac{1}{rn^2}\left(P_\theta - n\frac{\partial n}{\partial P_\theta}\right)$$

$$\frac{dP_r}{dt} = \frac{1}{n}\frac{\partial n}{\partial r} + P_\theta\frac{d\theta}{dt}$$

$$\frac{dP_\theta}{dt} = \frac{1}{r}\left(\frac{1}{n}\frac{\partial n}{\partial \theta} - P_\theta\frac{dr}{dt}\right)$$

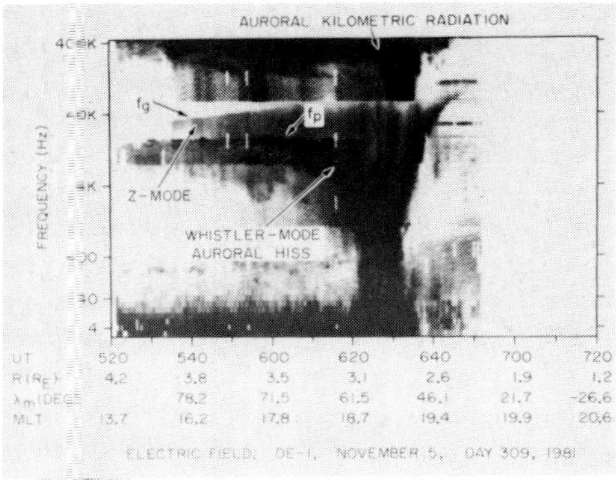

Figure 6. DE auroral zone pass with plasma frequency f_p much lower than gyrofrequency f_g. Note z-mode radiation. (Gurnett, et al, 1983)

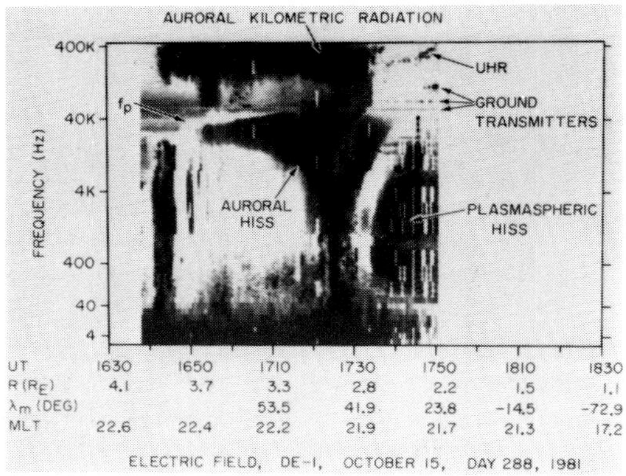

Figure 7. DE auroral zone pass with plasma frequency f_p greater than gyrofrequency f_g. Note UHR noise and VLF transmitters. (Gurnett, et al, 1983)

where r and θ are the radial distance and colatitude, n is the index of refraction, P_r and P_θ are the wave normal vector components with the magnitude of n and where t is the phase time along the ray path. These basic equations are also given for three dimensions which includes longitude and can be supplemented with equations for path length, group delay time and total growth or damping.

All the information about the plasma medium is contained in the expressions for the index of refraction which varies with spatial location (see Figure 3 for radial variation of parameters) and wave normal angle. Ray tracing programs consist of subroutines which contain at least analytical expressions for the plasma density N, the magnetic vector $\underline{B}$ and the ion composition and derivatives as a function of the spatial coordinates r, θ and longitude. The plasma dispersion expression n (f, $\underline{k}$) and various derivatives, and a solver routine for coupled partial differential equations. In practice many rays for different frequencies are started from the assumed source region and selected for analysis if they pass through the observation point. Results from a ray tracing analysis to understand the "funnel-shaped" auroral hiss in Figure 7 are presented in Figures 10 and 11 (Gurnett,et al, 1983). Figure 10 shows ray paths for different frequencies on auroral field lines for one source altitude. Figure 11 indicates that the best match to the observations occurs for a source at 1.7 to 1.9R_E. Impressive results for different modes of auroral kilometric radiation are given by Hashimoto (1984) and for AKR path growth by Omidi and Gurnett (1984), as other examples.

5. WAVE VECTOR AND WAVE DISTRIBUTION FUNCTION ANALYSIS

More information is needed and available from the wave fields than just amplitudes to characterize the waves as to nature--electrostatic or electromagnetic, to polarization--circular or linear, to direction of propagation and to the wave normal direction--$\underline{k}$. This information is needed to determine the wave mode, the probable source location since $\underline{k}$ is needed for ray tracing and wave-particle interaction mechanisms. A review of wave vector analysis is given by Shawhan (1983). Through the use of multiple antennas and receivers and cross-correlation of the waveforms these additional quantities can be obtained. A summary of some measurements to date are as follows:

FR-1	1964	Wave Normal Vector
Injun-5	1969	VLF Up/Down Direction
Voyager	1977	HF Wave Polarization
ISEE	1978	VLF Wave Normal and Up/Down Direction
GEOS	1979	ULF/ELF/VLF Wave Normal Vector
DE	1981	ELF/VLF Polarization and Up/Down Direction

Results from the DE-1 polarization measurement are shown in Figure 12 which gives the satellite-source geometry and in Figure 13 which gives an amplitude spectrogram and a polarization spectrogram. By correlating the long electric antenna EX when it is rotated perpendicular to the probable auroral zone source direction and EX which is perpendicular to the orbit plane, the polarization Right-hand (black) or Left-hand (white) can be obtained at each frequency step (Mellott, et al, 1984). Auroral hiss is known to be in the whistler mode or Right-hand mode in this kHz frequency range. Figure 13 illustrates that the strong AKR is also Right-hand but there is a weaker Left-hand component observable over a larger spatial extent which is consistent with the ray tracing analysis of Hashimoto (1984).

In reality the wave source regions are extensive in size and there exists many paths between the source region and the observing point. Consequently, there is not a unique $\underline{k}$-vector for a source but rather a distribution of $\underline{k}$-vectors. Lefeuvre and his co-workers have developed theoretical and computational techniques to derive this wave distribution function WDF from the available wave field information. At most there are six electromagnetic wave components--three electric and three magnetic. All autocorrelation and cross-correlation products provide 36 quantatives with which to estimate the WDF. In practice models of the WDF with adjustable parameters are fit to the correlation data to obtain a best-fit solution. Some sample results are presented in Figure 14 (Lefeuvre, et al, 1982). For this case the WDF is fairly well concentrated at a given direction. In other cases multiple sources are found. The output of these WDF analyses can provide the input for more realistic ray tracing analyses.

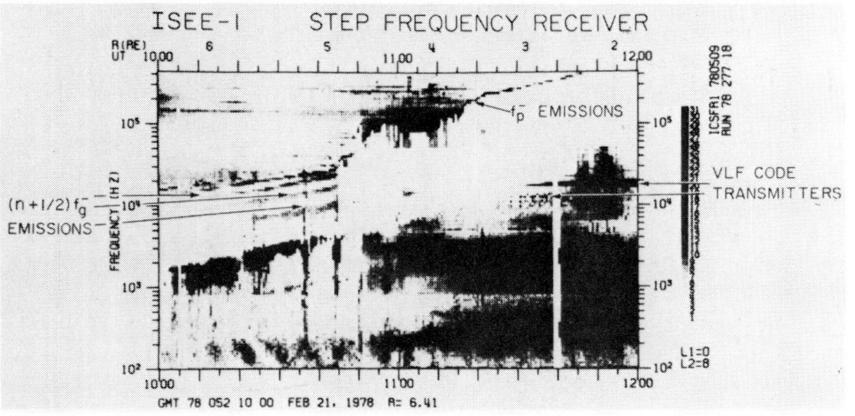

Figure 8. Spectrogram of emissions in the equatorial region from 2 to 6.5 earth radii from the International Sun-Earth Explorer (ISEE). Emission near the electron plasma frequency f_p called non-thermal continuum radiation are observed as are electrostatic emissions between harmonics of the electron gyrofrequency $(n+1/2)f_g$. (Shawhan, 1979b)

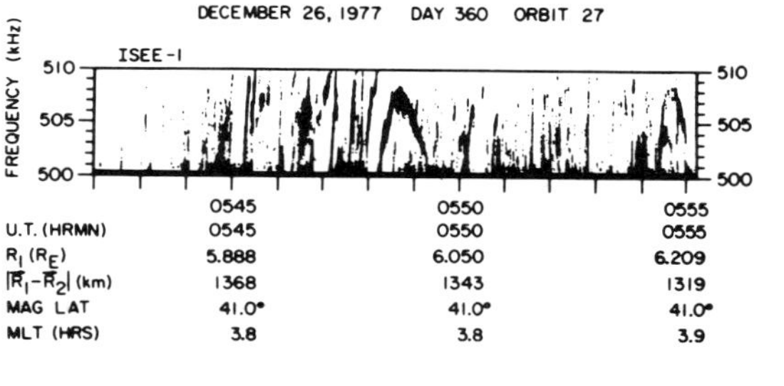

Figure 9. High time frequency resolution display of auroral kilometric radiation from the ISEE-1 wideband receiver. Signals are in the range of 500 to 510 kHz with 13 minutes of data displayed. The AKR is made up of many narrow band rising and falling tones. (Gurnett, et al, 1979)

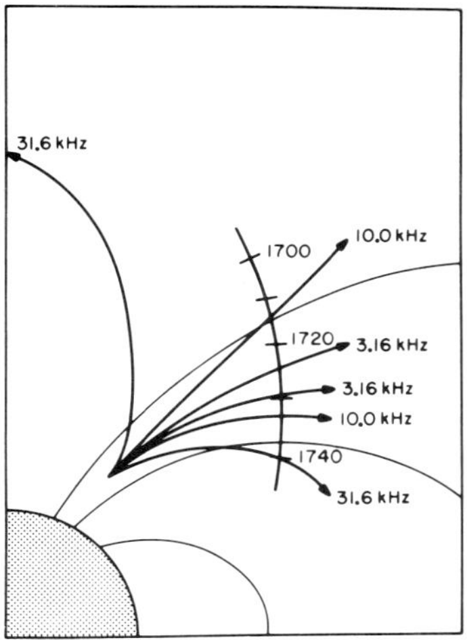

Figure 10. Ray paths computed for different frequencies from a source position on auroral field lines. (Gurnett, et al, 1983)

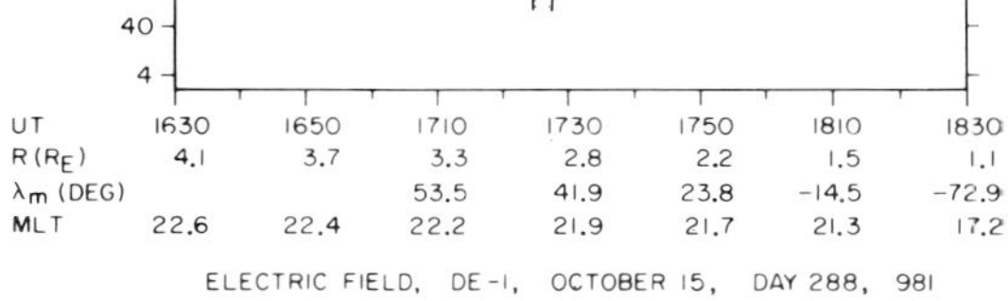

Figure 11. Frequency-time plots for the ray paths in Figure 11 for source positions located in the range of 1.3 to 2.1 Earth radii. Note that source positions of 1.7 to 1.9 R_E give the best fit to this funnel-shaped auroral hiss event shown on the spectogram in Figure 7. (Gurnett, et al, 1983)

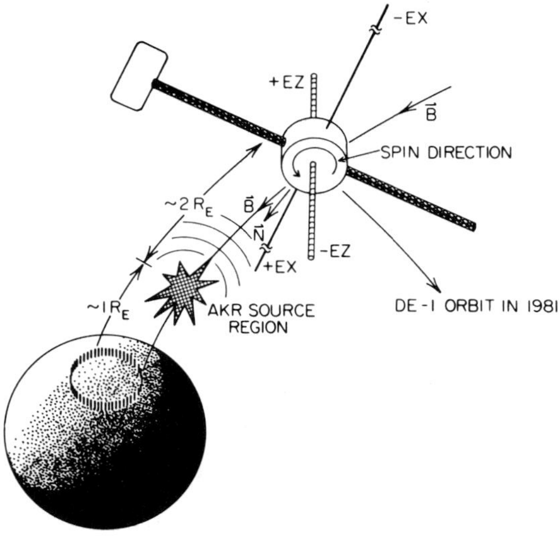

Figure 12. Geometry of the DE-1 spacecraft with respect to the auroral kilometric radiation source region.

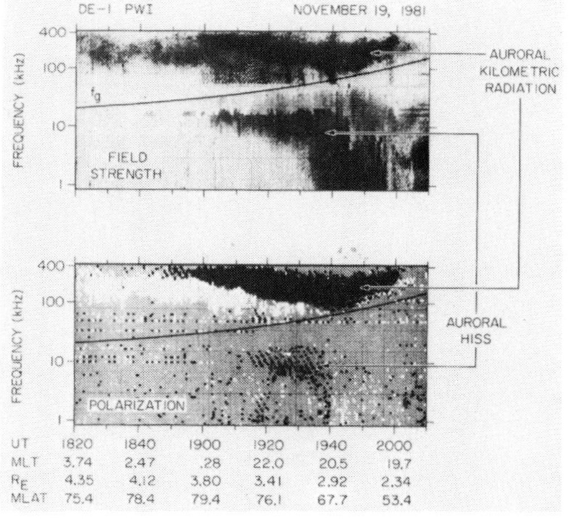

Figure 13. Frequency-time amplitude (top) and polarization (bottom) spectrograms of auroral kilometric radiation and auroral hiss as measured with DE-1. Electric field intensities are from the EX antenna and polarization is determined between EX and EZ (see Figure 12). Black indicates right-hand polarized waves and white, left-hand. The electron gyrofrequency is also plotted. Note that the intense AKR is right-hand polarized. (Mellott, et al, 1984)

TABLE 1. THE MENAGERIE OF GEOSPACE PLASMA WAVES BY FREQUENCY AND LOCATION

WAVE PHENOMENON	OBSERVED FREQUENCY	GEOSPACE LOCATION	WAVE PROPERTIES
Geomagnetic Pulsations and Magnetohydrodynamics Waves	0.001 Hz - 10 Hz	Polar Cusp Plasmapause	Transverse and compressional Alfven waves transmitted along geomagnetic field lines. Excited at time of geomagnetic storms on nightside. Excited by Kelvin-Helmholtz instability at quiet times on dayside. Indicators of changes in magnetosphere configuration.
Ion Cyclotron Waves	0.5 Hz - 100 Hz	Plasmapause Polar Cusp	Spectrum of waves just below the ion cyclotron frequency. Generally L.H. polarized. Causes proton precipitation?
Bow Shock Plasma Waves	20 Hz - 200 Hz	Bow Shock Magnetosheath	Electromagnetic turbulence comprised of bursts of whistler mode waves.
Lion Roars	90 Hz - 160 Hz	Magnetosheath	Packets of whistler-mode waves propagating along magnetic field. Associated with magnetic storms.
Magnetic Noise Bursts	10 Hz - 600 Hz Peak - 100 Hz	Auroral Field Lines Neutral Sheet	Superposition of intense bursts of magnetic noise. Associated with tail electrostatic noise and smiilar to lion roars.
Ion Cyclotron Whistlers	10 Hz - 750 Hz	Ionosphere Plasmasphere	L.H. polarized tones rising to the ion cyclotron frequency. Provides information on ion densities and temperatures.
ELF Hiss (Plasmaspheric)	10 Hz - 5 kHz	Plasmapause Plasmasphere Detached Plasma Regions	Whistler-mode turbulence at all local times. Fills plasmasphere. Causes electron precipitation from outer radiation zone.

WAVE PHENOMENON	OBSERVED FREQUENCY	GEOSPACE LOCATION	WAVE PROPERTIES
Chorus	10 Hz - 5 kHz	Outside Plasmasphere	Series of overlapping rising or falling tones. Quasi-periodic. Predominantly dayside equator.
VLF Hiss (Auroral and V-Shaped)	10 Hz - 100 kHz	Auroral Field Lines	V-shaped noise bands in frequency-time, often superimposed. Downgoing waves < 2500 km. Associated with electrostatic turbulence.
Saucers	500 Hz - 30 kHz	Auroral Field Lines	V-shaped noise bands in frequency-time, often superimposed. Upgoing waves > 1000 km.
Discrete Emissions	1 kHz - 12 kHz	Plasmapause	Rising, falling, or mixed tones, sporadic or quasi-periodic. Rarely seen with spacecraft.
LHR Noise	4 kHz - 18 kHz	Auroral Zone Plasmapause	Intense noise band above LHR freqiency, nearly electrostatic with $\underline{k} \perp \underline{B}$.
Upstream Whistler Mode Waves	.01 Hz - 4 Hz 20 kHz - 70 kHz	Upstream of Bow Shock in Solar Wind	Two bands; below ion cyclotron and below electron cyclotron frequency.
Trapped Non-Thermal Continuum Radiation	500 Hz - 20 kHz	Outside Plasmapause Inside Magnetosheath	Weak broadband noise trapped between plasmapause and magnetosheath. Generated on dayside 04-14 LT.
UHR Noise	100 - 600 kHz	Plasmapause	Intense noise band near upper hybrid resonance frequency.
Electron Whistlers	100 Hz - 1 MHz	Plasmasphere Plasmapause	R.H. polarized falling tones. Travel along magnetic field lines in ionization ducts or can be refracted across field lines.

WAVE PHENOMENON	OBSERVED FREQUENCY	GEOSPACE LOCATION	WAVE PROPERTIES
Tail Broadband Electrostatic Noise	10 Hz - 2 kHz	Magnetotail at Plasma Sheet Boundaries	Broadband emission consisting of discrete bursts with V-shaped structure, $\underline{k} \perp \underline{B}$. Associated with VLF hiss in same region and with AKR.
Auroral Field Line Turbulence	10 Hz - 10 kHz Peak 10 - 50 Hz	Auroral Field Lines All Local Times	Broadband emission consisting of discrete bursts at V-shaped structure. Associated with VLF hiss in same region with AKR.
Farley Instabilities	40 Hz - 10 kHz	Ionosphere E-Region	Narrowband ELF emission at $\sim$100 Hz. Broadband VLF emission < 10 kHz at higher altitude.
Electrostatic Electron Cyclotron Emissions	200 Hz - 50 kHz	Outside Plasmapause Near Plasma Sheet	Narrowband emission near $(n + 1/2)\ f_g$ with several harmonics observed simultaneously.
Bow Shock Turbulence	200 Hz - 30 kHz	Bow Shock Magnetosheath	Broadband electrostatic noise.
Bow Shock Plasma Oscillations	3 kHz - 50 kHz	Bow Shock	Narrowband electron plasma oscillations associated with electron heating.
Escaping Non-Thermal Continuum Radiation	20 kHz - 100 kHz	Outside Plasmapause Magnetosphere	Weak electromagnetic broadband noise escaping magnetosphere. Generated in morning and early afternoon. Associated with bands of electrostatic cyclotron noise at plasmapause.
Kilometric Radiation (Auroral or Terrestrial) (AKR or TKR)	20 kHz - 2 MHz	Auroral Field Lines	Broadband noise bursts pearked $\sim$ 200 kHz beamed into 2 sterad cone centered on auroral field lines at $\sim 2\ R_E$ emitting $\sim$10 W. Associated with VLF hiss, ELF hiss, auroral turbulence, and tail electrostatic noise. Generated $\sim$ 2200 LT in tail.

WAVE PHENOMENON	OBSERVED FREQUENCY	GEOSPACE LOCATION	WAVE PROPERTIES
Power System Harmonic Radiation	kHz	Plasmasphere Boundary $L \sim 4$	Narrowband radiation at harmonics of the rectified 60 Hz, spaced by 120 Hz and falling in kHz range. Affects other emissions at close frequencies.
Siple Stimulated VLF Emission	2 kHz - 16 kHz	Plasmasphere Boundary $L \sim 3$-5	Narrowband emissions, rising or falling tones, stimulated and modified by transmitted VLF waves.
Electron Beam Stimulated Emissions	DC - 12 MHz	Ionosphere	Waves stimulated by rocket-borne electron and argon guns; f_p, $2f_g$ and VLF and ELF whistler mode frequencies.

*Adapted from Shawhan (1979b)

f_g = electron gyrofrequency
f_p = electron plasma frequency
LHR = lower hybrid resonance frequency
UHR = upper hybrid resonance frequency
L.H. and R.H. = left- and right-hand polarization
$\underline{k}$ = wave normal vector direction
$\underline{B}_o$ = geomagnetic field direction

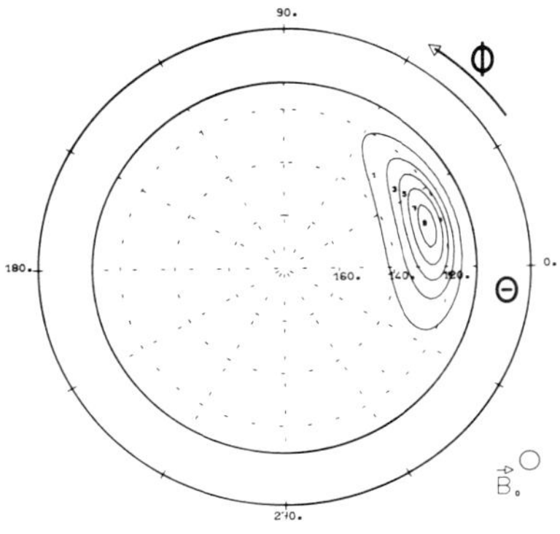

Figure 14. Wave Distribution Function (WDF) for a VLF transmitter signal received on GEOS derived from three wave magnetic components. (Lefeuvre, et al, 1982)

6. AREAS FOR FURTHER RESEARCH

Significant progress in plasma wave research to support the increasing efforts in theory, modelling and simulations could be achieved in the following areas:

- Develop waveform capture receivers so that samples of all six wave components can be analyzed for wave vector and WDF characteristics with high time and frequency resolution
- Enhance ray tracing analysis by improving the plasma parameter models and by carrying out growth rate calculations
- Improve WDF algorithms and apply the method to many more cases
- Carry out and analyze active wave propagation and wave-particle experiments under controlled conditions utilizing the Shuttle/Spacelab

7. REFERENCES

Al'pert, Ya. L., Waves and Satellites in the Near-Earth Plasma, Consultants Bureau: New York, 1974.

Anderson, R. R., Plasma Waves in Planetary Magnetospheres, Rev. Geophys. Space Phys., 21, 474-494, 1983.

Fraser, B. J., Observations of Ion Cyclotron Waves Near Geosynchronous Orbit and on the Ground, this volume, 1985.

Gendrin, R., Waves and Wave-Particle Interactions in the Magnetosphere; A Review, Space Sci. Rev., 18, 145-200, 1975.

Gendrin, R., Consequences of Hydromagnetic Waves on Magnetosphere Particle Dynamics, this volume, 1985.

Gurnett, D. A., The Earth as a Radio Source; Terrestrial Kilometric Radiation, J. Geophys. Res., 79, 4227-4238, 1974.

Gurnett, D. A., R. R. Anderson, F. L. Scarf, R. W. Fredricks and E. J. Smith, Initial Results from ISEE-1 and ISEE-2 Plasma Wave Investigation, Space Sci. Rev., 23, 103-777, 1979.

Gurnett, D. A., S. D. Shawhan and R. R. Shaw, Auroral Hiss, Z Mode Radiation, and Auroral Kilometric Radiation in the Polar Magnetosphere; DE-1 Observations, J. Geophys. Res., 88, 329-340, 1983.

Haselgrove, J., Ray Theory and a New Method for Ray Tracing, London Physical Soc.: Report on Conf. Phys. Ionosphere, pp. 355-364, 1955.

Hashimoto, Kozo, A Reconcilliation of Propagation Modes of Auroral Kilometric Radiation, J. Geophys. Res., 89, 7459-7466, 1984.

Helliwell, R. A., Whistlers and Related Ionospheric Phenomena, Stanford University Press; Stanford, 1965.

Kimura, I., Whistler Mode Propagation in the Earth and Planetary Magnetospheres and Ray Tracing Technique, this volume, 1985.

Lefeuvre, F., T. Neubert and M. Parrot, Wave Normal Directions and Wave Distrbution Functions for Ground-Based Transmitter Signals Observed With GEOS-1, J. Geophys. Res., 87, 6203-6217, 1982

Matsumoto, H., Coherent Non-Linear Effects on Electromagnetic Wave-Particle Interactions and Relevant Diagnostics in Simulations, this volume, 1985.

Mellott, M. M., W. Calvert, R. L. Huff, D. A. Gurnett and S. D. Shawhan, DE-1 Observations of Ordinary Mode and Extraordinary Mode Auroral Kilometric Radiation, Geophys. Res. Lett., 11, 1188-1191, 1984.

Omidi, N. and D. A. Gurnett, Path-Integrated Growth of Auroral Kilometric Radiation, J. Geophys. Res., 89, 10801-10812, 1984.

Ronnmark, K., Kinetic Theory of Plasma Waves, this volume, 1985.

S-300 Experimenters, Introduction to the S-300 Experiments Onboard GEOS; Space Sci. Rev., 22,327-332, 1978.

Scarf, F. L., Wave Observations in Outer Planet Magnetospheres, this volume, 1985.

Shawhan, S. D., The Use of Multiple Receivers to Measure the Wave Characteristics of Very-Low-Frequency Noise in Space, Space Sci. Rev., 10, 689-736, 1970.

Shawhan, S. D., Magnetospheric Plasma Waves, Solar System Plasma Physics, C. F. Kennell, L. J. Lanzerotti and E. N. Parker, (eds.), Volume III, 211-270, North-Holland; Amsterdam, 1979a.

Shawhan, S. D., Magnetospheric Plasma Wave Research 1975-1978, Rev. Geophys. Space Phys., 17, 705-724, 1979b.

Shawhan, S. D., Estimation of Wave Vector Characteristics, Adv. Sci Res., 2, 31-41, 1983.

Shawhan, S. D., D. A. Gurnett, D. L. Odem, R. A. Helliwell and C. G. Park, The Plasma Wave and Quasi-Static Electric Field Instrument (PWI) for Dynamics Explorer-A, Space Sci. Instrumentation, 5, 535-550, 1981.

Stix, T. H., Theory of Plasma Waves, McGraw Hill: New York, 1962.

OBSERVATIONS OF LOW ENERGY MAGNETOSPHERIC PLASMA OUTSIDE THE PLASMASPHERE

B. Hultqvist
Kiruna Geophysical Institute
P.O. Box 704
S-981 27 Kiruna
Sweden

ABSTRACT

After some introductory discussions about morphological concepts and limitations of various measurement techniques, existing low energy plasma data, orginating primarily from the GEOS, Dynamics Explorer, and Prognoz spacecraft, is described and discussed. The plasmasphere measurements are not included (but for some observations of plasmasphere refilling). It is finally concluded that we are very far from a complete picture of the low-energy plasma component in the magnetosphere and that this problem has to be given high priority in planning payloads of future space plasma physics missions.

1. INTRODUCTION

The distribution of density, characteristic energies, composition and flow of the plasma in the magnetosphere is the result of a complex interplay of source and loss processes, primarily at the outer (magnetopause) and inner (ionosphere) boundaries, as well as of transportation and energization/dissipation in various regions within the magnetosphere. One of the most important tasks of experimental magnetospheric research is to determine the main characteristics of the plasma in various parts of the magnetospheric system under different disturbance conditions. To know the plasma density in the magnetosphere is important because of the effect that plasma density has e.g. on wave-particle interactions, including those giving rise to auroral kilometric radiation and electrostatic ion cyclotron waves. The plasma density is also important for the propagation of waves of various kinds. So, besides being one of the main macroscopic characteristics of the magnetospheric plasma, the knowledge about the total plasma density is required for analysing most microphysical problems in the magnetosphere.

Space Science Reviews **42** (1985) 275–294. 0038–6308/85.15

Although a most important task, which has been recognized since before the beginning of the space era, we are still far from the goal of being able to determine density, temperature or energy distribution, composition and flow for all energy ranges of importance to the physical processes in the magnetosphere. This is because it is quite a difficult task to measure the plasma parameters in some energy ranges and an extensive set of sofisticated measuring instruments are required. Mostly the set is not complete even on satellites which are devoted to space plasma physics research. When most of the parameter space of interest is covered by a set of such instruments, there are generally still intercalibration problems to overcame in order to arrive at resonably accurate fluid parameters for the plasma in question.

In the 1960's and 1970's whistler measurements in combination with some direct satellite measurements provided a fairly comprehensive picture of the cold plasma density distribution in and close to the plasmasphere (e.g. Carpenter, 1966, 1970; Chappell, 1972), but temperature, pitch angle and density distributions along the magnetic field lines were not known. Until the late 70's the magnetospheric plasma was thought to be composed of two basic particle populations, one of solar wind origin, with high characteristics energies, and one cold ($T < 1$ eV), of ionospheric origin, found mainly in the plasmasphere. The change of this picture started in fact already at the end of the 1960's with the first flight of an energetic ion mass spectrometer on a low polar orbiting satellite by the Lockheed group (Shelley et al., 1972) which demonstrated the existence in the ring current of energetic O^+ ions, by necessity of ionospheric origin. Also observations of the cold plasma component provided quite early results which were inconsistent with the two component (hot and cold) magnetospheric plasma picture. Serbu and Maier (1970), on the basis of ion trap measurements on OGO 5, derived temperatures of 5-10 eV for the outer plasmasphere. Gurnett and Frank (1974) obtained a temperature of 8 eV in the dayside plasma trough and Bezrukikh and Gringauz (1975) reported temperatures above 2 eV in the outermost plasmasphere.

That the simple picture of the magnetospheric plasma being composed basically of a cold ("thermal") component originating in the ionosphere with temperatures below 1 eV and a hot component of solar wind origin was completely false was definitely demonstrated in the 1970's by means of the S3-3 and GEOS spacecraft for higher energies and by means of the plasma experiment on ATS 6 for the low, but superthermal, energies. It was found that the hot magnetospheric plasma in disturbed conditions may be dominated by ions of ionospheric origin (Shelley et al., 1976, Geiss et al., 1978 and others) and that the cold component has a long "tail" tying it together with the hot component along the energy axis (Lennartsson and Reasoner, 1978, Horwitz et al., 1978, Horwitz and Chappell, 1979, Horwitz, 1980 and others).

In this lecture I will concentrate on the thermal and superthermal parts of the energy distribution of the magnetospheric plasma. Others will discuss the hot component.

It should be emphasized that the term thermal plasma is not very well defined. If we mean fairly well Maxwell-distributed plasma, we may find such both at the cold end of the energy spectrum and in the hot part (plasmasheet). In this lecture we will use the term thermal plasma for the part that has a temperature of less than 1-2 eV (alternatively named the cold component). For the superthermal plasma T is in the range from 1-2 eV to about a hundred eV and for the hot plasma the characteristic energies are of the order of a hundred eV or more. These definitions are those mostly used in the literature, but differences occur.

The plasmasphere will not be dealt with here (but for when discussing phenomena where it is related to the magnetospheric regions outside it). It is a large subject of its own.

The material is organized regionally. That is a natural way of presenting the present knowledge of the cold and superthermal plasma in the magnetosphere, as the different regions of the magnetosphere have different histories of plasma research, different spacecraft making the measurements and, to some extent, different dominating physical processes.

As this is a tutorial lecture the goal is not to present a complete review of the field but rather to present the basic characteristics of the plasma in the magnetosphere outside the plasmasphere and the most important known physical processes affecting it.

2. SOME MORPHOLOGICAL CONCEPTS

In the parts of the magnetosphere outside the plasmasphere the magnetic field lines are supposed to be periodically or permanently open to the distant tail and the interplanetary space, whereas they are closed and dipole-like within the plasmasphere. How big the plasmasphere is depends on the magnetospheric disturbance level (see e.g. Chappell, 1972). During magnetic storms its size is strongly decreased to the extent that the L-value of the plasmapause in the midnight sector ($L \lesssim 4$) may be less than half of the value for very quiet conditions ($L \sim 8$). There is thus an outer part of the plasmasphere region of the magnetosphere which sometimes is part of the plasmasphere and sometimes not. In periods after storms this part is characterized by refilling processes, during which ionospheric plasma fills up the emptied region. The time scale for this replenishment of thermal ionospheric plasma has been found to be in the range 7-22 hours (e.g. Horwitz et al., 1984). The refilling of the plasmasphere is a field where we know very little of the basic physical processes and where very much remains to be investigated in the future.

Before going into the substance of this chapter we have to clarify another matter of nomenclature. The region of the magnetosphere next to the plasmasphere is usually called the plasma trough by researchers working in the field of thermal plasma. This name has come from observations that the ionosphere contains less electrons per unit volume in a region between the auroral oval and the plasmasphere than it does north and south of it. In the magnetosphere this trough region has been associated with a region in the dusk to midnight sector between the plasmapause and the inner edge of the plasmasheet (reported first by Vasyliunas, 1966). This intermediary region is difficult to understand on the basis of known magnetospheric processes. Some recent observations in fact indicate that it does not exist but the plasmapause and the inner edge of the plasma sheet, as seen in the electrons, coincide closely (Horwitz et al., 1982). On the dayside the trough region is located between the plasmapause and the magnetopause boundary layer (including the cusp region). It therefore appears likely - although not definitely demonstrated yet - that in the magnetosphere, as in the ionosphere, the trough is bounded on the outer side by the magnetic field lines connected with the auroral oval, i.e. by those in the plasma sheet boundary layer. In the trough there appears to occur generally a more or less intense electron precipitation associated with diffuse aurora (Horwitz et al., 1982) whereas the strong structured aurora is located on the field lines of the plasma sheet boundary layer.

The trough region of the researchers working with the thermal and superthermal components of the magnetospheric plasma, falls, for those working with the hot plasma, in the plasma sheet (on the nightside) and the ring current (on both day- and nightside). The ring current region overlaps with both the plasma sheet and the plasmasphere to some extent and it covers the trough region.

3. LIMITATIONS OF MEASUREMENT TECHNIQUES

To determine completely the characteristics of the magnetospheric plasma over the entire energy range from a tenth of an electron volt to hundreds of kiloelectron volts in the varying conditions found within the magnetosphere is, as mentioned in the introduction, a goal very difficult to reach and we are still in the early phase of approaching it. A number of different techniques have to be employed. Here we shall only briefly sketch the main characteristics and some problems associated with the different methods, which are important to understand in connection with the discussion of the measurement results presented below.

At the high energy end of the scale, i.e. in the hot region, the plasma is frequently far from Maxwell distributed. Therefore, the entire distribution function has to be determined for all major ion species in order to compute even the basic macroscopic variables, such as density, energy density, characteristic energies and flow vectors. At these energies each individual plasma particle entering the instrument is analysed and to cover all directions, energies and mass of interest in sufficiently short time requires so many different sensors that it has not yet been possible to measure in great detail but in limited parts of parameter space. Between or beyond the measured range interpolation and extrapolation have to be resorted to. Still, we have generally a better idea about density and characteristic energies of the hot plasma component than of the thermal and superthermal parts in most of the magnetosphere outside the plasmasphere.

The technique of measuring individual particles for determining density, temperature, composition and flow of the plasma has to be used also in the superthermal energy range (~1 eV - ~100 eV). It is, however, more difficult to apply such methods the lower the energy, mainly because of the spacecraft being charged up to several volts potential relative to the plasma in most of the magnetosphere outside the plasmasphere. Only very recently has it become possible to control the effect of an unknown spacecraft potential on the particle measurements reasonably well by varying the potential of the aperture of the measuring instrument (Sojka et al., 1983; Nagai et al., 1984). Still, the density values derived are subject to a number of assumptions which make the absolute accuracy of the measurements limited.

The best way to determine the total plasma number density (electron density) in the magnetosphere fairly accurately appears to be by deriving it from resonance effects in the surrounding plasma, which give rise to plasma waves detectable onboard the spacecraft. The technique may be passive or active, i.e. one may determine naturally excited waves or waves may be excited with a transmitter onboard. The passive method is generally more uncertain than the active methods because of difficulties under some conditions to identify the nature of the process producing the waves. Only from the late 1970's are observations in the magnetosphere, using active methods, available and only for limited regions (Decreau et al., 1982; Higel and Lei, 1984).

The relaxation sounding technique used for magnetospheric plasma density measurements first on the GEOS spacecraft is based on ionospheric plasma resonance phenomena discovered by the first topside sounder experiment (Calvert and Goe, 1963).

Passive methods of analysing plasma waves have provided magnetospheric plasma densities only in the 1970's and later (Gurnett and Shaw, 1973; Kurth et al., 1979; Persoon et al., 1983, and references therein).

The techniques using plasma waves for plasma density determination have the important limitation compared to the methods based on direct particle measurements that they only provide information about the density but not about the other characteristics of the plasma such as temperature, composition and flow.

The same is basically true for in situ probe techniques which have been used in the magnetosphere on several spacecraft since the middle of the 1970's but which have given rise to only little published results. The active wave mutual impedance probe onboard GEOS 1 and 2 have given electron densities from the geostationary orbit inwards (Decreau et al., 1982). The probe techniques have difficulties in hot low density plasma because of Debye length limitations and are therefore difficult to apply in many parts of the magnetosphere. A detailed comparison of most of the above techniques has been reported by Decreau et al. (1978).

After this summary of the limitations of the measurement techniques we proceed to the measurement results. The main problem everywhere outside the plasmasphere is to determine the density of a possibly present cold plasma component of ionospheric origion. This problem is bigger the lower the cold plasma density, the hotter the hot plasma and the higher the relative density of the hot plasma component.

4. THERMAL AND SUPERTHERMAL MAGNETOSPHERIC PLASMA ABOVE THE HIGH LATITUDE IONOSPHERE

A terrestrial contribution of cold plasma to the magnetosphere was hypothesized in 1961 by Dessler and Hanson. The theory of the "polar wind" (Axford, 1968) was initially discussed by Hanson and Patterson (1963), Dessler and Michel (1966), Axford (1968), Banks and Holzer (1968, 1969 a, b), Lemaire and Scherer (1970, 1973) and Marubashi (1970). The predictions of the decrease of the electron density with altitude above the ionosphere arrived at by Banks and Holzer (1969 b) from their polar wind theory are shown in Figure 1. For reasonable upper ionosphere temperatures only the light ions, H^+ and to a lesser degree He^+, flow out into the magnetosphere, whereas the O^+ ions do not overcome the gravitational force. The limiting flux is reached when the scale height of the escaping ion species is reduced to that of the friction-producing species at great altitudes.

Schunk and Watkins (1982) have developed a more advanced set of steady state solutions for the high-latitude plasma flow above 1500 km altitude from a 13-moment system of transport equations for a range of boundary conditions at 1500 km. Some of their predictions are shown in Figure 2 together with observational results from the Dynamics Explorer 1 satellite (Persoon et al., 1983).

Whereas the electron densities derived from whistler wave observations over the polar cap are below the values predicted by Banks and Holzer (1969 b) on the basis of equal electron and ion temperatures, the measured values fall within the range expected on the basis of Schunk's and Watkins' (1982) calculations with different electron and ion temperatures and temperature gradients.

The electron density height distribution obtained with the DE 1 satellite is related to the topside sounder data at lower altitudes in the polar ionosphere in Figure 3. The Alouette and ISIS density data shown do, however, not include the very low electron density values that were observed in the upper winter

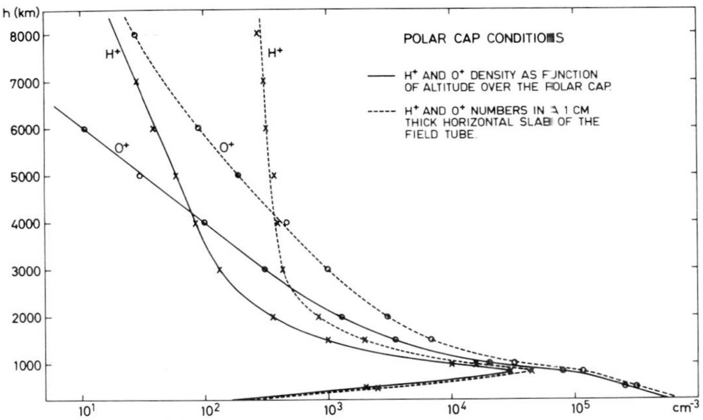

Figure 1. Altitude profiles for O^+ and H^+ ion densities under polar wind conditions, according to Banks and Holzer (1969b). In the figure are also given the corresponding variation with height of the numbers of O^+ and H^+ ions in a 1 cm thick horizontal slab of a typical auroral latitude field tube of 1 cm^2 cross section at 100 km (L=7) (after Hultqvist, 1983).

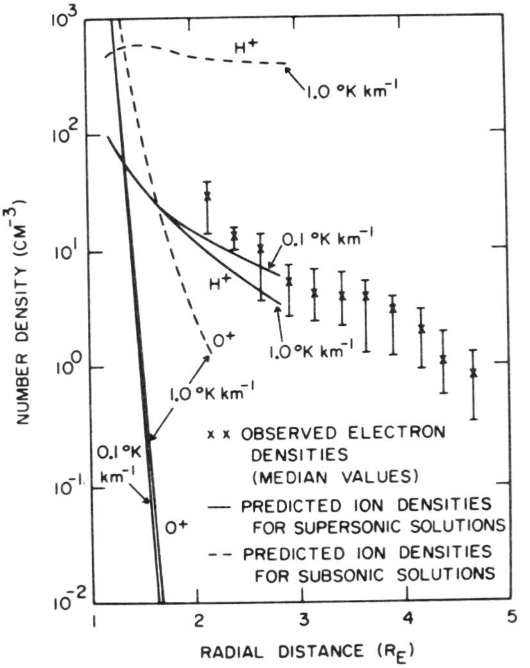

Figure 2. Predicted ion densities below 2.8 R_E based on solutions to the 13-moment transport equations (Schunk and Watkins, 1982). The subsonic outflow solutions correspond to higher H^+ and O^+ densities. These predicted densities exceed the DE 1 measured densities at 2 R_E by more than an order of magnitude. Supersonic outflow solutions correspond to lower H^+ and O^- densities. An increase in the boundary electron temperature gradient (from 0.1° K km^{-1} to 1° K km^{-1}) corresponds to a slight increase in the predicted O^+ densities and a larger decrease in the predicted H^+ densities. Both sets of predicted ion densities for the supersonic outflow solutions are within 25% of the median density values observed by DE 1 (after Persoon et al., 1983).

night ionosphere in about 12% of the passes (Hagg, 1967, Timleck and Nelms, 1969). The inclusion of these very low densities on Figure 3 would lower the Alouette/ISIS density curve in Figure 3 by an order of magnitude above 1.2 R_e.

Mozer et al. (1979) have derived electron densities in the range 4 to 40 cm^{-3} from lower-hybrid-frequency measurements onboard the S3-3 satellite. Calvert (1981), from Hawkeye wave measurements, and Temerin (1984), from S3-3 wave data, have derived electron densities as low as 1 cm^{-3} at about an earth radius altitude. These very low densities appear to be unusual, however, except possibly in outflowing ion beams.

The reasonable agreement between the DE 1 measurements and the polar wind theory predictions shown in Figure 2 indicates that the theory for the 'cold' polar wind is basically correct. This has also been confirmed by the first direct observations of a supersonic ion outflow outside the plasmasphere recently reported by Nagai et al. (1984). However, there are other results, besides the mentioned extremely low densities sometimes found over the polar cap which are not consistent with the polar wind theory, namely the observations on DE 1 of sizable fluxes of superthermal or even hot (i.e. energy above 100 eV with our nomenclature) field-aligned ions out of the polar cap ionosphere (Shelley et al., 1982; Moore et al., 1984; Lockwood et al., 1984; Waite et al., 1984 and Yau et al., 1984), sometimes composed of practically only O^+ ions. An example of such an observation is shown in Figure 4 after Shelley et al. (1982). This phenomenon has been observed in about 50% of the DE 1 passages over the polar caps. The total flux into the magnetosphere of ions due to this process has been estimated by Shelley et al. (1982) and Lockwood et al. (1984) to amount to $\sim 10^{25}$ ions/s, which is comparable in number to the estimated total polar wind outflow (see e.g. Cowley, 1980) and represents a much larger energy input to the magnetosphere than that of the polar wind.

Attempts have been made to modify the polar wind theory in such a way that also O^+ ions can overcome the gravitational force (Barakat and Schunk, 1983). This requires, however, so high electron temperatures in the upper ionosphere ($>$ 10000 K) that they become unrealistic. In addition, it is doubtful that even so high electron temperatures will suffice to have the O^+ ions penetrating the neutral hydrogen gas in the outermost ionosphere without loosing its charge to the hydrogen atoms (Moore, 1980, 1984). In any case such mechanisms seem unable to account for the ion beams consisting of only O^+. Therefore, other mechanisms, most likely involving field-aligned potential differences and/or ion cyclotron resonances, appear to be at work more or less continuously also over the polar caps, as they are in the auroral zones.

5. DAYSIDE PLASMA TROUGH

The total plasma density well outside the plasmapause in the day-

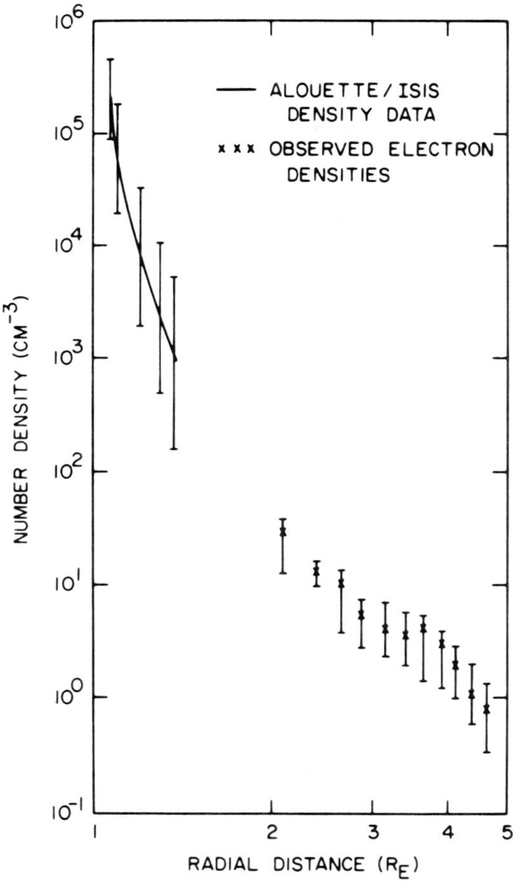

Figure 3. A log-log plot comparing low-altitude median electron densities obtained from the topside sounder data of Alouette II and ISIS 1 and high-altitude median electron densities obtained from the plasma wave instrument on DE 1 (after Persoon et al., 1983).

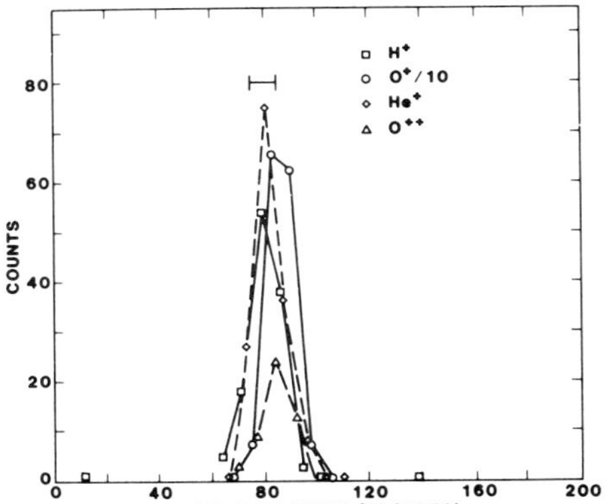

Figure 4. Counting rate as a function of mass and rotation angle for a portion of the DE 1 spin period beginning at 1241:00 UT on September 20, 1981. At this time the magnetic field direction was 88 degrees. The retarding potential difference was 7.2 V (after Shelley et al., 1982).

side of the magnetosphere has been determined for the first time with the active relaxation sounder and mutual impedance probe on GEOS 1 and 2 (Decreau et al., 1982; Higel and Lei, 1984). An example of the results obtained is shown in Figures 5 a, b. As can be seen the total electron density maximizes generally along the geostationary orbit on the dayside with values reaching ~10 electrons/cm^3. It builds up towards this value during the late morning hours from values around 1 cm^{-3} on night side. In the afternoon or early evening the spacecraft generally crosses the dusk-bulge of the plasmasphere containing appreciably higher electron densities.

What the energy and pitch angle distributions and the ion composition are cannot be determined with the wave methods used by Higel and Lei (1984) and Decreau et al. (1982). It turns out to be difficult to find answers in the litterature to the question what fractions of the plasma have energies in the thermal, superthermal and hot energy ranges. Sojka et al. (1983) report, from particle measurements on the DE 1 satellite, ion densities in the range 1-10 cm^{-3} both for ions of a few eV energy and for 10-15 eV ions, but Horwitz and Chappell (1979) derived much lower superthermal densities (0.1 - 0.4 cm^{-3}) from the ATS 6 measurements. Chappell (1983) stated that it is uncertain if there are <1 eV plasma in the plasma trough and Higel and Lei (1984) refer disagreements between wave measurements and particle measurements to a dominating cold component. The answer to the question formulated above is thus far from clear.

The DE 1 low energy particle experiment (RIMS) has recorded field-aligned ion beams (uni-or bidirectional) as well as conical distributions and isotropy.

The hot ion density for energies up to ~20 keV in the dayside trough region, i.e. in the outer ring current region from the hot plasma nomenclature point of view, can be seen from Figures 6 to be in the same density range (1-10 cm^{-3}) in storm conditions. As is evident in Figure 6, O^+ ions play a major role at keV energies in storms. In quiet conditions the H^+ ion density is not very much less than in Figure 6 but the O^+ density is an order of magnitude lower. Hultqvist (1983) has argued that the number densities of keV O^+ ions observed in the entire magnetosphere during disturbed conditions cannot have been produced by acceleration of cold O^+ ions existing in the magnetosphere when the disturbance set in. The amount of O^+ ions in the storm time magnetosphere is simply so large that it is equal to all cold O^+ ions above an altitude of the order of a thousand kilometers in the upper ionosphere in quite a wide latitude band on both hemispheres. It appears, therefore, that in disturbed magnetospheric conditions an effective extraction process works in the upper ionosphere over wide ranges in latitude and longitude. All ionospheric ion species are energized in this process, many to keV energies but others to lower (superthermal) energies.

It is still unclear in the literature what ion species dominate the trough region at lower energies during different distur-

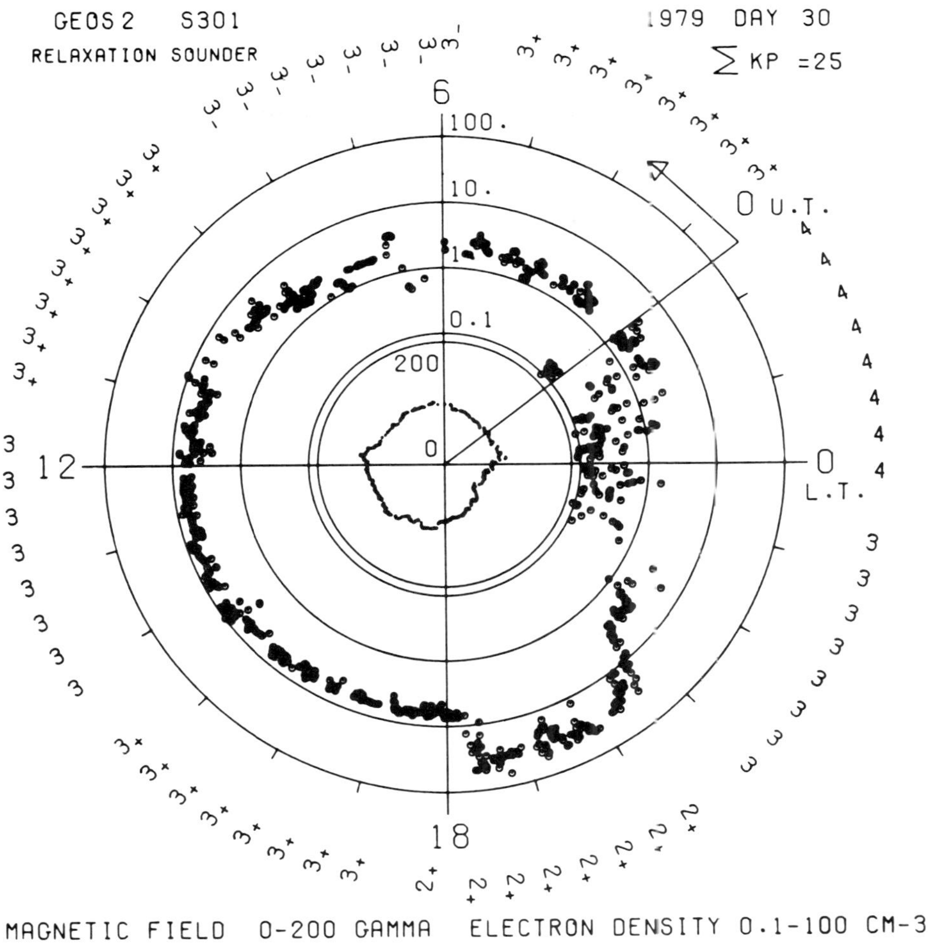

Figure 5a. Distribution versus local time of the magnetospheric plasma parameters measured by the GEOS 2 relaxation sounder (RS) along its 6.6 R_E orbit for one day (January 30, 1979). Electron density Ne and magnetic field strength B_0 are plotted radially, with time running as an angular coordinate starting from the 0000 UT radius. The electron density scale is logarithmic over the three decades (0.1–100 cm^{-3}) displayed by concentric rings. The magnetic field strength is plotted in the inner circle with a linear scale between 0 (center) and 200 nT (smaller circle). Ne and B_0 data are available in real time. Values of the three-hour Kp index are shown on the outer circumference (after Higel and Lei, 1984).

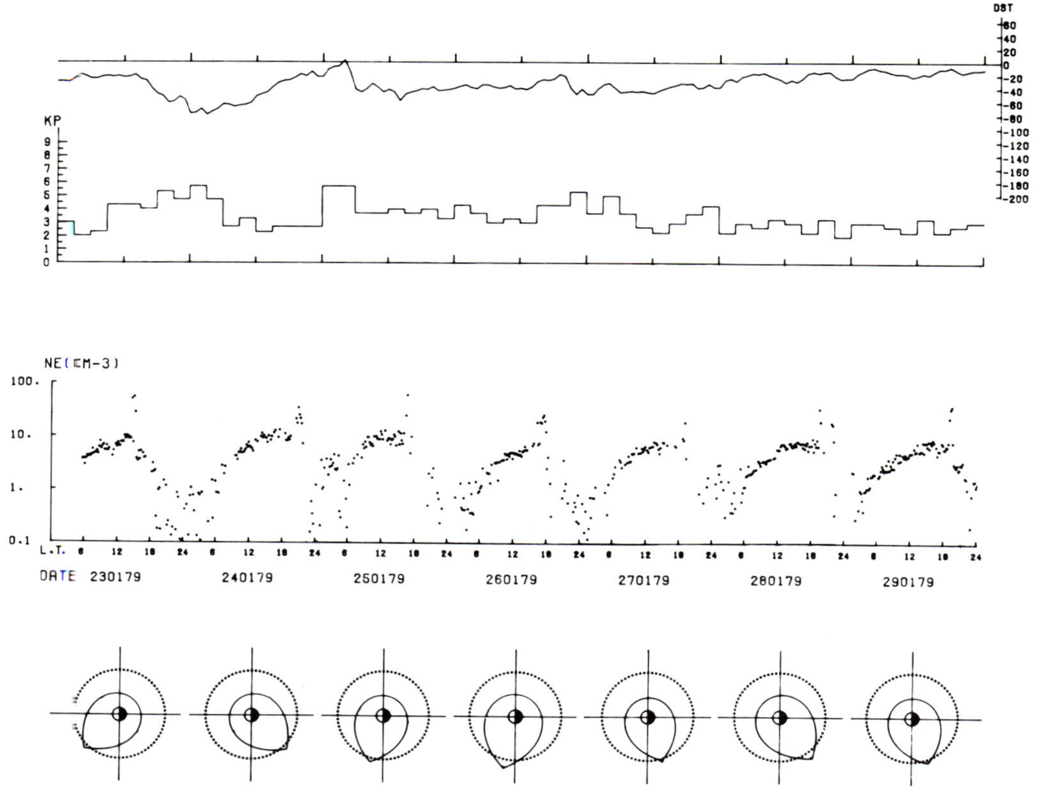

Figure 5b. Synoptic plot over seven consecutive days (January 23-29, 1979) of GEOS 2 RS electron density data. The center panel displays the density (vertical axis, log scale 0.1-100 cm^{-3}) versus local time (horizontal axis). Dates are given under the time axis. The top panel is devoted to geomagnetic activity indices (three-hour Kp with scale on the left, hourly Dst with scale on the right). The bottom panel shows, day by day, the orbit and the schematic equatorial section of the teardrop shaped model plasmapause that has the same bulge characteristics at 6.6 R_E as the actually measured profile displayed above (see text; after Higel and Lei, 1984).

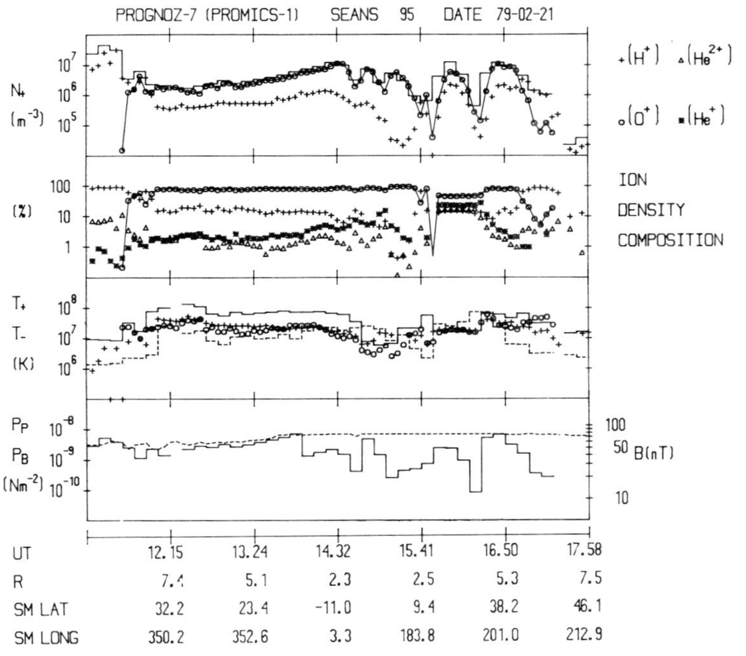

Figure 6. An example of observations in the dayside magnetosphere in magnetic storm conditions. At the lowest altitudes MeV electrons produce a dominating background which has not been eliminated from the data shown in the figure. The upper panel shows the ion number density (N_+) as deduced from the E/q spectrometers and assuming the ions were all protons (solid lines). Plus signs (+) represent the density of H^+ as deduced from the perpendicular ion mass spectrometers (assuming isotropy) and circles (o) represent the number density of O^+ using all mass spectrometers The second panel from the top represents the percentages of the four major ion constituents with respect to the total number density (logarithmic scale used). The third panel shows the temperatures of ions (solid line) and electrons (broken line) as deduced from the E/q electron and ion spectrometer data fitted onto Maxwellians. In the same panel the "perpendicular" H^+ (+) and O^+ (o) temperatures have been plotted. The fourth panel shows the ion plasma pressure (solid line) and magnetic field pressure (dotted line). The time and space coordinates (in Solar Magnetic, SM, coordinates) are given along the horizontal axis (after Hultqvist, 1983).

bance conditions. n_{H^+}/n_{He^+}=0.1-6 have been reported by Sojka et al. (1983) but no corresponding values for O^+ seem to have been published.

Chappell (1972) reported about detached plasma clouds in the dayside trough region and he suggested that they might originate in the dusk side bulge region and convect towards the dayside magnetopause. On the basis of the DE 1 measurements he concluded recently (Chappell, 1983) that these flux enhancements observed by the OGO 5 experiment may have been the equatorial manifestation of the out-flowing superthermal ion beams seen with DE 1 some ten years later.

6. THERMAL AND SUPERTHERMAL PLASMA IN THE PLASMAPAUSE REGION

The magnetospheric region just outside the plasmapause, as determined from the plasma density distribution, may in the recovery phase of disturbances in fact be located within a larger plasmasphere determined by the magnetospheric convection field at the moment in question. Horwitz et al. (1984) have investigated the formation of such a new outer plasmasphere near dusk. At the start the plasmapause was at L=4 (K_p=5-). K_p decreased then rapidly and the authors estimated that the new plasmapause was at L=8 during most of the period of investigation.

The time scale for filling up the outer plasmasphere was reported to be 7-22 hours. They saw $O^+/H^+>1$ at L=3-4 in the early period and a peak in the O^+ density just inside L=4, the so called "heavy ions torus" or "shell" or "cloak". The outer plasmasphere was O^+ enriched compared to the inner plasmasphere during the replenishment. There were indications that the O^+ ions were field-aligned at L~4 with 2-5 km/s speed along the field lines. The He^+/H^+ ratio was found to be in the range 0.2-0.5 in both the inner and the outer plasmasphere. The temperatures showed no overall evolution during the refilling of the outer part. The ion temperature they found was generally less than 2 eV between L=4 and L=8 and the various ion species had about the same temperatures. The "heavy ions torus" has also been discussed by Chappell (1983).

7. THE PLASMA SHEET

Higel and Lei (1984) have reported total electron density values from the region of the plasmasheet close to the plasmapause, the boundary of which is frequently crossed by the geostationary orbit in the late evening (except in very quiet periods when the entire geostationary orbit lies within the plasmasphere). Examples of measurement results from the midnight sector are also included in Figures 5a, b. The density values are lower and show a much larges spread than at other local times. As can be seen, the electron density is mostly between 0.1 and 1 cm^{-3} around midnight. Again it is difficult to say if there is any significant fraction of thermal or superthermal plasma in the inner plasmasheet. Considering that the hot plasma density in the plasmasheet generally is quoted to be in the density range mentioned, it seems likely that the low energy plasma components do not play a major role from a density point of view in this part of the magnetosphere, but better measurements are required before any firm conclusion can be drawn.

8. THE MAGNETOPAUSE BOUNDARY LAYERS AND THE CUSP

The dayside magnetopause boundary layer has been investigated in recent years by means of the ISEE 1,2 spacecraft (Peterson et al., 1982) and Prognoz 7 and 8 (Lundin and Dubinin, 1984a, b; Lundin, 1984).

The low latitude boundary layer appears to be a poor mix of solar wind and magnetosphere plasma. That is illustrated by Figure 7. Prognoz 7 crossed the magnetopause at 0619 on 9 February 1979 and encountered immediately inside it a magnetospheric kind of plasma without a low energy component. A few minutes later it saw a narrow beam of cold ionospheric ions, both H^+, He^+ and O^+, with a very narrow energy spectrum (at ~0625 in Figure 7). Three minutes later, within a region with a mixture of solar wind-like and magnetospheric plasma another beam of only He^+ was encountered. Another two minutes later a narrow beam composed of He^+ and O^+ ions was observed. Possibly it contained also ionospheric H^+ ions, but they could not be distinguished from those of solar wind origin. Figure 7 shows only the outermost part of the boundary layer. There were a number of transitions between magnetosphere-like and solar wind-like plasma regions similar to those shown in Figure 7. The solar wind-like "penetration" regions generally contain also ions of ionospheric origin as does the entire boundary layer. There are strong local increases of He^+ ion density in the solar wind penetration regions. These ions may be supposed to have reached the boundary layer from the plasmasphere. The reason that they are seen by the mass spectrometer on Prognoz 7 in the region of solar wind penetration is that the He^+ ions have been accelerated in those regions to energies above the lower threshold of the instrument.

These cold terrestrial ion beams in the boundary layer propagate perpendicularly to the magnetic field lines. The beam velocity is independent of ion mass. The thermal velocity is also roughly the same for all species and the beam velocity is generally higher than the thermal velocity by a factor of 2-4. The characteristic thermal energy falls in the superthermal energy range, but the lower threshold of the Prognoz-7 instrument of 200 eV did not make it possible to investigate in detail the thermal and superthermal plasma components, which may well be present in significant amounts. That thermal/superthermal ionospheric plasma exists in the boundary layer has, however, been clearly demonstrated by means of the PROMICS experiments on Prognoz 7 and 8. The observed independence of both beam velocity and thermal velocity on ion mass is a clear indication that the ions are accele-

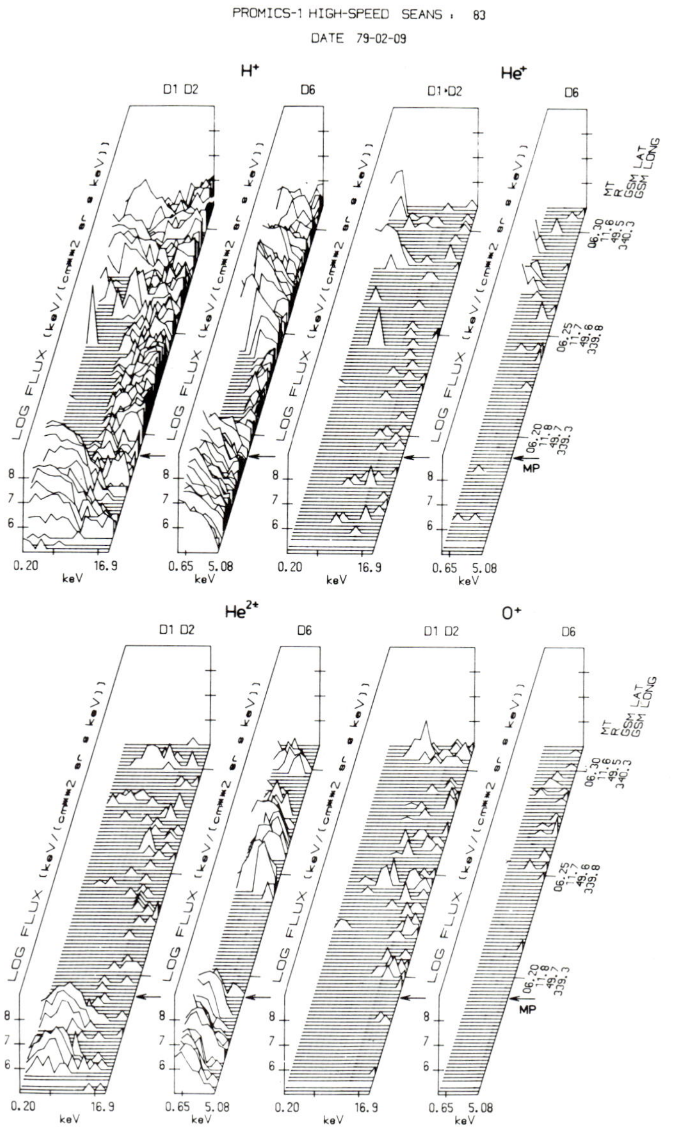

Figure 7. Spectrogram plot for H^+, He^{2+}, He^+ and O^+ during an inbound magnetopause crossing (marked MP) by Prognoz-7 on February 9, 1979. The D1D2-panels show energy spectra taken from the "perpendicular" spectrometers (scanning the YZ-ecliptic plane) whilst the D6 panels show energy spectra taken from the sunward oriented spectrometer. Time (MT = UT + 3) and orbital parameters are given along the inclined axis (after Lundin, 1984).

rated to an ExB velocity (Lundin and Dubinin, 1984b).

With the retarding potential ion mass spectrometer on DE 1 Chappell (1983) has observed outflowing H^+ and He^+ ions in the superthermal energy range in the cusp at altitudes of a few earth radii similar to what has been observed over the polar cap, but with more complex characteristics of the plasma energy distribution and flow than outside the cusp.

In discussing the boundary layers we have not emphasized density measurements as in previous sections. This is partly because it is more difficult to determine densities in a highly structured region such as a boundary layer and partly because the dominating physical processes in the boundary layers are likely to be more closely related to dynamical situations, structures and gradients than to averaged densities.

9. CONCLUDING REMARKS

Although the DE 1 satellite has provided us with very important new information about the low-energy plasma within a few earth radii altitude in the last few years, one general conclusion from the above presentation is that we still know very little about the magnetospheric plasma outside the plasmasphere at the very low end of the energy scale. The active wave instruments, flown for the first time in the late 1970's in the outer magnetosphere on the GEOS-satellites, which appear to be the best in including also the cold plasma component in the density measurements, have still only been used at and within the geosynchronous orbit. Systematic cold plasma density measurements remain to be carried out in the plasma sheet, boundary layers, lobes and tails. Even in regions where some measurements have been made a number of questions are unanswered.

The problem of how to make accurate measurements at or below a few eV of energy with particle detectors of various kinds in the presence of a potential difference of some volts between the spacecraft and the surrounding plasma is not a simple one, especially not in the presence of a large warm or hot plasma component. The good steps in the direction of solving that problem that have been taken by the Huntsville group (Chappell et al., 1981) are probably not the final ones needed. In any case an arsenal of different measurement techniques are required to cover the total range from cold to hot plasma in the magnetosphere and the low energy end requires particular care in planning the future space plasma physics payloads. Only when we master the determination of the density and temperature of the cold plasma component in those parts of the magnetosphere where there is a lot of high energy plasma will we have a basis for detailed investigations of microphysical processes, generation and propagation of many kinds of plasma waves and for answering questions about where some acceleration and heating processes take place. In other words, the answers to many fundamental questions in space plasma physics depend upon good measurements in the future of the low energy plasma in large parts of the magnetosphere.

10. REFERENCES

Axford, W.I., J. Geophys, Res., **73**, 6855, 1968.

Banks, P.M., and T.E. Holzer, J. Geophys. Res., **73**, 6846, 1968.

Banks, P.M., and T.E. Holzer, J. Geophys. Res., **74**, 6304, 1969a.

Banks, P.M., and T.E. Holqer, J. Geophys. Res., **74**, 6317, 1969b.

Banks, P.M., A.F. Nagy, and W.I. Axford, Planet. Space Sci., **19**, 1053, 1971.

Barakat, A.R., and R.W. Schunk, J.Geophys. Res., **88**, 7887, 1983.

Bezrukikh, V.V., and K.I. Gringauz, 'Hot zone in the outer plasmasphere of the earth', Techn. Rep. D-219, Acad. Sci. USSR, Space Res. Inst., Moscow, 1975.

Calvert, W., and G.B. Goe, J. Geophys. Res., **68**, 6113, 1963.

Calvert, W., Geophys. Res. Lett., **8**, 919, 1981.

Carpenter, D.L, J. Geophys. Res., **75**, 3887.

Carpenter, D.L., J. Geophys. Res., **71**, 693, 1966.

Chappell, C.R., Rev. Geophys. Space Phys., **10**, 951, 1972.

Chappell, C.R., C.R. Gaugher, and J.L. Horwitz, Rev. Geophys. Space Phys., **18**, 853, 1980.

Chappell, C.R., S.A. Fields, C.R. Baugher, J.H. Hoffman, W.B. Hanson, W.W. Wright, H.D. Hammack, G.R. Carignau, and A.F. Nagy, Space Sci. Instrum., **5**, 477, 1981.

Chappell, C.R., R.C. Olsen, J.L. Green, J.F.E. Johnson, and J.H. Waite, Jr., Geophys. Res. Lett., **9**, 937, 1982.

Chappell, C.R., in B. Hultqvist and T. Hagfors (Eds), High-Latitude Space Plasma Physics, p. 251, Plenum Press, N.Y., 1983.

Cowley, S.W.H., Space Sci. Rev., **26**, 217, 1980.

Decreau, P.M.E., J. Echeto, K. Knott, A. Pedersen, G.L. Wrenn, and D.T. Young, Space Sci. Rev., **22**, 633, 1978.

Decreau, P.M.E., C. Beghin, and M. Parrot, J. Geophys. Res., **87**, 695, 1982.

Dessler, A.J., and W.B. Hanson, Astrophys. J., **134**, 1024, 1961.

Dessler, A.J., and F.C. Michel, J. Geophys. Res., **71**, 1421, 1966.

de Ferandy, H., and B. Higel, Planet. Space Sci., **30**, 483, 1982.

Geiss, J., H. Balsiger, P., Eberhardt, H.P. Walker, L. Weber, D.T. Young, and H. Rosenbauer, Space Sci. Rev., **22**, 537, 1978.

Garrett, H.B., in W.P. Olson (Ed.), Quantitative Modeling of Magnetospheric Processes, p. 364, AGU, Washington, D.C., 1979.

Gurgiolo, C., and J.L. Burch, Geophys. Res. Lett., **9**, 945, 1982.

Gurnett, D.A., and R. Shaw, J. Geophys. Res., **78**, 8136, 1973.

Gurnett, D.A., and L.A. Frank, J. Geophys. Res., **79**, 2355, 1974.

Hagg, E.L., Can. J. Phys., **45**, 27, 1967.

Hanson, W.B., and T.N.L. Petterson, Planet. Space Sci., **11**, 1035, 1963.

Higel, B., and Wu Lei, J. Geophys. Res., **89**, 1583, 1984.

Hoffman, J.H., and W.H. Dodson, J. Geophys. Res., **85**, 626, 1980.

Horwitz, J.L., C.R. Baugher, E.G. Shelley, and D.T. Young, EOS, Trans. AGU, **59**, 1144, 1978.

Horwitz, J.L., and C.R. Chappell, J. Geophys. Res., **84**, 7075, 1979.
Horwitz, J.L., J. Geophys. Res., **85**, 2057, 1980.
Horwitz, J.L., J. Geophys. Res., **86** ,
Horwitz, J.L., Rev. Geophys. Space Phys., **20**, 929, 1982
Horwitz, J.L., W.K. Cobb, C.R. Baugher, C.R. Chappell. L.A. Frank, T.E. Eastman, R.R. Anderson, E.G. Shelley. and D.T. Young, J. Geophys. Res., **87**, 9059, 1982.
Horwitz, J.L., J. Atmos. Terr. Phys., **45**, 765, 1983.
Horwitz, J.L., R.H. Comfort, and C.R. Chappell, Geophys. Res. Lett., **11**, 701, 1984.
Hultqvist, B., Planet. Space Sci, **31**, 173, 1983.
Jackson, J.E., and E.S. Warren, Proc. IEEE, **57**, 861, 1969.
Kurth, W.S., J.D. Craven, L.A. Franck, and D.A. Gurnett, J. Geophys. Res., **84**, 4145, 1979.
Lemaire, J., and M. Scherer, Planet. Space Sci., **18**, 103, 1970.
Lemaire, J., and M. Scherer, Rev. Geophys. Space Phys., **11**, 427, 1973.
Lennartsson, W., and D.L. Reasoner, J. Geophys. Res., **83**, 2145, 1978.
Lockwood, M., J. Geophys. Res., **89**, 301, 1984.
Lockwood, M., J.H. Waite Jr., T.E. Moore, J.F.E. Johnson, and C.R. Chappell, 'A new source of suprathermal O^+ ions near the dayside polar cap boundary', Preprint, Space Sci. Lab. NASA Marshall Space Flight Center, Huntsville, 1984.
Lundin, R., Planet. Space Sci., **32**, 757, 1984.
Lundin, R., and Dubinin, E., Planet. Space Sci., **32**, 745, 1984a.
Lundin, R., and Dubinin, E.M., 'Solar wind energy transfer regions inside the dayside magnetopause-III Accelerated plasmaspheric ions as tracers for MHD-processes in the dayside boundary layer', Planet. Space Sci., in print, 1984b.
Marubashi, K., Rep. Ionos. Space Res., Japan, **24**, 322, 1970.
Moore, T.E., J. Geophys. Res., **85**, 2011, 1980.
Moore, T.E., Rev. Geophys. Space Phys., in print, 1984.
Moore, T.E., C.R. Chappell, M. Lockwood, and J.H. Waite, Jr., 'Superthermal ion signatures of auroral acceleration pro cesses', Preprint No. 85-101, Space Sci. Lab., NASA Marshall Space Flight Center, Huntsville, submitted to J. Geophys. Res., 1984.
Mozer, F.S., C.A. Cattell, M. Temerin, R.B. Torbert, S. von Glinski, M. Woldorff, and J. Wygant, J. Geophys.Res., **84**, 5875, 1979.
Nagai, T., J.H. Waite, Jr., J.L. Green, C.R. Chappell, R.C. Olsen, and R.H. Comfort, Geophys. Res. Lett., **11**, 669, 1984.
Persoon, A.M., D.A. Gurnett, and S.D. Shawhan, J. Geophys. Res., **88**, 10123, 1983.
Peterson, W.K., E.G. Shelley, H. Haerendel, and G. Paschmann, J. Geophys. Res., **87**, 2139, 1982.
Schunk, R.W., and D.S. Watkins, J. Geophys. Res., **87**, 171, 1982.
Serbu, G.P., and E.J.R. Maier, J. Geophys. Res., **75**, 6102, 1970.

Shelley, E.G., R.G. Johnson, and R.D. Sharp, J. Geophys. Res., **77**, 6104, 1972.
Shelley, E.G., R.D. Sharp, and R.G. Johnson, Geophys. Res. Lett., **3**, 654, 1976.
Shelley, E.G., W.K. Peterson, A.G. Ghielmetti, and J. Geiss, Geophys. Res. Lett., **9**, 941, 1982.
Sojka, J.J., R.W. Schunk, J.F.E. Johnson, J.H. Waite, Jr., and C.R. Chappell, J. Geophys. Res., **88**, 7895, 1983.
Temerin, M., J. Geophys. Res., **89**, 3945, 1984.
Timleck, P.L., and G.L. Nelms, Proc. IEEE, **57**, 1164, 1969.
Vasyliunas, V.M., 'Observations of low energy electrons with the OGO-A Satellite' Ph. D. Thesis, M.I.T., 1966.
Vasyliunas, V.M., (abstract) Trans. AGU, **47**, 142, 1966.
Waite,Jr. J.H., T. Nagai, J.F.E. Johnson, C.R. Chappell, J.L. Burch, T.L. Killeen, P.B. Hays, G.R. Carignan, W.K. Peterson, and E.G. Shelley, 'Escape of superthermal O^+ ions in the polar cap', J. Geophys. Res., in print, 1984.
Yau, A.W., P.H. Beckwith, W.K. Peterson, and E.G. Shelley, 'Long-term (solar-cycle) and seasonal variations of up-flowing ionospheric ion events at DE-1 altitudes', Preprint, Herzberg Inst. Astrophys., NRC, Canada, 1984.

SPACE SCIENCE REVIEWS

Volume 42 Nos. 3/4 1985

Space Science Reviews is published 6 times per annum, in double issues: February, April, June, August, October, and December.
Subscription prices, per volume: Institutions $ 101.00, Individuals $43.50.
Second-class postage paid at New York, N.Y. USPS No. 509–100.
U.S. Mailing Agent: Expediters of the Printed Word Ltd., 515 Madison Avenue (Suite 917), New York, NY 10022.
Published by D. Reidel Publishing Company, Spuiboulevard 50, P.O. Box 17, 3300 AA Dordrecht, Holland, and 190 Old Derby Street, Hingham, MA 02043, U.S.A.
Postmaster: please send all address corrections to: Kluwer Academic Publishers Group, c/o Expediters of the Printed Word Ltd., 515 Madison Avenue (Suite 917), New York, NY 10022, U.S.A.

FIELD-ALIGNED (BIRKELAND) CURRENTS

Thomas A. Potemra
The Johns Hopkins University
Applied Physics Laboratory
Laurel, Maryland 20707

ABSTRACT. Following earlier suggestions of Edmond Halley and Anders Celsius for the "magnetic behavior" of auroral phenomena, Kristian Birkeland discovered in his polar expeditions of 1902-03 that large-scale electric currents were associated with the aurora. He was also the first to suggest that these currents originated far from earth and that they flowed into the upper polar atmosphere and out of it along magnetic field lines; the existence of such field-aligned currents was widely disputed until satellite and rocket-borne instruments confirmed their permanent existence. The importance of these "Birkeland currents" to the coupling between the magnetosphere and the polar ionosphere is emphasized by their intensity, which ranges between 10^6 and 10^7 amperes, and by the energy which they dissipate in the upper atmosphere, which can exceed by a considerable factor the energy dissipated there by auroral particles. The large- and small-scale average properties of field-aligned currents, determined from spacecraft observations, are reviewed here.

1. HISTORICAL INTRODUCTION

The fact that the magnetic field is not absolutely steady but is continually changing and may experience violent changes was noted by several scientists; according to Chapman and Bartels (1940), the connection between the aurora and "magnetic storms" was first suggested by Halley in 1716. Later, in the same century, the Swedish physicist Celsius and his student Hiorter made similar discoveries. In a paper published at the Swedish Academy of Sciences in 1777, Hiorter reported the following (see Störmer, 1955):

> But who could have imagined that there was any connection between the aurora and the magnetic needle, and that the aurora, when passing over the zenith to the south or accumulating near the western or eastern horizon, would cause a considerable perturbation of the magnetic needle amounting to several degrees within a few minutes? The first time I saw an aurora in the south and simultaneously observed the greater

Space Science Reviews **42** (1985) 295–311. 0038–6308/85.15

> variation of the magnetic needle was on March 1, 1741 in the evening, after having several times noticed the irregular behavior of the needle but not seen the aurora owing to overcast sky. When I told the late professor [Celsius] of this he said that he had also observed a similar perturbation of the needle under the same circumstances, but that he had not mentioned it because he wished to see if I (such were his words) should happen to have the same idea.

One wonders how fair Celsius was being to his student!

In 1838, Gauss recognized that the magnetic field measured on the surface of the earth does not originate entirely within the earth, but that some of it may be of external origin due to currents flowing somewhere above the surface. Following the discovery of cathode rays, Gauss even suggested the possibility of "electric currents...sent out from the sun."

Inspired by his famous terrella experiments at the end of the 19th century and by his extensive studies of geomagnetic data recorded during magnetic storms, Kristian Birkeland suggested that the aurora was due to cathode rays or similar electric corpuscular rays sent out from the sun and deflected to the polar regions of the earth by the geomagnetic field. Birkeland recognized that the geomagnetic disturbances recorded on the earth's surface below the auroral region were due to intense currents flowing horizontally above. These are referred to today as auroral electrojets and are Hall currents resulting from large-scale electric fields directed perpendicular to the geomagnetic field. E. O. Hall received the second Ph.D. degree awarded by the Physics Department of The Johns Hopkins University in 1880 for his discovery, now referred to as the "Hall Effect" (see Potemra, 1980). In Birkeland's words:

> We consider it to be beyond doubt that the powerful storms in the northern regions, both those of long duration, and the short, well-defined storms that we have called elementary, are due to the action of electric currents above the surface of the Earth near the auroral zone. These currents, as far as the elementary storms are concerned at any rate, act, in the districts in which the perturbation is most powerful, as almost linear currents, that for a considerable distance are approximately horizontal.

He went on to compute the current strength to vary between 5×10^5 and 1×10^6 amperes and determined the altitude of these currents to be above 100 km. Present values are 3×10^5 amperes for the westward electrojet, which occurs between 100 and 120 km altitude, and 2×10^6 to 5×10^6 amperes for the total system of field-aligned currents.

Birkeland was concerned, however, about these horizontal currents, as is reflected in his writing:

> With regard to the further course of the current, there are two possibilities that may be considered. (1) The entire current-system belongs to the Earth. The current-lines are really lines where the current flows upon the Earth's surface, or rather at some height above it. (2) The current is maintained by a constant supply of electricity from without. The current will consist principally of vertical portions. At some distance from the Earth's surface, the current from above will turn off and continue for some time in an almost horizontal direction, and then either once more leave the Earth, or become partially absorbed by its atmosphere.

Birkeland apparently favored the second suggestion from his terrella experiments and Störmer's calculations, and proposed the field-aligned current system reproduced here as Figure 1. But his two suggestions became the source of a controversy concerning current systems that continued for a quarter of a century. Birkeland's first suggestion, concerning the containment of the current system within the earth's vicinity, was vigorously promoted by Sydney Chapman and his colleagues, and the model with the field-aligned currents was developed and advanced by Hannes Alfvén. Figure 2 shows Alfvén's diagram of those currents responsible for the aurora.

However, the reluctance to accept the concept of Birkeland's field-aligned currents is illustrated by the following quotations from papers by Chapman: "The question whether or not the currents flow wholly in the atmosphere is discussed, and it is concluded, though not decisively, that the evidence favors this view" (Chapman, 1935). The objections became stronger: "The electric current-system of Birkeland gives rise to a disturbance-field shown to be inconsistent with observations in several important respects" (Vestine and Chapman, 1938). The apparently unshakable hold on

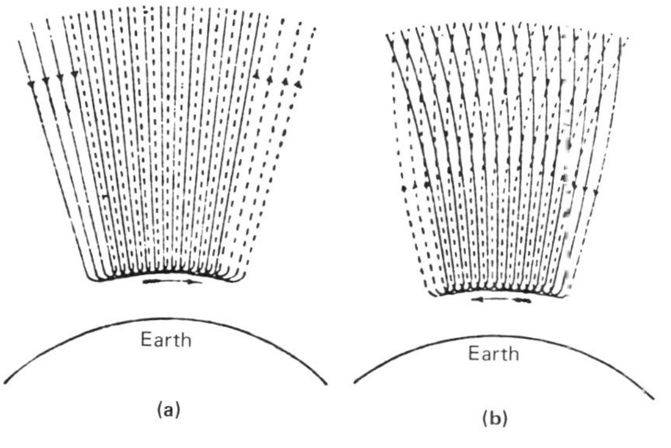

Figure 1. The system of field-aligned currents as published by Birkeland (1908). In consideration of the fact that auroral phenomena were often directly associated with solar activity, Birkeland believed that the horizontal currents had to couple to vertical currents flowing along geomagnetic field lines that, in turn, ultimately were connected to the sun.

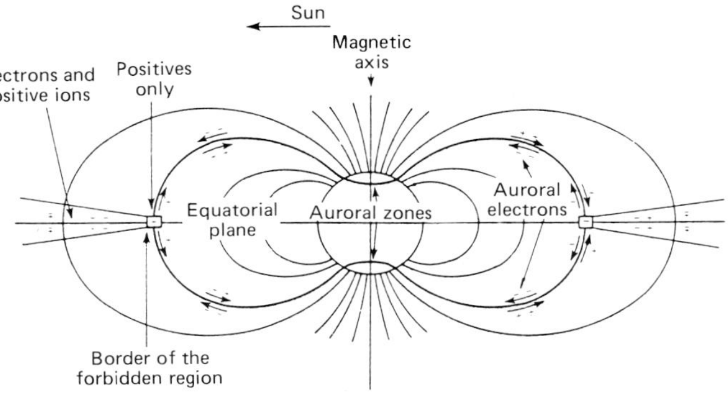

Figure 2. The complete system of field-aligned currents proposed by Alfvén (1939). He proposed that the interaction of solar wind plasma with the earth's magnetic field produced regions of positive and negative space charge (indicated in the figure) that were neutralized by currents flowing along geomagnetic field lines into and away from the auroral zones.

Birkeland's mind of his basic but invalid concept of intense electron beams, mingled error inextricably with truth in the presentation of ideas and experiments on auroras and magnetic storms" (Chapman, 1968).

The existence of field-aligned currents was disputed because it is not possible to distinguish unambiguously between current systems that are field-aligned and those that are completely ionospheric from a study of surface magnetic field measurements only. The first satellite measurements of Birkeland currents were made by A. J. Zmuda et al. (1966, 1967) with a magnetometer on board the polar orbiting satellite 1963-38C at approximately 1100 kilometers altitude. The magnetic disturbances observed by Zmuda et al. were initially interpreted as hydromagnetic waves. But it was soon realized that their latitudinal extent was too small for waves of the appropriate wavelength, and they were interpreted as being due to field-aligned or Birkeland currents (Cummings and Dessler, 1967). Many observations of Birkeland currents have been reported during the last few years by virtue of their associated magnetic variations detected by satellite-borne magnetometers.

The importance of the Birkeland currents to the coupling between the magnetosphere and the auroral atmosphere and ionosphere is demonstrated by their total intensity, which ranges between 10^6 and 10^7 amperes, and by the energy that they dissipate in the upper atmosphere, which can exceed by a considerable factor the energy deposition associated with the dramatic visual auroral forms.

2. MAGNETIC DISTURBANCES AND BIRKELAND CURRENTS

The density and flow direction of Birkeland currents associated with magnetic disturbances, δB, are determined from the relationship

$$J = \frac{1}{\mu_0} \nabla \times \delta B. \tag{1}$$

The magnetic disturbances associated with Birkeland currents occur predominantly in the geomagnetic east-west direction (B_y) so that

$$J_{\parallel} = J_z = \frac{1}{\mu_0} \frac{\partial}{\partial x} B_y, \tag{2}$$

where x is directed toward the north, y toward the east, and z is positive downward and parallel to the main geomagnetic field. This formulation also presumes that the Birkeland current sheet is uniform in the east-west (y) direction; (i.e., the Birkeland current sheets are aligned in the east-west direction and are "infinite").

Since $\mu_0 = 4\pi \times 10^{-7}$ henry/meter,

$$J_{\parallel} = 8 \times 10^{-4} \frac{\partial B_y}{\partial x} \text{ A/m}^2, \tag{3}$$

where B_y is in nanoteslas (nT) and x is in meters.

Since the speed of low altitude satellites like TRIAD and HILAT is approximately 8 km/second,

$$J_{\parallel} = \frac{1}{\mu_0} \left(\frac{\partial x}{\partial t}\right)^{-1} \frac{\partial B_y}{\partial t} = 0.1 \frac{\partial B_y}{\partial t} \mu\text{A/m}^2, \tag{4}$$

where $\partial B_y/\partial t$ is in nT per second.

A 200 nT gradient over a 160 km distance (covered in 20 secs by a low altitude satellite) corresponds to a typical Birkeland current sheet density of 1.0 μA/m^2.

3. CHARACTERISTICS OF BIRKELAND CURRENTS

Observations of perturbations transverse to the geomagnetic field from low altitude satellites have been used in numerous studies to determine characteristics of Birkeland currents (see for example, Saflekos et al., 1982; Potemra, 1983a,b; Potemra and Zanetti, 1985). These characteristics may be arbitrarily separated into two major categories based on the spatial size (or extent) of the currents. Birkeland currents with scales larger than 50 km (equivalent to about 0.5° of latitude) are considered to be "large-scale" and are believed to comprise a part of the global, three-dimensional current system which connects the auroral ionosphere to the most distant regions of the magnetosphere which interact with the solar wind. Birkeland currents with spatial scales less than 50 km (0.5°) are usually more intense, exhibit complicated spatial and time variations and are believed to be associated with plasma processes within the magnetosphere. These characteristics are summarized in Tables 1 and 2 below.

TABLE 1

Characteristics of Large-Scale Birkeland Currents > 50 km (0.5°)

- Stable patterns in auroral zone, cusp, polar cap (Region 1 and 2, cusp, and NBZ systems)
- Close association with large-scale convection and electric field patterns
- Close relationship to IMF
- Directly coupled to large-scale ionospheric currents detected on surface
- Carried by precipitating low-energy electrons (≤ few 100 eV) for upward flowing currents
- Carried by upward flowing ionospheric electrons (thermal) for downward flowing currents
- Generated by solar wind/magnetospheric processes

TABLE 2

Characteristics of Small-Scale Birkeland Currents < 50 km (0.5°)

- Irregular patterns embedded in large-scale Region 1, Region 2, cusp and NBZ systems
- Close association with visible and UV auroral forms
- Close association with AKR, micropulsations, and ionospheric instabilities
- Associated with plasma processes within the magnetosphere and ionospheric processes
- Carried by high energy (10's of keV) electrons or ions

Some characteristics will be discussed in more detail in the following sections.

4. STATISTICAL DISTRIBUTIONS OF LARGE-SCALE BIRKELAND CURRENTS

The large-scale Birkeland currents are concentrated in two principal areas which encircle the geomagnetic pole (Zmuda and Armstrong, 1974a,b; Iijima and Potemra, 1976a,b). These are shown in Figure 3 (from Iijima and Potemra, 1976b, their Figure 6). The Birkeland current flow regions have been arbitrarily designated by Iijima and Potemra (1976a) as "Region 1" located at the poleward side, and "Region 2" located at the equatorward side. The Region 1 currents

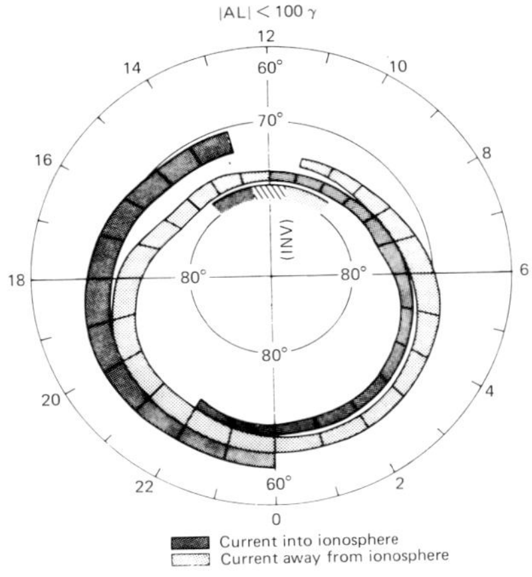

Figure 3. A summary of the distribution and flow directions of large-scale field-aligned currents determined from data obtained on 493 passes of TRIAD during weakly disturbed conditions (|AL| < 100 γ) (from Iijima and Potemra 1976b).

flow into the ionosphere in the morning sector and away from the ionosphere in the evening sector. The Region 2 currents flow in the opposite direction at any given local time.

The densities of the Region 1 and Region 2 currents are not necessarily equal at given local time (i.e., the upward and downward flowing currents are not always completely coupled by ionospheric Pedersen currents flowing along a geomagnetic meridian). The regions of greatest imbalance between the densities of the Region 1 and Region 2 systems occur near 1400 MLT (magnetic local time) and 0800 MLT (see Figure 14 of Iijima and Potemra, 1978). The largest Region 1 currents flow into the ionosphere near 0800 MLT and away from the ionosphere near 1400 MLT producing the "net" or "unbalanced" field-aligned currents discussed by Sugiura and Potemra (1976). The region near 1400 MLT may have a special significance with regard to particle precipitation, auroral emissions, and topology of the magnetosphere (Potemra, 1983a).

A three-region flow pattern of Birkeland currents exists in the 2200 to 2400 MLT sector (sometimes referred to as the "Harang discontinuity region") and may be thought of as an "overlapping" of the two-region flow patterns usually observed in the surrounding MLT sectors; it is not possible to distinguish between Region 1 and Region 2 Birkeland currents in this MLT sector. The basic flow pattern of Birkeland currents is the same during geomagnetically inactive and active periods, although the regions widen and shift to lower latitudes during disturbed periods (similar to the variations in the statistical "auroral oval" determined by Feldstein, 1966). Complicated small-scale structures are superimposed upon the large-scale Birkeland current features, especially near the high-latitude noon region associated with the cusp and on the nightside during substorm events.

5. CUSP-REGION BIRKELAND CURRENTS

Complicated patterns of magnetic disturbances have been observed with various low altitude satellites in the high latitude region near magnetic noon (see the review of Potemra and Zanetti, 1985). The disturbances at the poleward edge of this region were believed to be due to the flow of Birkeland currents in the cusp because of their location (for example, between 78° and 80° invariant latitude and between 1000 MLT and 1400 MLT during quiet geomagnetic conditions) and because of their relationship to the IMF B_z component (larger intensities were usually observed during periods of southward B_z). These cusp-region Birkeland currents were determined by Iijima and Potemra (1976b) to be flowing away from the ionosphere in the pre-noon hours (MLT) and into the ionosphere in the post-noon hours (as shown in Figure 3), from a data base that included both sector orientations of the IMF. Subsequent analysis of transverse disturbances over the north and south polar regions acquired by the TRIAD, ISIS-2, and S3-2 spacecraft (Wilhjelm et al., 1978; McDiarmid et al., 1978b, 1979; Iijima et al., 1978; Doyle et al., 1981) have revealed that one flow direction is observed more often than the other, depending upon the polarity of B_y. This dependence is predicted from theoretical and simulation considerations (e.g., Gizler et al., 1977; Levitin et al.,1977; Reiff et al., 1978; Ogino and Walker, 1984; Ogino et al., 1984, and the proceedings of this conference).

An attempt to schematically diagram the complicated relationships between Birkeland current flow patterns and the IMF orientation is shown in Figure 4 (from Potemra and Saflekos, 1979). The region of cusp Birkeland currents is shown as the crescent-shaped area poleward of Regions 1 and 2. The demarcation line between upward and downward flowing Birkeland currents is depicted to shift to the afternoon or morning side of the cusp, depending upon the polarity of B_y. This interpretation retains the general flow pattern of downward flowing Birkeland currents on the afternoon side

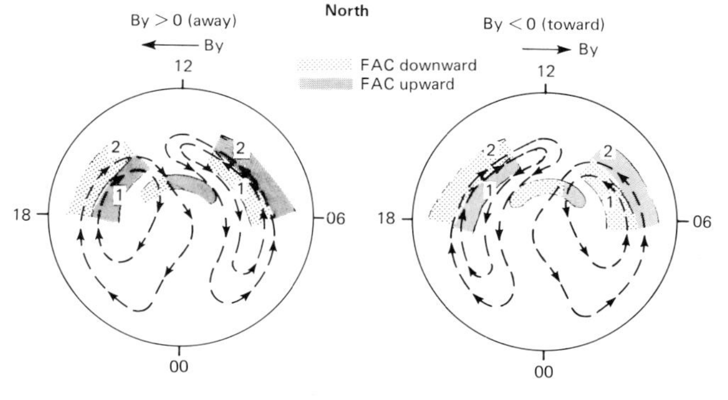

Figure 4. A schematic diagram of convective flow patterns and Birkeland current patterns in the north polar regions during conditions of positive and negative B_y (from Potemra and Saflekos, 1978, their Figure 1).

of the cusp and upward flowing Birkeland currents on the morning side, while also incorporating the observation of dominant upward current flow for positive B_y and dominant downward current flow for negative B_y. The dashed arrows in this diagram are intended to show the convection flow pattern for each B_y polarity, and the DPY ionospheric current is believed to flow opposite to this flow near noon between the cusp and Region 1.

A summary of the patterns of Birkeland currents, ionospheric currents, and convection patterns is provided in Table 3 (from Potemra et al., 1979). All of these patterns apply to conditions of negative B_z only.

TABLE 3

Summary of B_y-Dependent Current Patterns (for $B_z < 0$)

Positive B_y, Northern (Southern)[1] Hemisphere:

1. Cusp Birkeland current flow at the most poleward latitudes is predominantly upward[2,3] (downward)[4]

2. Region 1 and 2 Birkeland current flow directions are not affected by B_y and are contained within regions of sunward convection[5,6]. Larger Region 1 and 2 current intensities occur on the dawn side of the auroral zone[7] (dusk side)[8].

3. DPY ionospheric current is located between two Birkeland current sheets and flows eastward[2] (westward?).

4. Convective flow near noon is directed predominantly westward (eastward)[6,9]. The "throat" is shifted to the afternoon (morning).

5. Maximum convective flow in the polar cap (and associated E-fields) is located near dawn (dusk)[10].

All patterns and flow directions are systematically reversed for negative B_y.

[1] For each item the flow direction or spatial location in the Southern Hemisphere is given in parenthesis.
[2] Wilhjelm et al. (1978).
[3] McDiarmid et al. (1978b).
[4] Iijima et al. (1978).
[5] Iijima and Potemra (1976b).
[6] Saflekos et al. (1979).
[7] McDiarmid et al. (1978a).
[8] Saflekos and Potemra (1980).
[9] Heelis et al. (1976).
[10] Heppner (1972, 1973).

6. BIRKELAND CURRENTS DURING POSITIVE B_z

Iijima (1983) discovered the presence of a large-scale Birkeland current system in the northern summer polar region during a period of strongly northward IMF. He determined this from his analysis of a remarkably stable pattern of transverse magnetic disturbance which appeared in data acquired during 8 consecutive MAGSAT orbits over the northern hemisphere on May 10, 1980 (see his Figure 4). Araki et al. (1984) came to the same conclusion from their analysis of MAGSAT data acquired on the same day. This large-scale Birkeland current system occupies the entire region poleward of the dayside Region 1 Birkeland current system, and has the opposite flow direction. Namely, the polar region Birkeland current system flows into the ionosphere on the dusk side and away from the ionosphere on the dawn side.

Iijima et al. (1984) examined magnetic field data acquired during 146 orbits of MAGSAT over the south polar regions during November, 1979 to January, 1980. The characteristics of these polar disturbances include the following: (1) They occur at latitudes poleward of the Region 1 Birkeland current system at daytime magnetic local times (0600 MLT through noon to 1800 MLT). (2) The spatial distribution along the dawn-dusk direction resembles the "W" shaped distribution of electric fields observed in the polar cap during periods of positive B_z (Burke et al., 1979). (3) These patterns show remarkable stability showing little change from orbit to orbit, up to 7 orbits of MAGSAT (equivalent to a period of 10 hours). (4) As shown in Figure 5 (from Iijima et al. 1984), the

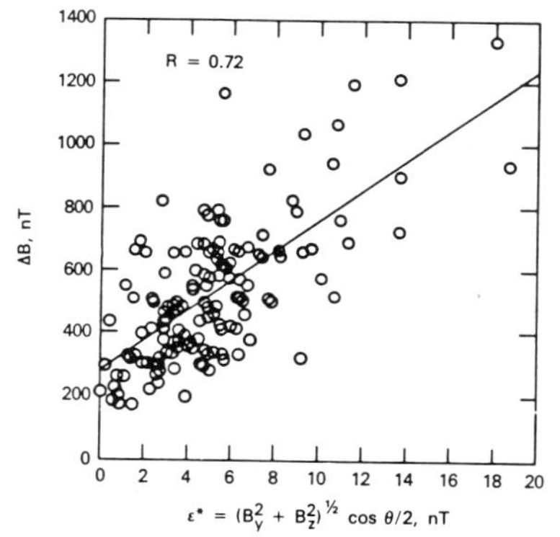

Figure 5. The magnitude (ΔB) of peak triangular Δ S disturbance plotted versus the "complementary transmission function" ϵ * for 146 MAGSAT events, analyzed by Iijima et al. (1984, their Figure 10).

magnitude of the peak disturbance, ΔB, correlates with a "complementary" magnetospheric transmission function of the form: $\varepsilon^* = (B_y^2 + B_z^2)^{1/2} \cos \theta/2$, where θ is the angle between the positive z-axis and the IMF.

If the magnetic disturbances are interpreted in terms of

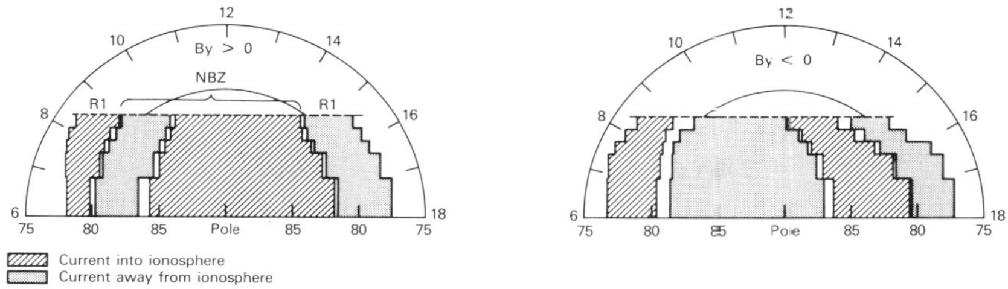

Figure 6. The spatial distribution of the flow directions of the large-scale NBZ and Region 1 Birkeland currents in the south polar region during $B_y > 0$ and during $B_y < 0$.

Birkeland currents, they flow downward on the dusk side and flow away on the morning-side (identical to the cusp current flow reported by Iijima and Potemra, 1976b). During periods of negative B_y the region of morning-side (upward flowing) currents is much larger than the evening-side (downward flowing) current region in the southern hemisphere (as shown schematically in Figure 6 from Iijima et al., 1984). The density of the currents in the smaller spatial region is larger than the density of the currents in the bigger region. This pattern systematically reverses during periods of positive B_y. These observations are interpreted as evidence for a large-scale, stable Birkeland current system in the polar region which is associated with merging on field lines in the geomagnetic tail. This current system intensifies and is more stable as B_z becomes more northward (reminiscent of the behavior of the Region 1 current system with increasing southward values of B_z). The new stable polar cap current system is referred to as the "NBZ" Birkeland current system for "northward B_z" and is important because of its relationship to a variety of other "northward B_z" phenomena such as polar cap auroral arcs ("theta aurora") and multi-cell convective flow patterns (see, for example, Potemra et al., 1984).

Zanetti et al. (1984) used the radial component of magnetic field measurements from MAGSAT to deduce horizontal currents flowing in the ionosphere during periods of positive B_z. They compared the locations, intensities, and flow directions of the ionospheric currents and Birkeland currents deduced from the same MAGSAT orbits. They discovered a B_y dependence in the characteristics of the ionospheric currents that matched a B_y dependence in the Birkeland currents noted by Iijima et al. (1984) above.

7. SMALL-SCALE BIRKELAND CURRENTS

Evidence for the relationship between Birkeland currents and plasma waves in the magnetospheres of the Earth and Jupiter has been growing in the last few years (see Barbosa et al. 1981 and references contained therein). Birkeland currents have been associated with auroral kilometric radiation (AKR) (Green et al., 1979), ionospheric scintillations (Vickrey et al., 1980), Pc5 micropulsations (Poulter et al., 1982), electrostatic ion cyclotron waves

(Cattell, 1981), and Pil irregular micropulsations (Engebretson et al., 1984).

The detection at low altitude of very high density Birkeland currents has recently become a topic of great interest due, for the most part, to plasma instabilities that may result from their presence. To date, only a few low altitude satellites have been instrumented to measure particles, electric fields, and magnetic perturbations (to determine Birkeland currents). Small-scale Birkeland currents of both signs (flowing into and away from the ionosphere) have been detected within and nearby auroral arc formations (Burke et al., 1980; Burke et al, 1983; Sugiura et al., 1982, 1983). In almost all cases, observed currents with densities of ~ 100 $\mu A/m^2$ have been directed away from the ionosphere and were associated with a net inward flux of energetic electrons. Intense Birkeland currents directed into the ionosphere have been observed with densities of ~ 10 to 20 $\mu A/m^2$ (Burke et al., 1983). These currents were usually paired with more intense upward flowing currents.

The most intense earthward flowing Birkeland current yet detected was ~ 94 $\mu A/m^2$ (detected by the HILAT satellite, Bythrow et al., 1984). This current was less than 4 km in extent and was located at the equatorward edge of a large-scale earthward flowing current in the late evening sector. Simultaneous UV images acquired with an onboard auroral imager showed that the large-scale Birkeland current was embedded within the diffuse aurora.

Figure 7 (from Bythrow et al., 1984, their Figure 1) shows plots of transverse magnetic field, particle flux of precipitating electrons in the energy range between 20 eV and 20 keV, energy flux computed from the observed precipitating flux, and the horizontal component of ion drift. The sharp gradient in magnetic field indicates the presence of the intense (97 $\mu A/m^2$) downward flowing Birkeland current. A gradient in the electron flux occurred about 1/4 second before the onset of the intense Birkeland current. The horizontal component of ion drift underwent a series of large excursions in an extended region surrounding the intense field-aligned current.

From these observations, Bythrow et al. (1984), concluded that the intense Birkeland current was associated with a sharp gradient in the height-integrated Pedersen conductivity (produced by the precipitating energetic electrons) in the presence of an electric field, i.e., a result of the E_x $(\partial \Sigma_p / \partial x)$ term of the current continuity relation. The highly structured fluctuations in ion drift may have resulted from current driven instabilities excited by the downward flowing Birkeland current (presumably carried by a high flux of upward flowing thermal ionospheric electrons).

8. BIRKELAND CURRENTS AND AURORAL EMISSIONS

The presence of Birkeland currents have been correlated with observations of auroral forms from the ground and from separate satellites (see the review of Potemra, 1979). Figure 8 shows a bright UV

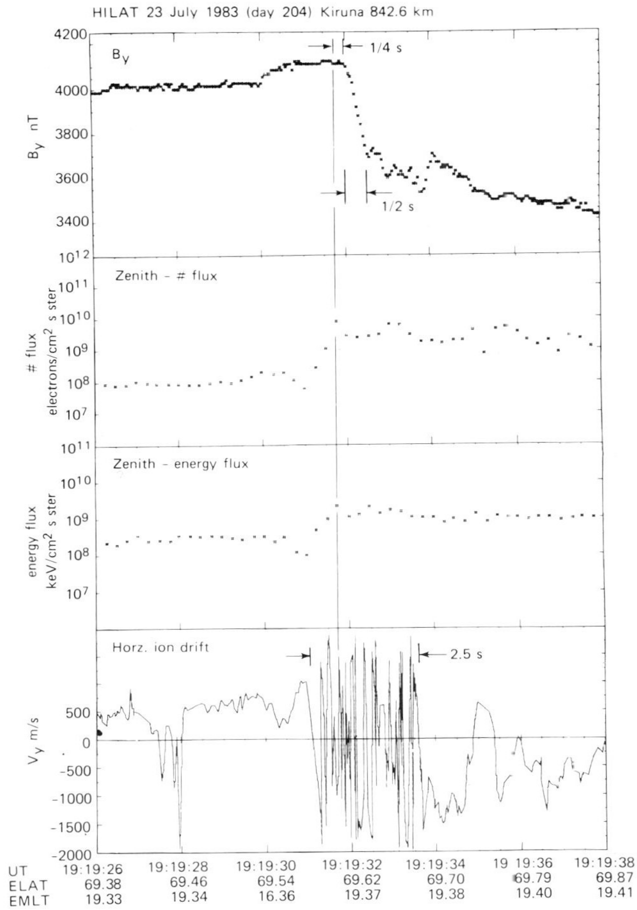

Figure 7. A stack plot of simultaneous y component magnetic fields, zenith number flux, zenith energy flux, and y component ion drift velocity from 1919:26 UT to 1919:38 UT on July 23, 1983.

auroral form in the daylit atmosphere and magnetic perturbations acquired with instruments on the same low-altitude satellite (HILAT, from Potemra, 1983b). This a black and white version of his color Figure 11). The spacecraft is traveling in complete daylight from high to low latitudes over the dawn sector near 0500 magnetic local time. The magnetometer trace along the right side of the figure shows a large perturbation extending from 63° to 69° magnetic latitude, which is due to a large-scale pair of Birkeland currents. A current of 1.1 $\mu A/m^2$ flows away from the ionosphere in the region of the intense UV auroral arc. The downward flowing electrons producing the UV emissions carry the upward flowing current (the Region 2 system). Poleward of this current is a downward flowing Birkeland current of 0.7 microampere per square meter, which is the corresponding Region 1 system.

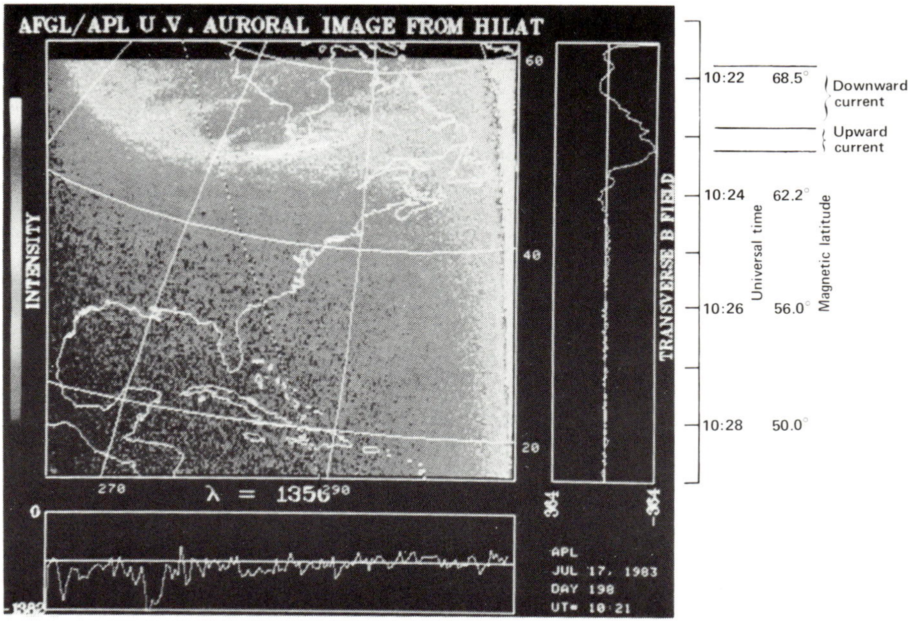

Figure 8. A bright vacuum ultraviolet (1356 Å) auroral form observed in the daylit atmosphere below Hudson Bay by the Auroral Ionospheric Mapper instrument on the HILAT satellite. The magnetic field observed by HILAT in approximately the east-west geomagnetic direction is shown in the line plot on the right. In this plot, the vertical axis is the distance along HILAT's trajectory (which occurs down the middle of the image map), and the horizontal axis is the magnetic disturbance. The large magnetic disturbance near the top of the plot (first to the right and then to the left) reveals the presence of a pair of field-aligned currents. The current at the poleward side flows downward, while the current at the equatorward side flows upward and is located at the same position of the arc. The downward flowing electrons that carry the current also produce the arc. (The Auroral Ionospheric Mapper image was provided through the courtesy of R. E. Hoffman, AFGL, and C. I. Meng, APL.)

9. SUMMARY

Large-scale Birkeland currents flow into and away from the auroral regions, dayside cusp, and polar cap. These Birkeland currents are intimately associated with ionospheric currents, convection electric fields, auroral emissions, and plasma instabilities. The locations, intensities, and flow directions of the Birkeland currents correlate with the amplitude and orientation of the interplanetary magnetic field. These relationships suggest that the Birkeland currents are the important link between the interplanetary medium and the high latitude ionosphere. This Birkeland current system can account for the close correlation between surface geomagnetic variation in the polar regions and the sector direction of the interplanetary magnetic field (Svalgaard, 1973; Mansurov, 1969).

Although the Birkeland current system in the auroral zone diminish in intensity during periods of northward interplanetary magnetic field (as do many of the standard geomagnetic indices), a current system develops and intensifies in the polar cap. This "NBZ" system of Birkeland and ionospheric currents is important because it demonstrates that a considerable amount of energy flows into the magnetosphere even during periods of northward IMF. These various systems of Birkeland currents comprise important elements in the interplanetary-magnetosphere-ionosphere network during almost all conditions.

ACKNOWLEDGEMENT

This work was supported by the Atmospheric Research Section of the National Science Foundation, the Office of Naval Research, and the Defense Nuclear Agency. I am grateful for the active discussions and help from L. J. Zanetti, P. F. Bythrow, and A. T. Y. Lui.

REFERENCES

Alfvén, H. 1939, Kungl. Sv. Vet.-Akademiens Handl., **18**, No. 3.
Araki, T., Kamei, T., and Iyemori, T. 1984, Geophys. Res. Lett., **11**, 23.
Barbosa, D. D., Scarf, F. L., Kurth, W. S., and Gurnett, D. A. 1981, J. Geophys. Res., **86**, 8357.
Birkeland, K. 1908, The Norwegian Aurora Polaris Expedition, 1902-03, Aschehoug, Oslo, Norway.
Burke, W. J., Kelley, M. C., Sagalyn, R. C., Smiddy, M., and Lai, S. T. 1979, Geophys. Res. Lett., **6**, 21.
Burke, W. J., Hardy, D. A., Rich, F. J., Kelly, M. C., Smiddy, M., Shuman, B., Sagalyn, R. C., Vancour, R. P., Wipman, P. J. L., and Lai, S. T. 1980, J. Geophys. Res., **85**, 1179.
Burke, W. J., Silevitch, M., and Hardy, D. A. 1983, J. Geophys. Res., **88**, 3127.
Bythrow, P. F., Potemra, T. A., Hanson, W. B., Zanetti, L. J., Meng, C.-I., Huffman, R. E., Rich, F. J., and Hardy, D. A. 1984, J. Geophys. Res., **89**, 9114.
Cattell, Cynthia 1981, J. Geophys. Res., **86**, 3641.
Chapman, S. 1935, Terr. Magn. Atmos. Electr., **40**, 349.
Chapman, S., and Bartels, J. 1940, Geomagnetism, **Vols. I and II**, Oxford Press.
Chapman, S. 1968, Ann. Geophys., **24**, 497.
Cummings, W. D., and Dessler, A. J. 1967, J. Geophys. Res., **72**, 1007.
Doyle, M. A., Rich, F. J., Burke, W. J., and Smiddy, M. 1981, J. Geophys. Res., **86**, 5656.
Engebretson, M. J., Cahill, L. J., Potemra, T. A., Zanetti, L. J., Arnoldy, R. L., Mende, S. B., and Rosenberg, T. J. 1984, J. Geophys. Res., **89**, 1602.

Feldstein, Y. I. 1966, Planet. & Space Sci., **14**, 121.

Gizler, V. A., Semenov, V. S., Troshichev, O. A. 1977, EOS, (Abstract) **58**, 719.

Green, J. L., Gurnett, D. A., Saflekos, N. A., and Potemra, T. A. 1979, J. Geophys. Res., (submitted).

Heelis, R. A., Hanson, W. B., and Burch, J. L. 1976, J. Geophys. Res., **81**, 3803.

Heppner, J. P. 1972, J. Geophys. Res., **77**, 4877.

Heppner, J. P. 1973, Radio Sci., **8**, 933.

Iijima, T., and Potemra, T. A. 1976a, J. Geophys. Res., **81**, 2165.

Iijima, T., and Potemra, T. A. 1976b, J. Geophys. Res., **81**, 5971.

Iijima, T., and Potemra, T. A. 1978, J. Geophys.Res., **83**, 599.

Iijima, T., Fujii, R., Potemra, T. A., and Saflekos, N. 1978, J. Geophys. Res., **83**, 5595.

Iijima, T. 1983, in Magnetospheric Physics, Geophys. Monograph **28**, 115.

Iijima, T., Potemra, T. A., Zanetti, L. J., and Bythrow, P. F. 1984, J. Geophys. Res., **89**, 7441.

Levitin, A. E., Belov, B. A., Afonina, R. G., Faermark, D. S. and Feldstein, Y. I. 1977, USSR Academy of Sciences IZMIRAN Preprint N 17a.

Mansurov, S. M. 1969, Geomagn. Aeron., **9**, 622.

McDiarmid, I. B., Burrows, J. R., and Wilson, M. D. 1978a, J. Geophys. Res., **83**, 5753.

McDiarmid, I. B., Burrows, J. R., and Wilson, M. D. 1978b, J. Geophys. Res., **83**, 681.

McDiarmid, I. B., Burrows, J. R., and Wilson, M. D. 1979, J. Geophys. Res., **84**, 1431.

Ogino, T. and Walker, R. J. 1984, EOS, **65**, 1057.

Ogino, T., Walker, R. J., and Ashour-Abdalla 1984, EOS, **65**, 1063.

Potemra, T. A., and Saflekos, N. A. 1979, Proc. of Magnetospheric Boundary Layers Conf., Alpbach, June 1979, 193.

Potemra, T. A. 1979, Rev. Geophys. Space Phys., **17**, 640.

Potemra, T. A., Iijima, T., and Saflekos, N. A. 1979, Dynamics of the Magnetosphere, ed. by S.-I. Akasofu, D. Reidel Pub. Co., 165.

Potemra, T. A. 1980, in The Hall Effect and Its Applications, C. L. Chein and C. R. Westgate, eds., Plenum Pub. Co., N. Y., pp. 399-415.

Potemra, T. A. 1983a, High-Latitude Space Plasma Physics, Plenum Publishing Co.

Potemra, T. A. 1983b, Johns Hopkins APL Technical Digest, **4**, 276.

Potemra, T. A., Zanetti, L. J., Bythrow, P. F., Lui, A. T. Y., and Iijima, T. 1984, **89**, 9753.

Potemra, T. A. and Zanetti, L. J. 1985, Proc. of the NATO Advanced Workshop on Morphology and Dynamics of the Polar Cusp, ed. by J. A. Holtet and A. Egeland, D. Reidel Pub. Co.

Poulter, E. M. Nielsen, E. and Potemra, T. A. 1982, J. Geophys. Res., **87**, 2331.

Reiff, P. H., Burch, J. L., and Heelis, R. A. 1978, Geophys. Res. Lett., **5**, 391.

Saflekos, N. A., Potemra, T. A., Kintner, Jr., P. M., and Green, J. Lauer 1979, J. Geophys. Res., **84**, 1391.
Saflekos, N. A., and Potemra, T. A. 1980, J. Geophys. Res., **85**, 1987.
Saflekos, N. A., Sheehan, R. E., and Carovillano, R. L. 1982, Rev. of Geophys. & Space Phys., **20**, 709.
Störmer, C. 1955, The Polar Aurora, Oxford Press.
Sugiura, M., and Potemra, T. A. 1976, J. Geophys. Res., **81**, 2155.
Sugiura, M., Maynard, N. C., Farthing, W. H., Heppner, J. P., Ledley, B. G., and Cahill, Jr., L. J. 1982, Geophys. Res. Lett., **9**, 985.
Sugiura, M., Iyemori, T., Hoffman, R. A., Maynard, N. C., Burch, J. L., and J. D. Winningham 1983, in Magnetospheric Currents, Geophysical Monograph **28**, edited by T. A. Potemra, 96.
Svalgaard, L. 1973, J. Geophys. Res., **78**, 2064.
Vestine, E. H., and Chapman, S. 1938, Terr. Magn. Atmos. Electr., **43**, 351.
Vickrey, James F., Rino, Charles L., and Potemra, Thomas A. 1980, Geophys. Res. Lett., **7**, 789.
Wilhjelm, J. E., Christensen, E. Friis, and Potemra, T. A. 1978, J. Geophys. Res., **83**, 5586.
Zanetti, L. J., Potemra, T. A., Iijima, T., Baumjohann, W., and Bythrow, P. F. 1984, J. Geophys. Res., **89**, 7453.
Zmuda, A. J., Martin, J. H., and Heuring, F. T. 1966, J. Geophys. Res., **71**, 5033.
Zmuda, A. J., Heuring, F. T., and Martin, J. H. 1967, J. Geophys. Res., **72**, 1115.
Zmuda, A. J., and Armstrong, J. C. 1974a, J. Geophys. Res., **79**, 2501.
Zmuda, A. J., and Armstrong, J. C. 1974b, J. Geophys. Res., **79**, 4611.

LARGE ELECTRIC FIELDS IN THE MAGNETOSPHERE

F. S. Mozer,* M. H. Boehm,* C. A. Cattell, M. Temerin
and J. R. Wygant
Space Sciences Laboratory
University of California
Berkeley, CA 94720
* also Physics Department

ABSTRACT. In several regions of the magnetosphere, perpendicular and/or parallel electric fields are found to be orders-of-magnitude larger than expected from simple considerations. Problems associated with these large fields that may be amenable to study through computer simulations are discussed. Regions in which large electric fields are observed include: a) The auroral ionosphere, where Langmuir soliton-like structures have been measured to contain plasma frequency oscillations as large as 500 mV/m, the envelopes of which have parallel electric fields of ~100 mV/m lasting for fractions of a millisecond; b) The auroral acceleration region, where electrostatic shocks have been observed to contain perpendicular fields as large as 1000 mV/m and parallel fields as large as 100 mV/m, and where double layers having parallel fields up to 10 mV/m have been observed; c) The high latitude boundary of the plasma sheet, where turbulent electric fields as large as 100 mV/m have been seen along with quasi-static fields of 5-10 mV/m; d) Inside the plasma sheet, where fields of 5-10 mV/m have frequently been observed; e) The bow shock, where turbulent fields as large as 100 mV/m and d.c. fields of ~5 mV/m normal to the shock have been seen.

1. INTRODUCTION

The purpose of this paper is to discuss large electric fields observed in the magnetosphere. By "large electric fields" are meant d.c. and low frequency fields having amplitudes which are large compared to those contained in static convection electric field models that assume a total potential of ~60 kilovolts across the magnetosphere. Such models do not have parallel electric fields and

Space Science Reviews **42** (1985) 313–335. 0038–6308/85.15

contain perpendicular fields of tens of mV/m at low altitudes and ~0.1 mV/m in the tail.

Parallel and perpendicular d.c. and low frequency electric fields orders-of-magnitude larger than those associated with the static convection model have been observed at altitudes below 1000 kilometers, in the auroral acceleration region at an altitude of ~1 R_e, in the plasma sheet and plasma sheet boundary layer in the tail, and at the bow shock. Observational characteristics of the observed fields and some of their problems that may be studied by computer simulations will be described.

2. DISCUSSION

2.1. Large Electric Fields at Altitudes Below 1000 Kilometers in the Auroral Zone

Recent Berkeley sounding rocket payloads have carried instrumentation capable of collecting bursts of electric field data with a time resolution of ~10 microseconds. On two recent flights into visible auroras, temporal structures that appear to be Langmuir solitons have been observed (*Boehm et al.*, 1984a; *Boehm et al.*, 1984b). Similar structures have also been seen on another rocket (*Kellogg et al.*, 1984). Since these observations are discussed elsewhere (*Boehm et al.*, 1985), it is the purpose here to include only one example of such data will be given in order to suggest the possible study of such phenomena by computer simulations. Figure 1 contains 7 milliseconds of data obtained at an altitude of about 750 kilometers on a rocket launched from Poker Flat, Alaska. The top panel gives the RMS amplitude of electric field oscillations parallel to the local magnetic field at frequencies above 300 kHz. From spectral measurements, it is known that these oscillations occurred predominantly at frequencies near the local plasma frequency of about 900 kHz. The amplitude of the parallel field oscillations exceeded 500 mV/m under the assumption that the wavelength was long compared to the receiving antenna length of 1.5 meters. If the wavelengths were shorter than the antenna length, the field strengths would be correspondingly larger.

The bursts of large plasma frequency oscillations lasted typically for ~100 microseconds, or for about 100 plasma periods. As shown in the third panel, the envelope of these Langmuir wave pulses contained parallel electric field fluctuations whose mean was non-zero and whose magnitudes were as large as 100 mV/m. Over the interval of Figure 1, the average parallel electric due to Langmuir pulses was about 15 mV/m. The formation of such fields and their influence on the ambient plasma might be topics which are amenable to simulation studies.

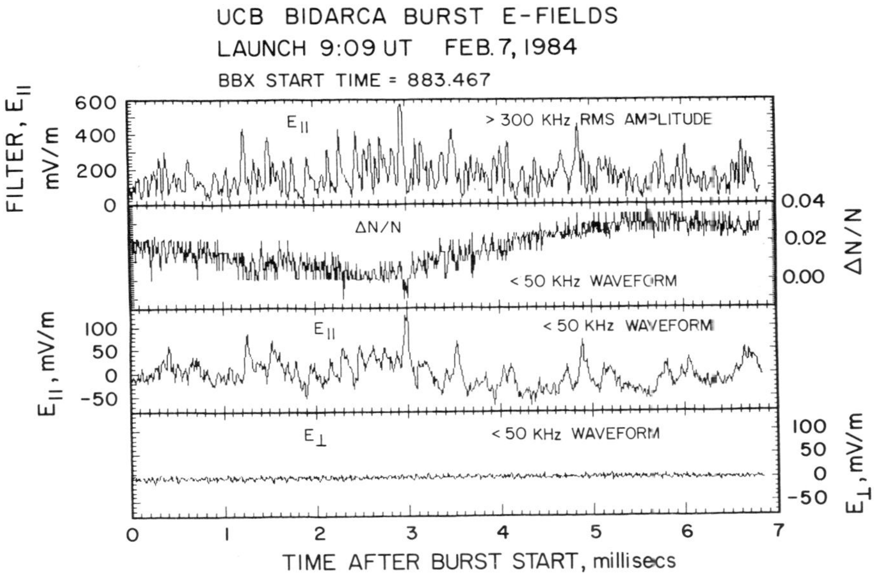

Figure 1. High time resolution electric field and density fluctuations observed on a sounding rocket.

2.2. The Auroral Acceleration Region

Observations from the S3-3 satellite (*Mozer et al.*, 1980; *Temerin et al.*, 1982) have shown that large perpendicular and parallel electric fields exist on auroral magnetic field lines at altitudes of about 8,000 km. These fields cause the acceleration of both downgoing electrons which produce the visible aurora when they impinge on the upper atmosphere, and upgoing ions which populate the magnetosphere with plasma of atmospheric origin. That such processes occur at the observed altitudes can be understood from the fact that, for a given field-aligned current, the ratio of the electron drift velocity to its thermal velocity as a function of altitude, is a maximum near these altitudes (*Mozer et al.*, 1980). Hence, current driven instabilities should occur first at the altitudes of the observed large parallel and perpendicular electric fields.

The two classes of large electric fields that have been observed in the auroral acceleration region are called "electrostatic shocks" and "double layers." An example of an electrostatic shock is given in Figure 2, which presents two minutes of measured electric field components in a non-rotating, magnetic-field-aligned coordinate system. At about 0121:13, 0121:25 and 0121:40, the spacecraft passed through three large ($\gtrsim$400 mV/m), mainly perpendicular, electric field structures. Each structure was observed for 3-10 seconds, which, for the satellite velocity of about 4 km/sec through an assumed stationary structure, corresponds to a thickness of 10-40 km at the

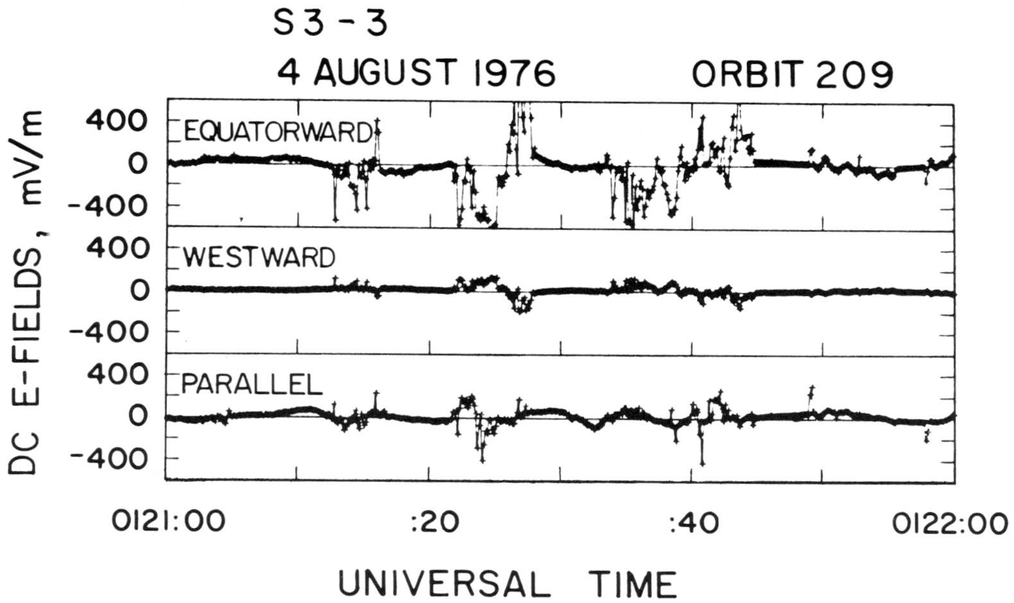

Figure 2. Electric fields in electrostatic shocks measured in the auroral acceleration region.

spacecraft altitude of 8000 km, or about 3-12 km at a mapped altitude of 100 km. This example illustrates many of the observed features of electrostatic shocks, which are:

1. Electrostatic shocks contain quasi-static electric fields of 100-1000 mV/m.

2. The electric fields are mostly perpendicular to the magnetic field and in the north-south direction.

3. Paired, oppositely directed, electric fields are observed (V-shocks), but less frequently than are unidirectional fields (S-shocks).

4. The potentials associated with the shock structures are often narrow in latitude ($<0.1°$), and broad in longitude.

5. Shock structures are repetitive in latitude.

6. Shocks are imbedded in a few degree wide latitudinal region of low frequency turbulence.

7. Field-aligned currents, ion cyclotron waves, upward accelerated ions, and downward accelerated electrons appear both in the electrostatic shocks and through the neighboring low frequency turbulence.

8. Electrostatic shocks occur on auroral zone field lines at all altitudes from 1000 km to the satellite apogee of 8000 km and at all local times, with an occurrence frequency similar to that of discrete auroral arcs.

9. Because of the above properties and because of detailed correlations found in case studies, it is concluded that the particle acceleration responsible

for discrete auroras and inverted-V electron precipitation occurs, at least in part, in the parallel electric fields associated with electrostatic shocks.

An example of an S-shaped electrostatic shock and the particle acceleration associated with it, is given in Figure 3. The top panel of this figure shows the presence of a unidirectional, perpendicular, electric field larger than 600 mV/m for about one second at a time near 1619:16 UT. Following this electrostatic shock, ~15 mV/m low frequency turbulence was observed for about 100 seconds (see the second panel, which is a plot of the same field component as given in the top panel, except on a magnified scale). The lower seven panels give the electron and ion fluxes measured at different energies by detectors that rotated with the spacecraft and that measured particles coming up the magnetic field line from the ionosphere at the times given by the upward pointing arrows above the plots for each particle species. To understand particle acceleration in relation to the electrostatic shock, equipotentials of the electrostatic shock are modelled in Figure 4, in a frame in which the magnetic field lines are vertical. Prior to the satellite passage through the shock, the parallel potential is above the spacecraft in this model. In this region one expects that secondary electrons with energies less than the shock potential will be mirrored by the shock above and the converging magnetic field below, so upgoing and downgoing low energy electrons should be observed, as they are (see the peaks labelled "F" in Figure 3). After spacecraft passage through the shock, the parallel potential is below the spacecraft, so low energy upgoing electrons will be reflected by the potential before they reach the spacecraft.

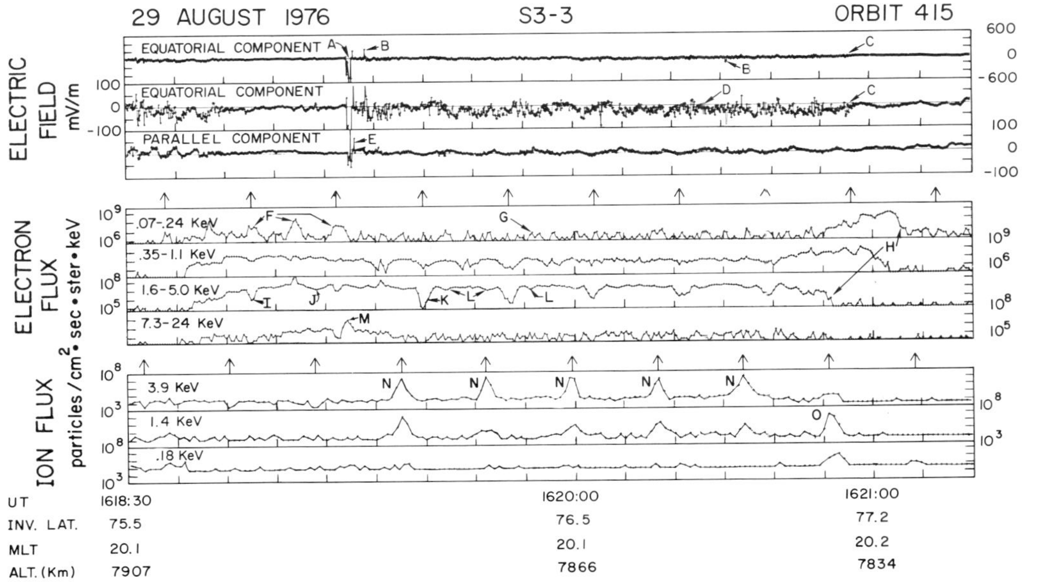

Figure 3. Electric fields and particle fluxes observed near an electrostatic shock.

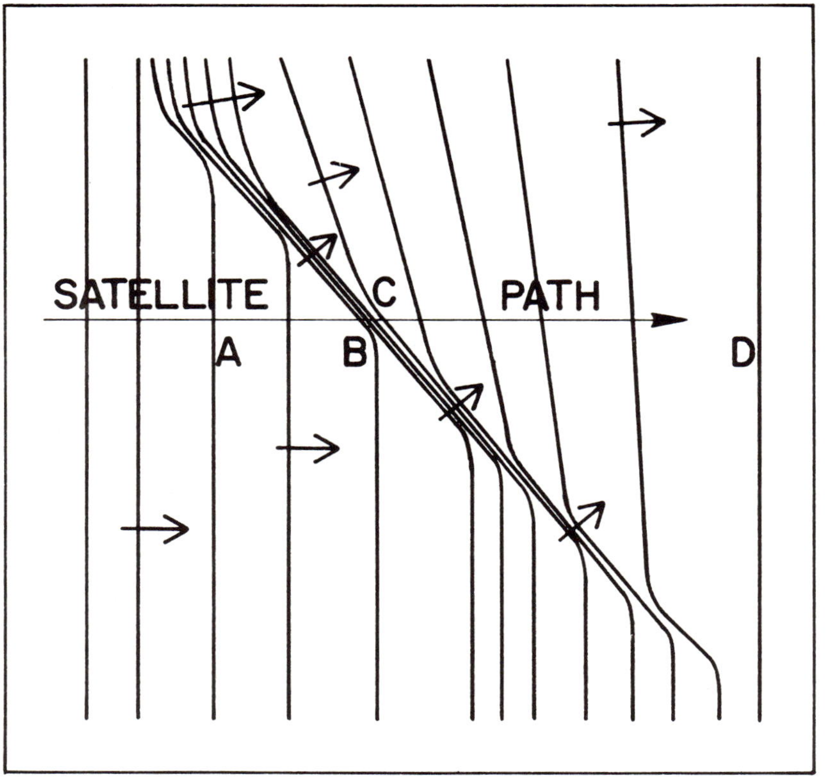

Figure 4. Model of the S-shaped electrostatic shock of Figure 3.

Hence, the satellite measured flux of low energy electrons should decrease when the shock is crossed, as it does in Figure 3. Similarly, the flux of the highest energy downgoing electrons should be maximum just before or as the shock is traversed, and it should decrease after the shock is crossed, in agreement with Figure 3 (M). After the shock is traversed, the parallel potential below the spacecraft should accelerate ambient ions upwards past the satellite to produce upgoing ion beams (N). The loss cones of the electron distribution should also broaden after the shock is traversed because the potential below the spacecraft causes precipitation into the ionosphere of electrons that would otherwise mirror. This effect is also observed in Figure 3 (K). Because so many features of the model of Figure 4 fit the data of Figure 3, there is little doubt but that potential drops on auroral field lines play a major role in accelerating both the electrons that make discrete aurora and the ionospheric ions that populate the magnetosphere.

Before discussing questions associated with electrostatic shocks, the properties of double layers will be described. An example of double layers observed in the auroral acceleration region is given by the 0.4 seconds of electric field data illustrated in Figure 5. The perpendicular components of the electric field show coherent electrostatic ion cyclotron waves and low frequency turbulence, while the parallel component shows 2-10 milliseconds unidirectional spikes, which are the double layers. The dominant parallel field direction in the double layers is such as to accelerate ions upwards and electrons downwards.

The occurrence frequency of double layer events is difficult to estimate, both because searching a significant data set with a one millisecond time resolution is tedious and because the data are often dominated (and sometimes saturated) by electrostatic ion cyclotron waves and by low frequency turbulence. Nevertheless, based on a search of some dozen orbits, the probability of finding an event containing tens to hundreds of double layers on a single satellite pass through the dusk auroral zone at an altitude between 6000 and 8000 kilometers is estimated as greater than 0.5. Thus, double layers occur frequently in the auroral acceleration region.

Since a typical parallel electric field in a double layer is ~10 mV/m, and because double layers are observed about 5% of the time in examples such as that of Figure 5, the average parallel electric field associated with double layers in such regions is ~0.5 mV/m, which produces an upward parallel potential drop of about 1 kilovolt over an altitude interval of 2000 kilometers. In the approximately 12 events that have been examined in the dusk region, the average potential had approximately this magnitude and the net parallel electric field pointed upwards. Thus, the double layers have sufficient net potential, have parallel electric fields in the proper direction, and occur sufficiently often to account for the parallel acceleration of particles in the

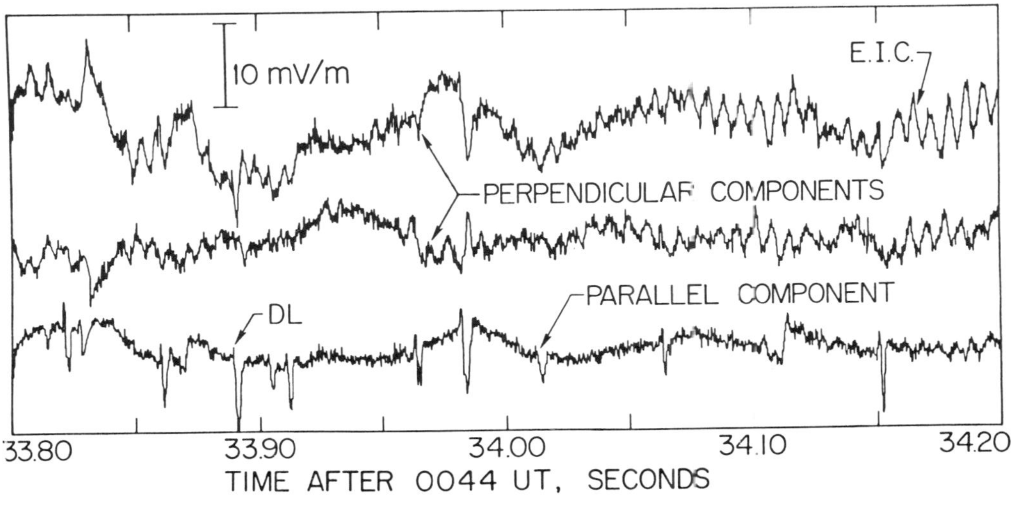

Figure 5. Perpendicular and parallel electric fields measured in a region containing many double layers.

auroral acceleration region.

Observed features of double layers are:

1. Double layers are pulses of upward pointing parallel electric field that pass by the spacecraft in 2-10 milliseconds and that have amplitudes of $\sim$10 mV/m.

2. Double layers occur in regions of upgoing field-aligned currents, upgoing ion beams, downgoing electron beams, and coherent electrostatic ion cyclotron wave emission.

3. Electromagnetic energy is converted to particle kinetic energy in double layers, since $\bar{j} \cdot \bar{E} > 0$.

4. The many double layers on a given magnetic field line contain sufficient potential to account for the electron and ion acceleration.

5. Double layers move along the magnetic field line at $v \gtrsim 50$ km/sec. Thus, their thickness and spacing are $\gtrsim$200 m and 4 km, respectively. For reasonable estimates of the plasma temperature and density, this thickness is about 20 Debye lengths and this spacing is about 400 Debye lengths.

6. For the same estimated temperature and density, $e\phi/kT \sim 1$.

7. If the observed double layers were moving *up* the magnetic field line, they were preceded by a negative potential rarefractive pulse. If they were moving *down* the magnetic field line, they were preceded by a positive potential compressive pulse.

8. Auroral electrons and upgoing ions are accelerated by the double layers.

Many of the above features are illustrated in Figure 6, which contains 10 minutes of particle and field data obtained while the satellite traversed an invariant latitude interval of about 5° at an altitude near 7000 kilometers, at a local time near 1600. A single, small, electrostatic shock was observed near 0029:20 UT as illustrated at the top of the figure. Three bursts of ion cyclotron waves, two of which lasted over latitudinal intervals of $\sim$0.5°, were also observed (second bar plot from the top of the figure). During at least a part of each of these ion cyclotron events, double layers were observed. The solid regions of the third bar plot indicate times of known double layer occurrence, while the cross-hatched regions indicate the probable presence of double layers, confirmation of which is difficult to obtain due to the detector saturation discussed above. The field-aligned currents obtained from the 17 second spin period averaged magnetometer data (the highest time resolution available) are presented in the top panel, which shows that the double layers were all observed in upward-directed field-aligned current regions. When mapped to ionosphere heights, the observed upward currents had magnitudes of $\sim$1 microampere/m^2, which is a rather small field-aligned current. The currents may be much larger on smaller scale sizes, judging from studies of auroral field-aligned currents from high-time resolution magnetometers (*Burke et al.*, 1980; *Sugiura et al.*, 1984).

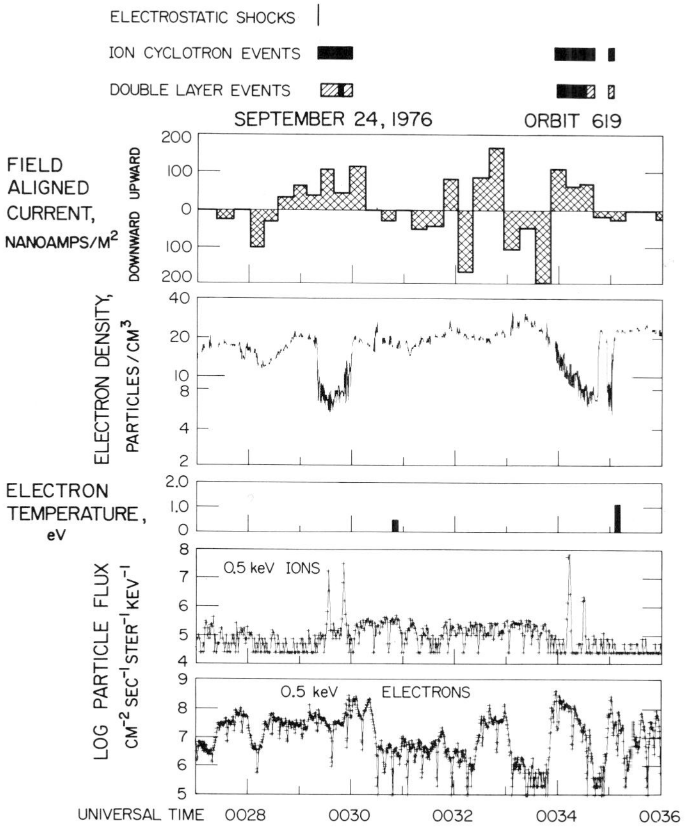

Figure 6. Field-aligned currents and particle fluxes measured near regions containing double layers.

That particles are accelerated by the double layers (and in the absence of electrostatic shocks) is most clearly seen by the factor of 100 increases in the flux of upgoing ions in the two double layer regions (second panel from the bottom). Signatures of accelerated electrons are also present in the loss cones and fluxes of the bottom panel.

Several questions associated with electrostatic shocks and double layers in the auroral acceleration region may be amenable to further study by computer simulations. They include:

1. Why are double layers observed sometimes, while electrostatic shocks are observed at other times?

2. What is the self-consistent physics of many double layers along a single magnetic field line?

3. What is the three dimensional structure of double layers?

4. What is the three dimensional structure of electrostatic shocks in a strong magnetic field?

5. What is the perpendicular acceleration mechanism that produces ion conic distributions?

Each of these questions will be discussed briefly below.

2.2.1. *Why Are Double Layers Observed Sometimes While Electrostatic Shocks Are Observed At Other Times?* In the dozen or so orbits of double layer data that have been studied, the best example of near-simultaneous observations of double layers and electrostatic shocks is that given in Figure 6. Even in this case, it is probable that there were no double layers during the few second interval of the electrostatic shock. Thus, the data suggest that one but not both of the electric field structures may appear at a given location and time.

A suggestion as to why and when double layers or electrostatic shocks appear is obtained from consideration of the intensities of each type of event. As discussed earlier, the double layers of Figure 6 appeared in regions of relatively weak field-aligned currents. They produced ions with an energy at the peak ion flux of ~300 eV. They occurred during a period of low K_p. These characteristics appear to hold for most of the double layer events, so, with some qualifications, double layers appear to be associated with weaker events while electrostatic shocks accompany the more energetic events. An important qualification is that it might be difficult to see double layers in energetic events if they occurred, because detector saturation in the larger ion cyclotron and low frequency turbulence might mask them.

That double layers might be associated with weaker events and electrostatic shocks might accompany stronger events is consistent with the view that the observed double layers are the weak double layers of current-driven computer simulations and that electrostatic shocks are the potential structures found in voltage-driven simulations. Thus, one might imagine that the boundary conditions imposed on the auroral acceleration region by magnetotail processes appear to be current boundary conditions for weak events which evolve into voltage boundary conditions as the events become larger. It would be interesting if simulations could test these conjectures.

2.2.2. *What Is The Self-Consistent Physics of Many Double Layers Along A Single Magnetic Field Line?* The distribution function of the auroral plasma in the neighborhood of a given double layer is altered by the plasma's interaction with the other double layers on the same magnetic field line. Thus, it is not obvious that an initial distribution that supports the evolution of a single double layer will also cause hundreds of double layers to be formed along the same magnetic field line, as is required if the sum of the double layer potentials is sufficient to produce the observed particle acceleration. It would be interesting to simulate the presence of many double layers in a self-consistent fashion.

2.2.3. *What Is The Three Dimensional Structure Of Double Layers?* The simplest picture of double layers is that they are planar one-dimensional structures that propagate exactly along the magnetic field line. Experimental data, such as that in Figure 5, is in disagreement with this simple picture, since apparent signatures of double layers also exist in the perpendicular electric field components at times such as 0044:33.88 or 0044:33.98. Furthermore, as illustrated in Figure 7, double layer signatures in different antenna pairs are not as similar as one would expect if double layers were planar structures propagating past the spacecraft. Comparisons such as these suggest that the perpendicular scale size of double layers may be comparable to their parallel scale. This suggestion is borne out by recent two dimensional simulations (*Barnes, Hudson, and Lotko,* 1984).

2.2.4. *What Is The Three Dimensional Structure of Electrostatic Shocks In A Strong Magnetic Field?* Since the latitudinal thickness of many observed electrostatic shocks is comparable to the energetic ion gyroradius, complete simulations of shocks require a three dimensional treatment in a strong, converging, magnetic field.

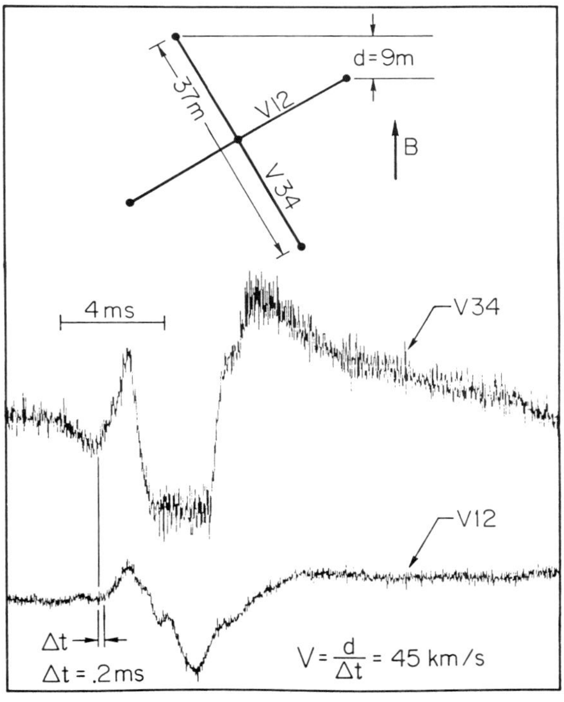

Figure 7. Detailed structure of a single double layer.

2.2.5. *What Is The Perpendicular Acceleration Mechanism That Produces Ion Conic Distributions?* The lower panels of Figure 8 give ion fluxes over three satellite spins, including times when the ion detector looked down the magnetic field line at upcoming ions (during the minima in the saturated sine wave of the top panel). At the first two such observations, upcoming beams were observed, while the last observation showed an ion conic distribution (upcoming ions at pitch angles between 90° and 180°). Conic distributions are thought to be produced by perpendicular acceleration below the spacecraft and the consequent folding of the distribution along $\bar{B}$ as the particles move up the field line into a weaker magnetic field. Both the beams and the conics in this example are associated with passage of the spacecraft through an electrostatic shock (see panels immediately above the particle data).

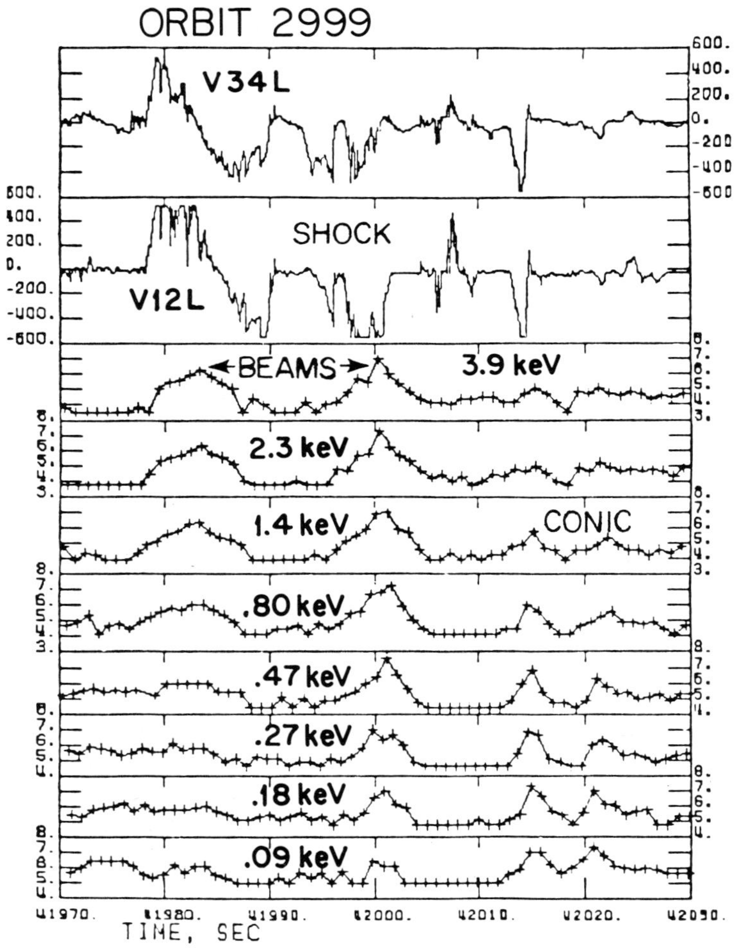

Figure 8. Ion pitch angle distributions in the vicinity of electrostatic shocks.

Perpendicular particle acceleration can be produced by electrostatic ion cyclotron waves. However, the correlation between the presence of conic distributions and electrostatic ion cyclotron waves is poor (*Cattell,* 1984). On the other hand, turbulence, which is often large enough to be classified as an electrostatic shock, is generally present in regions of conic distributions. Furthermore, the spatial distribution of shocks containing conics is different from that of shocks containing ion beams (*Redsun et al.,* 1984). These results suggest that there may be more than one type of electrostatic shock and that conics may be accelerated by the low frequency turbulence in the type of shocks that accompany them.

2.3. The Geomagnetic Tail

Figure 9 illustrates measurements of field and plasma parameters on ISEE-1 in the geomagnetic tail during a two hour interval when the plasma β (third panel from the bottom) indicated spacecraft passage from the plasma sheet to the southern lobe near 0240 UT, several crossings into and out of the plasma sheet before about 0300 UT, and final entry of the spacecraft into the plasma sheet at that time. The y-component of the electric field (third panel from the top) was large (~5-10 mV/m) and mostly positive for the 40 minute interval encompassing the lobe-plasma sheet traversals. An expanded view of 90 seconds of this data is given in Figure 10, which is unusual in that it contains lobe, high latitude plasma sheet boundary, plasma sheet, and neutral sheet data. Prior to about 0256:45, the spacecraft was in the lobe, as is confirmed by the spacecraft potential of ~20 volts (the negative of the quantity plotted in the top panel). At about 0256:45 the spacecraft crossed the high latitude boundary and passed into the plasma sheet where the higher density of thermal plasma caused the spacecraft potential to decrease to about 10 volts. The diamagnetic effect of the higher plasma density is indicated by the decrease of the magnitude of the magnetic field near 0256:45 (third panel from bottom). Shortly after 0257:00 UT the spacecraft was near the neutral line since B_x was very small (second panel from bottom) and the magnitude of $\bar{B}$ was a minimum. The raw electric field measurement by the spherical double probe sensor is shown in the middle panel of the figure. The three second sinusoidal variation of this data is caused by rotation of the sensor through the quasistatic external electric field at the satellite spin frequency. The GSE x and y components of electric field, obtained from sine wave least-squares fits of the raw data every spin period are given in the second and third panels of Figure 10. Two facts are evident from this electric field data: 1) Large amplitude (~20 mV/m), low frequency turbulence was present at the high latitude boundary crossing near 0256:45 UT; 2) Throughout the time interval, in the lobes, in the plasma sheet, and at the neutral line, the quasistatic electric field was large (~10 mV/m), and mainly in the $+y$ direction. These features, which are often observed in the tail, raise questions that are discussed below.

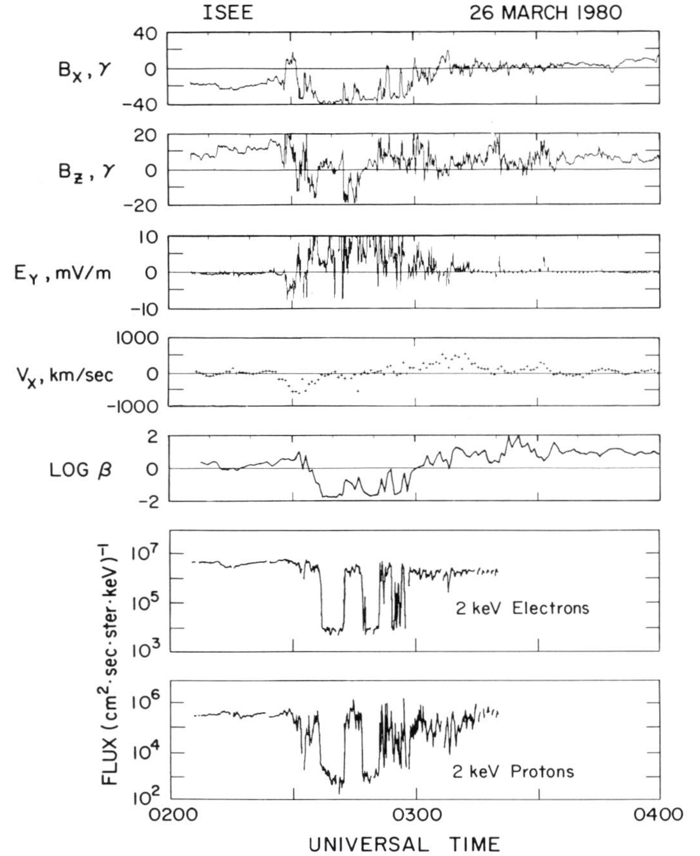

Figure 9. Fields and particles observed in the geomagnetic tail.

The low frequency turbulence at the high latitude boundary of the plasma sheet has been studied through analyses of five months of ISEE-1 tail data in 1980 (*Cattell et al.*, 1982; *Levin et al.*, 1983). One purpose of this study was to distinguish between the two following possibilities: 1) The low frequency turbulence is the high altitude mapping of electrostatic noise produced in the auroral acceleration region; in this case, the average amplitude of the turbulence should decrease with altitude; 2) The low frequency turbulence is created by fundamental processes occurring in the tail. In this case, the average amplitude of the turbulence should be largest in the tail source region.

In Figure 11 a measure of the five month average amplitude of the low frequency turbulence is plotted versus geocentric distance from the earth. The decrease of amplitude with radial distance is approximately that expected for a static electric field that maps along equipotential, diverging, magnetic field lines. Hence, the low frequency turbulence at the high latitude boundary of

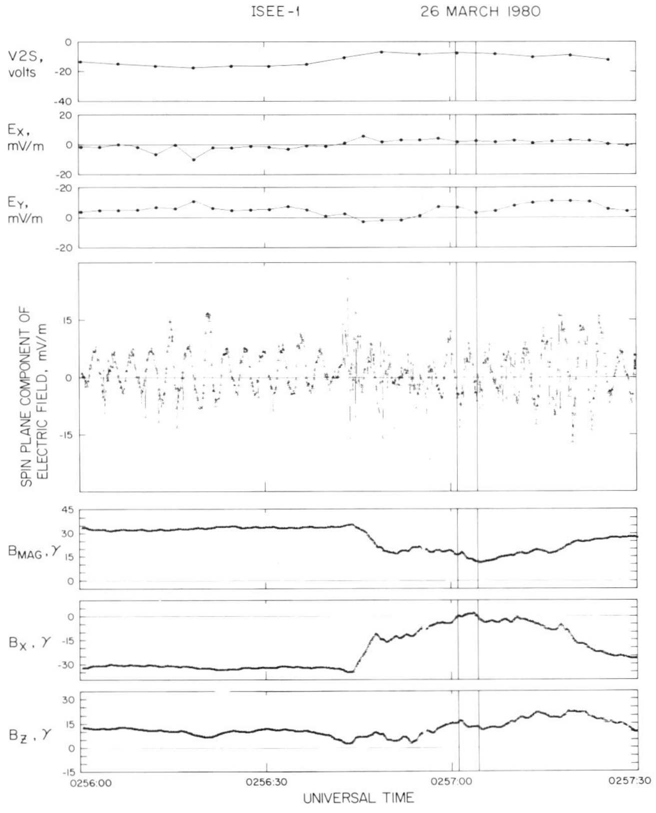

Figure 10. Electric and magnetic fields observed during a 90 second interval during which the spacecraft passed from the tail lobe to the neutral sheet.

the plasma sheet may come from low altitude processes on the same field lines. This result is evidence that the low altitude auroral zone and the high latitude boundary of the plasma sheet exist instantaneously on the same magnetic field line, since the low altitude turbulence occurs in the auroral acceleration region and the high altitude, apparently mapped, turbulence occurs in the high latitude boundary of the plasma sheet.

This instantaneous mapping of low altitude auroral zone magnetic field lines into the high latitude boundary layer has a topological problem. The boundary layer in the tail is thought to contain the last open magnetic field lines. Hence, magnetic field lines through it should map into the auroral zone-polar cap boundary at low altitudes. However, it is known that the auroral zone does not straddle this boundary at low altitudes (*Torbert et al.*, 1981). Thus, if the turbulence measurements correctly identify low altitude auroral zone magnetic field lines as mapping into the high latitude boundary layer of the plasma sheet, then the high latitude boundary must lie well

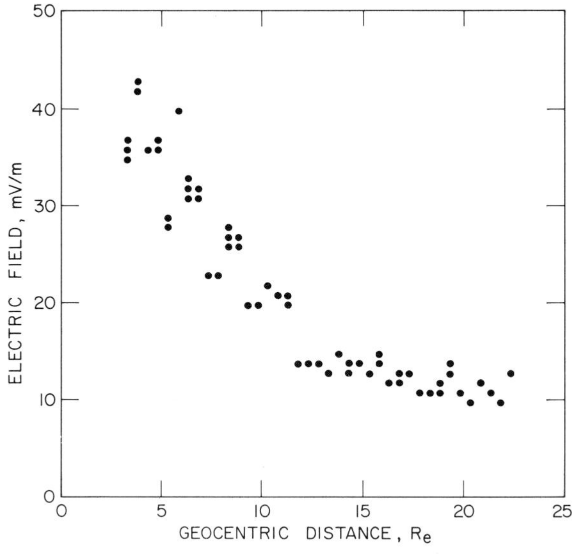

Figure 11. Amplitude of nightside electric field turbulence as a function of radial distance from the earth.

equatorward of the last open field line on the average. Perhaps simulation studies could further test this conclusion and thereby contribute to our knowledge of the magnetospheric topology.

The data of Figure 12 present two further examples of the large (~5-10 mV/m), primarily dawn-to-dusk, quasi-static electric fields that are frequently seen in the plasma sheet for typical periods of several minutes (*Cattell and Mozer*, 1984; *Cattell et al.*, 1985). A search of tail electric field data from February, 1980, to May, 1980, has found approximately 60 such events that satisfied criteria that $(\bar{E} \times \bar{B}/B^2)_x > 100$ km/sec, that $|B_z/B_x| > 0.2$, and that $|B_z/B_y| > 0.2$. The criterion on B_z were imposed in order that E_z and the components of $\bar{E} \times \bar{B}/B^2$ could be estimated with sufficient accuracy from the requirement that $\bar{E} \cdot \bar{B} = 0$. These B_z criteria cause most of the observed earthward flow events to be located inside the plasma sheet. The average probability of finding events that satisfied these criteria was between 3 and 4%, and the average dawn to dusk electric field in these events was 5 mV/m.

These large electric fields cannot be potential fields that extend across the magnetotail because the crosstail potential associated with them would be too large (~2,000,000 volts). The potential would be reasonable if the large fields were spatially confined to a y-dimension of ~1 R_e. However, the measured occurrence frequency is too large for a model in which individual large field regions extended across only a few percent of the tail.

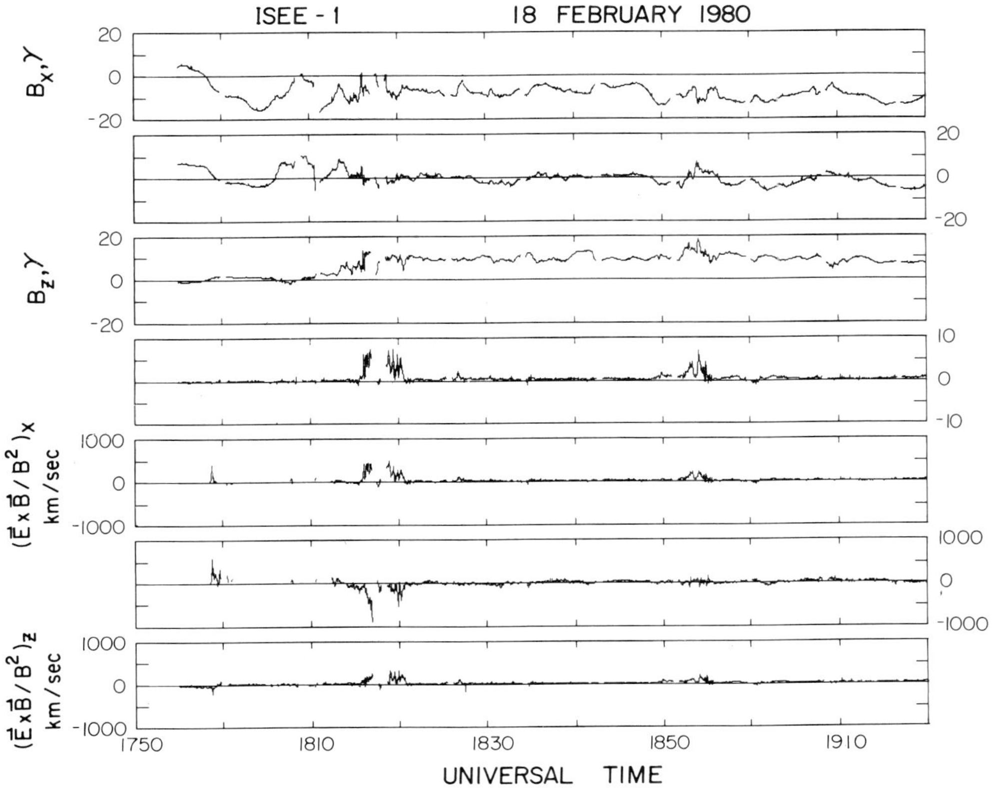

Figure 12. Geomagnetic tail measurements of enhanced quasi-static electric fields.

The observed field magnitudes are consistent with their being inductive electric fields, as the following estimate shows. Suppose that E_y varied with z on a spatial scale of 1 R_e. This $\nabla \times \bar{E}$ would be consistent with Maxwell's equations if B_x changed by about 1 γ/second in the region of the curl. As Figure 10 illustrates, such rapid magnetic field variations are often observed.

Further testing of the hypothesis that the observed fields are inductive requires a model of the magnetic and electric field configurations since Maxwell's equations cannot be solved without boundary conditions. The model that the observed electric fields are associated with the flapping of an otherwise time-stationary tail past the satellite does not work since, in this model, E_y is negative as often as positive, in disagreement with the observations.

Another model holding more promise is that the observed electric fields are associated with enhanced reconnection in the tail and with the associated earthward plasma convection. Simulations of such models yield predominantly positive dawn-to-dusk electric fields with reasonable time scales (*Birn and Hones,* 1981; *Sato et al.,* 1982; *Birn,* 1984), in agreement with experiment. However, there are important differences between the simulations and the data. These include: 1) The magnitudes of the observed fields are larger than those predicted by simulations; 2) The fields are observed over a wider region of the tail than expected from simulations.

The simulation by *Birn and Hones* (1981) yielded electric fields that were typically a few tenths of a mV/m, while that of *Sato et al.* (1982), gave somewhat bigger fields. In both cases, on the earthward side of the X-point, the fields were large only within ~10 R_e of the X-point. Thus, both simulations predict typical electric fields an order of magnitude smaller than those observed, especially if the X-point is beyond 30 R_e down the tail.

In Figure 13, the probability of observing a large electric field event during the 1980 tail search is summed over y and plotted versus x, and vice-versa. While the statistical fluctuations in the probability estimate are large, there is little evidence that the probability of observing a large electric field depends on the location of the satellite in the near-earth tail, especially in the y-direction. This result is in disagreement with a 3-D reconnection simulation (*Birn and Hones,* 1981) which predicted large fields only within a few R_e of

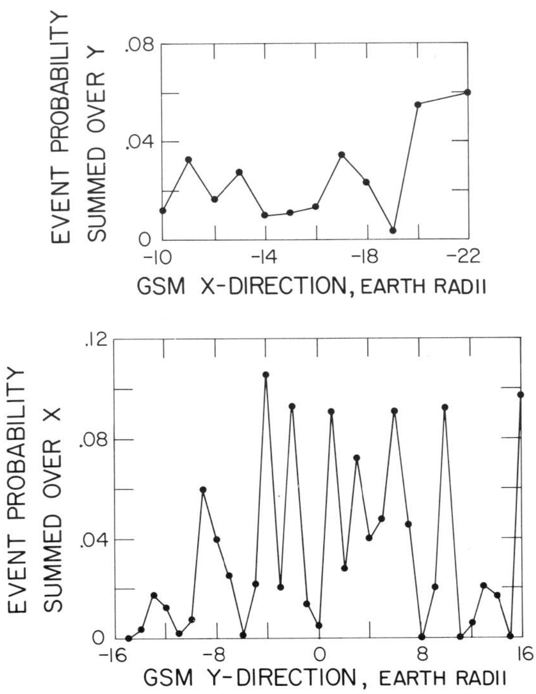

Figure 13. Probabilities of observing large electric fields in the tail.

$y = 0$. Such inconsistencies between the simulation results and the tail electric field observations might be clarified by further simulation studies, as well as data analysis.

2.4. The Relationship Of The Auroral Acceleration Region And Tail Processes

The above discussed electric field data as well as many plasma observations support the views that auroral primaries are accelerated near geocentric altitudes of $\sim 2\ R_e$, and that reconnection with the formation of neutral lines and plasmoids occurs in the tail. An important question to magnetospheric physics is the self-consistent relationship between processes occurring in the auroral acceleration region and the tail. A complete understanding of the magnetosphere cannot be obtained until these two phenomena are interrelated and understood together.

2.5. The Bow Shock

Large amplitude quasi-static electric fields and low frequency fluctuations are both observed at the bow shock (*Bensadoun et al.*, 1984). An example of a 6 mV/m quasi-static electric field normal to the shock is given in the top panel of Figure 14. Its correlation with B_z (bottom panel) is understood in terms of the precurser whistler associated with dispersive, sub-critical shocks. Less well understood is the low frequency turbulence that is often found in sub-critical shocks. Figure 15 presents 21 seconds of raw electric field data obtained at a bow shock crossing in which turbulent electric fields larger than 100 mV/m were observed. Although one cannot unfold single component measurements to determine the electric field vector on a time scale faster than the satellite spin period, the data of Figure 15 suggest an intriguing possibility. Many of the ~ 0.1 second electric field spikes are preceded or followed by a spike of the opposite sign. This signature, which is observed in auroral double layers (see Figure 5), suggests the possibility that the turbulence observed in the bow shock may be composed of double layers. Since the amplitude of this turbulence in sub-critical shocks depends on the ratio of the electron to ion temperature, as shown by the data of Figure 16, these conjectured double layers might arise from ion acoustic turbulence. It would be useful to obtain three component electric field measurements in the bow shock and to compare them with double layer simulations of the bow shock to study this possibility further.

2.6. The Sub-Solar Magnetopause

In view of the observation of large amplitude low frequency turbulence in the auroral acceleration region, the tail, and the bow shock, it might be reasonable to suppose that such turbulence would also be present at the sub-solar magnetopause, for example, in association with observations of flux transfer events. A survey of the electric fields associated with FTE's (*Dailey et al.*, 1984) found no cases with large turbulent fields. This may be because the densities are usually higher so that the drift velocities associated with observed events are smaller than in other regions.

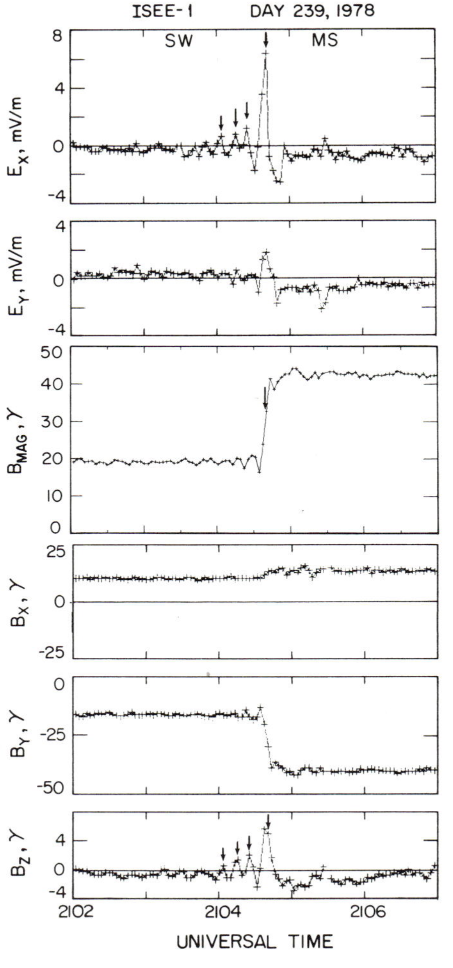

Figure 14. Quasi-static electric and magnetic fields measured at a sub-critical bow shock crossing.

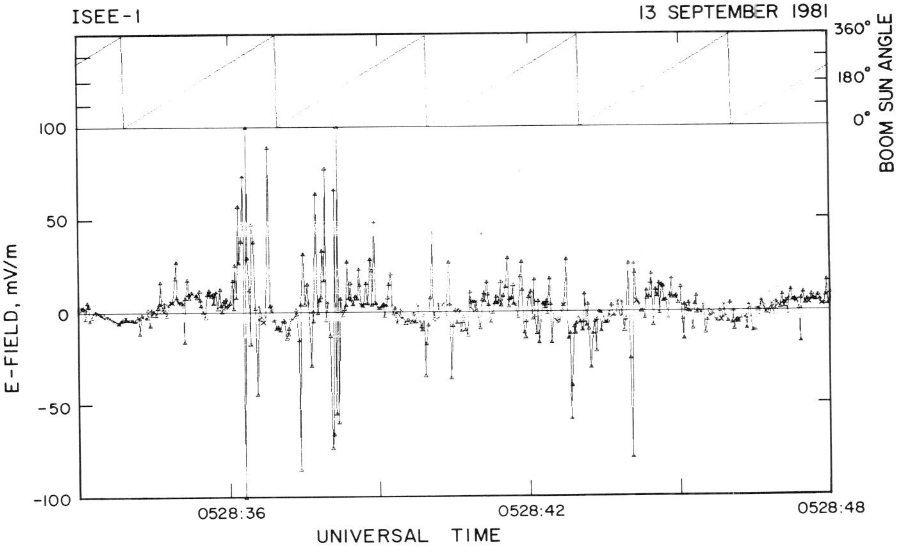

Figure 15. Fluctuating electric fields measured at a sub-critical bow shock crossing.

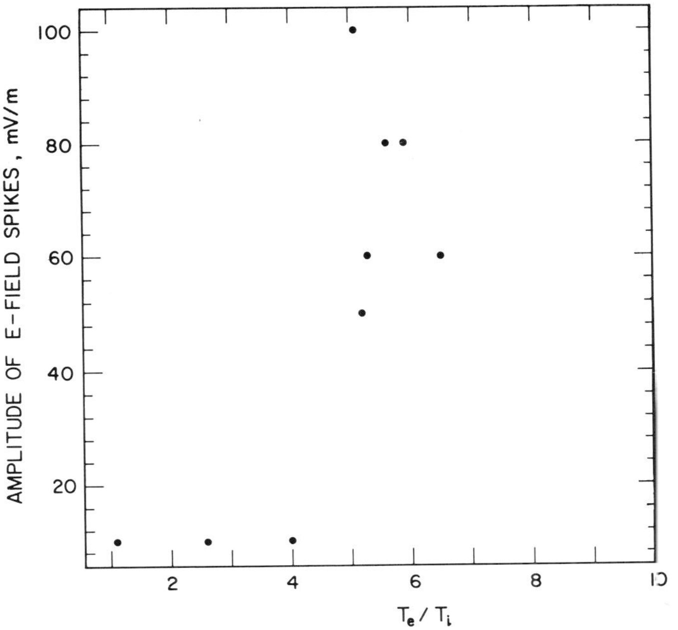

Figure 16. Dependence of the amplitudes of the fluctuating electric fields measured at sub-critical bow shock crossings on the ratio of the electron to ion temperatures.

ACKNOWLEDGEMENTS

The authors wish to thank Jeff Dhont, Brian Lorber and Winston Teitler for assistance in data reduction and analyses, and C.-G. Fälthammar for many helpful discussions. This work was performed under ONR contract N00014-81-C-0006 and NASA grant NAG5-375.

REFERENCES

Barnes, C., M. K. Hudson and W. Lotko, 'Weak double layers in ion acoustic turbulence,' submitted to *Phys. Fluids,* 1984.

Bensadoun, M., J. R. Wygant and F. S. Mozer, 'Measurement of large amplitude fluctuating electric fields at the earth's bow shock,' E⊕S *Trans. Am. Geophys. Union,* **65**, 1073, 1984.

Birn, J., 'Three-dimensional computer modeling of dynamic reconnection in the magnetotail: Plasmoid signatures in the near and distant tail,' in *Magnetic Reconnection in Space and Laboratory Plasmas, Geophysical Monograph* **30**, edited by Edward W. Hones, Jr., p. 264, American Geophysical Union, Washington, D.C., 1984.

Birn, J. and E. W. Hones, Jr., 'Three-dimensional computer modeling of dynamic reconnection in the geomagnetic tail,' *J. Geophys. Res.* **86**, 6802, 1981.

Boehm, M. H., C. W. Carlson, J. McFadden and F. S. Mozer, 'Observations of double layer-like and soliton-like structures in the ionosphere,' *Geophys. Res. Lett.* **11**, 511, 1984a.

Boehm, M. H., C. W. Carlson, J. McFadden and F. S. Mozer, 'Sounding rocket observations of intense high frequency waves,' E⊕S *Trans. Am. Geophys. Union,* **65**, 258, 1984b.

Boehm, M. H., C. W. Carlson, J. McFadden and F. S. Mozer, 'Observations of non-linear effects in high frequency waves in the ionosphere,' Second International School On Space Simulations, Kauai, Hawaii, 1985.

Burke, W. J., D. A. Hardy, F. J. Rich, M. C. Kelley, M. Smiddy, B. Schumann, R. C. Sagalyn, R. P. Vancour, P. J. Widman and S. T. Lai, 'Electrodynamic structure of the late evening sector of the auroral zone,' *J. Geophys. Res.* **85**, 1179, 1980.

Cattell, C. A., 'Association of field-aligned currents with small-scale auroral phenomena,' in *Magnetospheric Currents, Geophysical Monograph* **28**, edited by T. A. Potemra, p. 304, American Geophysical Union, Washington, D.C., 1984.

Cattell, C. A., M. Kim, R. P. Lin and F. S. Mozer, 'Observations of large electric fields near the plasma sheet boundary by ISEE-1,' *Geophys. Res. Lett.* **9**, 539, 1982.

Cattell, C. A. and F. S. Mozer, 'Substorm electric fields in the earth's magnetotail, in *Magnetic Reconnection in Space and Laboratory Plasmas, Geophysical Monograph* **30**, edited by Edward W. Hones, Jr., p. 208, American Geophysical Union, Washington, D.C., 1984.

Cattell, C. A., F. S. Mozer, E. W. Hones, Jr., R. R. Anderson and W. Lennartssen, 'Electric and magnetic fields, waves, particles, and plasma measurements, during crossings of the plasma sheet, plasma sheet boundary and neutral sheet,' in preparation, 1985.

Dailey, R., C. Cattell, F. S. Mozer and J. Berchem, 'Observations of electric fields and $\vec{E} \times \vec{B}/B^2$ velocities during flux transfer events, E⊕S *Trans. Am. Geophys. Union,* **65**, 1039, 1984.

Kellogg, P. J., S. J. Monson and B. A. Whalen, 'Observations of double layer-like structures at rocket altitudes,' *Geophys. Res. Lett.* **11**, 515, 1984.

Levin, S., K. Whitley and F. S. Mozer, 'A statistical study of large electric field events in the Earth's magnetotail,' *J. Geophys. Res.* **88**, 7765, 1983.

Mozer, F. S., C. A. Cattell, M. K. Hudson, R. L. Lysak, M. Temerin and R. B. Torbert, 'Satellite measurements and theories of low altitude auroral particle acceleration,' *Space Sci. Rev.* **27**, 155, 1980.

Redsun, M. S., M. Temerin and F. S. Mozer, 'Classification of auroral electrostatic shocks by their ion and electron associations,' submitted to *J. Geophys. Res.*, 1984.

Sato, T., H. Matsumoto and K. Nagai, 'Particle acceleration in time-developing magnetic reconnection process,' *J. Geophys. Res.* **87**, 6089, 1982.

Sugiura, M., T. Iyemari, R. A. Hoffman, N. C. Maynard, J. L. Burch and J. D. Winningham, 'Relationships between field-aligned currents, electric fields, and particle precipitations as observed by DE-2,' in *Magnetospheric Currents, Geophysical Monograph* **28**, edited by T. A. Potemra, p. 96, American Geophysical Union, Washington, D.C., 1984.

Temerin, M., K. Cerny, W. Lotko and F. S. Mozer, 'Observations of double layers and solitary waves in the auroral plasma,' *Phys. Rev. Lett.* **48**, 1175, 1982.

Torbert, R. B., C. A. Cattell, F. S. Mozer and C.-I. Meng, 'The boundary of the polar cap and its relation to electric fields, field-aligned currents, and auroral particle precipitation,' in *Physics of Auroral Arc Formation, Geophysical Monograph* **25**, edited by S.-I. Akasofu and J. R. Kan, p. 143, American Geophysical Union, Washington, D.C., 1981.

PLASMA OBSERVATIONS IN THE AURORAL AND POLAR CAP REGION

J. F. Fennell
Space Sciences Laboratory
The Aerospace Corporation
P. O. Box 29257
Los Angeles, CA 90009

ABSTRACT. The last decade has seen a period of rapid growth in our understanding of the processes which occur in the auroral regions. Much of our understanding is based on the copious new observations which have been made available in the auroral community. The present work is a short overview of the plasma conditions which obtain throughout much of the auroral region. It covers the diffuse and discrete auroral electron precipitation in the morning and evening oval, cusp, and polar cap. The ionospheric ion outflow throughout the high latitude regime is also described and related to the electron observations.

1. INTRODUCTION

This overview will attempt to summarize our present knowledge of the plasma distributions which obtain in the auroral and polar cap regions. As such, little will be said about the theory of auroral plasmas. The intent is to present a reference description of the plasma regime that the theory and simulations must contend with. Given the shortness of this report, the topics are not covered in detail. A fairly detailed reference list has been included so that the reader can easily find the most recent and detailed information.

The report treats the electron observations first (Section 2) and then the ion observations (Section 3). Each major section is broken down into subsections with the electron subsections covering the different auroral regions according to local time and latitude and type of precipitation. The ion section emphasizes ionospheric ion outflow and distinguishes between the energetic and polar wind ion outflow. The energetic ion outflow discussion is further divided into sections on conics and field-aligned beams.

2. ELECTRON PRECIPITATION

2.1. Diffuse Aurora

The diffuse aurora appears as a broad band of weak auroral luminosity extending around the auroral zone. Simultaneous observations of the

Space Science Reviews **42** (1985) 337–355. 0038–6308/85.15

diffuse auroral precipitation and the equatorial central plasma sheet fluxes show that the central plasma sheet is the source of the precipitating particles. No acceleration is required to cause the spectral characteristics of the central plasma sheet and auroral precipitation to match. Meng et al. (1979) found good agreement between the spectral shapes in the two regions, as is shown in Figure 1. The electron intensities were similar. They concluded that the precipitation did not require a potential drop, only pitch angle transport.

The equatorward edge of the diffuse aurora generally maps to the inner edge of the plasma sheet (Horwitz et al., 1982; Gussenhoven et al., 1983). This requires that pitch-angle scattering processes are acting to continuously dump electrons into the loss cone throughout the central plasma sheet region and down to the innermost edge of the plasma sheet. It is speculated that the wave particle interaction which caused the electron precipitation is at the strong scattering limit.

Fairfield and Viñas (1984) have argued that the plasma sheet electrons are frequently not in strong diffusion. They and others (e.g., Ashour-Abdalla and Thorne, 1978) have noted that the anisotropies resulting from single particle motion in the outer magnetosphere are the most likely source of free energy for the waves which precipitate the electrons. Belmont et al. (1983) argues that the observed electrostatic cyclotron harmonic wave amplitudes in the equatorial plasma sheet seldom reach the levels required to cause strong pitch angle scattering. They suggest that other mechanisms may play important roles in generating the electron precipitation from the plasma sheet required to generate the diffuse aurora.

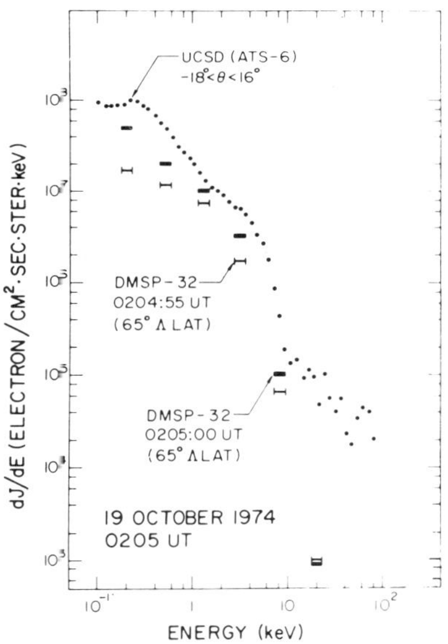

Figure 1. Coordinated observation of plasma sheet electrons at geosynchronous orbit by ATS-6 and the diffuse aurora electron precipitation at low altitudes near the conjugate field lines by DMSP-32 (after Meng et al., 1979).

Thus we are left with the impression that the current wisdom concerning the source and mechanism for generating the electron precipitation for the diffuse aurora is generally correct but some questions concerning the details remain. For example, is strong pitch angle diffusion required to provide the precipitating electron fluences observed at low altitude or are the observed wave amplitudes and spatial distributions consistent with those required to supply the precipitation?

2.2. Night Side Precipitation

The night side auroral arc precipitation is dominated by discrete arc precipitation and 'inverted-V' precipitation regions combined with the diffuse aurora precipitation (Akasofu and Kan, 1980; Reiff, 1983). The discrete arc and 'inverted-V' precipitations show signatures of field-aligned acceleration of plasma sheet electrons by 'parallel' electric fields (see Burke, 1982; Reiff, 1983; Fennell, 1985 and references therein). The 'inverted-V' structures span the range of latitudinal scale lengths from ~ 0.2° to 3° and have a mean width of ~ 0.5° in invariant latitude (Lin and Hoffman, 1982 and references therein). These are wider than discrete arcs or electrostatic shocks. Generally, the inverted-V region also corresponds to a region of accelerated ion outflow (see Sec. 3) and intense wave turbulence (Mozer et al., 1980). Sometimes the inverted-Vs have embedded in them bursts of intense electron precipitation associated with the electrostatic shocks (Lin and Hoffman, 1979b; Mizera et al., 1981).

The kinds of particle distribution functions seen in the nightside auroral zones have been summarized by several authors (see Fennell, 1984 and references therein). Examples of these distribution functions are shown in Figure 2a. Basically, at satellite altitudes the inverted-V electrons show the peaked energy spectra indicative of a potential drop existing at $\geq$ few thousand km altitude on the field line. In rare instances the inverted-V signature has been seen at very high altitudes (~ 13 R_e, Huang et al., 1984). If the inverted-V observations are in the acceleration region, as is the case in Figure 2a, the resulting enhanced loss cone, field-aligned beam and a "trapped" component of the electrons are observed (see Chiu et al., 1983a and references therein). This indicates that the particle motion is sufficiently adiabatic at the observation altitude to maintain signatures of the forces which are acting on the particles. These signatures are not always maintained down to low (few hundred kilometers) altitudes. Often at low altitudes the particle distributions have been much smoothed by scattering processes and at very low altitudes by the interaction with the topside of the atmosphere (Pulliam et al., 1981; Prasad et al., 1983).

The electron distribution functions measured in the inverted-V regions are generally consistent with the picture presented by Evans (1974) in which a hot Maxwellian source distribution is assumed to fall through a potential drop and interact with the atmosphere producing secondaries and degraded primaries. The upgoing secondaries and primaries, with energies less than the potential barrier, are reflected downward to interact with the atmosphere again (see Pulliam et al., 1981).

One question which is often asked is how stable is the inverted-V acceleration process? Auroral arcs are known to persist for tens of minutes, based on photographic studies (see Reiff, 1983). But it is very difficult to determine whether the spatial and temporal character of the precipitating electron spectra remains stable for such periods. Rocket flights are too short and the time between successive polar traversals by a satellite is too long. It is known that the altitudinal

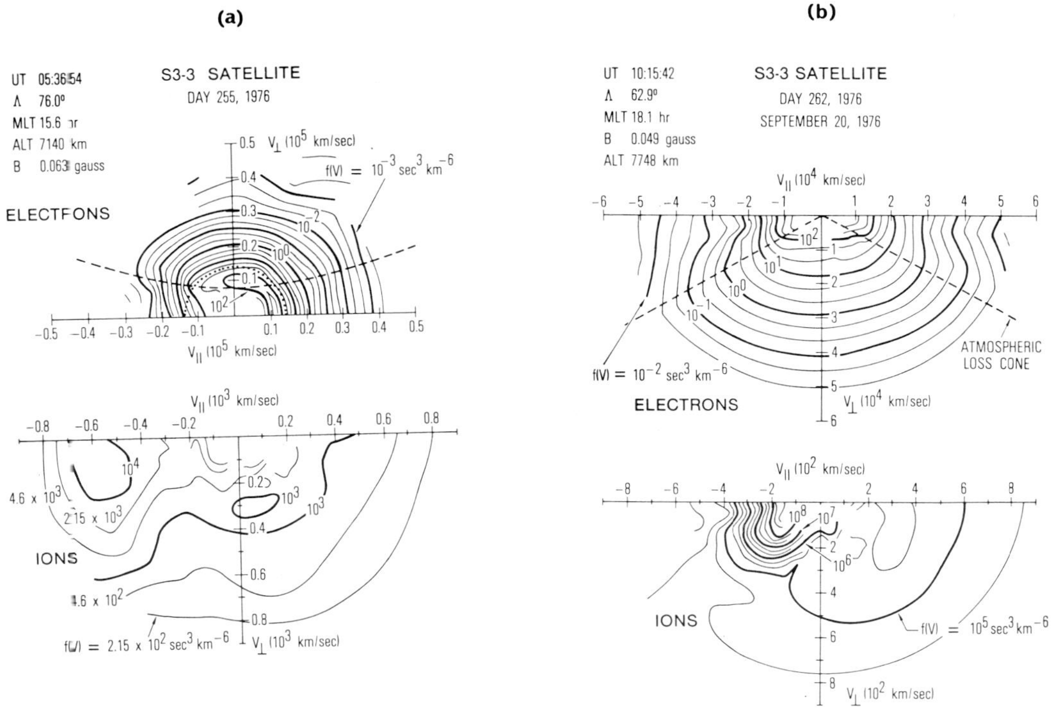

Figure 2. Typical electron and ion distribution functions observed in inverted-V regions (a) and in the return current region (b). The electron and ion distribution in (a) are consistent with a potential drop of 0.5 kV above and 2.2 kV below the observation altitude. Panel (b) shows the simultaneous occurrence of bidirectional electron beams and ion conics (after Chiu et al., 1983).

distribution of the potential drop varies within an inverted-V (see Fig. 6 of Fennell et al., 1981). It is not known whether this distribution is time-stationary. Using the recent combined DE-1 and -2 coplanar observations, Thieman and Hoffman (1985) argue that inverted-Vs can remain reasonably stable for at least a few minutes and up to 18 minutes. At this point in time there is no satisfactory answer to the stability question.

There is a relatively large collection of plasma observations, taken predominantly in the nightside auroral region, which do not appear to fit the simple Evans (1974) picture. For example, intense field-aligned electrons with steep, unpeaked energy spectra are often observed at the edges of inverted-Vs usually at lower altitudes (Bryant, 1978; Wahlen and Daly, 1979; Fennell et al., 1981, Collin et al., 1982; Bryant, 1983). Also, at times the inverted-V precipitating electrons just below the peak of the energy spectrum are field-aligned, not isotropic, as is expected for secondary electrons. Thus it is argued they are primary downgoing particles (see Bryant, 1983). Finally, there are double peaked inverted-V precipitating electron energy spectra (Hoffman and Lin, 1981; Arnoldy, 1981). Examples of both of these latter characteristics are shown in Figure 3. These data were taken at relatively low altitudes. Generally, wave-particle interactions are called upon to explain such auroral electron distributions (Lin and Hoffman, 1979b; Bryant, 1983; Bingham et al., 1984; and references therein). When the wave observations are examined it is often found that the waves do not carry enough energy to provide the necessary acceleration (Gurnett and Frank, 1977) to explain such particle observations. Part of the difficulty lies with insufficient simultaneous high-resolution wave and particle data. Thus we are left with a set of nightside auroral observations for which there are no generally accepted explanations.

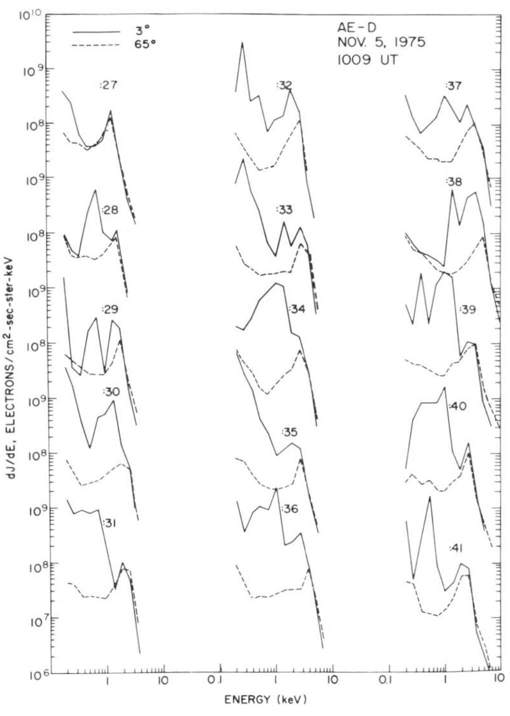

Figure 3. Energy spectra showing multiple peaks and strong field alignment of the low energy component of inverted-V electron precipitation (after Lin and Hoffman, 1979).

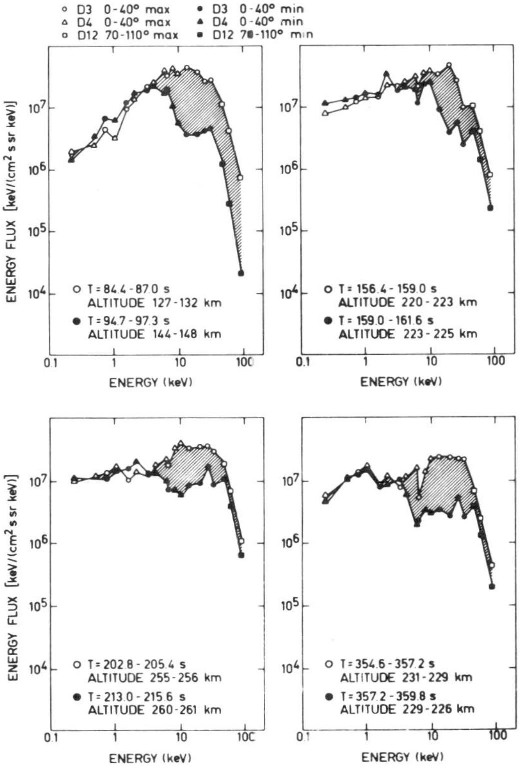

Figure 4. Electron spectra taken at maximum and minimum of a pulsating auroral patch (after Sandahl, 1984).

The general picture for the nightside that one gets from examining the wealth of precipitating electron data available and the DC and AC electric field results is; the narrow discrete auroral arcs correspond to the intense latitudinally narrow regions of electron precipitation and the broader 'arcs' correspond to the inverted-V precipitation. The narrow arcs may correspond with electrostatic shocks. At least, the shock scale lengths are comparable to that of the narrow arcs. Recently, Kletzing et al. (1983) has shown evidence that the electrostatic shocks are indeed correlated to discrete arcs.

2.3. Morning Side Aurora

The morning side aurora appears to be a relatively broad region of emission composed of different forms. There are the rare higher latitude arcs which often extend from the dayside or nightside (Akasofu and Kan, 1980; Akasofu, 1981) and which are caused by relatively soft and structured electron precipitations. The source region electron spectra of the high latitude arc precipitation can be characterized as magnetosheath or boundary-plasma-sheet like. There are also the hard precipitating electrons at lower latitudes, which are usually uniform but can take a structured appearance at the recovery phase of the substorms. The recovery is often associated with the hard and intense mantle aurora precipitation which spreads through local morning from the nightside (Hoffman and Burch, 1973; Meng and Akasofu, 1983). Often these low-latitude precipitations are patchy in nature. The patches are often associated with pulsations (Johnstone, 1978; Roywick and Davis, 1977). Although the morning discrete arcs are rare, when they occur they appear, optically, much as other discrete arcs. Of course, the diffuse auroral precipitation also occurs on the morningside.

Because the morning side also contains the pulsating aurora it has been the focus of investigations to understand how the pulsating electron precipitation is generated (Berkey, 1980; Sandahl et al., 1980; Sandahl, 1984). There is evidence that these pulsating aurora undergo E x B drift motion (Scourfield et al., 1983) and drift with the background plasma. Many of the features of pulsating aurora have been summarized by Roywik and Davis (1977) and in Sandahl's (1984) thesis. It is known to occur, typically, at the recovery phase of substorms. The variation of the electron precipitation tends to be only at the higher energies (> few keV and extending to many tens of keV). An example of the electron spectra at the maximum and minimum precipitation intensity is shown in Figure 4. At the peak intensity the angular distribution becomes isotropic or slightly field-aligned over the downgoing hemisphere.

Generally the flux changes are ascribed to modulation of the pitch angle scattering in the equatorial plane via Doppler shifted gyro resonance with whistler mode waves (see Kennel and Petschek, 1966; Johnstone, 1978, 1983). Recently, Romich and Deehr (1984) have shown a relationship between the pulsation period and the intensity of the ion precipitation. They also noted the duration of the pulses was relatively constant but that the time between pulses varied thus causing the period to vary. Several modulation mechanisms have been put forward (see Sandahl, 1984) but so far they all leave unanswered questions when compared to the observations.

The morning side discrete auroral arcs embedded in the diffuse precipitation generally do not show the monoenergetic peak in the electron precipitation that is associated with the night side arcs (ref. Fig. 3 of Chiu et al., 1983). Thus the morning side auroral zone arcs are not due to potential drops. This has left the question of how the arc forms unanswered. Chiu et al. (1983b) attempted to answer this question by considering the spatial modulation of the plasma-sheet warm plasma density by the mirror instability. This is still a topic that could use some theoretical work. More observations of the character of electron precipitation in these particular morning side arcs would also be helpful.

Inverted-V type precipitation does occur, rarely, on the poleward edge of the morning side aurora. Questions remain as to whether these precipitations always map to the boundary layer of the plasma sheet or whether they are ever magnetosheath plasma (Hoffman and Lin, 1981; Chiu et al., 1985). These points are discussed in Section 2.5.

2.4. Dayside Precipitation

The dayside electron precipitation can be identified as two general types; the soft cusp precipitation and the harder dayside auroral zone. The cusp electron precipitation coincides with a magnetosheath like ion precipitation and is generally found at higher latitudes than the dayside auroral zone (see Reiff, 1983; Fennell, 1985 and references therein). The cusp electron precipitation is intense and of relatively low energy (< 1 keV). The source of the cusp precipitation is generally stated to be the magnetosheath. The mechanisms for the plasma entry to the cusp have been described as direct entry on interconnected magnetosheath-magnetosphere field lines, diffusive entry and impulsive entry.

Using ion time of flight and dispersion effects the entry altitude has been estimated to be ~ 8-20 Re (see Burch et al., 1982; Reiff, 1983). There are also indications that the cusp ions are accelerated during entry to the cusp (Torbert and Carlson, 1980). The signature of the convection of plasma on cusp field lines has even been observed in the upflowing ion signatures (Chappell, 1984; Moore, 1985) and in the precipitating ion energy dispersion with latitude (Reiff et al., 1980). This latter signature is present when the IMF is southward and is expected in the 'open magnetosphere' model. During northward IMF the cusp ion access appears to be diffusive (Reiff et al., 1980).

The cusp electron precipitation is normally described as unaccelerated, isotropic and Maxwellian (Heikkila and Winningham, 1971; Frank and Ackerson, 1972; see also Frank, 1975 and references therein). Yet, there are observations which deviate from this picture. In particular, Zanetti et al. (1981) describe the occurrence of field-aligned electron distributions in the cusp. These distributions occurred in about half the cusp crossings and tended to lie near the equatorward edge. They did not appear to be consistent with acceleration by parallel electric fields (e.g., see Fig. 3 and Table 2 of Zanetti et al., 1981). While the authors discussed wave particle interactions as a possible mechanism for forming the field-aligned electron distributions, they came to no firm conclusions concerning such possibilities.

2.5. Polar Cap Electron Precipitation

The latitudinal region above the auroral zone is the polar cap region. It is generally thought to be a region of open field lines which maps into the magnetotail lobes. Most of the time, a relatively structureless precipitation of low energy electrons, "polar rain" (Winningham and Heikkila, 1974), is observed in this region (see Gussenhoven et al., 1984 and references therein). The polar rain is usually weak (10^{-2} to 10^{-3} erg cm^{-2} s^{-1}) and soft (average energy $\lesssim$ 100 eV precipitation. The relative intensity of the polar rain over the two polar caps is controlled by the Interplanetary Magnetic Field (IMF) direction. The precipitating electron intensity over the northern (southern) polar cap is greater than that over the southern (northern) polar cap for an antisunward (sunward) directed IMF (Mizera and Fennell, 1978; Gussenhoven, 1982). This combined with IMF control of the tail lobe fluxes (Yeager and Frank, 1976) lead to the conclusion that the polar rain was a component of the solar wind which gained unimpeded access to the tail lobes and polar caps (see Mizera and Fennell, 1978). The IMF control was related to the existence of an anisotropy of the solar wind electrons, or 'strahl'. This idea has been recently reinforced by Fairfield and Scudder (1985). They observed the 'strahl' electrons in the magnetotail lobes. The observed distributions were consistent with unimpeded and preferential entry of the strahl into the polar cap most directly connected magnetically to the sun. Figure 5 schematically shows this idea. Not everyone agrees with this picture of polar rain (Gussenhoven et al., 1984). It remains to examine the angular distributions of the polar rain in great detail to see if the weak pitch angle anisotropy (~ 5% at 1 Re) which is predicted in the strahl access picture is seen at lower altitude. DE-1 is probably the best platform for such a study.

More intense electron precipitation is also seen in the polar caps. Winningham and Heikkila (1974) called these "polar showers" and "polar squalls". Usually these structured precipitations are embedded in or near the boundaries of the polar rain. The peak precipitation can be relatively intense (~ 0.1 to 1 erg cm^{-2} s^{-1}) and have average energies of 100-200 eV. Hardy (1984) has studied these precipitations and found that they occur predominantly during positive IMF B_Z. This kind of relationship has also been noted for the high latitude and polar arcs, which are also often sun aligned (Gussenhoven, 1982 and references therein). These sun aligned polar arcs are relatively rare (~ 1% occurrence probability) and occur most often during magnetic quiescence (Ismail et al., 1977).

Examination of the electron precipitation over the sun aligned arcs has shown that the electrons have been accelerated by potential drops (see Fig. 4 of Murphree et al., 1983; Meng, 1981b). Chiu and Gorney (1983) find that the polar arcs have the same formation processes as the evening discrete arcs. Murphree et al. (1983) suggest that the polar rain particles have been accelerated to a few hundred eV to generate the arc precipitation. Meng (1981b) first suggested these were arcs on the

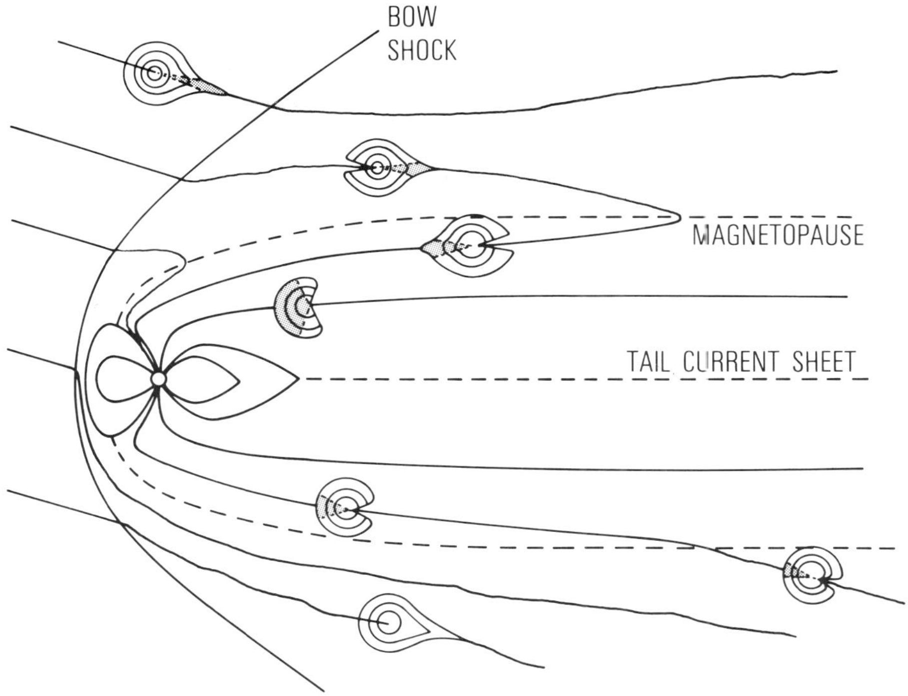

Figure 5. Illustration showing electron distributions in various regions near and in the earth's magnetosphere (after Fairfield and Scudder, 1985).

poleward edge of expanded, quiet auroral oval. Recently, Makita and Meng (1984) have stated that the source region for these quiet time arcs is not completely understood. They argue that near the dayside it may be the low latitude boundary layer plasma and at midnight the boundary plasma sheet plasma.

Frank et al. (1982) have reported the observation of a polar arc which extends from the nightside oval to the midday auroral zone. They have called this a Θ aurora. The precipitating electrons associated with this aurora are similar to those of the oval arcs (see Plate 2 and Fig. 3 of Frank et al., 1982). They argue this feature may result from arcs, which are also often sun aligned (Gussenhoven, 1982 and references therein). These sun aligned polar arcs are relatively rare 1% a two cell convection in the polar cap and might require a bifurcation of the magnetotail lobe. Peterson and Shelley (1984) used the ion composition data in an attempt to distinguish the source plasma for the Θ aurora. While their results provide strong support for the idea that the arc occurs on field lines closing through the distant plasma sheet or plasma sheet boundary layer, it does not substantiate either the model of a bifurcated magnetotail or the expanded oval with an arc at the poleward edge. Thus the discussion of source plasma for the sun aligned polar arcs and the corresponding magnetospheric topology is not yet complete.

2.6. Low Energy Field-Aligned Electron Distributions

Narrowly collimated field-aligned electron distributions at relatively low energies (< 500 eV) have been observed throughout the auroral regions. These distributions are usually to be distinguished from the electron beam associated with inverted-Vs which is field aligned only at energies near the peak of the precipitating electron spectrum. Here we are referring to upgoing, downgoing and counter-streaming field aligned electrons which generally have soft monotonic spectra (Fennell et al., 1979; Sharp et al., 1980; Klumpar, 1981; Collin et al., 1982; Lin et al., 1982, 1984a and references therein). The counter streaming electrons are least understood. They have been observed at middle altitudes (Lin et al., 1982, 1984a) and at the equator (Richardson et al., 1981 and references therein). Figure 2b (upper panel) is an example of counter streaming electrons in the auroral region.

The upflowing electron beams are often observed at the low latitude edge of inverted-Vs, and as part of the Region I Birkeland current (Klumpar, 1981; Burch et al., 1983; Gorney et al., 1985 and references therein). An example of upflowing electrons is shown in Figure 6. Often these upflowing electron beams are associated with the conic ion beams (see Section 2.1 and Gorney et al., 1985). Generally, these upflowing electrons have been accelerated through relatively small ($\lesssim$ tens of eV) potential drops. Also, the probability of observing these beams with existing instruments increases with increasing altitude, up to ~ 6000 km (Collin et al., 1982). This would indicate an extended potential drop. The evidence for the potential drops has come from analysis of the distribution functions of the electrons (Klumpar and Heikkila, 1982; Burch et al., 1983) or both electrons and ions (Gorney et al., 1985). At this point, no direct measurement of a downward electric field has been made although Mozer (1980) has inferred its existence from the measurements of the perpendicular electric field.

Downward electron beams are observed almost uniformly in altitude over the range 3000-8000 km and occur more frequently than upstreaming electrons (Collin et al., 1982). The limited DE-1 observations indicate that when the plasma current measured by plasma instruments is compared to the current measured by the magnetometer that they agree. When the component carried by the low energy beams is compared to the total, it is found that these beams can carry as much as 40 to 50% of the total current, when they occur. Generally, the beams are capable of carrying a more significant fraction of the downward current (Lin et al., 1984a).

The counter streaming electrons observed in the auroral regions have been classified into two types by Lin et al. (1982). Type 1 are those associated with an isotopic high energy ($\gtrsim$ 1 keV) precipitating electron population. They generally have a two component Maxwellian distribution function and are fairly stable, existing over a small spatial region. The type 1 may involve wave particle interactions (Lin et al., 1982). Type 2 counter streaming electrons are beams parallel and antiparallel to the magnetic field. They are transient and possibly consistent with acceleration and trapping by parallel electric fields. Ion conics are often observed with the counter streaming electrons by the S3-3 satellite (Sharp et al., 1980). At the geomagnetic equator the counter streaming electrons are always associated with field-aligned ion distributions (see Fig. 8 and Richardson et al., 1981). At this point, the generation mechanisms for these counter streaming electrons is not totally clear.

3. IONOSPHERIC ION OUTFLOW

3.1. Conics

Over the past several years it has become clear that ionospheric ions are often a major source of the plasmasheet and magnetospheric plasma. It has also become clear that transverse energization of these ions at varying altitudes plays a major role in their being supplied to the outer magnetosphere with significant energies and fluences (Chiu et al., 1983, 1984; Rosner et al., 1984 and

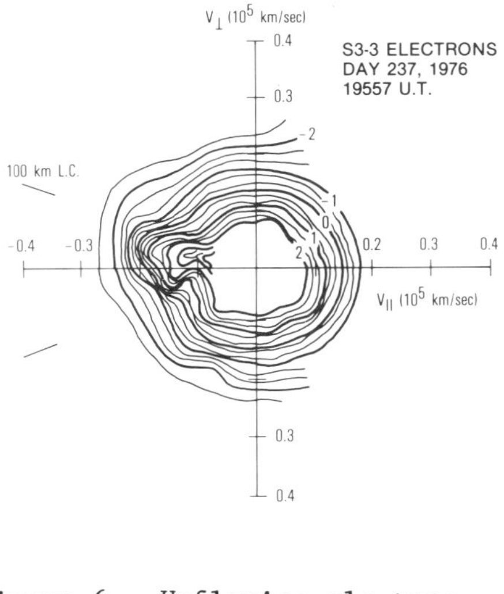

Figure 6. Upflowing electron distribution observed on the S3-3 satellite. This distribution was observed simultaneously with a conic ion distribution (after Gorney et al., 1985).

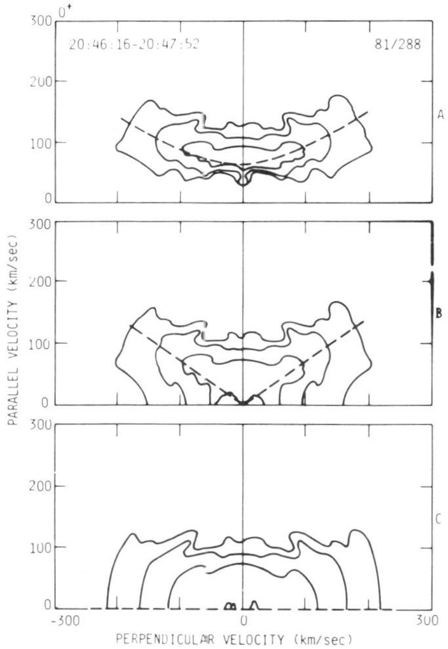

Figure 7. Oxygen ion distribution function from DE-1 on 15 October 1981. The three panels show: (A) measured distribution; (B) distribution transformed through 310 V potential drop; and (C) distribution (B) adiabatically transformed to region where magnetic field intensity is a factor of 1.49 that at the satellite (after Klumpar et al., 1984).

references therein). Figure 2b (bottom panel) shows a distribution function for ions that have undergone transverse energization. This kind of distribution has been called an ion conic.

The transverse energization of ions on high latitude field lines has been observed from a few hundred kilometers (Wahlen et al., 1978; Yau et al., 1983) to 18,000 kilometers (Klumpar et al., 1984; Yau et al., 1984; Winningham and Burch, 1985) altitude on auroral field lines and even in the far magnetotail (Sharp et al., 1981, 1983). Initially, it was thought that the transverse acceleration occurred predominantly below a few thousand kilometers (Gorney et al., 1981) and on separate flux tubes from regions of parallel acceleration of ions (Cattell et al., 1979; Gorney et al., 1981).

The upward flowing "ion beams" (see below) had a perpendicular temperature greater than the temperature of en in the regions where parallel electric fields were known to exist. Thus, it appears a mass selective acceleration is process acting.

Recently, Klumpar et al. (1984) and Winningham and Burch (1985) have observed ion distributions which clearly show the effects of both parallel and transverse to B accelerations in the process of being transported upwards into the magnetosphere (see Fig. 7). The parallel acceleration signature was consistent with an inverted-V structure, indicating both types of acceleration can occur throughout the auroral zone. Similar observations have been made at low altitudes (few x 10^3 km) with transverse heating of thermal and suprathermal ions being clearly associated with O^+ outflows (Chappell, 1984; Lockwood et al., 1984; Moore et al., 1985). These latter data showed that the transverse heating can involve the entire thermal plasma to the point that ion "ring" or "torus" distributions are generated. These transversely heated ion outflows occur in low density ($n < 100\ cm^{-3}$) regions.

In the dayside auroral region the suprathermal ion source appears as a field aligned flow dispersed tailward across the polar cap by the convective flow as sketched in Fig. 8 (Horwitz, 1984). On the night side the ion beams are often bounded by ion conics with the cone angle increasing with distance from the beam region (see Moore et al., 1985).

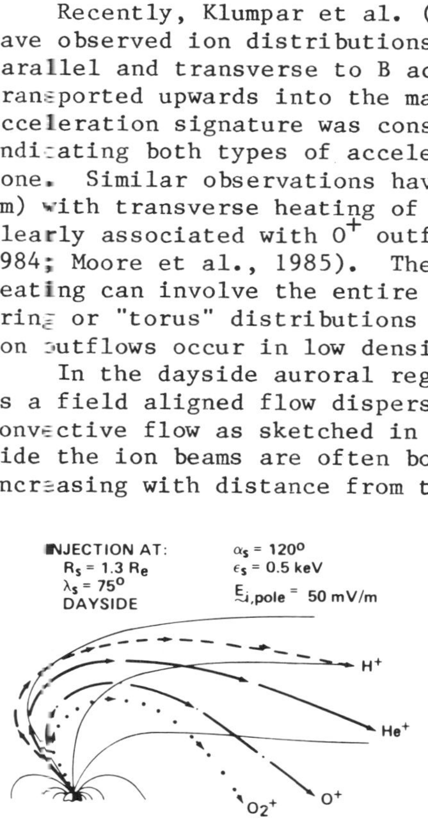

Figure 8. Trajectories of upflowing ions in the polar magnetosphere (after Horwitz, 1984).

The combination of these results leaves us with the impression that the night side ion acceleration may go through multiple stages of transverse and parallel acceleration starting with transverse acceleration at ionospheric altitudes and passing into parallel and transverse acceleration at different points up the field line. We also see that the ion transport is affected by the convection electric fields such that the ions are disperse throughout the polar region on tail lobe and plasma sheet field lines (Horwitz, 1984).

Examination of disturbed time data, from polar orbiting satellites, shows that conics are seen throughout the central plasma sheet (Fennell, 1985; Winningham and Burch, 1985) down to the plasma pause. The transverse acceleration of ions has also been observed near the outer regions of the plasmasphere at the geomagnetic equator (see Horwitz, 1980, 1982; and references therein). When measurements are made with high angular resolution and down to very low (< 10 eV) energies, then the combined transverse energization and net parallel and perpendicular drifts can be separated (Winningham and Burch, 1985).

In rare instances it appears that the accelerated upflowing ions are capable of carrying significant current. For example, Hellis and Winningham (1984) have reported a DE-2 observation of upward thermal ion fluxes as high as 10^{10} cm^{-2} sec^{-1} at the edges of inverted-V events. The upwards field aligned currents were ~ 100 μ A m^{-2} with the upgoing ions contributing more than 10 μ A m^{-2} and at times currents comparable to the observed field aligned current. The observations were made at 900 km near the ionosphere-magnetosphere interface region and may be unique to this region in very disturbed times.

A recent statistical study at 8-23 x 10^3 km altitude by Yau et al. (1984) has shown that the upflowing ions occur more frequently near the equatorward edge of the oval and have higher energies there. At these altitudes the conic ions occurrence shows a dawn-dusk asymmetry favoring the dusk sector. The occurrence frequency of O^+ conics decreases with altitude relative to H^+ conics. In the polar cap the upflowing ions are of relatively low energy but the outflow rates are significant, ranging from 2-3 x 10^{-24} s^{-1} to 3-10 x 10^{24} s^{-1} for quiet and disturbed periods, respectively.

The overall impression that transverse energization of ionospheric ions is a fundamental process occurring throughout the higher latitude regions where significant currents are flowing is succinctly stated in a paper by Winningham and Burch (1985). These authors state, "it is possible that conics represent a very natural.....state of the magnetosphere...As such they are a part of a more generalized polar wind and should be theoretically treated in a more realistic, expanded polar wind theory." This seems to represent our current understanding very nicely.

3.2. Beams

The term "ion beam" has been used by the S3-3 experimenters and some others to distinguish upflowing ion distributions which show indications of acceleration by a potential drop below the observation point (i.e., a peak in $f(v_\parallel)$). The statistical distribution of these upflowing ion beams shows a preference for the nightside auroral zones and inverted V regions (see Gorney et al., 1981; Yau et al., 1984 and references therein) as the regions where ion beams are seen. As discussed above, the artificial separation of upflowing ions into beams and conics may be breaking down as observations push to lower energy. Nevertheless, the existence of ion beams as defined here, no matter how low their energy, are indicators that field-aligned potential drops exist on the flux tubes which map to their remote source point. They can be used to provide information about physical mechanisms which cause ion outflow, but they must be interpreted carefully.

The altitudinal distribution of the upflowing ion beams (Ghielmetti et al., 1978; Gorney et. al., 1981; Yau et al., 1984) has been used to predict the lower bound of the potential drop associated with inverted-V events and electrostatic shocks. This is generally quoted as 3000-4000 km. This lower bound depends somewhat on the threshold energy used to define the ion beams, although the results for a 90 eV and a 500 eV threshold are nearly identical (Ghielmetti et al., 1978; Gorney et al., 1981; Collin et al., 1981).

Downward flowing ion beams have rarely been observed by the S3-3 satellite (Fennell et al., 1979; Ghielmetti et al., 1979). At the equator (Richardson et al., 1981; Olsen, 1982 and references therein) and in the high altitude magnetosphere (Lundin et al., 1982) field-aligned ion beams are observed fairly often during magnetically disturbed periods. Recently, downward flowing ion beams have been observed on DE (Winningham et al., 1984). These have been interpreted in terms of the convection electric field acting on ions flowing out from the ionosphere in the opposite hemisphere. The latitudinal (L-shell) displacement caused by inward convection acts as a 'velocity-filter' separating the different ion energies and species. Thus, at the equator, at high altitudes and at the opposite hemisphere from the source, the ions would always appear as beams (i.e., peaked $f(v_{\parallel})$). Therefore, once the observing platform is far enough from the source/acceleration region that the convection electric field plays a significant role in forming the observed ion f(v) (i.e., see Fig. 5 of Winningham and Burch, 1985), the ion beams must be carefully interpreted. These kinds of studies are just beginning to occur.

3.3. Polar Wind

The escape of thermal ions from the ionosphere along field lines outside the plasmasphere has been predicted by theory and inferred from density measurements for some time (see Reiff, 1983 and references therein). The flow on closed field lines outside the plasmasphere exists because of the dynamic emptying and refilling which occurs there in response to magnetic activity and magnetospheric convection. The outer part of the magnetosphere contains flux tubes which open into the magnetotail. On these open field lines the polar wind becomes a source of ions for the plasma sheet or looses ions to the interplanetary medium. (See reviews by Horwitz, 1982 and Young, 1983 and references therein.) We will discuss here only those results pertaining to the auroral oval and polar cap regions.

Direct measurements of the upward flowing thermal ions, that make up the polar wind, were only recently made (Hoffman and Dodson, 1980; Gurgiolo and Burch, 1982). The initial DE satellite results showed that a heated and unheated component of the polar wind could be observed. They also showed that the average energy of the upflowing ions was higher than theoretically expected, and that the accelerated component of the polar wind was observed on each polar cap traversal (Chappell et al., 1982; Gurgiolo and Burch, 1982). It was also noted that the polar cap photoelectrons showed anomalous distributions indicative of a parallel electric field which could provide the polar wind ions escape energy (see Fig. 2 of Winningham and Gurgiolo, 1982).

More recently, it has been shown that the polar wind is supersonic with an average Mach number of 3, corresponding to a flow velocity of 16-25 km/sec (Nagai et al., 1984). The H^+ flow matched classical polar wind theory. Thus, we know that the light ion component of the polar wind is supersonic. We also know that a large fraction, if not all, of the time the polar regions experience significant ion heating and acceleration sufficient for the heated ions, both light and heavy, to obtain escape velocity.

4. SUMMARY

We see that the electron and ion acceleration by parallel electric fields is substantiated for the inverted-V precipitation regions and many discrete arcs. The detailed processes by which the potential drops are generated are still being discussed. There is some evidence that the discrete arcs may be associated with electrostatic shocks, except in the dawn oval. We also see that pitch angle diffusion is still a good explanation for the diffuse aurora but the existing models may need some refinements of detail. The polar cap precipitation is still an area where much work needs to be done to define the source plasmas for the polar arcs, especially for IMF $B_z > 0$. The polar rain appears to be of solar origin, although there is not a consensus on this point. The dawn side discrete oval arcs have not been studied very well at this point. These arcs plus the pulsating aurora precipitation are still topics for fruitful research both observationally and theoretically.

The ion outflow studies have been most fruitful in the last decade. We now know that transverse acceleration of ions is observed throughout the auroral zone and down to the plasmasphere. It appears that nearly all ionospheric ion outflow, other than the classic polar wind, has its origin in transverse acceleration which provides the escape energy (especially for oxygen). The transverse acceleration occurs most often in the downward current regions but is not restricted to these regions.

Once the ions leave the ionosphere they can be further accelerated either perpendicular or parallel to the magnetic field and they are transported in latitude in the combined magnetospheric electric and magnetic fields. Thus, the ionospheric ions are transported throughout the outer magnetosphere.

5. References

Akasofu, S. -I., Physics of Auroral Arc Formation, Akasofu and Kan eds., p 1 Am. Geophysical Union, Washington, D.C., 1981.

Akasofu, S. -I. and J. R. Kan, Geophys. Res. Lett., **7**, 753, 1980.

Arnoldy, R. L., in Physics of Auroral Arc Formation, Akasofu and Kan eds., p. 56, Am. Geophysical Union, Washington, D.C., 1981.

Ashour-Abdalla, M. and R. M. Thorne, J. Geophys. Res., **83**, 4775, 1978.

Belmont, G., D. Fontaine and P. Canu, J. Geophys. Res., **88**. 9163, 1983.

Berkey, F. T., J. Geophys. Res. **85**, 6075, 1980.
Bingham, R., D. A. Bryant and D. S. Hall, Geophys. Res. Lett., **11**, 327, 1984.
Bryant, D. A., D. S. Hall and D. R. Lepine, Planet. Space Sci., **26**, 81, 1978.
Bryant, D. A., High-Latitude Space Plasma Physics, Hultqvist and Hagfors eds., p. 295, Plenum Press, New York, 1983.
Burch, J. L., P. H. Reiff, R. A. Heelis, J. D. Winningham, W. B. Hanson, C. Gurgiolo, J. D. Menietti, R. A. Hoffman, J. N. Barfield, Geophys. Res. Lett., **9**, 921, 1982.
Burch, J. L., P. H. Reiff and M. Sugiura, Geophys. Res. Lett., **10**, 753, 1983.
Burke, W. J., Rev. Geophys., **20**, 685, 1982.
Cattell, C., R. Lysak, R. B. Torbert and F. S. Mozer, Geophys. Res. Lett., **6**, 621, 1979.
Cattell, C., Magnetospheric Currents, T. A. Potemra, ed., Am. Geophys. Union, Washington, D.C., 1984.
Chappell, C. R., Adv. Space Res., to be published, COSPAR, 1984.
Chappell, C. R., J. L. Green, J. F. E. Johnson and J. H. Waite, Jr., Geophys. Res. Lett., **9**, 933, 1982.
Chiu, Y. T., R. Anderson, J. Fennell, L. Frank, R. Hoffman, M. Hudson, L. Lyons, P. Palmadesso, E. Ungstrup, R. Vondrak, D. Williams and R. Wolf, Solar Terrestrial Physics: Present and Future, NASA R. P. **1120**, p. 7-3, 1984.
Chiu, Y. T., J. M. Cornwall, J. F. Fennell, D. J. Gorney and P. F. Mizera, Space Sci. Rev., **35**, 211, 1983a.
Chiu, Y. T., N. U. Crooker and D. J. Gorney, J. Geophys. Res., to be published, 1985.
Chiu, Y. T. and D. J. Gorney, Geophys. Res. Lett., **10**, 463, 1983.
Chiu, Y. T. and A. M. Kishi, J. Geophys. Res., **89**, 5531, 1984.
Chiu, Y. T., M. Schulz, J. F. Fennell and A. M. Kishi, J. Geophys. Res., **88**, 4041, 1983b.
Collin, H. L., R. D. Sharp, E. G. Shelley and R. G. Johnson, J. Geophys. Res., **86**, 6820, 1981.
Collin, H. L., R. D. Sharp and E. G. Shelley, Geophys. Res., **87**. 7504, 1982.
Collin, H. L., R. D. Sharp and E. G. Shelley, J. Geophys. Res., **89**, 2185, 1984.
Evans, D. S., J. Geophys. Res., **79**, 2853, 1974.
Fairfield, D. H. and J. D. Scudder, J. Geophys. Res., to be published, 1985.
Fairfield, D. H. and Vinas, A. F., J. Geophys. Res., **89**, 841, 1984.
Fennell, J. F., Proceedings of the Symposium on Achievements of the IMS, ESASP-, Report in press, 1984.
Fennell, J. F., D. J. Gorney and P. F. Mizera, Physics of Auroral Arc Formation, Akasofu and Kan, eds., p. 91, Am. Geophys. Union, Washington, D.C., 1981.
Fennell, J. F., P. F. Mizera and D. R. Croley, Jr., Proceedings of Magnetospheric Boundary Layers Conference, ESA SP-148 1979.
Frank, L. A., Rev. of Geophys. and Space Phys., **13**, 974, 1975.
Frank, L. A. and K. Ackerson, J. Geophys. Res., **77**, 4116, 1972.

Frank, L. A., J. D. Craven, J. L. Burch and J. D. Winningham, Geophys. Res. Lett., **9**, 1001, 1982.
Ghielmetti, A. G., R. G. Johnson, R. D. Sharp and E. G. Shelley, Geophys. Res. Lett., **5**, 59, 1978.
Ghielmetti, A. G., R. D. Sharp, E. G. Shelley and R. G. Johnson, J. Geophys. Res., **84**, 5781, 1979.
Gorney, D. J., A. Clarke, D. Croley, J. Fennell, J. Luhmann and P. Mizera, J. Geophys. Res., **86**, 83, 1981.
Gorney, D. J., Y. -T. Chiu and D. R. Croley, Jr., J. Geophys. Res., **90**, in press, 1985.
Gurgiolo, C. and J. L. Burch, Geophys. Res. Lett., **9**, 945, 1982.
Gurnett, D. A. and L. A. Frank, J. Geophys. Res., **82**, 1031, 1977.
Gussenhoven, M. S., J. Geophys. Res., **87**, 2401, 1982.
Gussenhoven, M. S., D. A. Hardy and N. Heinemann, J. Geophys. Res., **88**, 5692, 1983.
Gussenhoven, M. S., D. A. Hardy, N. Heinemann and R. K. Burkhardt, J. Geophys. Res., **89**, 9785, 1984.
Hardy, D. A., J. Geophys. Res., **89**, 3883, 1984.
Heelis, R. A. and J. D. Winningham, J. Geophys. Res., **89**, 3893, 1984.
Heikkila, W. J. and J. C. Winningham, J. Geophys. Res., **76**, 883, 1971.
Hoffman, R. A. and J. L. Burch, J. Geophys. Res., **78**, 2867, 1973.
Hoffman, J. H. and W. H. Dodson, J. Geophys. Res., **85**, 626, 1980.
Hoffmann, R. A. and C. S. Lin, Physics of Auroral Arc Formation, p. 80, Am. Geophysical Union, Washington, D.C., 1981.
Horwitz, J. L., J. Geophys. Res., **85**, 2057, 1980.
Horwitz, J. L., Rev. Geophys. Space Phys., **20**, 929, 1982.
Horwitz, J. L., W. K. Cobb, C. R. Baugher, C. R. Chappell, L. A. Frank, T. E. Eastman, R. R. Anderson, E. G. Shelley and D. T. Young, J. Geophys. Res., **87**, 9059, 1982.
Horwitz, J. L., Geophys. Res. Lett., **11**, 1111, 1984.
Huang, C. Y., L. A. Frank and T. E. Eastman, J. Geophys. Res., **89**, 7423, 1984.
Ismail, S., D. D. Wallis and L. L.Cogger, J. Geophys. Res., **82**, 4741, 1977.
Johnstone, A. D., Nature, **274**, 119, 1978.
Johnstone, A. D., Annales Geophysical, **1**, 397, 1983.
Kennel, C. F. and H. E. Petschek, J. Geophys. Res., **71**, 1, 1966.
Kletzing, C., C. Cattell, F. S. Mozer, S. -I. Akasofu and K. Makita, J. Geophys. Res., **88**, 4105, 1983.
Klumpar, D. M., Physics of Auroral Arc Formation, Akasofu and Kan, eds., p. 22, Am. Geophys. Union, Washington, D.C., 1981.
Klumpar, D. M. and W. J. Heikkila, Geophys. Res. Lett., **9**, 873, 1982.
Klumpar, D. M., W. K. Petersen and E. G. Shelley, J. Geophys. Res., **89**, 10779, 1984.
Lin, C. S., J. L. Burch, J. D. Winningham, J. D. Menietti and R. A. Hoffman, Geophys. Res. Lett., **9**, 925, 1982.
Lin, C. S. and R. A. Hoffman, J. Geophys. Res., **84**, 1514, 1979a.
Lin, C. S. and R. A. Hoffman, J. Geophys. Res., **84**, 6547, 1979b.
Lin, C. S. and R. A. Hoffman, Sp. Sci. Rev., **33**, 415, 1982.
Lin, C. S., M. Sugiura, J. L. Burch, J. N. Barfield and E. Nielsen, J. Geophys. Res., **89**, 8907, 1984a.

Lin, C. S., J. L. Burch, S. D. Shawhan and D. A. Gurnett, J. Geophys. Res., **89**, 925, 1984b.
Lockwood, M., J. Geophys. Res., **89**, 301, 1984.
Lundin, R., B. Hultquist, E. Dubinin, A. Zackarov and N. Pissarenko, Planet. Space Sci., **30**. 715, 1982.
Makita, K. and C. -I. Meng, J. Geophys. Res., **89**, 2861, 1984.
Meng, C. -I., in Physics of Auroral Arc Formation, Akasofu and Kan, eds., p. 67, Am. Geophys. Union, Washington, D.C., 1981a.
Meng, C. -I., Geophys. Res. Lett., **8**, 273, 1981b.
Meng, C. -I. and S. -I. Akasofu, Planet. Sp. Sci., **31**, 889, 1983.
Meng, C. -I. and H. W. Kroehl, J. Geophys. Res., **16**, 2305, 1977.
Meng, C. -I., B. Mauk and C. E. McIlwain, J. Geophys. Res., **84**, 2545, 1979.
Mizera, P. F. and J. F. Fennell, Rev. Geophys. and Sp. Sci., **16**. 147, 1978.
Mizera, P. F., J. F. Fennell, D. R. Croley, Jr., A. Vampola, F. S. Mozer, R. B. Torbert, M. Temerin, R. L. Lysak, M. K. Hudson, C. A. Cattell, R. G. Johnson, R. D. Sharp, P. M. Kintner and M. C. Kelley, J. Geophys. Res., **86**, 2329, 1981.
Moore, T. E., C. R. Chappel, M. Lockwood and J. H. Waite, J. Geophys. Res., to be published, 1985.
Mozer, F. S., Geophys. Res. Lett., **7**. 1097, 1980.
Mozer, F. S., C. A. Cattell, M. K. Hudson, R. L. Lysak, M. Temerin and R. B. Torbert, Space Sci. Rev., **27**, 155, 1980.
Murphree, J. S., S. Ismail, L. L. Cogger, D. D. Wallis, G. G. Shepherl, R. Link and D. M. Klumpar, Planet. Space Sci., **31**, 161, 1983.
Nagai, T., J. H. Waite, Jr., J. L. Green, C. R. Chappell, R. C. Olsen and R. H. Comfort, Geophys. Res. Lett., **11**, 669, 1984.
Olsen, R. C., J. Geophys. Res., **87**, 3481, 1982.
Peterson, W. K. and E. G. Shelley, J. Geophys.Res., **89**, 6729, 1984.
Prasad, S. S., D. J. Strickland and Y. T. Chiu, J. Geophys. Res., **88**, 4123, 1983.
Pulliam, D. M., H. R. Anderson, K. Stamnes and M. H. Rees, J. Geophys. Res., **86**, 2397, 1981.
Reiff, P. H., J. L. Burch and R. W. Spiro, J. Geophys. Res., **85**, 5997, 1980.
Reiff, P. H., Rev. Geophys. Space Phys., **21**, 418, 1983.
Richardson, J. D., J. F. Fennell and D. R. Croley, Jr., J. Geophys. Res., **86**, 10105, 1981.
Romik, G. J. and C. S. Deehr, in Achievements of the IMS, Abstract, 25th COSPAR Meeting, 1984.
Rosner, R., E. L. Chupp, G. Gloeckler, D. J. Gorney, S. M. Krimigis, Y. Mok, R. Ramaty, D. W. Swift, L.Vlahos and E. G. Zweibel, Solar Terrestrial Physics: Present and Future, NASA R. P. **1120**, 1984.
Roywick, O. and T. N. Davis, J. Geophys. Res., **82**, 4720, 1977.
Sandahl, I., Kiruna Geophysical Institute Report #**185**, Kiruna, Sweden, 1984.
Sandahl, I., L. Eliasson and R. Lundin, Geophys. Res. Lett., **7**, 309, 1980.
Scourfield, M. W. J., J. G. Keys, E. Nielsen, C. K. Goertz and H. Collin, J. Geophys. Res., **88**, 7983, 1983.

Sharp, R. D., E. G. Shelley, R. G. Johnson and A. G. Ghielmetti, J. Geophys. Res., **85**, 92, 1980.
Sharp, R. D., D. L. Carr, W. K. Peterson and E. G. Shelley, J. Geophys. Res., **86**, 4639, 1981.
Sharp, R. D., R. G. Johnson, W. Lennartsson, W. K. Peterson and E. G. Shelley, in Energetic Ion Composition in the Earth's Magnetosphere, R. G. Johnson, ed., **231**. Terra Sci. Publ. Co., Tokyo, 1983.
Thieman, J. R. and R. A. Hoffman, J. Geophys. Res., to be published, 1985.
Torbert, R. B. and C. W. Carlson, J. Geophys. Res., **85**, 2909, 1980.
Wahlen, B. A., W. Bernstein and P. W. Daly, Geophys. Res. Lett., **5**, 55, 1978.
Wahlen, B. A. and P. W. Daly, J. Geophys. Res., **84**, 4175, 1979.
Winningham, J. D. and J. L. Burch, J. Geophys. Res., to be published, 1985.
Winningham, J. D., J. L. Burch and R. A. Frahm, J. Geophys. Res., **89**. 1749, 1984.
Winningham, J. D. and W. J. Heikkila, J. Geophys. Res., **79**, 949, 1974.
Winningham, J. D. and C. Gurgiolo, Geophys. Res. Lett., **9**, 977, 1982.
Yau, A. W., B. A. Wahlen, A. G. McNamara, P. J. Kellogg and W. Bernstein, J. Geophys. Res., **88**, 341, 1983.
Yau, A. W., B. A. Wahlen, W. K. Petersen and E. G. Shelley, J. Geophys. Res., **89**. 5507, 1984.
Yeager, D. M. and L. A. Frank, J. Geophys. Res., **81**, 3966, 1976.
Young, D. T., Rev. Geophys. Space Phys., **21**, 402, 1983.
Zanetti, L. J., T. A. Potemra, J. P. Doering, J. S. Lee and R. A. Hoffman, J. Geophys. Res., **86**, 8957, 1981.

6. ACKNOWLEDGMENTS

The author would like to thank Drs. Y. T. Chiu, D. J. Gorney, L. R. Lyons and P. F. Mizera for many useful discussions. This work was supported in part by the U. S. Air Force System Command's Space Division under Contract F04701-84-C-0085.

OBSERVATIONS OF ION CYCLOTRON WAVES NEAR SYNCHRONOUS ORBIT AND ON THE GROUND

B. J. Fraser
Department of Physics
University of Newcastle
N.S.W. 2308, Australia

ABSTRACT. Ion cyclotron waves (hereafter ICW's) generated in the magnetosphere by the ion cyclotron instability of 10-100 keV protons are now known to be the origin of short-period (0.1-5 Hz) electromagnetic field oscillations observed by synchronous spacecraft and on the earth's surface. Observations of the various wave characteristics, including spectral and polarization properties that lead to the identification of generation and propagation mechanisms and regions in the magnetosphere are described with reference to ATS-6, GEOS and ground-based wave data and interpreted using cold plasma propagation theory. The presence of heavy ions (O^+, He^+) dramatically modifies ICW magnetospheric propagation characteristics giving rise to spectral slots and polarization reversals. These properties may be used in plasma diagnostics. Finally satellite-ground correlations and techniques for determining the magnetospheric source position of ICW's not seen at synchronous orbit but observed on the ground as structured Pc1 pulsations are considered.

1. INTRODUCTION

Observations of ultra-low-frequency (ULF) waves in the Pc1 (0.2-5 Hz) and Pc2 (0.1-0.2 Hz) frequency ranges have been seriously undertaken since the beginning of the IGY in 1958 using both single stations and multistation arrays (see for example reviews by Troitskaya, 1967; Jacobs, 1970; Orr, 1973; Fraser-Smith, 1982; Fukunishi, 1984). These observations provided the basis for the theoretical work (Kennel and Petschek, 1966; Cornwall, 1966; Liemohn, 1967) which suggested that the waves were generated in the equatorial L = 4-8 region of the magnetosphere by the proton cyclotron instability involving an anisotropic (T > T) and energetic proton population (10-100 keV). The waves then propagate from the generation region in a field-aligned direction in the left-hand (LH) polarized mode down to the ionosphere. Here some of the energy is coupled to the right-hand (RH) mode and propagates by way of the F2 region ionospheric waveguide from the magnetospheric exit region to higher and lower latitudes (Manchester, 1968). The remaining energy is reflected back into the magnetosphere and propagates to the

Space Science Reviews **42** (1985) 357–374. 0038–6308/85.15

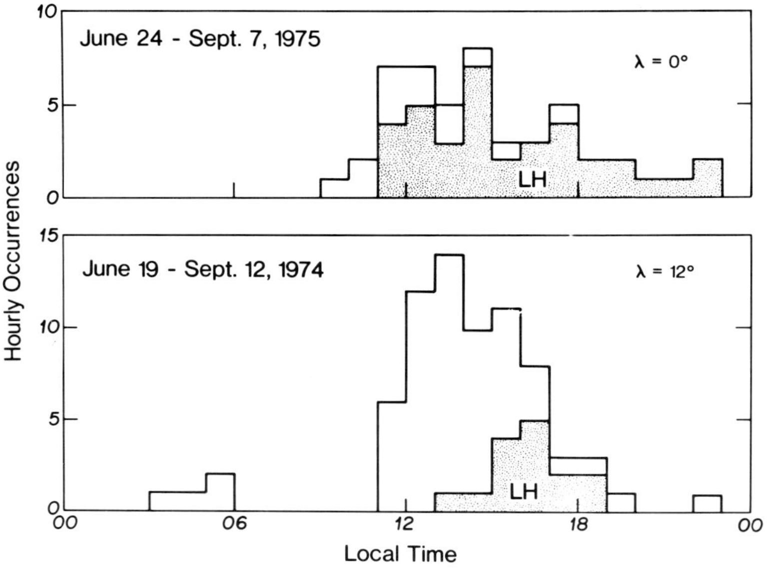

Fig. 1. Diurnal variation in hourly occurrences of ICW's observed by ATS-6 on the equator ($\lambda = 0^o$) and north of the equator ($\lambda = 12^o$). Unshaded areas indicate RH and linear polarization.

conjugate ionosphere. This gives rise to a wave packet bouncing between hemispheres. These basic generation and propagation mechanisms are currently accepted but, as will be shown later, the profound effects produced by the presence of heavy ions in the magnetospheric plasma must be taken into consideration in some frequency ranges.

Although many types of Pc1-2 waves have been identified by ground-based spectral observations (Fukunishi et al., 1981) waves may be most simply defined in two categories: structured and unstructured (Hayashi et al., 1981; Fraser et al., 1984). Structured emissions include the classic regularly fine structured "pearl" type periodic emissions, hydromagnetic whistlers, chorus and sometimes IPDP (intervals of pulsations of diminishing period). These pulsations are associated with individual wave packets propagating between hemispheres. The unstructured pulsations generally correspond to non-periodic emissions. It should be noted that observations of the various wave types may depend on the latitude of the recording station.

In the magnetospheric plasma Pc1-2 wave frequencies are generally between 0.1 to 0.6 times the proton cyclotron frequency and are electromagnetic ICW's (Stix, 1962). Early in situ observations of ICW's were made by elliptically orbiting satellites in studies of pitch angle diffusion processes associated with ring current protons (Taylor and Lyons, 1976). Observations of ICW phenomena are most easily studied from geostationary satellites and early observations were made on the ATS-1 fluxgate magnetometer by Bossen et al. (1976a,b). Large amplitude non-periodic LH polarized ICW's associated with sub-storm activity

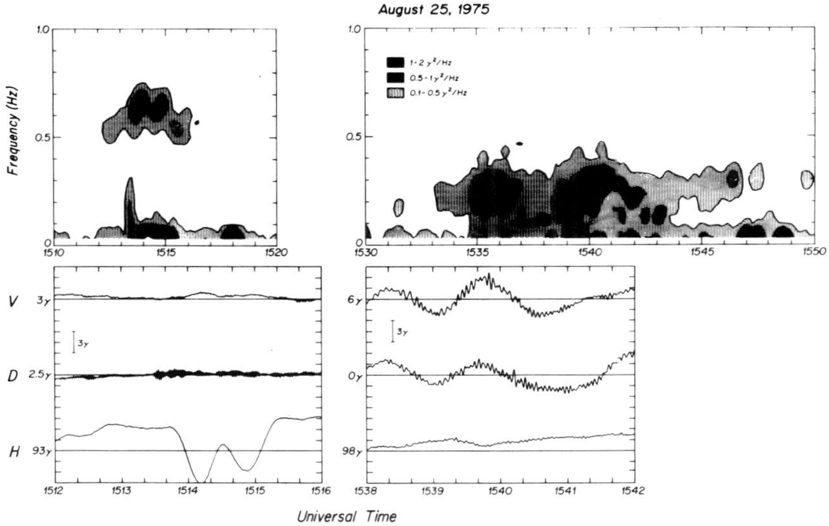

Fig. 2. Dynamic spectra and waveforms of the two distinct types of ICW's seen by ATS-6. The high frequency waves on the left are essentially azimuthal while the lower frequency waves on the right are LH circularly polarized.

were observed. More recently ATS-6 and GEOS-1 and -2 wave and particle data have been used to verify directly that ICW-particle interaction is in fact the source of ICW's (Mauk and McPherron, 1980; Roux et al., 1982).

After the discovery of heavy ion species in the thermal plasma (M/Q > 1) by Geiss et al (1978) it soon became obvious that the assumption of ICW's propagating in a simple H^+ - e^- two component plasma was not always valid. The profound effect of the presence of heavy ions in the cool plasma population was evident in both wave and particle observations (Perraut et al., 1978; Young et al., 1981; Fraser and McPherron, 1982). Helium (He^+) and oxygen (O^+) ions play a dominant role in the determination of wave spectra by introducing additional resonance and cutoff frequencies and polarization reversals into the propagation process. Energy transfer between different particle species by particle 1-wave-particle 2 interactions also occurs (Gendrin, 1983a,b).

The understanding of ICW propagation processes will be complete when consistent experimental and theoretical descriptions are available for waves propagating along geomagnetic field-aligned paths between hemispheres and transmitting energy into the ionospheric waveguide and ultimately to the ground as electromagnetic waves. It is therefore important to relate the observation of individual ICW emissions in the magnetosphere to Pc1-2 pulsation events seen at conjugate ground locations. Direct comparisons have been made but the results are not always easy to interpret.

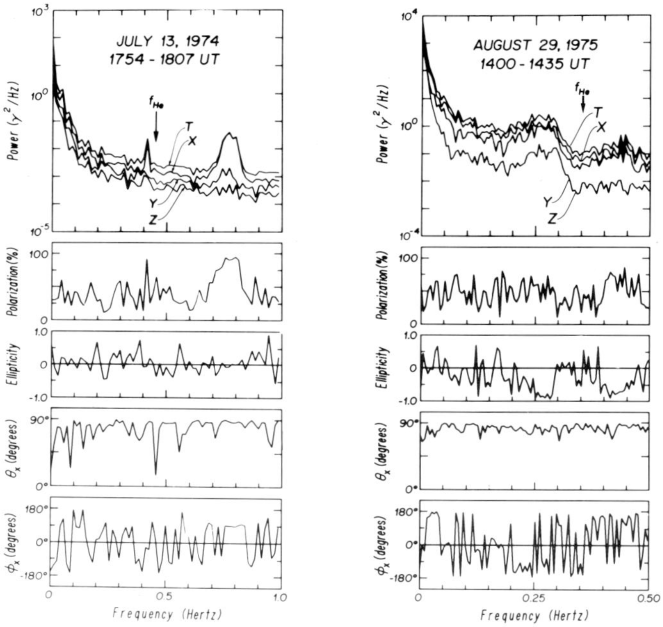

Fig. 3. Spectral and polarization characteristics for ICW events observed off the equator at $\lambda = 12^{\circ}$ (left panels) and on the equator $\lambda = 0^{\circ}$ (right panels). f_{He} is the equatorial cyclotron frequency.

The purpose of this tutorial review paper is to present the properties of ICW's seen at synchronous orbit and interpret them in terms of simple multicomponent cold plasma propagation theory. This leads to the application of wave diagnostic techniques to determine relative heavy ion concentrations. The relationship of Pc1-2 pulsations observed on the ground to the spacecraft waves and their association with the plasmapause will be discussed. This involves ionospheric wave transmission and groundbased source location techniques.

2. ION CYCLOTRON WAVES AT SYNCHRONOUS ORBIT

The ICW results presented were recorded by magnetometers and particle detectors on the ATS-6 (Mauk and McPherron, 1980) and GEOS-1 and -2 (Perraut et al., 1978) geostationary spacecraft at L = 6.6. These instrumentation packages were designed to study wave-particle interactions occurring in the equatorial region of the magnetosphere. Additionally, the GEOS spacecraft provided plasma composition measurements (Geiss et al., 1978). Results of the GEOS studies are well documented (Perraut et al., 1978; 1984; Gendrin, 1980; 1983a,b; Young et al., 1981; Roux et al., 1982; Perraut, 1982; Korth et al., 1984) and emphasis here will be placed on ATS-6 results.

The ATS-6 spacecraft was situated at 94°W longitude and about 12°N off the magnetic equator in 1974 and at 35°E on the magnetic equator in

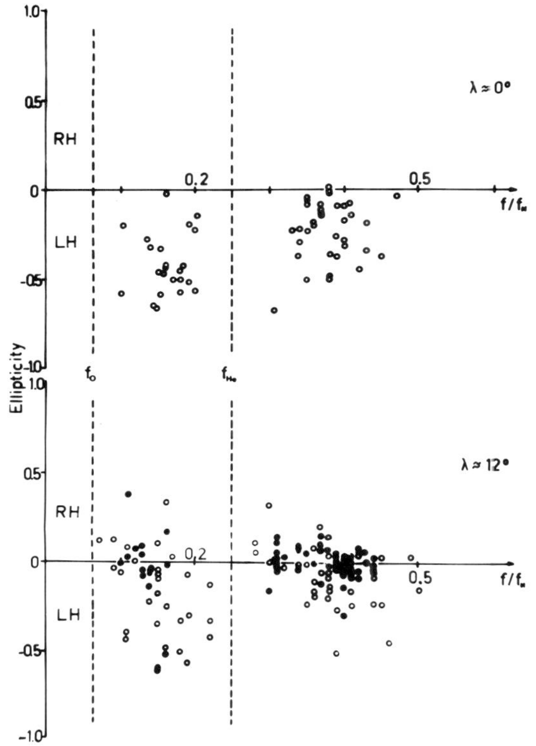

Fig. 4. Wave ellipticity variation with normalized frequency over the data recording intervals indicated in Figure 1. The solid data points indicate the first two months of data off the equator.

1975. GEOS-1 did not achieve synchronous orbit and observations were essentially off the equator whereas GEOS-2 was truly geostationary and situated on the equator. With these orbits both the ATS and GEOS experiments were capable of detecting on- and off-equator wave phenomena. As it turns out this was an important aspect of the studies since wave regime properties in the two different locations are significantly different.

At synchronous orbit under conditions of moderate magnetic activity spacecraft are typically situated in the plasma trough region from local midnight until midafternoon and within the bulge region of the plasmasphere for a few hours between afternoon and midnight (Higel and Lei, 1984). The midtime position of the bulge varies between 14 and 21 hr LT with a bulge time width of between 1 and 4 hr. In general, data recorded prior to noon and after midnight will be associated with the plasma trough while midafternoon to midevening data will be recorded in the plasmasphere. The occurrences of ICW's at ATS-6 both on the equator (λ = 0°) and off the equator (λ = 12°) are shown in Figure 1. Waves occur between about 09-23 hr LT indicating a predominant plasmasphere

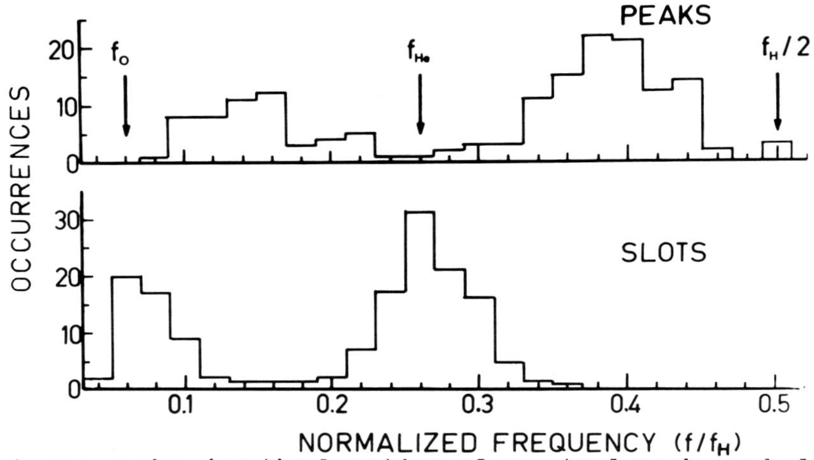

Fig. 5. Histograms showing the location of spectral peaks and slots with respect to f_O and f_{He} (Fraser and McPherron, 1982).

bulge region association. Some of the waves occurring prior to noon may be associated with detached plasma regions (Chappell, 1972).

Two distinct regimes of waves have been identified by ATS and GEOS and their spectral properties are illustrated in Figure 2. The emission on the right is a typically left-hand (LH) polarized transverse wave superimposed on a longer period wave with a compressional component. The emission on the left, at frequencies above 0.5 Hz is predominantly azimuthal (D) and essentially linearly polarized. The accompanying decrease in the H component is often seen and is probably a diamagnetic effect associated with the presence of keV protons. ATS-6 data is displayed in the V-D-H coordinate system where V is radially outwards from the centre of the earth, D is azimuthally east and H is antiparallel to the dipole axis of the Earth. The lower frequency LH waves are observed both on and off the equator and correspond to the shaded areas in Figure 1. The higher frequency azimuthally polarized events predominate off the equator.

Power spectra and polarization parameters for wave events seen on and off the equator are plotted in Figure 3 as a function of frequency. Four parameters are plotted: the trace of the spectral matrix (T), and the major (X), intermediate (Y) and minor (Z) eigen values (McPherron et al., 1972). Polarization parameters are also plotted with the ellipticity defined as the ratio of the minor to major axis of the ellipse with +1 for right circular polarization, -1 for left circular polarization and zero for linear polarization. θ_X and ϕ_X are major eigenvector polar and azimuth angles respectively with respect to the V-D-H system. The emission on July 13, 1974 in Figure 3, recorded off the equator is linearly polarized over the bandwidth 0.70-0.83 Hz. However, the August 29, 1975 event recorded on the equator is LH polarized below and above the equatorial He^+ cyclotron frequency (f_{He}). It has a minimum in spectral power density at f_{He}. This minimum

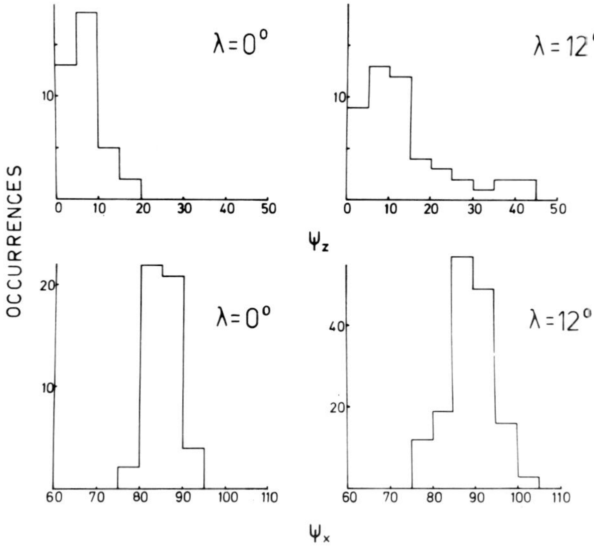

Fig. 6. Distributions of wave normal angle ψ_Z and the polarization ellipse major axis angle to the ambient field ψ_X at ATS-6 on and off the equator.

constitutes a slot in the wave spectrum which is direct evidence of the presence of thermal He^+ in the magnetospheric plasma (Mauk et al., 1981; Mauk, 1982; 1983) and has been well documented in GEOS results (Young et al., 1981; Roux et al., 1982).

Off-equator Poynting flux measurements by Mauk and McPherron (1980) show that ICW's are in fact travelling as opposed to standing waves and propagate away from the equator. It is therefore of interest to directly compare the on and off equator wave properties. Figure 4 shows polarization ellipticity plotted as a function of wave frequency normalized to the equatorial proton cyclotron frequency (f_H). The $\lambda = 12^o$ data are mapped back to the equator using the Olsen-Pfitzer 1977 model. The He^+ slot is apparent in both plots by the absence of data points around f_{He}. For the $\lambda = 0^o$ data most waves are LH polarized with some almost linear cases observed. Off the equator the wave regime above f_{He} changes character to almost completely linear polarization. Waves are considered linear for $|e| < 0.2$. Below f_{He} the waves remain LH with the exception of a small linear group in the frequency range (0.1-0.2) f_H, which is just above the oxygen cyclotron frequency (f_O). These linearly polarized waves observed off the equator indicate the presence of O^+ in the thermal plasma (Fraser and McPherron, 1982). O^+ with concentrations comparable to or greater than He^+ has been observed in the thermal plasma at synchronous orbit (Geiss et al., 1978). Unfortunately the GEOS spacecrafts spin at a frequency of 0.17 Hz which is in the f_O frequency band and the S-300 wave experiment cannot normally observe O^+ effects. Further support for the presence of O^+ propagation effects is shown in Figure 5 where off-equator ($\lambda = 12^o$) ATS-6 distributions of the mean normalized wave frequencies of spectral slots and

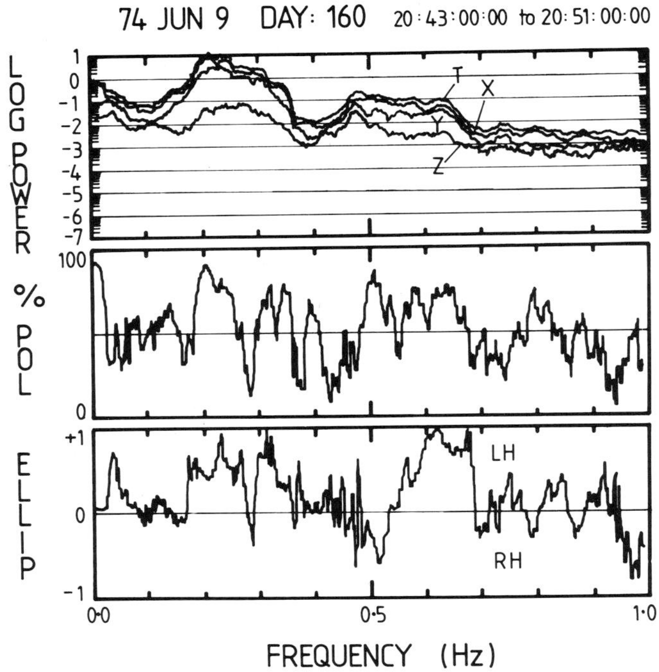

Fig. 7. ICW spectra recorded off the equator showing the spectral slot and a polarization reversal at the satellite (f_{xo} = 0.54 Hz). Note positive ellipticity indicates LH polarization and negative RH polarization.

peaks are plotted. Equatorial frequencies f_O and f_{He} are indicated and only data below $f_H/2$ and with $|e| < 0.2$ are plotted. Spectral slots occur only in association with a heavy ion cyclotron frequency whereas peaks are seen only between two cyclotron frequencies.

Orientation of the polarization ellipse with respect to the d.c. magnetic field is important in establishing wave propagation characteristics. The distributions of ψ_X, the angle between the ambient field and the major polarization axis (X), and ψ_Z, the angle between the ambient field and the normal to the plane of the ellipse (Z), are plotted in Figure 6. ψ_Z is indicative of the wave normal direction and shows that waves with $|e| \geqslant 0.2$ observed both on and off the equator are essentially aligned along the ambient field at angles less than 15°. For linear polarization ($|e| < 0.2$) the plane of polarization degenerates into a line and ψ_Z cannot be determined. In the bottom panels the polarization ellipse major axis is shown to be normal to the field at 90° as would be expected for transverse waves. ψ_X cannot be determined for circular polarization and only data with $|e| < 0.5$ are included.

It has been pointed out that waves observed off the equator often show linear polarization at frequencies above f_{He} (Figures 3 and 4). There are also many emissions in which a polarization reversal is observed above f_{He}. Figure 7 illustrates an event with a polarization reversal with increasing frequency from RH to LH at 0.54 Hz. The spectral slot is seen between 0.36 Hz and 0.46 Hz and wave polarization at frequencies below the slot is LH.

The properties described indicate that electromagnetic ICW's are transverse waves propagating at small wave normal angles. Both ATS-6 and GEOS data show that on the equator waves of all frequencies are predominantly LH polarized while off the equator polarization is LH below f_{He} and mixed but mainly linear above f_{He} with polarization reversals often seen. Many of these properties are consequences of wave propagation in a multicomponent magnetospheric plasma.

3. ION CYCLOTRON WAVE PROPAGATION

For wave propagation in a cold, infinite, uniform, homogeneous, collisionless plasma the dispersion relation (Stix, 1962; Smith and Brice, 1964; Gurnett et al., 1965; Booker, 1984) is given by

$$An^4 - Bn^2 + C = 0 \tag{1}$$

$$A = S \sin^2\theta + P \cos^2\theta$$

$$B = RL \sin^2\theta + PS\,(1 + \cos^2\theta)$$

$$C = PRL$$

θ is the angle between the static magnetic field and the wave normal direction.

The wave polarization ρ is given by

$$\rho = i\frac{B_y}{B_x} = \frac{iDn^2\cos^2\theta}{Sn^2 - RL} \tag{2}$$

where

$$S = 0.5\,(R+L) \quad , \quad D = 0.5\,(R-L)$$

For propagation in a three ion H^+, He^+ and O^+ plasma with relative concentrations α, β and γ respectively ($\alpha + \beta + \gamma = 1$) the R and L parameters are given by

$$\begin{matrix} R \\ L \end{matrix} = 1 + \frac{f_{Ne}^2}{f_{He}^2} \pm \frac{f_{Ne}^2}{f_H f_{He}}\left[\frac{1}{\Lambda} - \frac{\alpha}{\Lambda(1\pm\Lambda)} - \frac{\beta}{\Lambda(1\pm4\Lambda)} - \frac{\gamma}{\Lambda(1\pm16\Lambda)}\right] \tag{3}$$

where f_{Ne} is the electron plasma frequency and the normalized frequency

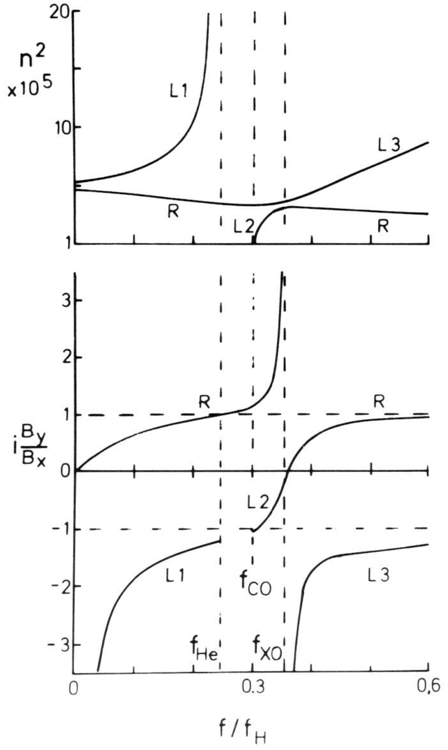

Fig. 3. Refractive index and ellipticity for ICW's propagating in a cold H^+-He^+ plasma with 7% He^+ and wave normal angle $\theta = 24^o$.

$\Lambda = f/f_H$.

The refractive index squared in the dispersion relation in Eq. 1 and the wave polarization from Eq. 2 are plotted as a function of normalized frequency in Figure 8. For simplicity only two ions are included with 93% H^+ and 7% He^+ ($\gamma = 0$). Non-parallel wave propagation is assumed with $\theta = 24^o$. It can be seen from the upper panel that the dispersion characteristics of the LH polarized wave mode are significantly modified by the presence of the He^+ ions and are broken into three modes (L1, L2 and L3). A resonance ($n^2 = \infty$) occurs at f_{He} and a new cutoff frequency ($n^2 = L = 0$, Eq. 3) is defined by $f_{co} = (1 + 3\beta)f_{He}$. Waves cannot propagate between f_{He} and f_{co} in the LH mode. Above f_{co} the dispersion curves for the LH and RH waves are no longer decoupled and a polarization reversal is seen at a crossover frequency $f_{xo} = (1 + 15\beta)^{\frac{1}{2}}f_{He}$ given by $D = 0$, since $\rho = 0$ in Eq. 2. This occurs for $\theta \neq 0$. If $\theta = 0$ a crossover of modes still occurs at f_{xo} but the polarization reversal does not (Young et al., 1981). At f_{xo} wave polarization is linear and azimuthal ($B_X = 0$) for the L3 branch and radial ($B_Y = 0$) for the L2 branch. For transverse propagation ($\theta \approx 90^o$) a hybrid resonance between the two ion species (H^+, He^+) occurs at the bi-ion frequency, $f_{bi} = [(1 + 3\beta)/(1 + 3\beta/4)]f_{He}$.

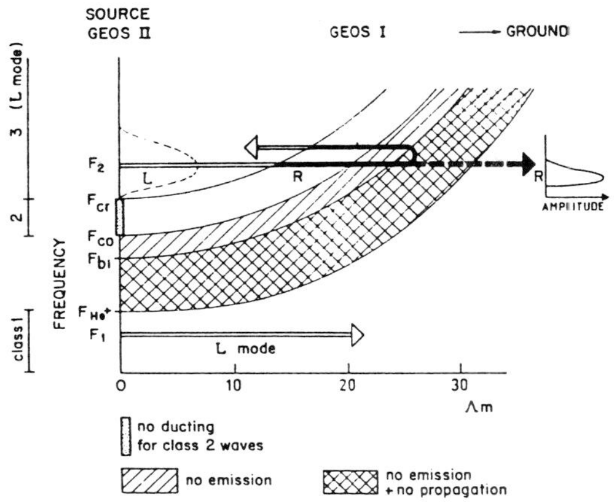

Fig. 9. Schematic diagram showing propagation effects from the equatorial region to higher geomagnetic latitudes including wave reflection and transmission by tunnelling. F_{cr} is the crossover frequency, F_{cf} the cutoff frequency, F_{bi} the bi-ion hybrid frequency, F_{He} the He^+ cyclotron frequency (Perraut et al., 1984).

The proton cyclotron instability produces LH polarized waves and the L modes shown in Figure 8 are therefore of greatest importance. Essential features of the waves have been well documented by Young et al., (1981), Roux et al., (1982) and Rauch and Roux (1982). The L1 mode propagation at $f < f_{He}$ is essentially unmodified by the presence of He^+ and is well guided along the field direction. The L2 waves ($f_{co} < f < f_{xo}$) are unguided and propagate away from the earth. The L3-R waves ($0 < f < f_H$) are unguided below f_{bi} and guided above it. According to Rauch and Roux (1982) the L3-R waves can be reflected near f_{bi} due to a rapid increase in θ.

Current models suggest that ICW amplification is initated off the equator (Mauk, 1982; Gendrin et al., 1984) and waves propagate with increasing f/f_H towards the equator and then with decreasing f/f_H away from the equator into the opposite hemisphere. A schematic model for a wave commencing at the equator and propagating with increasing latitude is shown in Figure 9. Due to the increasing magnetic field all critical frequencies gradually increase with latitude. Wave regimes L1, L2 and L3 are represented by classes 1, 2 and 3 respectively. It can be seen that L1 waves (F_1) are LH polarized both on and off the equator and their propagation mode is unaffected by the presence of He^+. This is supported by the observations for $f < f_{He}$ in Figure 4. The L3 waves (F_2) leave the equator as LH waves. However, as they propagate into regions of increasing magnetic field the normalized frequency f/f_H decreases and propagation in the L3 branch moves towards the crossover frequency f_{xo} (Figure 8) where it changes polarization from LH through linear azimuthal to RH. The R mode then propagates wave energy through the L wave stop band to higher latitudes. The R waves then encounter the bi-ion frequency and are reflected. This reflection eventually gives rise to waves bouncing between low latitude mirror points at $\theta \approx 25^{o}$ (Rauch and Roux, 1982). Thus, an off-equator observation may show LH, linear or RH waves depending on the location of the satellite with respect to f_{xo}. This can be seen in the off-equator polarization results for $f > f_{He}$ in Figure 4. The predominance of linear polarization here could indicate that the f_{xo} latitude is often close to the satellite. However, Rauch and Roux assume that this linear polarization results from propagation at large wave normal angles. Both mechanisms are probably operative but further study is required to ascertain their relative contributions.

The spectral slot seen in wave spectra near f_{He} is the result of two mechanisms. In addition to the L mode propagation stop band mechanism, non-linear off-resonant He^{+} cyclotron absorption also plays a role (Mauk et al., 1981; Mauk, 1982; 1983).

In addition to He^{+} there is a significant population of thermal O^{+} in the plasmapause region. Limited evidence is available suggesting that O^{+} propagation effects are seen in ATS-6 wave data thus supporting the statistical evidence in Figure 5. For O^{+} the f_O, f_{co} and f_{xo} frequencies occur in the $(0.05-0.15)f/f_H$ range (Fraser, 1982). Figure 4 indicates a small but significant linearly polarized group of waves above f_O at $(0.1-0.2)f/f_H$ which shows similar properties to the linear group above f_{He}. Polarization reversals have also been seen in this frequency range but further analysis is required to conclusively establish properties of the O^{+} effect. Recently Inhester et al. (1984) have observed LH polarized Pc2 waves on GEOS-2 with frequencies that are below but closely correlated with f_O.

Heavy ion effects on ICW propagation are also seen in the Io plasma torus of Jupiter. Here the heavy ion concentration exceeds that of the protons resulting in an extremely wide propagation stop band. Energy from obliquely propagating L mode ion cyclotron waves excited at intermediate latitudes reaches the equator by undergoing a polarization reversal to the R mode at the crossover frequency (Thorne and Moses, 1983).

4. PLASMA DIAGNOSTICS

It has been indicated earlier that three classes of critical frequencies are associated with a plasma containing two or three ions: the crossover, cutoff and bi-ion frequencies. One of each of these frequencies is associated with each heavy ion. In a three ion plasma knowledge of any two of these critical frequencies uniquely determines the relative concentrations α, β and γ (Shawhan and Gurnett, 1966). This technique, utilizing the two crossover frequencies, was invoked by Fraser (1982) to determine simultaneous He^{+} and O^{+} relative concentrations from an ATS-6 wave event. In the two ion (H^{+}, He^{+}) situation measurement of a single critical frequency determines α and β. The most accurately determined critical frequency from wave spectra is the crossover frequency defined by the polarization reversal ($\rho = 0$). In

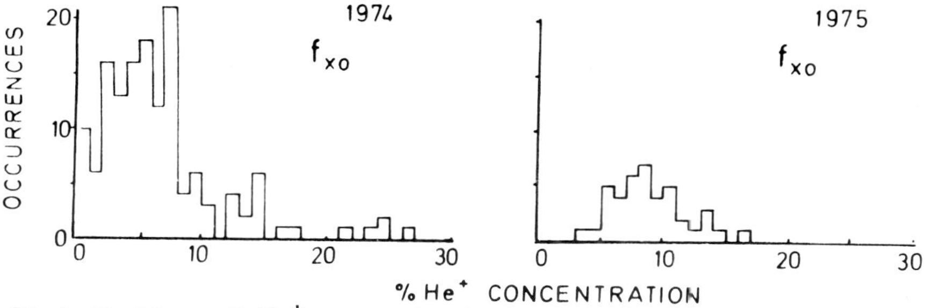

Fig. 10. Distribution of He^+ concentrations determined from the crossover frequency (f_{xo}) off the equator (1974), and on the equator (1975).

Eq. 2 this yields $\beta = (16\Lambda_{xo} - 1)/15$ where Λ_{xo} is the normalized crossover frequency. Direct spectral measurements may now be used to determine β. For example, in the wave event illustrated in Figure 7 $f_{xo} = 0.54$ Hz and $\Lambda_{xo} = 0.30$ for a 120 nT ambient magnetic field measured at the satellite. This yields a relative He^+ concentration of $\beta = 3\%$. The distribution of He^+ concentrations determined by measuring f_{xo} from ATS-6 ICW's observed off the equator in 1974 and on the equator in 1975 are shown in the histograms in Figure 10. He^+ concentrations are generally < 10%, a result which is in agreement with linear theoretical calculations of wave spectra (Roux et al., 1982). It also appears that very low He^+ concentrations are seen less frequently at the equator. There is still a lot of quantitative work to be done and it is anticipated that further understanding of wave propagation models will lead to the extension of diagnostic studies.

5. SATELLITE-GROUND CORRELATIONS

The direct comparison of ICW's at their source with groundbased observations is a crucial test of the global propagation processes. The first correlative study of individual ICW events at synchronous orbit with Pc1-2 pulsations in the conjugate region was undertaken by Bossen et al. (1976a,b) using ATS-1 magnetometer data. Large amplitude LH polarized waves with $f = (0.1-0.2)f_H$ were seen at synchronous orbit simultaneously with IPDP events on the ground. More recent studies show that almost all wave events observed at GEOS-1 and -2 with $f < f_{He}$ propagate to the ground while only half the waves with $f > f_{He}$ are seen at the ground (Perraut et al., 1984). The absence of waves on the ground at $f > f_{He}$ is attributed to the reflection of wave energy at the bi-ion frequency (Figure 9). The waves that do reach the ground often show lower frequencies than seen at the satellite. For example, in Figure 11(a) the emission at GEOS-1 peaks at 1.0 Hz while on the ground a lower frequency of 0.8 Hz is seen. Perraut et al. (1984) show that if He^+ concentrations are relatively low (< 5%) then wave tunnelling through the reflection region can occur for frequencies above f_{xo} (Figure 9). For wave transmission this region acts as a low pass filter thereby lowering the peak frequency seen at the ground. Poor conjugacy could also contribute to this effect.

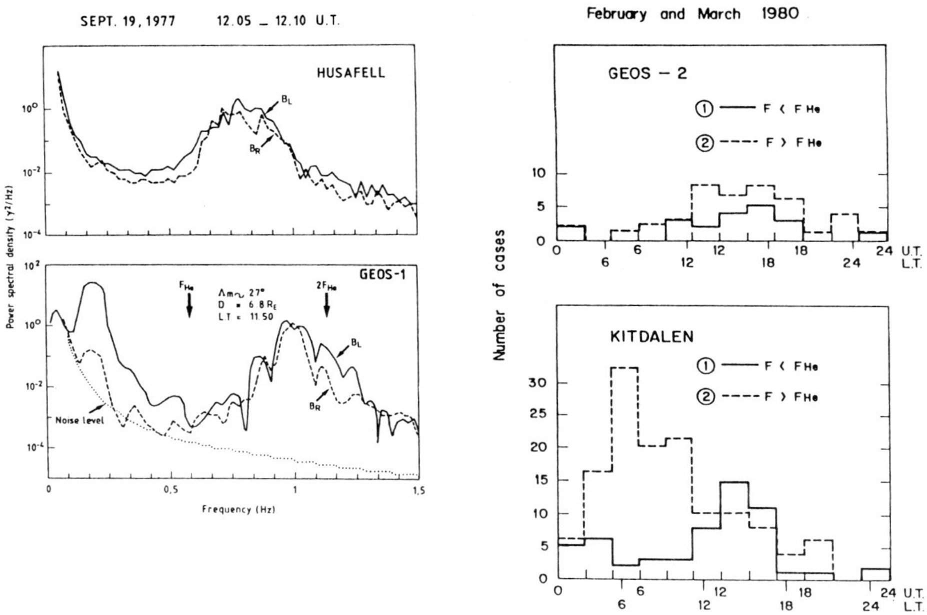

Fig. 11. (a) Power spectra for a strong ICW emission observed on GEOS-1 and linearly polarized above f_{He} with a spectral peak at 1 Hz. The corresponding emission on the ground peaks at 0.8 Hz. (b) Diurnal distribution of ICW events observed above and below f_{He} at GEOS-2 and on the ground (Perraut et al., 1984).

The diurnal variations in wave occurrence at synchronous orbit and at the conjugate ground station for simultaneous data are plotted in Figure 11(b). The GEOS-2 observations show daytime occurrence and are similar to those from ATS-6 shown in Figure 1. The waves with $f < f_{He}$ on the ground show a similar pattern in agreement with results of Bossen et al. (1976b). However, there is a large group of high frequency waves (up to 1 Hz) observed in the early morning hours. These are the structured emissions with bounce periods of the order of 1 mn and which propagate from lower latitudes via the ionospheric waveguide. This illustrates the fact that ground stations may observe Pc1-2 pulsations from a wide range of field lines whereas guided waves observed in space are confined to a single field line.

6. Pc1 WAVES OBSERVED ON THE GROUND

The waves discussed so far have been associated with the afternoon-evening bulge sector of the plasmasphere and plasma trough regions and are predominantly unstructured emissions (Section 2). The large group of structured pulsations observed on the ground are rarely seen at synchronous orbit (Figure 11(b)) since they are generated at lower altitudes in the nighttime and morning plasmapause-plasmasphere sectors. The He^+ propagation effects are not important here since f_{He} is generally above the wave frequency. In this case the propagation

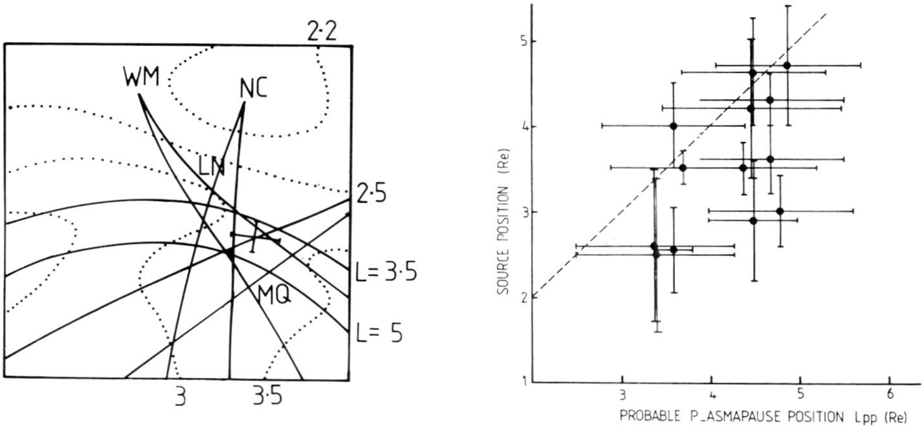

Fig. 12. (a) Polarization ellipse major axis projections from three stations (continuous lines) and source position from the time delay method on October 10, 1978, 1545-1710 hr UT. Dotted lines are contours of constant f_oF2 in MHz. Lpp = 3.6±0.4. (b) Measured source position plotted as a function of probable plasmapause position (Lpp). Dashed line represents sources on the plasmapause (Fraser et al., 1984).

mechanism is the simple bouncing wave packet phenomenon in the L1 mode. Although O^+ critical frequencies at these altitudes may occur in the Pc1-2 band no effects have been identified in structured pulsations seen on the ground. However, at higher latitudes Inhester et al. (1984) have observed unusually large amplitude Pc2 waves at 0.14 Hz on the ground and on GEOS-2 which have been interpreted as ICW's destabilized by high suprathermal O^+ concentrations.

It is only recently that serious attempts have been made to specifically locate the source of structured Pc1-2 using ground data. Four techniques have been employed and are summarized by Fraser et al. (1984), Fukunishi (1984) and Kemp and Fraser (1985). The latitude profile of the wave amplitude may be used to identify the source location as the L shell conjugate to the station observing the largest signal. Hayashi et al. (1981) and Baransky et al., (1981) used this technique at middle and high latitudes and on a statistical basis found the source of structured Pc1-2 to be near the plasmapause. On the other hand, unstructured emissions relate to the detached plasma regions. Other techniques use the dispersion characteristics of ICW's bouncing between hemispheres in a simple H^+-e^- plasma (Watanabe, 1965) and the wave packet propagation time along field lines (Feygin et al., 1979).

The fourth method is a more definitive technique to determine the exit region of Pc1 wave energy from the magnetosphere into the ionosphere, using the ionospheric waveguide propagation properties applied to a network of stations situated at low latitudes. Source determinations for individual pulsation events have been made using time delay, ULF goniometer and polarization techniques (Fraser et al., 1984; Webster and Fraser, 1984; 1985; Kemp and Fraser, 1985). In this method the projection of the major axis of linearly polarized waves in the plane of the earth's surface indicates the direction of arrival of the signal in the waveguide. Ray paths are plotted taking into account the statistical spread in major axis azimuth angles and refraction from f_oF2 gradients as shown in Figure 12(a). The overlap region of the three wedge shaped projections defines a small source region. The geometry of the elliptical wavefront expanding horizontally outwards from the source is determined from the full correlation technique of Briggs (1976) used in ionospheric drift analysis. Taking into account the wavefront geometry in conjunction with interstation differences in Pc1 signal arrival times calculated by correlation and cross-spectral techniques (typically 0-3 s) allows a unique source location to be determined using iteration methods. The cross in Figure 12(a) indicates a particular Pc1 emission source and the associated errors in latitude and longitude. It can be seen that the small source region determined by the polarization projections overlaps the error range of the time delay determined source. The probable plasmapause position determined by the statistical method of Roth and Orr (1975) is Lpp = 3.6 ± 0.4. The location of Pc1 pulsation source positions for 13 events is plotted in Figure 12(b) against the corresponding statistical plasmapause position. Eight of the measured sources overlap the plasmapause position with the mean position of only one event being significantly outside by 0.5 R_e. The remaining five positions are inside by up to 2 R_e.

7. CONCLUSION

It has been shown how the detailed measurements of ICW properties in space and their ground signatures as Pc1-2 pulsations have contributed to our knowledge of the physical processes involved in the propagation of ULF plasma waves. ICW emissions not only involve interaction with energetic protons but are also associated with changes in the cold plasma composition, heavy ion heating, and electron acceleration. The propagation of ICW's in a simple H^+, e^- plasma is no longer valid and the profound effects produced by the presence of small concentrations of He^+ and O^+ in the cold plasma must be taken into consideration. Heavy ions give rise to spectral slots, polarization reversals and wave reflection in the magnetosphere. All these properties can be explained by simple multicomponent cold plasma theory. As a consequence diagnostic techniques provide information on He^+ concentrations in the presence of ICW's. The increased knowledge of magnetospheric wave processes has rekindled interest in groundbased observations. The transmission of waves from the magnetosphere to the ground is now better understood.

Much work remains to be done in studying temporal variations in space and on the ground, over both long time scales of hours with respect to plasmapause dynamics and short time scales of minutes with respect to modulation effects on wave processes.

ACKNOWLEDGEMENTS

Thanks are due to R.L. McPherron for his many contributions to the ATS-6 studies and to D.J. Webster and W.J. Kemp for their work on ground-based data. This research was supported at the University of Newcastle, N.S.W. by the Australian Research Grants Scheme and at UCLA by contracts ONR N00014-75-C-0396 and NAS 5-11679 and the University of California.

REFERENCES

Baransky, L., Golikov, Yu., Feygin, F., Harchenko, I., Kangas, J., and Pikkarainen, T.: 1981, J. Atmos. Terr. Phys., **43**, 875.
Booker, H.G.: 1984, Cold Plasma Waves, M. Nijhoff, Dordrecht, 349pp.
Bossen, M., McPherron, R.L., and Russell, C.T.: 1976a, J. Geophys. Res., **81**, 6083.
Bossen, M., McPherron, R.L., and Russell, C.T.: 1976b, J. Atmos Terr. Phys., **38**, 1157.
Briggs, B.H.: 1976, Radio Sci., **11**, 817.
Chappell, C.R.: 1972, Rev. Geophys. Space Phys., **10**, 951.
Cornwall, J.M.: 1966, J. Geophys. Res., **71**, 2185.
Feygin, F.Z., Kurchashov, J.P., Troitskaya, V.A., Fligel, D.S., and Dobes, K.: 1979, Planet. Space Sci., **27**, 151.
Fraser, B.J.: 1982, Planet. Space Sci., **30**, 1229.
Fraser, B.J., and McPherron R.L.: 1982, J. Geophys. Res., **87**, 4560.
Fraser, B.J., Kemp, W.J., and Webster, D.J.: 1984, Proc. Conf. Achievements of the IMS, ESA SP-217, p609.
Fraser-Smith, A.C.: 1982, Rev. Geophys. Space Phys., **20**, 497.
Fukunishi, H.: 1984, Proc. Conf. Achievements of the IMS, ESA SP-217, p437.
Fukunishi, H., Toya, T., Koike, K., Kuwashima, K., and Kawamura, M.: 1981, J. Geophys. Res., **86**, 9029.
Geiss, J., Balsinger, H., Eberhardt, P., Walker, H.P., Weber, L., and Young, D.T.: 1978, Space Sci. Rev., **22**, 537.
Gendrin, R.: 1980, in Exploration of the Polar Upper Atmosphere, C.S. Deehr and J.A. Holtet eds., D. Reidel, Dordrecht, p337.
Gendrin, R.: 1983a, in High Latitude Space Plasma Physics, B. Hultqvist and T. Hagfors eds., Plenum N.Y., p415.
Gendrin, R.: 1983b, Space Sci. Rev., **34**, 271.
Gendrin, R., Ashour-Abdalla, M., Omura, Y., and Quest, K.: 1984, J. Geophys. Res., **89**, 9119.
Gurnett, D.A., Shawhan, S.D., Brice, N.M., and Smith, R.L.: 1965, J. Geophys. Res., **70**, 1665.
Hayashi, K., Kokubun, T., Oguti, T., Tsuruda, K., Machida, S., Kitamura, T., Saka, O., and Watanbe, T.: 1981, Can. J. Phys., **59**, 1097.
Higel, B., and Lei, W.: 1984, J. Geophys. Res., **89**, 1583.
Inhester, B., Wedeken, U., Korth, A., Perraut, S., and Stokholm, M.: 1984, J. Geophys., **55**, 134.

Jacobs, J.A.: 1970, Geomagnetic Micropulsations Springer-Verlag, Berlin, 176pp.
Kemp, W.J., and Fraser, B.J.: 1985, Planet. Space Sci., submitted.
Kennel, C.F., and Petschek, H.E.: 1966, J. Geophys. Res., **71**, 1.
Korth, A., Kremser, G., Perraut, S., and Roux, A.: 1984, Planet. Space Sci., **32**, 1393.
Liemohn, H.B.: 1967, J. Geophys. Res, **72**, 39.
Manchester, R.N.: 1968, J. Geophys. Res., **75**, 3549.
Mauk, B.H.: 1982, J. Geophys. Res., **87**, 9107.
Mauk, B.H.: 1983, Geophys. Res. Lett., **10**, 635.
Mauk, B.H. and McIlwain, C.E., and McPherron, R.L.: 1981, Geophys. Res. Lett., **8**, 103.
Mauk, B.H., and McPherron, R.L.: 1980, Phys. Fluids **23**, 2111.
McPherron, R.L., Russell, C.T., and Coleman, P.J.: 1972, Space Sci. Rev., **13**, 411.
Orr, D.: 1973, J. Atmos. Terr. Phys., **27**, 943.
Perraut, S.: 1982, Planet. Space Sci., **30**, 1219.
Perraut, S., Gendrin, R., Roux, A., and de Villedary, C.: 1978, Space Sci. Rev., **22**, 347.
Perraut, S., Gendrin R., Roux, A., and de Villedary, C.: 1984, J. Geophys. Res., **89**, 195.
Rauch, J.L., and Roux, A.: 1982, J. Geophys. Res., **87**, 8191.
Roth, B., and Orr, D.: 1975, Planet. Space Sci., **23**, 993.
Roux, A., Perraut, S., de Villedary, C., Gendrin, R., Kremser, G., Korth, A., and Young, D.T.: 1982, J. Geophys. Res., **87**, 8174.
Shawhan, S.D., and Gurnett, D.A.: 1966, J. Geophys. Res., **71**, 47.
Smith, R.L., and Brice, N.: 1964, J. Geophys. Res., **69**, 5029.
Stix, T.H.: 1962, The Theory of Plasma Waves, McGraw Hill, N.Y., 283pp.
Taylor, W.W.L., and Lyons, L.R.: 1976, J. Geophys. Res., **81**, 6177.
Thorne, R.L., and Moses, J.: 1983, Geophys. Res. Lett., **10**, 631.
Troitskaya, V.A.: 1967, in Solar Terrestrial Physics, J.W. King and W.S. Newman eds., Academic Press, London, p213.
Watanabe, T.: 1965, J. Geophys. Res., **70**, 5839.
Webster, D.J., and Fraser, B.J.: 1984, Planet. Space Sci., **32**, 935.
Webster, D.J., and Fraser, B.J.: 1985, Planet. Space Sci., to appear.
Young, D.T., Perraut, S., Roux, A., de Villedary, C., Gendrin, R., Korth, A., Kremser, G., and Jones D.: 1981, J. Geophys. Res., **86**, 6755.

DYNAMICS OF THE EARTH'S RING CURRENT: THEORY AND OBSERVATION

D. J. Williams
The Johns Hopkins University
Applied Physics Laboratory
Johns Hopkins Road
Laurel, Maryland 20707
U.S.A.

ABSTRACT. The development of currents within an arbitrary distribution of particles trapped in the geomagnetic field is described. These currents combine to form the earth's ring current and thus are responsible for the worldwide depressions of surface magnetic field strength during periods of magnetic activity known as magnetic storms. Following a brief review of trapped particle motion in magnetic fields, ring current development is described and presented in terms of basic field and particle distribution parameters. Experimental observations then are presented and discussed within the theoretical framework developed earlier. New results are presented which, in the area of composition and charge state observations, hold high promise in solving many long standing ring current problems. Finally, available experimental results will be used to assess our present understanding as to ring current sources, generation, and dissipation.

1. INTRODUCTION

The Earth's ring current is the current system responsible for observed worldwide decreases in the surface magnetic field, known as geomagnetic storms. It is an excellent example of a general class of basic plasma physics problems which can be pursued within the Earth's environment, namely the creation and dynamics of electrical currents in magnetized plasmas of cosmological interest.

Since the early pioneering studies of Chapman and Ferraro (1931, 1932, 1933), theoretical and experimental studies have established the location of and driving force for the Earth's ring current system (Alfvén, 1955; Parker, 1957; Singer, 1957; Dessler and Parker, 1959; Akasofu and Chapman, 1961; Apel et al., 1962; Hoffman and Bracken, 1965, 1967; Frank, 1967a, b; Smith et al., 1979; Williams, 1980, 1981, 1983). This work has shown that the ring current is generated by the motion of charged particles trapped in the geomagnetic field. However, the source, energization, and injection processes for these particles remain to be determined.

Space Science Reviews **42** (1985) 375–396. 0038–6308/85.15

The most useful approach in tracking charged particles in the Earth's magnetosphere has been the guiding center approximation and subsequent development of adiabatic invariant concepts (Alfvén, 1950; Spitzer, 1956; Northrup, 1963). In the guiding center approximation, the instantaneous position of a charged particle moving in a magnetic field is broken down into its circular motion about the magnetic field and the displacement of the center of this motion, the guiding center, with respect to the field. This allows charged particle motion in the earth's magnetic field, the magnetosphere, to be separated into three components: gyration about a field line, bounce back and forth along a field line between mirror points, and a slow longitudinal drift around the earth. These motions are illustrated in Figure 1. Although these motions are not separate in a strict sense, the large differences in their respective time scales makes a mathematical separation possible and leads directly to consideration of the adiabatic invariants.

Adiabatic invariants are those quantities which may be considered constants of the particle's motion under the condition that magnetic field variations are small compared to the time and spatial scales associated with the particle's motion. There are three such invariants applicable to particles trapped in the geomagnetic field, each associated with one of the characteristic motions illustrated in Figure 1, gyration, bounce, and drift.

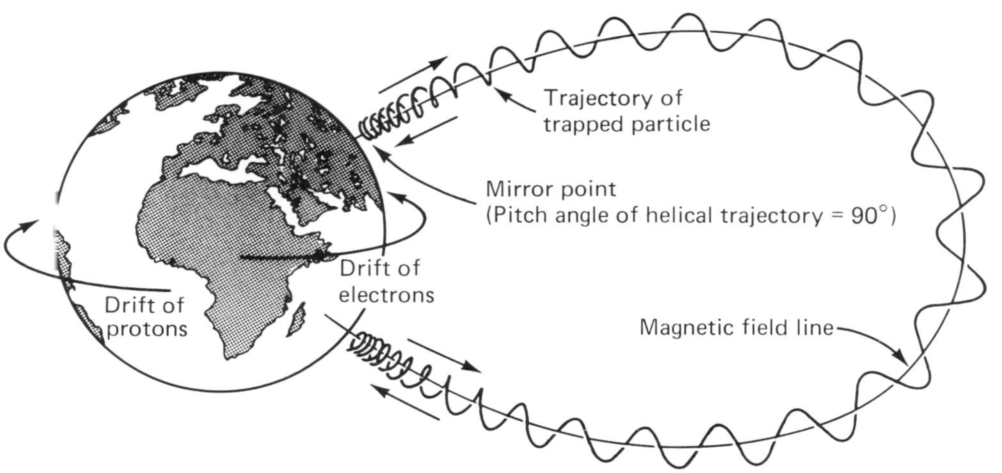

Figure 1. The motion of a charged particle trapped in the earth's magnetic field.

The first adiabatic invariant is associated with the particle's gyration about the magnetic field and can be estimated by calculating the magnetic moment produced by a current equivalent to the particle's circular motion about the field, $i=qv_\perp/2\pi\rho$ where $\rho=m\gamma_r v_\perp/qB$ is the gyroradius. Here q is the charge, $v_\perp$ is the particle velocity perpendicular to the magnetic field, B is the magnetic field magnitude, m is the particle rest mass, and $\gamma_r = (1 - v^2/c^2)^{-1/2}$. We have

$$\mu_B = iA = \frac{mv_\perp^2}{2B}\gamma_r \tag{1}$$

If magnetic field changes are small over a gyroradius and gyroperiod, then $\mu_B \simeq$ constant. Formal calculations of the effects on the particle trajectory of small temporal and spatial changes in B show that the correct relativistic adiabatic invariant associated with the particle's gyration about the field line is $\mu = p_\perp^2/B$ where $p_\perp$ is the particle's transverse momentum (Alfvén, 1950; Alfvén and Fälthammar, 1963; Spitzer, 1956; Roederer, 1970). We see that $\gamma_r \mu_B = \mu =$ constant and that μ_B is not a relativistically correct invariant. However, non-relativistically μ_B is constant and historically has been identified as the first adiabatic invariant:

$$\mu_B = \frac{mv_\perp^2}{2B} \equiv \frac{K_\perp}{B} = \text{constant (non-relativistic)} \tag{2}$$

From (2) we obtain the mirror equation describing the reflection of a charged particle in a region of converging magnetic field. In a static field, the particle's kinetic energy is constant and we have

$$\mu_B = \frac{K\sin^2\alpha}{B} = \text{constant} \tag{3}$$

and

$$\frac{\sin^2\alpha_1}{B_1} = \frac{\sin^2\alpha_2}{B_2} \tag{4}$$

where α is the particle pitch angle (angle between $\overline{B}$ and the particle's velocity vector). As a particle moves into a converging magnetic field, α increases to $\pi/2$ at which point the particle reverses its motion and travels back along the field line in the direction from which it came. The gradient of $\overline{B}$ parallel to the field direction produces the force necessary to drive the particle (of moment μ_B) away from regions of increasing field strength.

If magnetic field variations are small compared to the time scale of bounce motion (Figure 1), a second adiabatic invariant can be defined from the action integral associated with the bounce motion (Northrup and Teller, 1960; Alfvén and Fälthammar, 1963). It is given by

$$J = \int p_\parallel d\ell = 2\int_{M_1}^{M_2} \gamma_r m v_\parallel d\ell \tag{5}$$

where the integral is taken along the field line between the two conjugate mirror points M_1 and M_2. Using $v_{\parallel} = v(1-\sin^2\alpha)^{1/2}$ and (3), we can rewrite (4) to define the integral invariant I as

$$I \equiv \frac{J}{2mv\gamma_r} = \int_{M_1}^{M_2} \left(1 - \frac{B}{B_M}\right)^{1/2} d\ell \qquad (6)$$

The convenience of using I as the invariant quantity is that it is conserved in a static magnetic field and depends only on the field configuration. If $\partial B/\partial t \neq 0$, the actual invariant J, given by (4), must be used to describe the particle's motion.

The third and last adiabatic invariant is associated with the particle's drift in azimuth in the geomagnetic field (Figure 1). This drift is due to magnetic field gradients and field line curvature and in an ideal dipole field will produce drift surfaces which are simply figures of revolution of lines of force around the dipole axis. The drift velocity produced by these effects can be expressed as

$$\bar{V}_{DB} = \frac{K}{qB^2}(1+\cos^2\alpha)\bar{B}\times\bar{\nabla}B \qquad (7)$$

The third invariant is the magnetic flux linked in the particle's drift,

$$\Phi_M \equiv \int B ds \qquad (8)$$

Because of the large spatial and time scales of the particle's drift in azimuth, Φ_M is the most easily violated of the three adiabatic invariants. Table I shows characteristic times of charged particle motions in the geomagnetic field which are associated with the adiabatic invariants (1st, gyration; 2nd, bounce; 3rd, drift).

2. PARTICLE DISTRIBUTIONS AND CURRENTS

From the preceding discussion of the motion of particles trapped in the geomagnetic field, we may develop a description of the effects of such particles on the field itself and the establishment of currents within the trapped particle distribution. Let us begin simply with a single charged particle trapped in a dipole field at the magnetic equator and with a pitch angle of 90° (Figure 2). First consider the magnetic field effect at the origin of the dipole due to the particle drift velocity caused by the main dipole field. At the magnetic equator $\underline{B}=M/r^3 \lambda$ where λ is a unit vector in magnetic latitude and M is the dipole moment. With this and $\alpha=\pi/2$ we have from equation (7)

$$\bar{V}_{DB} = -\frac{3K}{qM} r^2 \hat{\Psi}$$

Table I

CHARACTERISTIC TIMES ASSOCIATED WITH PARTICLES TRAPPED IN THE EARTH'S MAGNETIC FIELD FOR $R_0 = 4$ AND $\alpha_0 = \pi/2$

		τ_c (seconds) $\propto R_0^3$	τ_B (seconds) $\propto R_0$	τ_{DB} (hours) $\propto \frac{1}{R_0}$
Electrons	1 keV	7.4×10^{-5}	4.0	184
	10	7.5×10^{-5}	1.3	18.4
	100	8.8×10^{-5}	0.46	1.5
	1000	2.2×10^{-4}	0.27	0.061
Protons	1 keV	0.14	172	184
	10	0.14	54.5	18.4
	100	0.14	17.2	1.8
	1000	0.14	5.45	0.18

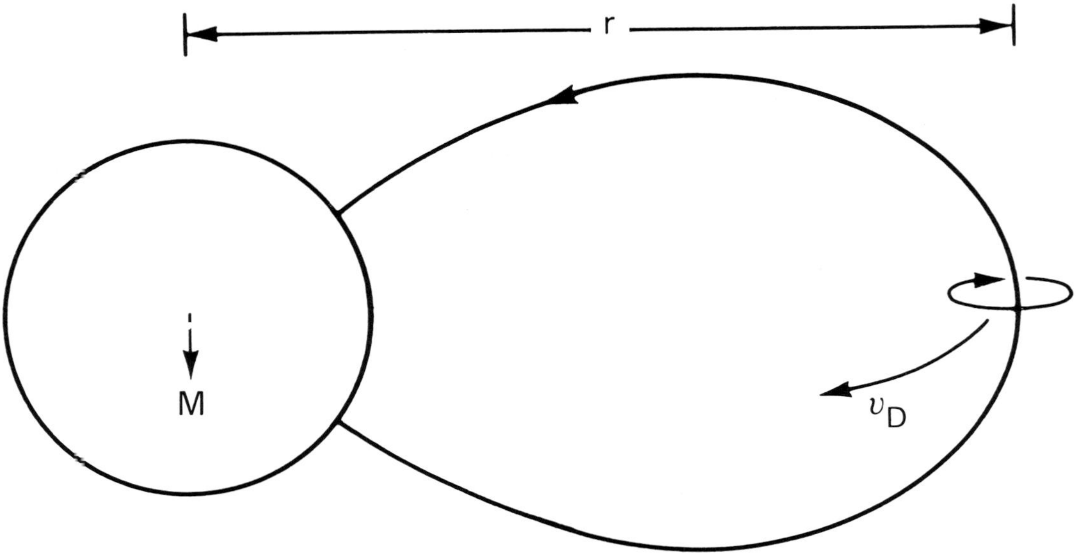

Figure 2. Earth's dipole field. Field line direction and proton gyration and drift are indicated.

where $\hat{\Psi}$ is a unit vector in magnetic longitude. With $i=qV_{DB}/2\pi r$, an application of the Biot-Savat law yields for the field, B_D, at the origin due to the particles' drift

$$B_D = \frac{\mu_o i}{4\pi} \int_o^{2\pi r} \frac{ds}{r^2} = -\frac{\mu_o i}{2r} = -\frac{\mu_o}{4\pi}\frac{3K}{M} \tag{9}$$

The minus sign indicates that B_D is opposite to the main dipole direction near the origin. Next consider the magnetic effect at the origin due to the particle's motion around the field line. Here the particle can be considered a small dipole of moment $\mu_B = K/B$ (equation (3)) giving a magnetic field at the origin of

$$B_\mu = \frac{\mu_o}{4\pi}\frac{\mu_B}{r^3} = \frac{\mu_o}{4\pi}\frac{K}{B} \tag{10}$$

Thus the total field perturbation at the origin due to a single particle trapped at the magnetic equator is $\Delta B = B_D + B_\mu = -(\mu_o K/2\pi M)$.

Considering the earth's field to be nearly dipolar we have at the surface $B_s = M/r_s^3$ and $\Delta B/B_s = -(\mu_o K/2\pi B_s^2 r_s^3)$. Noting that the total energy of the earth's dipole field above the earth's surface ($r=r_s$) is $U_M=(4\pi B_s^2 r_s^3/3\mu_o)$ we get

$$\frac{\Delta B}{B_s} = -\frac{2K}{3U_M} \tag{11}$$

for the magnetic perturbation at the center of the earth due to a single trapped particle at the geomagnetic equator. This is a general result known as the Dessler-Parker relation (Dessler and Parker, 1959) and has been shown to be accurate for any distribution of trapped particles in the earth's magnetic field if K is the total kinetic energy of the distribution (Sckopke, 1966; Carovillano and Maguire, 1968; Olbert et al., 1968; Carovillano and Siscoe, 1973). It requires $\sim 4(10)^{13}$ joules of particle kinetic energy to produce a surface magnetic field depression of $\sim 10^{-5}$ G. In cases where the particle kinetic energy density is comparable to or greater than the magnetic field energy density, the accuracy of this relation deteriorates by about a factor of 2.

By summing and averaging over individual particle motions in a magnetic field, Parker (1957) established a hydromagnetic formalism in which the currents due to the particle distribution are expressed in terms of particle pressures normal, $P_\perp$, and parallel, $P_\parallel$, to the magnetic field line. Defining

$$P_\perp = \frac{1}{2}\int f(v)mv^2 \sin^2 \alpha \, d^3v; \quad P_\parallel = \int f(v,\alpha)mv^2 \cos^2 \alpha \, d^3v \quad (12)$$

the current due only to gradient and curvature drifts driven by the magnetic field can be expressed as

$$\bar{i}_D = \frac{\bar{B}}{B^2} \times \left(\frac{P_\perp}{B}\,\bar{\nabla}B + \frac{P_\parallel}{B^2}\,(\bar{B}\cdot\bar{\nabla})B\right). \quad (13)$$

The first term on the right-hand side of (13) is due to the particle drift driven by the field gradient, and the second term is due to field line curvature driven drift.

The current due only to gyration effects within the particle distribution is

$$\bar{i}_c = \frac{\bar{B}}{B^2} \times \left(\bar{\nabla}P_\perp - \frac{P_\perp}{B}\,\bar{\nabla}B - \frac{P_\perp}{B^2}\,(\bar{B}\cdot\bar{\nabla})B\right). \quad (14)$$

The three terms on the right-hand side of (14) represent, respectively, currents caused by the particle pressure gradient, the magnetic field gradient, and magnetic field line curvature.

The total current is

$$\bar{i} = \bar{i}_D + \bar{i}_c = \frac{\bar{B}}{B^2} \times \left(\bar{\nabla}P_\perp + \frac{P_\parallel - P_\perp}{B^2}\,(\bar{B}\cdot\bar{\nabla})B\right). \quad (15)$$

Note that the drift and gyration current terms driven by the magnetic field gradient are equal and opposite so that the resultant current due to a particle distribution trapped in a magnetic field does not depend on gradients in that field. In addition, for the case of an isotropic distribution ($P_\parallel = P_\perp$), magnetic field inhomogeneites play no role in establishing currents in the trapped particle population. In this case the current system is established solely by particle pressure gradients.

Of the three pressure terms on the right-hand side of (15), the first term is due to a particle pressure gradient, the second term is due to field line curvature driven drift, and the third term is due to the crowding of gyro-orbits inside a curved field line. The first and third terms are illustrated in Figure 3. In the next section we shall discuss the importance of these terms for measured ring current distributions.

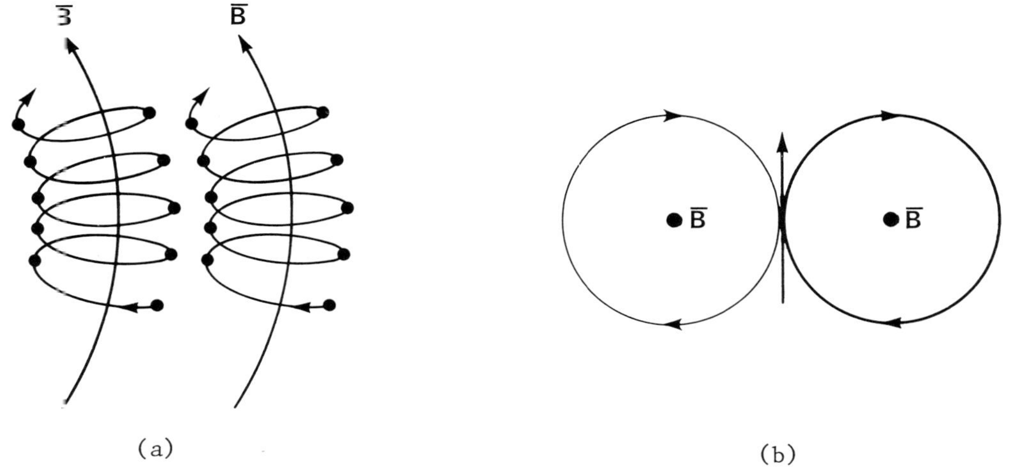

Figure 3. Source of current due to gyration effects on curved field lines. (a) Eastward current between the two curved field lines is due to the crowding of gyro-orbits on the concave side of the lines; (b) source of current due to particle pressure gradient. Particle intensity is indicated by darkness of gyration circles. It is seen that a positive gradient (intensity increasing away from earth) leads to an eastward current.

The magnetic field produced by the above current expression changes the magnetic field geometry in which the current was calculated, thus in turn changing the current. Therefore an iterative process is required to converge on an accurate evaluation of the field perturbations produced by a given particle distribution (Akasofu and Chapman, 1961; Hoffman and Bracken, 1967). Figure 4 shows the results of using particle measurements plus an iterative process in fitting the measured field perturbation in the recovery phase of the magnetic storm of June 17-18, 1972 (Berko et al., 1975). The agreement is quite good for the third-order iteration. The discrepancies seen may be due to measurement inaccuracies and the neglect of electric field effects. Note that while $\overline{E}\times\overline{B}$ drift produces no net current, the electric field may change the geometry of the current paths in the magnetic field and thereby change expected effects at a particular point in space.

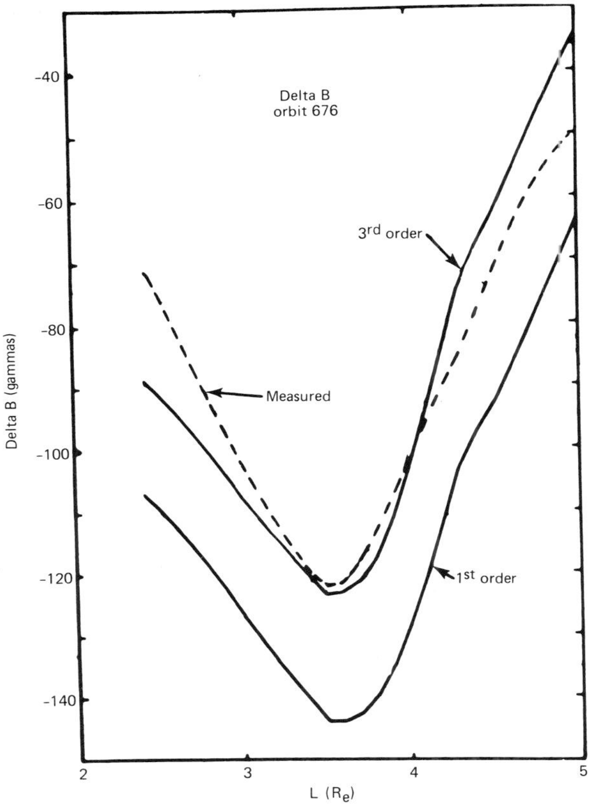

Figure 4. Measured field perturbation, ΔB, and 1st and 3rd order calculated ΔB curves based on Explorer 45 particle measurements (protons assumed) in recovery phase of June 17-18, 1972, magnetic storm (from Berko et al., 1975).

3. THE RING CURRENT

Since we showed earlier that the magnetic field perturbation at the earth's surface is proportional to the total kinetic energy of the particle population, it is appropriate to consider the ring current in an energy density perspective. Figure 5 shows ring current energy density versus altitude plots obtained during the 17 December 1971 magnetic storm (Smith and Hoffman, 1973). The energy density calculation assumes that the measured ions are protons and is carried out over the 1-138 keV energy range. Orbit 97 shows prestorm levels, orbit 101 shows the asymmetric main phase development, and orbits 102 and 103 show storm recovery phase. Of interest in Figure 5 are the pressure gradients during the storm. The steep decrease at low altitudes represents an eastward current which is opposite to the westward current required for the ring current. However, the steady decrease towards high altitudes gives a westward current and since the pressure gradient term is weighted as L^3 (equation (15)), it is this portion of the particle distribution that will provide the ring current system causing the observed surface field depression.

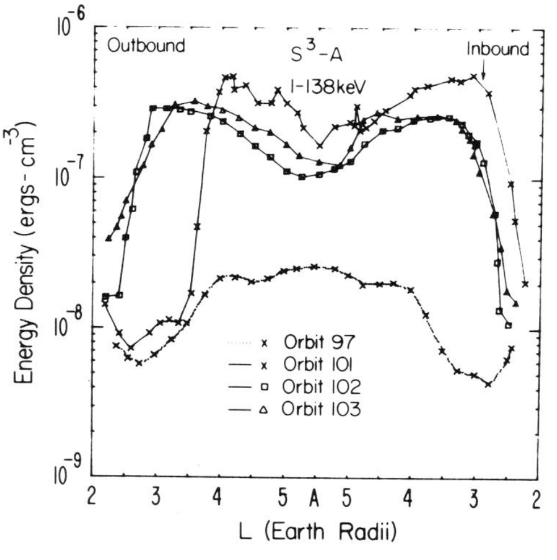

Figure 5. Proton energy density radial profiles as measured by Explorer 45 (S^3-A) during pre-storm quiet (orbit 97) main phase injection (orbit 101), and recovery phase (orbits 102 and 103) (from Smith and Hoffman, 1973).

For the same magnetic storm, Figure 6 shows the measured differential energy density of the ring current (Smith and Hoffman, 1973). The solid data points (orbit 102) were obtained in storm recovery phase and at the altitude ($\sim 3.6 R_E$) of maximum ring current intensity. A pre-storm plot of the quiet time energy density (orbit 97) is shown for comparison. Very similar results have been reported for a magnetic storm in November 1977 (Williams, 1981a).

It is seen from Figure 6 that the main injection of ring current particles is at energies $\lesssim 200$ keV while particles at energies $\gtrsim 200$ keV show a decrease in intensity. This result has been reported and discussed by Lyons and Williams (1976) and Williams (1981a) who show that the high energy particle decrease is due to an adiabatic response of the energetic particles to the new magnetic field configuration caused by the low energy particle injection (the ring current).

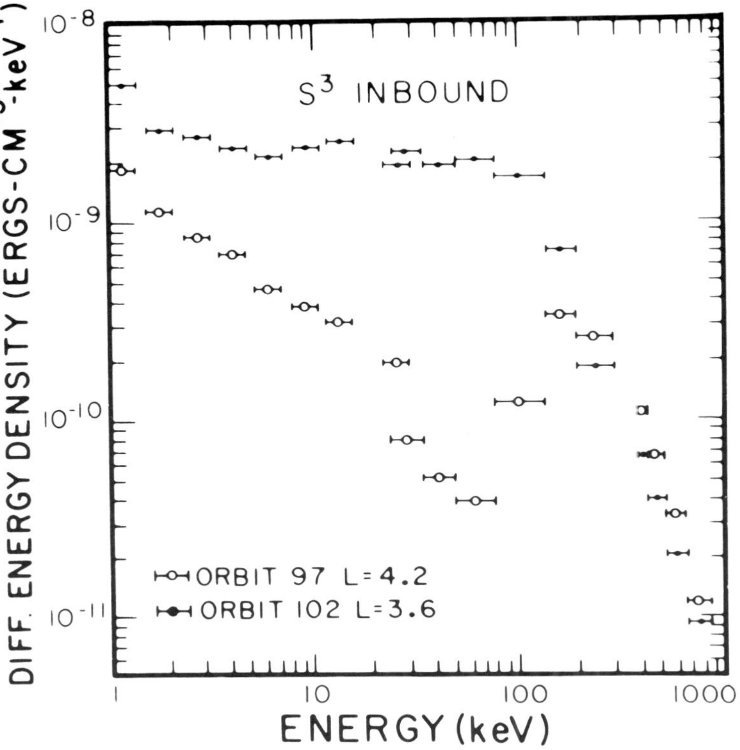

Figure 6. Ion energy density spectra (protons assumed) measured by Explorer 45 during pre-storm quiet (orbit 97) and recovery phase (orbit 102) for December 17, 1971, magnetic storm (from Smith and Hoffman, 1973).

Integration of the curves of Figure 6 yields the results shown in Figure 7 (Williams, 1980, 1981b). Here the accumulated percentage of ring current energy density is plotted against energy. Quiet time and recovery phase are shown. The recovery phase solid curve assumes the ring current particles are all protons while the dashed curve assumes all O^+ ions below 50 keV and all protons above 50 keV. Assuming protons, the median energy in the integral energy density curve is ~85 keV. Energies for which composition measurements have been made in the magnetosphere (prior to recent Charge Composition Explorer, CCE, observations) are indicated.

Figure 7 shows that while we know the energy range of the ring current particles, we have not had composition observations for the bulk (80-90%) of these particles. This effectively has prevented us from identifying the source of the ring current particles. The low energy tail ($\lesssim$20 keV) of the ring current distribution is thought to have an ionospheric origin whereas the high energy tail ($\gtrsim$600 keV) is thought to have a solar wind origin. How, or if, these two sources mix to produce the bulk ring current is unknown (Williams, 1980, 1981b, 1983).

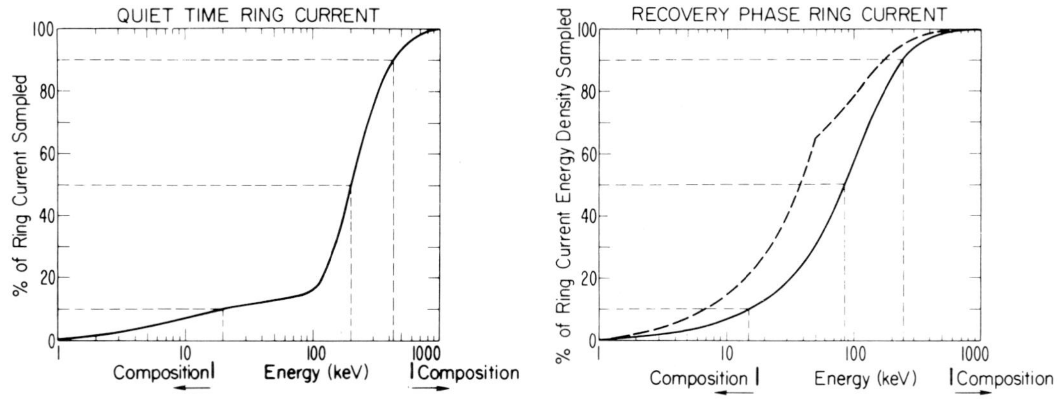

Figure 7. Accumulated percentage of ring current energy density vs. energy. Quiet-time and recovery phase curves are from data shown in Figure 3.14. Solid curves assume all ions are protons. The dashed curve in recovery phase shows case for all O^+ ions below 50 keV and all protons above 50 keV. Energy ranges of available direct composition observations are indicated (from Williams, 1980, 1981b).

Happily this situation is now changing. The U.S. Charge Composition Explorer (CCE) was launched in August 1984 as part of the joint three-satellite Active Magnetospheric Particle Tracer Explorers (AMPTE) program conducted by the U.S., the Federal Republic of Germany and the United Kingdom. A major objective of the AMPTE program was the release of gases in the solar wind and distant geomagnetic tail and the observation of the entry and transport of the resulting ions to the lower magnetosphere. To perform these latter observations, the CCE was placed in an elliptical, nearly equatorial orbit having an 8.8 R_E apogee and 15.7 hour period. It carried into space the first comprehensive set of composition and charge state instrumentation spanning the entire ring current energy range. Initial CCE ring current results have been published for the 4 September 1984 magnetic storm in a set of seven papers contained in the May 1985 Special Issue of the Geophysical Research Letters. While these results are preliminary and pertain to just one magnetic storm, they are exciting and hold high promise to solve several long standing ring current problems. We briefly highlight these results in the following paragraphs:

Figure 8 shows measured oxygen charge states summed over an inbound pass on 5 September 1984 during the magnetic storm (Gloeckler et al., 1985). The two source regions are identified by the two charge state groups -- high charge states being associated with the solar wind and low charge states with the ionosphere.

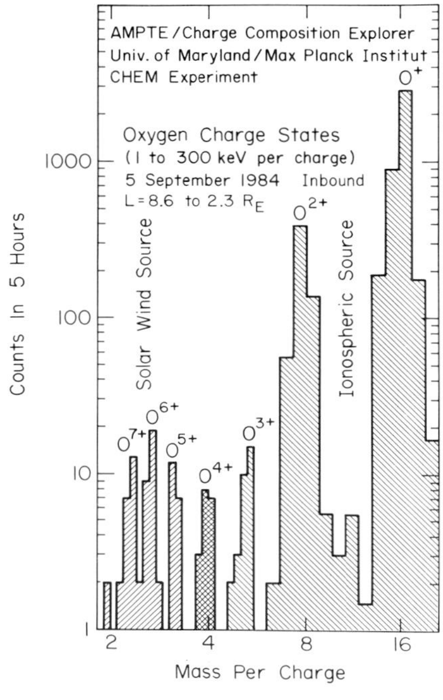

Figure 8. Charge-state histogram of ~1 to 300 keV/e oxygen in the earth's ring current during the geomagnetic storm of Sept. 4-7, 1984. Contributions from the ionosphere ($O^{+1,2}$), the solar wind ($O^{+5,6,7}$) and from charge exchange ($O^{+3,4}$) are evident (from Gloeckler et al., 1985).

While Figure 8 shows a dominant ionospheric oxygen source as summed over L values 8.6 to 2.3 R_E, Figure 9 (Gloeckler et al., 1985) shows that protons dominated the ring current energy density at all altitudes over the 5.2-315 keV/e range. It also can be seen that relative strength of the ionospheric source decreases with respect to the solar wind source with increasing altitude.

One of the main results concerning the 4 September 1984 storm was the observation that protons were the dominant ion throughout the bulk of the ring current (Krimigis et al., 1985). Figures 10a and b show composite intensity and energy density spectra from the CCE particle instruments, respectively for a pre-storm "quiet" period (3 September 1984) and during storm main phase (5 September 1984) (Krimigis et al., 1985). From these data, Krimigis et al. (1985) show that i) the pre-storm ring current primarily consisted of protons and was peaked in energy at ~200 keV (see Figure 7), ii) the main ring current intensity enhancement occurred in the ~20-200 keV, and iii) protons $\gtrsim$50 keV dominated the storm time ring current energy density with oxygen ions contributing ~27%.

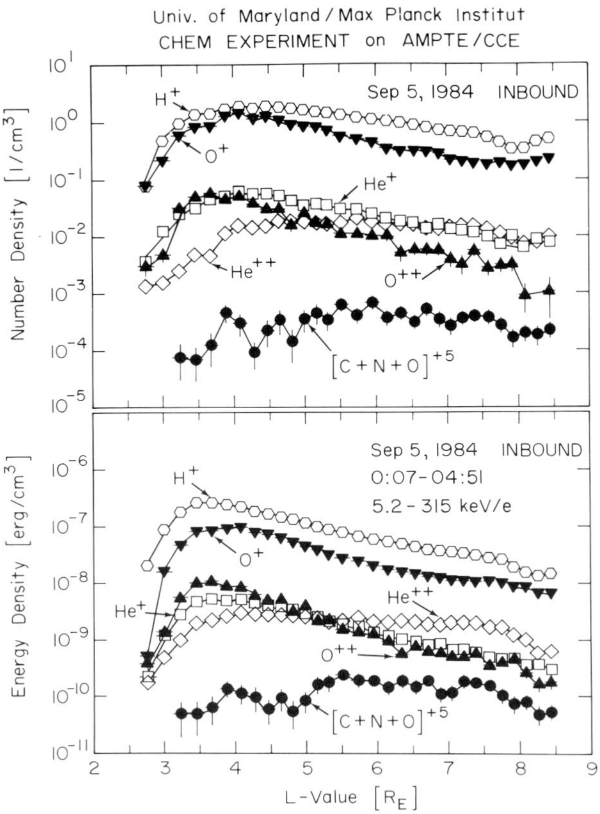

Figure 9. Radial profiles of (a) the number density and (b) the energy density of 5-315 keV/e H^+, O^+, He^+, He^{++}, O^{++} and $[CNO]^{+5}$ during the inbound pass on Sept. 5. The e-folding values for the number (energy) densities between L=4 and 6.5 are 3.3 (1.5), 1.7 (1.3), 1.2 (1.0), 2.0 (1.5) R_E for H^+, O^+, O^{++} and He^+ respectively (from Gloeckler et al., 1985).

Using the result that the dominant ring current ion in this storm was H^+, McEntire et al. (1985) have computed the current density (equation 15) resulting from the 25-1000 keV proton distribution. Their result is shown in Figure 11. It can be seen that the main increase in current density is due to the pressure gradient which occurs at higher altitudes (see Figure 9), as discussed earlier.

While many more magnetic storms have to be analyzed before any comprehensive picture of ring current sources, energization, and generation will emerge, these present CCE results do show an exciting new capability to solve several age old space plasma physics problems.

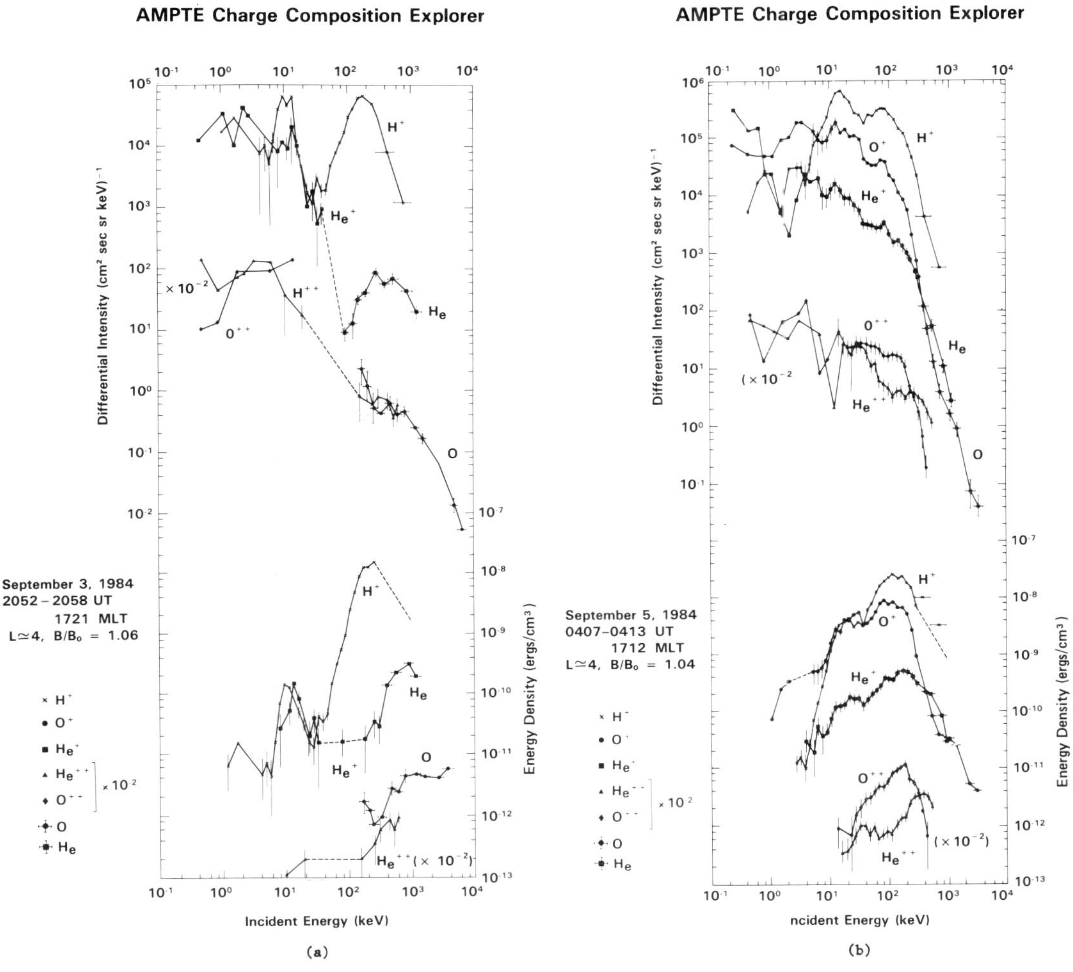

Figure 10. Differential energy spectra and densities for hydrogen, helium and oxygen ions, at L=4; O^{2+} and He^{2+} data are offset by a factor $\times 10^{-2}$ for clarity. (a) On September 3, 1984, some eleven hours prior to the storm sudden commencement. (b) During storm main phase on September 5; note the dramatic increase in oxygen and protons and substantial increase in helium; peak energy densities are in the range of 100-200 keV (bottom part of figure) (from Krimigis et al., 1985).

4. RING CURRENT GENERATION

Since the composition and thus the source of the ring current particles remains unknown, it is not possible to discuss quantitatively generation mechanisms. However, we present in this section a brief listing and qualitative description of various mechanisms which have been popular for many years.

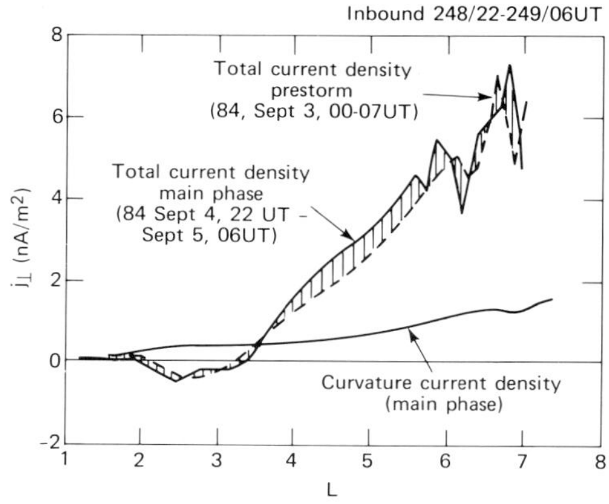

Figure 11. Radial profiles of the current densities derived from the MEPA measurements. Positive value indicates a westward current and negative value an eastward one. Enhancement in the current density is seen in the L-range of 3.5 to 6 for the main phase of the magnetic storm. The total current density is found to be dominated by the pressure gradient current (from McEntire et al., 1985).

(1) Earthward convection of the plasma sheet. Here an enhanced dawn-dusk electric field quickly convects ($\overline{E}\times\overline{B}$) the plasma sheet particles towards the Earth on the nightside hemisphere. The convecting particles experience betatron and Fermi acceleration as they are transported into low altitudes (~2.5–4.0R_E) by the enhanced field to form the ring current. (See for example discussion by Frank (1967b) and Axford (1969).)

(2) Acceleration of ionospheric ions by electric fields parallel to auroral field lines. Since the observation of energetic ($\lesssim$10 keV) ion beams streaming out of the ionosphere in auroral regions (Shelley et al. (1976) it has been attractive to consider these particles as a source for the ring current. Additional mechanisms are required to form an equatorial $\sin^n\alpha (n \geqslant 0)$ pitch angle distribution which would then be displaced to low altitude by an enhanced dawn-dusk electric field and accelerated to ring current energies by betatron and Fermi processes.

(3) Acceleration at a source boundary. To explain the complex energy-time signature displayed by energetic ($\lesssim$30 keV) particles at geostationary orbit McIlwain (1974) invoked the concept that enhanced electric fields cause the appearance of fresh energetic particles at locations outside a sharp, well-defined boundary in the nightside hemisphere. After the transient, enhanced electric field, the particles move in the ambient, quiescent magnetic-electric field configuration. Using a realistic magnetic field and tailoring an electric field model inferred from particle data, McIlwain (1974) was able to explain many of the particle energy-time signatures. This hypothesis has been successfully tested by Mauk and McIlwain (1974) and Konradi et al. (1975). While the particle observations in the dusk-midnight sector also are consistent with particle motion in the existing magnetic and

electric field configuration (Kivelson, 1976; Kivelson and Southward, 1975; Kivelson et al., 1980), the inclusion of midnight-dawn sector observations and observations of particle 'forbidden' zones presents further evidence supporting the source boundary concept (Mauk and Meng, 1983a, b).

(4) Inward motion of the outer radiation zone. The same enhanced electric field responsible for rapid convection of the plasma sheet towards the Earth will also cause an inward motion of previously trapped particles in the Earth's outer radiation zone. Lyons and Williams (1980) quantitatively tested this concept using data from two main phase magnetic storms in 1971 and 1972. They found that the main phase ring current could be accounted for by the inward motion of the pre-storm outer zone trapped particles. Ions and electrons simultaneously were consistent with this hypothesis. Electric fields giving inward motions consistent with rough plasmapause position measurements were required.

It is clear that these mechanisms bear a resemblance to each other and may represent different aspects of the overall particle transport and acceleration process responsible for the creation of the ring current. For example, mechanism (2) may be closely related to (3), although McIlwain has associated (3) with the plasmapause boundary. Mechanism (4) clearly seems related to all the others with the recognition that (1), (2), and (3) may be providing a fresh supply of trapped particles for the outer zone, which would be transported into low altitudes at the next major electric field enhancement thereby forming the ring current. Much more quantitative work is required to sort through the present qualitative possibilities for ring current generation.

5. RING CURRENT DECAY

Singer (1957) was the first to suggest that the main phase ring current was due to the motion of particles trapped in the Earth's magnetic field. He suggested the decay of the ring current was due to scattering of the particles either by field changes or atomic collisions. His estimate of an ~1 day lifetime was essentially a strong diffusion scattering limit. Stuart (1959), Dessler and Parker (1959), and Wentworth et al. (1959a, b), were the first to note the importance of charge exchange in removing ring current particles. Charge exchange has since been used extensively in magnetospheric studies. However, since we do not yet know the time and spatial dependencies of ring current composition, it is not possible to predict the charge exchange decay time in advance because of the species and energy dependence of the charge exchange cross sections. We show in Figure 12 the results of a charge exchange calculation to explain ring current decay after the 24 February 1972 magnetic storm (Smith et al., 1981). They performed a best fit charge exchange decay analysis to the observed decay by assuming a multi-iron particle population. Parameters extracted from the fits were the relative abundance and charge exchange lifetime of each ion species. The lifetimes obtained from the best fit analysis were essentially the charge exchange lifetimes of H^+, He^+, and O^+. Figure 12 shows one set

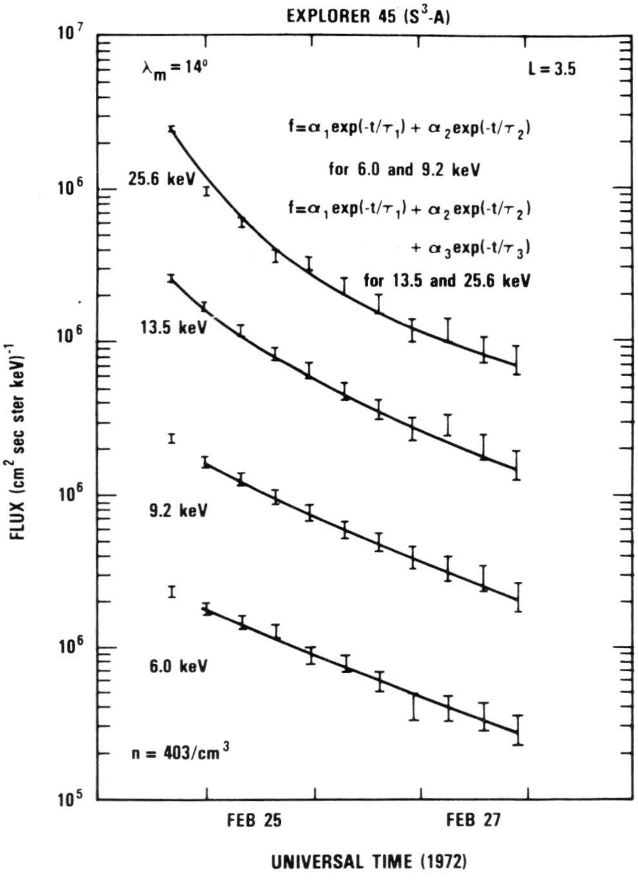

Figure 12. Ion fluxes measured by Explorer 45 at indicated energies during recovery phase of February 24, 1972, magnetic storms. Best fit curves using multi-ion fit to data are shown as solid lines. Best fit charge exchange lifetimes, τ_1, τ_2, τ_3, are essentially those of H^+, O^+, and He^+. The fits require an admixture of H^+ and O^+ at 6.0 and 9.2 keV and an admixture of H^+, O^+, and He^+ at 13.5 and 25.6 keV (from Smith et al., 1981).

of results for energies from 6 to ~26 keV. Thus proper accounting for a multi-ion ring current shows charge exchange to be a major cause of the overall decay in recovery phase.

However charge exchange is not the sole dissipation mechanism operating on the ring current. Cornwall et al. (1970) predicted intense ion cyclotron wave turbulence generated by the interaction of the ring current particles with the cold plasmasphere plasma.
The resulting pitch angle diffusion of the ring current particles would represent a major loss process. Williams and Lyons (1974a, b) showed that the pitch angle evolution of ring current particles in the plasma-

pause region was consistent with the generation of ion cyclotron waves, although wave growth and resulting particle loss was weak, not strong as predicted.

Even though ring current pitch angle diffusion was weak instead of strong, sufficient energy was lost from the ring current to cause observable atmospheric effects. Integrating the energy lost from the ring current due to pitch angle distribution changes observed in the plasmapause region, Williams et al. (1976) estimated the energy deposition per cm^2 into the ionosphere as a function of L value. Comparison of their results with simultaneously observed double SAR arc structures in the ionosphere showed that the energy lost from the ring current was sufficient to sustain the observed SAR arcs, consistent with the prediction of Cornwall et al. (1971). Furthermore, the equatorial location of the two high altitude peaks in the ring current loss were consistent with the observed SAR arc positions when the ring current inflated magnetic field is taken into account.

Thus it seems that the general ring current decay is due primarily to charge exchange with weak pitch angle scattering occurring in the plasmapause region, as discussed by Williams (1974). The pitch angle scattering is sufficient to cause observable atmospheric effects.

6. SUMMARY

The calculation of magnetic field perturbations due to an arbitrary distribution of particles trapped in the Earth's magnetic field apparently is well in hand. However the generation mechanism(s) responsible for the ring current particle distribution remain unknown. Ring current decay is due primarily to charge exchange and secondarily to ion cyclotron wave generation. The composition of the bulk of the ring current ions and the manner in which sources (ionosphere, solar wind) mix throughout the ring current presently are not known. However, results from the Charge Composition Explorer should clarify these characteristics in the near future.

ACKNOWLEDGEMENT. This work was supported in part by a National Aeronautics and Space Administration contract to The Johns Hopkins University Applied Physics Laboratory and the Department of the Navy under Task I2UOS10 contract N00024-85-C-5301.

REFERENCES

Akasofu, S.I. and S. Chapman, "The ring current, geomagnetic disturbance, and the Van Allen radiation belts", J. Geophys. Res., 66, 1321 (1961).
Alfven, H., "Cosmical Electrodynamics", 1st ed., Oxford Univ. Press, London and New York (1950).
Alfven, H., "On the electric field theory of magnetic storms and aurorae", Tellus, 7, 54, (1955).
Alfven, H. and C.G. Falthammar, "Cosmical Electrodynamics", 2nd ed., Oxford Univ. Press, London and New York (1963).
Apel, J.R., S.F. Singer, and R.C. Wentworth, "Effects of trapped particles on the geomagnetic field", Advances in Geophysics 9, 131 (1962).
Axford, W.I., "Magnetospheric convection", Rev. of Geophys., 7, 421 (1969).
Berko, F.W., L.J. Cahill, Jr., and T.A. Fritz, "Protons as the prime contributors to storm time ring current", J. Geophys. Res., 80, 3549 (1975).
Carovillano, R.L. and J.J. Maguire, "Magnetic energy relationships in the magnetosphere", in Physics of the Magnetosphere, R.L. Carovillano, J.F. McClay, and H.R. Radoski (eds.), D. Reidel Publ. Co., Dordrecht, Holland, 290 (1968).
Carovillano, R.L. and G.L. Siscoe, "Energy and momentum theorems in magnetospheric processes", Rev. of Geophys. and Space Phys., 11, 289 (1973).
Chapman, S. and V.C.A. Ferraro, "A new theory of magnetic storms", Terr. Magn. Atmos. Elec., 36, 77-97, 171-186 (1931).
Chapman, S. and V.C.A. Ferraro, "A new theory of magnetic storms", Terr. Magn. Atmos. Elec., 37, 147-156, 421-429 (1932).
Chapman, S. and V.C.A. Ferraro, "A new theory of magnetic storms", Terr. Magn. Atmos. Elec., 38, 79-96 (1933).
Cornwall, J.M., F.B. Coroniti, and R.M. Thorne, "Turbulent loss of ring current protons", J. Geophys. Res., 75, 4699 (1970).
Cornwall, J.M., F.V. Coroniti, and R.M. Thorne, "A unified theory for SAR arc formation at the plasmapause", J. Geophys. Res., 76, 4428 (1971).
Dessler, A.J. and E.N. Parker, "Hydromagnetic theory of geomagnetic storms", J. Geophys. Res., 64, 2239 (1959).
Frank, L.A., "Several observations of low energy protons and electrons in the earth's magnetosphere with OGO 3", J. Geophys. Res., 72, 1905 (1967a).
Frank, L.A., "On the extraterrestrial ring current during geomagnetic storms", J. Geophys. Res., 72, 3753 (1967b).
Gloeckler, G., B. Wilken, W. Studemann, F.M. Ipavich, D. Hovestadt, D.C. Hamilton, and G. Kremser, "First Composition Measurements of the Bulk of the Storm Time Ring Current (1 to 300 keV/e), submitted to Geophysical Research Letters, February 1985.
Hoffman, R.A. and P.A. Bracken, "Magnetic effects of the quiet-time proton belt", J. Geophys. Res., 70, 3541 (1965).

Hoffman, R.A. and P.A. Bracken, "Higher order ring currents and particle energy storage in the magnetosphere", J. Geophys. Res., 72, 6039 (1967).

Kivelson, M.G., "Magnetospheric electric fields and their variation with geomagnetic activity", Rev. Geophys. Space Phys., 14, 189 (1976).

Kivelson, M.G. and D.J. Southwood, "Approximations for the study of drift boundaries in the magnetosphere", J. Geophys. Res., 80, 3528 (1975).

Kivelson, M.G., S.M. Kaye, and D.J. Southwood, "The physics of plasma injection events", in Dynamics of the Magnetosphere, S.-I. Akasofu (ed.), D. Reidel Publ. Co., Dordrecht, Holland, 385 (1980).

Konradi, A., C.L. Semar, and T.A. Fritz, "Substorm-injected protons and electrons and the injection boundary model", J. Geophys. Res., 80, 543 (1975).

Krimigis, S.M., G. Gloeckler, R.W. McEntire, T.A. Potemra, F.L. Scarf, E.G. Shelley, "Magnetic Storm of 4 September 1984: A Synthesis of Ring Current Spectra and Energy Densities", submitted to Geophysical Research Letters, February 1985.

Lyons, L.R. and D.J. Williams, "Storm associated variations of equatorially mirroring ring current protons, 1-800 keV, at constant first adiabatic invariant", J. Geophys. Res., 81, 216 (1976).

Lyons, L.R. and D.J. Williams, "A source for the geomagnetic storm main phase ring current", J. Geophys. Res., 85, 523 (1980).

Mauk, B.H. and C.E. McIlwain, "Correlation of Kp with the substorm-injected plasma boundary", J. Geophys. Res., 79, 3193 (1974).

Mauk, B.H. and C.I. Meng, "Dynamical injections as the source of near geostationary, quiet-time particle spatial boundaries, J. Geophys. Res., 88, 10011-10024 (1983a).

Mauk, B.H. and C.I. Meng, "Characterization of geostationary particle signatures based on the 'injection boundary' model", J. Geophys. Res., 88, 3055-3071 (1983b).

McEntire, R.W., A.T.Y. Lui, S.M. Krimigis and E.P. Keath, "Energetic Particle Composition Measurements During the September 4, 1984 Magnetic Storm", submitted to Geophysical Research Letters, February 1985.

McIlwain, C.E., "Substorm injection boundaries", in Magnetospheric Physics, B.M. McCormac (ed.), D. Reidel Publ. Co., Dordrecht, Holland, 143 (1974).

Northrup, T.G. and E. Teller, "Stability of the adiabatic motion of charged particles in the earth's field", Phys. Rev., 117, 215 (1960).

Northrup, T.G., "The adiabatic motion of charged Particles", Wiley (Interscience), New York, 109 pp. (1963).

Olbert, S., G. L. Siscoe, and V.M. Vasyliunas, "A simple derivation of the Dessler-Parker-Sckopke relation", J. Geophys. Res., 73, 1115 (1968).

Parker, E.N., "Newtonian development of the dynamical properties of ionized gases at low density", Phys. Rev., 107, 924 (1957).

Roederer, J.G., "Dynamics of geomagnetically trapped radiation", Springer-Verlag, New York, Heidelberg, Berlin, pp. 166 (1970).

Sckopke, N., "A general relation between the energy of trapped particles and the disturbance field near the earth", J. Geophys. Res., 71, 3125 (1966).

Shelley, E.G., R.D. Sharp, and R.G. Johnson, "Satellite observations of an ionospheric acceleration mechanism", Geophys. Res. Letters, 3, 654 (1976).

Singer, S.F., "A new model of magnetic storms and aurorae", Trans. Am. Geophys. Union, 38, 175 (1957).

Smith, P.H., N.K. Bewtra, and R.A. Hoffman, "Motions of charged particles in the magnetosphere under the influence of a time-varying large scale convection electric field", in Quantitative Modelling of Magnetospheric Processes, ed. W.P. Olsen, Geophysical Monograph 28, American Geophysical Union, Washington, D.C., 513-535, (1959).

Smith, P.H. and R.A. Hoffman, "Ring current particle distribution during the magnetic storms of Dec. 16-18, 1971", J. Geophys. Res., 78, 4731 (1973).

Smith, P.H., N.K. Bewtra and R.A. Hoffman, "Inference of the ring current ion composition by means of charge exchange decay", J. Geophys. Res., 86, 3470, (1981).

Spitzer, L., "Physics of fully ionized gases", Wiley (Interscience), New York, 105 pp. (1956).

Stuart, G.W., "Satellite measured radiation", Phys. Rev. Letters, 2, 417 (1959).

Wentworth, R.C., W.M. MacDonald, and S.F. Singer, "Lifetimes of trapped auroral radiation belt particles", Bull. Am. Phys. Soc. Ser 2 4, 7, (1959).

Wentworth, R.C., W.M. MacDonald, and S.F. Singer, "Lifetimes of trapped radiation belt particles determined by Coulomb scattering", Phys. of Fluids, 2, 499, (1959b).

Williams, D.J., "Magnetospheric Proton Dynamics", EOS, 55, 1021 (1974).

Williams, D.J., "Ring current composition and sources", in Dynamics of the Magnetosphere, S.-I. Akasofu (ed.), D. Reidel Publ. Co., Dordrecht, Holland, 407 (1980).

Williams, D.J., "Phase space variations of equatorially mirroring ring current ions", J. Geophys. Res., 86, 189 (1981a).

Williams, D.J., "Ring current composition and sources: An update", Planet. Space Sci., 29, 1195 (1981b).

Williams, D.J. and L.R. Lyons, "The proton ring current and its interaction with the plasmapause: Storm recovery phase", J. Geophys. Res., 79, 4195, (1974a).

Williams, D.J. and L.R. Lyons, "Further aspects of the proton ring current interaction with the plasmapause: Main and recovery phases", J. Geophys. Res., 79, 4791, (1974b).

Williams, D.J., G. Hernandez, and L.R. Lyons, "Simultaneous observations of the proton ring current and stable auroral red arc", J. Geophys. Res., 81, 608 (1976).

Williams, D.J., "The earth's ring current: causes, generation and decay", Space Science Reviews, 34, 223-234, 1983.

PART 3 - THEORY

SPACE PLASMA TURBULENT DISSIPATION: REALITY OR MYTH?

F. V. Coroniti
Department of Astronomy and Department of Physics
University of California
Los Angeles, California 90024

ABSTRACT. A prevalent approach to understanding magnetospheric dynamics is to combine a hydromagnetic description of the large scale magnetic structure and convection flows with a locally determined anomalous dissipation which develops in boundary layers. Three problems - nose and tail reconnection, auroral field-aligned currents, and diffuse auroral precipitation - are critically examined to test the validity of this theoretical philosophy. Although the expected plasma wave turbulence is observed for each case, the concept of local anomalous dissipation fails to provide an adequate or complete description of the phemomenae.

1.0 INTRODUCTION

About 15 to 20 years ago satellite observations established the magnetic structure and plasma populations of the terrestrial magnetosphere and identified many of the modes of plasma wave turbulence. With this observational base, a theoretical philosophy or framework developed in order to describe the magnetosphere's interaction with the solar wind and internal dynamics. Since magnetospheric scale lengths are large compared to particle gyroradii, inertial lengths or any dissipative lengths, ideal hydromagnetics or MHD should provide an accurate lowest order description of the global configuration. Convection flows should obey the ideal Ohm's law $\underline{E} + \underline{v} \times \underline{B}/c = 0$, and should interact with each other and with the ionosphere by large scale Reynolds, pressure, and Maxwell stresses.

Ideal MHD, however, generally breaks-down at boundaries which separate different plasma regions. Usually, dissipation must be added to MHD in order to permit the flow and magnetic field to change smoothly through the boundary layers. Except in the ionosphere, Coulomb collisions are unimportant, so that magnetospheric dissipation must be anomalous. Within the boundary regions, plasma flow and magnetic stresses inevitably create velocity and configuration space free energy which can drive local plasma wave instabilities; the resulting nonlinear wave turbulence should provide the necessary anomalous dissipation.

The inclusion of local anomalous transport coeffficients into the

Space Science Reviews **42** (1985) 399–410. 0038–6308/85.15

equations of hydromagnetics is a prevalent theoretical framework for describing space plasmas. With the comprehensive plasma and wave measurements obtained by recent satellites, and with the development of modern plasma numerical simulations, we can now evaluate whether this theoretical approach successfully describes magnetospheric processes. In this paper we will examine the problems of magnetopause and tail reconnection (section 2.0), the field-aligned current interaction between the magnetosphere and ionosphere (section 3.0), and the convection-induced diffuse auroral electron precipitation (section 4.0). For each problem we will argue that the description based on coupling hydromagnetics and local plasma turbulent dissipation is at best only partially correct and, in some cases, is quite likely to be wrong.

2. RECONNECTION AND ANOMALOUS RESISTANCE

2.1. Petschek Reconnection

The hydromagnetic description of fast reconnection was first proposed by Petschek (1964), and has subsequently been confirmed by both analysis and simulation (see Sonnerup, 1979). Oppositely directed magnetic field lines $\underline{E} \times \underline{B}$ drift into a thin boundary layer; at a single stagnation point, both the flow velocity and the magnetic field must vanish, leaving the electric field to be balanced only by ohmic dissipation. Actually resistivity decouples the field and plasma throughout a diffusion region which surrounds the neutral line and has a dimension transverse to (along) the boundary layer of r_m (r_m/M_A); r_m is the magnetic Reynolds length (see below) and M_A is the Alfven Mach number of the incoming flow. Away from the diffusion region slow switch-off shocks stand in the incoming flow; the tangential (upstream) component of the magnetic field is resistively dissipated across the slow shock, and downstream the plasma is tangentially accelerated to the Alfven speed. The basic prediction of hydromagnetic reconnection models is that the plasma flows away from the neutral line perpendicular to the weak magnetic field component which is normal to the boundary layer.

2.2. Anomalous Resistivity

For collisionless plasmas, the hydromagnetic reconnection model requires an anomalous resistance to the flow of current in both the diffusion region and the slow shocks. Since the current drift velocity should be large within the thin diffusion region, current driven instabilities should readily be excited. The ideal instability would be the ion acoustic mode since it exists in a field-free plasma; unfortunately, the high ion to electron temperature ratio found throughout most of the magnetosphere makes the critical drift velocity for exciting ion acoustic turbulence much higher than the observed current drift speeds.

An alternative anomalous resistance candidate is the lower hybrid drift (LHD) instability (Huba, et. al, 1978) which is driven by the anti-parallel density (i.e., current density) and magnetic field gradients that are expected in the diffusion region. The LHD mode has frequencies near the lower hybrid ($\Omega_{LH} = (\Omega_i \Omega_e)^{1/2}$), and is strongly unstable when the magnetic gradient length (λ) is comparable to the ion gyro-radius ($R_i = a_i/\Omega_i$). Since the mode exists only for magnetized electrons, the LHD is primarily excited near the low-β edges of the diffusion region, and is strongly damped by the electron grad-B drift resonance in the high-β region near the neutral line (Huba, et. al, 1980.)

The LHD anomalous resistivity arises from turbulent perpendicular electric fields ($\delta \underline{E}$) which create $\delta \underline{E} \times \underline{B}$ drifts that transport the current carriers across the magnetic field. Roughly, the cross-field diffusion coefficient is $D \sim (c\delta E/B)^2 \Delta t$ where Δt is the correlation time. For strongly driven LHD turbulence, Δt is of order the growth time or Ω_{LH}^{-1}. We can define an anomalous collision frequency (ν_*) for cross-field transport by setting $D = \nu_* a_e^2/\Omega_e^2$ to find

$$\nu_* = \frac{\omega_{pe}^2}{\Omega_{LH}} \frac{(\delta E)^2}{8\pi n T_e} \tag{1}$$

If we ignore the difficulty that the LHD wave turbulence may not penetrate into the low field region around the neutral line (Tanaka and Sato, 1981; Brackbill, et. al, 1984), we can use ν_* to obtain the magnetic Reynolds length, which we then equate to the thickness of the resistive diffusion region

$$r_m = \frac{c^2}{\omega_{pe}^2} \frac{\nu_*}{M_A C_A} = \frac{cB}{4\pi n V_D} = \lambda \tag{2}$$

where V_D is the current drift speed. By combining (1) and (2) we can define a critical turbulent electric field energy density for which r_m will equal the observed value of λ

$$\frac{(\delta E)^2_{cr}}{8\pi n T_e} = M_A \left(\frac{m_i}{m_e}\right)^{1/2} \frac{C_A^2}{V_D} \frac{C_A}{c^2} = M_A \frac{C_A}{c} \frac{\Omega_{LH}}{\omega_{pi}} \left(\frac{\lambda \omega_{pi}}{c}\right) \tag{3}$$

We now examine whether the electric field amplitudes which are observed in magnetospheric reconnection regions are comparable to δE_{cr}.

2.3 Dayside Magnetopause Reconnection

The plasma flow signature of hydromagnetic reconnection is occasionally observed within the magnetopause boundary layer (Sonnerup, et. al, 1981). The tangential flow velocity is comparable to the Alfven speed, and the magnetic field transition is roughly consistant with the theoretically expected (Levy, et. al, 1984) rotational discontinuity

(Berchem and Russell, 1984). Hence there is evidence that fast ($M_A \sim 0.1$) hydromagnetic reconnection does occur at the magnetopause.

Gurnett, et. al (1979) have shown that the dayside magnetopause contains an intense level of electric and magnetic wave turbulence which extends from 10 HZ to above 10^3 HZ. At 10 HZ the average electric field power spectral density is typically 10^{-7} V^2/m^2-HZ and decreases above 10 HZ as f^{-2} (Anderson, et. al, 1982; Tsurutani, et. al, 1981). Since the lower hybrid frequency is usually in the range 20 - 40 HZ, we can integrate the observed spectral density for f > 10 HZ to obtain an estimate of the electric field energy density $(\delta E)^2_{ob} \approx 10^{-6}$ V^2/m^2. We will attribute all of $(\delta E)^2_{ob}$ to the LHD mode even though some (or even all) of the observed electric fields may be due to whistler turbulence, which is simultaneously observed at the magnetopause.

For typical magnetopause conditions, the plasma density $n \sim 40$ cm^{-3}, the electron temperature $T_e \sim 40$ ev, and the magnetosheath southward magnetic fields $B \sim 30$ nT (Paschmann, et. al, 1982); note that these quantities vary significantly from day-to-day. Near the magnetic equator, the average magnetopause thickness is $\lambda \sim 500$ km (Berchem and Russell, 1982). Substituting these values into (4) and assuming $M_A = 0.1$, we find $(\delta E)^2_{ob}/(\delta E)^2_{cr} \sim 10^{-3}$; the observed LHD turbulence appears to be inadequate to provide the anomalous resistance which is required for fast reconnection ($M_A \sim 0.1$).

Another potential difficulty with the hydromagnetic reconnection model was discovered by Russell and Elphic (1979). They observed temporally impulsive flux transfer events in which spatially localized flux tubes are stripped away from the magnetopause. At present it is unclear whether flux transfer events are describable as unsteady Petschek reconnection or as non-hydromagnetic, tearing-like instabilities.

2.4 Tail Reconnection

With the recent ISEE-3 discovery of substorm-associated tailward flowing plasmoids (Hones, et. al, 1984), little doubt remains that substorm break-up is caused by the onset of strong reconnection dissipation in the near earth plasma sheet (McPherron, et al., 1973; Hones, 1973). Sato and Hayashi (1979) and Ugai (1982) have used hydromagnetic simulations to study substorm reconnection. An initially thick current sheet collapses until the current drift exceeds the LHD instability threshold; the sudden increase of LHD anomalous resistance allows the rapid development of the Petschek reconnection flow.

Since LHD anomalous resistance is expected in the tail (Huba, et. al, 1978), we can ask are the waves observed? The answer is not entirely clear since the frequency response of satellite wave detectors usually does not extend to the 10 - 20 HZ lower hybrid frequency. Gurnett, et. al (1976) reported intense electric field waves with an average power spectral density at 40 HZ of 10^{-8} v^2/m^2 - HZ and a f^{-1} spectrum above 40 HZ. These waves peaked in intensity at the edge of plasma sheet as expected for the LHD mode. If we extend the f^{-1} spectrum to 10 HZ (below the local lower hybrid) and integrate over frequency, we can obtain an estimate of the wide-band electric field amplitude of $(\delta E)^2_{ob} = 10^{-6}\ V^2/m^2$.

Typically, the plasma sheet density and electron temperature are $n \sim 0.3\ cm^{-3}$ and Te = 500 eV, respectively. At the edge of the near earth plasma sheet, the field strength is of order $B \sim 30$ nT. The thickness of the plasma sheet is quite variable, but has never been observed to be less than $\lambda = 10^3$ km. Using these values in (4) with $\dot{M}_A = 0.1$ for fast reconnection, we find $(\delta E)^2_{ob}/(\delta E)^2_{cr} \approx 10^{-3}$; once again LHD anomalous resistance fails to provide the necessary dissipation for strong reconnection.

During substorms strong plasma flows with velocities comparable to the Alfven speed are detected flowing both toward and away from the earth. However, in contrast to the prediction of the hydromagnetic reconnection model, the plasma is most often detected flowing parallel to the magnetic field rather than across the lines of force (De Coster and Frank, 1979). Parallel Alfvenic flows are predicted by Eastwood's (1972) model of the reconnection outflow region which is based on single particle dynamics rather than on MHD.

2.5 Slow Shocks

In the Petschek model, most of the magnetic field dissipation occurs in the standing slow switch-off shocks. Perhaps the absence of anomalous resistivity in the diffusion region is actually not that important, and we should really inquire about the dissipative structure of slow shocks. In the near earth plasma sheet, the expected slow shocks have been very difficult to identify, perhaps because the large ion gyroradius prevents a sharp shock discontinuity.

Recently, Feldman, et. al (1984) interpreted several crossings of the plasma sheet boundary in the distant tail as slow shocks. They argued that the change in plasma density, electron temperature, and flow velocity was consistent with the slow shock Rankine-Hugoniot relations. In a study of one slow shock event, Scarf, et. al (1984) reported the detection of low-frequency electric field turbulence which maximized in the magnetic gradient; the wave spectrum peaked near 300 HZ with a 200 HZ bandwidth, and had a peak amplitude of 3×10^{-4} v/m - (HZ)$^{1/2}$.

Since the ion plasma frequency was about 60 HZ at the slow shock, Scarf, et. al (1984) tested whether the observed turbulence could represent ion acoustic anomalous resistance; doppler shifting could easily extend the ion acoustic spectrum to 500 HZ. For an estimated slow shock thickness of 3-6 c/ω_{pi}, they found that the observed $(\delta E)^2_{ob}$ failed by a factor of 10^3 to provide the necessary resistivity. As an alternative they suggested that the LHD instability, which would not be detected by ISEE-3 wave instrument, might provide the anomalous resistance if the electric field amplitude was 2×10^{-3} v/m - (HZ)$^{1/2}$, which is not implausible.

However there is potential difficulty with this suggestion.

Theories of slow shocks predict that the electron Hall current should cause the magnetic field to rotate out of the V - B plane (Coroniti, 1971; Swift, 1983). The slow shocks observed by Feldman, et al. (1984) exhibited no rotation of the field. In order to supress the Hall current field rotations, the anomalous resistance must be so strong that the magnetic Reynolds length equals c/ω_{pi}. For $r_m = c/\omega_{pi}$, the anomalous collision frequency $\nu_* = \Omega_e$, thus decoupling the electrons from the magnetic field. Theories for the nonlinear saturation of the LHD instability predict $\nu_* \lesssim \Omega_{LH}$ (Sotnikov, et. al, 1981; Drake, et. al, 1984). Hence LHD anomalous resistance probably does not explain the observed slow shock structure.

3. Auroral Field-aligned Currents and Anomalous Resistance

The collision-dominated plasma in the ionosphere represents a resistive load which opposes magnetospheric convection. In order to sustain an ionospheric convection electric field, the magnetosphere must drive a Pedersen current across the ionospheric field lines; the Pedersen current is fed by a system of field-aligned currents which communicate the magnetospheric J x B stresses to the ionospheric plasma. At quiet times the convective flow is slow, and the field-aligned currents are readily carried by the thermal plasma. During substorms, however, the magnetosphere is highly stressed, and the field-aligned currents become sufficiently intense that the drift velocity should exceed the threshold for exciting current-driven instabilities in the topside ionosphere (Coroniti and Kennel, 1972); for the plasma temperatures and composition in the topside, the electrostatic ion cyclotron (EIC) wave has the lowest drift threshold (Kindel and Kennel, 1971). We therefore expect that EIC turbulence should provide an anomalous resistance to the current flow. Since the convection stresses demand that the ionospheric electric field be maintained, a parallel electric field should develop to derive the field-aligned current through the anomalous resistor.

The ionosphere can also act as a generator of field-aligned currents. The poleward edge of auroral oval connects magnetically to the outer boundary of the plasma sheet. The precipitation of hot plasma sheet electrons creates an ionospheric density discontinuity which separates polar cap and plasma sheet field lines. The corresponding discontinuity in the ionospheric Hall current, which is the convective flow of electrons, requires that a field-aligned current flow out of the ionosphere at the poleward edge. If the field-aligned current exceeds an instability threshold, an upward parallel electric field should develop in order to maintain the current against the anomalous resistivity.

Abundant evidence exists that parallel electric fields do develop in the far topside auroral ionosphere. The inverted-V electron precipitation structures (Frank and Ackerson, 1971) are certainly formed by the acceleration of electrons through a V-shaped electrostatic potential. The electrostatic shocks (Mozer, et al., 1977) with their spatially localized perpendicular electric field reversals indicate parallel potential drops in the kilovolt range. In association with the electrostatic shocks, Kintner, et. al (1978) also observed the anticipated EIC waves with potential fluctuations comparable to the ion temperature as predicted by strong turbulence theory (Hudson, et. al, 1978). The EIC waves are so powerful that cold ionospheric oxygen ions are accelerated to ~ keV's by the perpendicular wave electric fields (Mizera, et. al, 1981).

Is the observed parallel electric field the result of EIC anomalous resistance? A theoretical argument due to Cornwall and Chiu (1982) indicates probably not. The difficulty is that anomalous resistance is too powerful and too rapid a dissipation process. Typically wave turbulent effective collision frequencies are of order $\nu_* \sim 10^{-2}$ to 10^{-4} ω_{pe}; ν_* for EIC turbulence should be comparable to Ω_i. Suppose we calculate the field-line integrated Joule heating rate

$$\int J_\parallel^2/\sigma_* \, d\ell \sim nm_e v_D^2 \nu_* \ell \tag{4}$$

where ℓ is the length of the anomalous resistance region and $J_\parallel = nev_D$. The total Joule dissipation must be bounded by the rate at which the magnetosphere can supply new energy. The largest, and most likely, source of energy for the inverted-V auroral arcs is the flow of 1-10 keV field-aligned ions which are observed at the outer edge of the plasma sheet (DeCoster and Frank, 1979). Bounding (4) by $nm_i v_i{}^3$, where v_i is the ion flow speed, and setting $m_e V_D^2 \sim m_i v_i^2$ by energy conservation, we find $\nu_* < v_i/\ell$, the ion transit time, which is much smaller than any reasonable anomalous collision frequency.

Local plasma instabilities dissipate energy much more rapidly than the global magnetospheric configuration and convection can supply it. If an anomalous resistivity were to develop on the auroral field lines, the wave turbulence and parallel electric fields could not be sustained and would quickly die out. The observed large-scale (inverted-V) parallel potential probably results from the magnetic mirror force acting on electron and ion distributions with different pitch-angle anisotropies (Chiu and Cornwall, 1980); the field-aligned ion flows observed in the plasma sheet boundary layer are very likely the source of anisotropy (Serizawa and Sato, 1985). Small scale parallel potentials may also develop across ion acoustic solitons or "weak" double layers (Hudson, et. al, 1983).

4. Diffuse Auroral Precipitation

Convection transports hot electrons and ions from the plasma sheet into the dipolar region of the nightside auroral zone. In flowing earthward, the length of a field line decreases more slowly than the equatorial field strength increases. Since convection conserves the electron's first two adiabatic invariants, electrons gain energy preferentially perpendicular to the field direction. Hence the electron distribution becomes anisotropic with more particles having larger than smaller pitch-angles (α); the equatorial loss-cone at pitch-angle $\alpha_o^2 \approx 1/2L^3$ also weakly contributes to the anisotropy since particles with $\alpha < \alpha_o$ are lost to the atmosphere.

The velocity-space anisotropy should destabilize plasma waves which, upon growing to a finite amplitude, scatter the excess pitch-angle particles towards the loss-cone (Kennel and Petschek, 1966). If the excited wave spectrum is broad in frequency (ω) and/or wave vector (k), the pitch-angle scattering should be adequately described by the quasi-linear diffusion theory (Kennel and Engelmann, 1966). At large wave amplitudes, the pitch-angle diffusion can be sufficiently strong that the distribution is maintained isotropic within the equatorial loss-cone (Kennel, 1969).

In addition to discrete arcs and inverted-V type structures, the nightside auroral zone exhibits a latitudinally diffuse luminosity which is excited by the precipitation of ~ keV plasma sheet electrons. Low altitude satellites typically observe that the diffuse auroral precipipitation is isotropic with a down-going flux which is comparable to ~ keV electron fluxes measured at the magnetic equator (Winningham, et. al, 1975). Thus strong pitch-angle diffusion driven by near-equatorial plasma wave turbulence should provide an excellent description of the diffuse aurora (Ashour-Abdalla and Kennel, 1976).

The OGO-5 satellite discovered a class of electrostatic emissions which occur at odd half-harmonics (n + 1/2) of the electron cyclotron frequency with $\omega = 3/2\ \Omega_e$ being the most prevalent (Kennel, et. al, 1970). These emissions are quasi-continuously detected on auroral field-lines and are typically confined to within ± 3° of the magnetic equator (Belmont, et. al, 1983). Linear instability theory has demonstrated that the $3/2\ \Omega_e$ wave can be excited by a weak loss-cone distribution; the unstable waves have parallel resonant velocities $v_\parallel = (\omega - \Omega_e)/k_\parallel \sim a_e$, and propagate obliquely to the magnetic field with $k_\parallel/k_\perp \sim 0.1$ and $k_\parallel a_e/\Omega_e \sim 0.5$ (see Kennel and Ashour-Abdalla (1982) for an excellent review). The $3/2\ \Omega_e$ wave is essentially non-convective since the group velocities are typically comparable to the cold electron thermal speed (Ashour-Abdalla and Kennel, 1978).

The properties of the $3/2\ \Omega_e$ waves are apparently ideal for explaining the diffuse auroral electron precipitation. The only question is whether or not the $3/2\ \Omega_e$ wave amplitudes are sufficiently intense for the electrons to be on strong diffusion. For low-pitch electrons, the quasi-linear pitch-angle diffusion coefficient is approximately

$$D_{\alpha\alpha} = \frac{\pi}{2} \frac{e^2(\delta E)^2}{m^2 v_\parallel^2 \Omega} \left(\frac{k_\parallel}{\Delta k_\parallel}\right) \tag{5}$$

where δE is the wide-band amplitude and $\Delta k_\parallel$ is the width of the unstable $k_\parallel$ spectrum (an accurate calculation of $D_{\alpha\alpha}$ is given by Lyons (1974) and Belmont, et. al (1983)). Strong diffusion occurs if an electron is scattered across the loss-cone (α_o) during a transit time (Δt) across the the unstable equatorial region; for a latitudinal half-width λ, $\Delta t \approx$

$\lambda LR_E/v_\parallel$. Setting $D_{\alpha\alpha} = \alpha_o^2/2\Delta t$, we can estimate the critical wave amplitude necessary for strong diffusion as

$$(\delta E)_{cr} = \frac{1.5\ W_1^{3/4}}{L_6^{7/2}} \left(\frac{\Delta k_\parallel}{K_\parallel}\right)^{1/2} \frac{mv}{m} \tag{6}$$

where W_1 is the electron energy in units of 1 kev and $L_6 = L/_6$. Linear instability theory indicates that $\Delta k_\parallel \sim k_\parallel/2$ (Ashour-Abdalla and Kennel, 1976).

In a statistical study using the GOES-2 wave data, Belmont, et. al (1983) found that on the nightside the 3/2 Ω_e wave amplitude at the equator was less than 1.0 mv/m for 85% of the time and that $\delta E < 0.1$ mv/m was observed for 74% of the time. Since the 3/2 Ω_e amplitudes exceed $(\delta E)_{cr}$ for only 15% of the observation time, Belmont, et. al (1983) question whether the 3/2 Ωe waves are responsible for the diffuse auroral precipation.

The disagreement between the expected strong diffusion amplitudes and the GEOS observations is very puzzling. Although it is always tempting to criticize the quasilinear theory, the observed frequency and expected wave number bandwidths of the 3/2 Ω_e waves strongly indicate that quasilinear diffusion should be an excellent description of the pitch-angle scattering process. Belmont, et. al (1983) suggest that other wave modes and/or enhanced non-normal mode fluctuations may be responsible for the diffuse auroral precipitation, but there is no observational evidence for either possibility.

The GEOS-2 amplitude statistics at the magnetic equator were obtained only for a two month interval; no information on simultaneous observations of the diffuse auroral precipitation was reported. The two month interval was not particularly disturbed; the planetary index K_p was less than 3 for 80% of the time. At quiet times the diffuse electron precipitation is often anisotropic, and the equatorward boundary of the diffuse aurorae is at higher invariant latitudes than correspond to the geosynchronous L-shell (Winningham, et. al, 1975). Hence if GEOS-2 was near the inner edge of the plasma sheet most of the time, and if the diffuse auroral precipitation was anisotropic most of the time, the GEOS-2 and 3/2 Ω_e amplitudes might be reconciled with the quasilinear predictions. Clearly, further data analysis which includes simulanteous wave amplitude and precipitation fluxes is needed.

5. DISCUSSION

The coupling of hydromagnetics with locally determined anomalous transport is such a reasonable theoretical description of magnetospheric dynamics that its validity has rarely been questioned. In this paper we have attempted a quasi-rigorous evaluation of three processes for which the application of this theoretical philosophy is well-developed. For each case the theoretically expected plasma wave turbulence is indeed observed. However, either the strength of the anomalous dissipation is too weak (reconnection, diffuse aurora) or the anomalous dissipation description is inappropriate (auroral field-aligned currents).

We should conclude, however, not that plasma wave turbulence is always unimportant, but only that the local description of collisionless dissipation is often inadequate. In the hot magnetospheric plasmas, particles rapidly transverse local interaction regions and ballistically communicate information about these interactions over large spatial distances. Since particle thermal velocities are comparable to hydromagnetic wave and flow speeds, collisionless wave turbulent dissipation interacts with the global structure and must adjust to the hydromagnetic space and time scales. In space plasmas collisionless dissipation is therefore fundamentally nonlocal. For the future, the theoretical challenge is to develop analytical models and plasma simulation experiments which self-consistently describe the global coupling of dissipation processes to the hydromagnetic configuration of the magnetosphere.

ACKNOWLEDGEMENTS

This work was partially supported by the National Science Foundation under Grant ATM-81-19544 and by NASA under Grant NGL-05-007-190.

REFERENCES

Anderson, R. R., C. C. Harvey, M. M. Hoppe, B. T. Tsuratani, T. E. Eastman, and J. Etcheto, J. Geophys. Res., 87, 2087, 1982.
Ashour-Abdalla, M., and C. F. Kennel, in Magnetospheric Particles and Fields, ed. by B. M. McCormac, pg. 181, D. Reidel, Highman, Mass, 1976.
Ashour-Abdalla, M., and C. F. Kennel, J. Geophys. Res., 83, 1531, 1978.
Belmont, J. and D. Fontaine, and P. Canu, J. Geophys. Res., 88, 9163, 1983.
Berchem, J. and C. T. Russell, J. Geophys. Res., 87, 2108, 1982.
Berchem, J. and C. T. Russell, J. Geophys. Res., 89, 127, 1984.
Brackbill, J. U., D. W. Forslund, K. B. Quest, and D. Winske, Phys. Fluids, 27, 2682, 1984.
Chiu, Y. T., and J. M. Cornwall, J. Geophys. Res., 85, 543, 1980.
Cornwall, J. M., and Y. T. Chiu, J. Geophys. Res., 87, 1517, 1982.
Coroniti, F. V., Nucl. Fusion, 11, 261, 1971.
Coroniti, F. V., and C. F. Kennel, J. Geophys. Res. 77, 2835, 1982.
DeCoster, R. J. and L. A. Frank, J. Geophys. Res., 84, 5099, 1979.
Drake, J. F., P. N. Guzdar, A. B. Hassam, and J. D. Huba, Phys. Fluids, 27, 1147, 1984.
Eastwood, J. W., Planet. Space Sci., 20, 1555, 1972.
Feldman, W. C., S. J. Schwartz, S. J. Bame, D. M. Baker, J. Birn, J. T. Gosling, E. W. Hones, Jr., D. J. McComas, J. A. Slavin, E. J. Smith, and R. D. Zwickl, Geophys. Res. Lett., 11, 599, 1984.
Frank, L. A. and K. Ackerson, J. Geophys. Res., 76, 3612, 1971.
Gurnett, D. A., L. A. Frank and R. P. Lepping, J. Geophys. Res., 81, 6059, 1976.
Gurnett, D. A., R. D. Anderson, B. T. Tsurutani, E. J. Smith, G. Paschmann, G. Haerendel, S. J. Bame, and C. T. Russell, J. Geophys. Res., 84, 7043, 1979.

Hones, E. W., Jr., Radio Sci., 8, 879, 1973.
Hones, E. W., Jr., D. N. Baker, S. J. Bame, W. C. Feldman, J. T. Gosling, D. J. McComas, R. D. Zwickl, J. A. Slavin, E. J. Smith, and B. T. Tsurutani, Geophys. Res. Lett., 11, 5, 1984.
Huba, J. D., N. T. Gladd, and K. Papadopoulos, Geophys. Res. Lett., 83, 52217, 1978.
Huba, J. D., J. F. Drake, and N. T. Gladd, Phys. Fluids, 23, 522, 1980.
Hudson, M. K., R. L. Lysak, and F. S. Mozer, Geophys. Res. Lett., 5, 143, 1978.
Hudson, M. K., W. Lotko, I. Roth, and E. Witt, Geophys. Res. Lett., 88, 916, 1983.
Kennel, C. F., Rev. Geophys. 7, 379, 1969.
Kennel, C. F. and F. E. Engelmann, Phys. Fluids, 9, 2377, 1966.
Kennel, C. F. and H. E. Petschek, J. Geophys. Res., 71, 1, 1966.
Kennel, C. F., F. L. Scarf, R. W. Fredricks, J. H. McGehee, and F. V. Coroniti, Geophys. Res. Lett., 75, 6136, 1970.
Kennel, C. F. and M. Ashour Abdalla, in Magnetospheric Plasma Physics, 04, ed. by A. Nishida, pg. 245, (CAPJ-Reidel), 1982.
Kindel, J. and C. F. Kennel, Geophys. Res. Lett., 76, 3055, 1971.
Kintner, P. M., M. C. Kelly, and F. S. Mozer, Geophys. Res. Lett., 5, 139, 1978.
Levy, R. H., H. E. Petschek, and G. L. Siscoe, AIAA, 2, 2065, 1964.
Lyons, L. R., J. Geophys. Res., 79, 575, 1974.
McPherron, R. L., C. T. Russell and M. P. Aubry, J. Geophys. Res., 78, 3131, 1973.
Mizera, P. F., J. F. Fennel, D. R. Croley, Jr., and D. J. Gurney, J. Geophys. Res., 86, 7566, 1981.
Mozer, F. S., C. W. Carlson, M. K. Hudson, R. B. Torbert, B. Parady, J. Yatteau and M. C. Kelly, Phys. Rev. Lett., 38, 292, 1977.
Pashmann, G., G. Haerendel, I. Papamastorakis, N. Sckopke, S. J. Bame, J. T. Gosling and C. T. Russell, J. Geophys. Res., 87, 2159, 1982.
Petschek, H. E., in Proceedings of the AAS-NASA Symposium on the Physics of Solar Flares, NASA SP-50, p. 425, 1964.
Russell, C. J., and R. C. Elphic, J. Geophys. Res., 6, 33, 1979.
Sato, T., and T. Hayashi, Phys. Fluids, 22, 1189, 1979.
Scarf, F. L., F. V. Coroniti, C. F. Kennel, E. J. Smith, J. D. Slavin, B. J. Tsurutani, S. J. Bame, and W. C. Feldman, Geophys. Res. Lett., 11, 1050, 1984.
Serizawa, Y. and T. Sato, these Proceedings, 1985.
Sonnerup, B. U. O., in Solar System Plasma Physics III, ed. by L. J. Lanzerotti, C. F. Kennel, and E. N. Parker, p. 45, North-Holland, 1979.
Sonnerup, B. U. O., G. Paschmann, I. Papamastoraski, N. Schopke, G. Haerendel, S. J. Bame, J. R. Ashbridge, J. T. Gosling and C. T. Russell, J. Geophys. Res., 86, 10,049, 1981.
Sotnikov, V. I., V. D. Shapiro, and V. I. Shechenko, Physica 2D, 170, 1981.
Swift, D. W., Geophys. Res. Lett., 8, 5185, 1983.
Tanaka, M. and T. Sato, J. Geophys. Res., 86, 4451, 1981.

Tsurutani, B. T., E. J. Smith, R. M. Thorne, R. R. Anderson, D. A. Gurnett, G. K. Parks, C. S. Liu, and C. T. Russell, Geophys. Res. Lett., 8, 183, 1981.
Ugai, M., Phys. Fluids, 25, 1027, 1982.
Winningham, J. P., F. Tasuharo, S.-I. Akasofu and W. J. Heikkila, J. Geophys. Res., 80, 3148, 1975.

KINETIC THEORY OF PLASMA WAVES

K. G. Rönnmark
Kiruna Geophysical Institute
University of Umeå
S-901 87 Umeå
Sweden

ABSTRACT. In this paper, I outline the solution of Vlasov-Maxwell's equations with given initial conditions. When transients have died out, the temporal evolution of each spatial Fourier component is completely determined by a dispersion relation. I derive the electrostatic dispersion relation and discuss the resonant interaction between particles and electrostatic waves. A new derivation of the wave energy density in a plasma with arbitary dissipation is given. The numerical solution of the dispersion relation of waves in a Maxwellian plasma is discussed, and finally I show some examples of numerically evaluated dispersion surfaces.

1. FORMAL SOLUTION OF VLASOV-MAXWELL'S EQUATIONS

1.1. The basic equations

Waves in plasmas can be described by several different mathematical models at various levels of approximation. All treatments of plasma waves are based on Maxwell's equations

$$\underline{\nabla}\times\underline{E} + \partial_t\underline{B} = 0 \tag{1a}$$

$$\underline{\nabla}\cdot\underline{B} = 0 \tag{1b}$$

$$\varepsilon_o\underline{\nabla}\cdot\underline{E} = \rho \tag{1c}$$

$$\mu_o^{-1}\underline{\nabla}\times\underline{B} - \varepsilon_o\partial_t\underline{E} = \underline{j} \ , \tag{1d}$$

and the differences are mainly manifested in the way the source terms ρ and $\underline{j}$ are calculated from the plasma properties. In the kinetic theory, which I will develop here, the charge and current densities are calculated from particle distribution functions $\hat{f}_s(\underline{v},\underline{r},\underline{t})$ as

$$\rho = \sum_s q\int\hat{f}_s d\underline{v} \tag{2a}$$

Space Science Reviews **42** (1985) 411–428. 0038–6308/85.15

and

$$\underline{j} = \sum_s q \int \underline{v}\hat{f}_s d\underline{v} \ , \tag{2b}$$

where the sums are over all species of particles. The distribution function of each species satisfies the Vlasov equation

$$[\partial_t + \underline{v}\cdot\underline{\nabla} + q/m(\underline{E}+\underline{v}\times\underline{B})\cdot\partial_{\underline{v}}]\hat{f}_s = 0 \tag{3}$$

A detailed discussion of the Vlasov equation is beyond the scope of this paper, but it should be pointed out that since the plasma is described by continuous functions $\hat{f}_s$, discrete particle effects are not included in this theory. The definition of the distribution function implies that $\hat{f}_s(\underline{v},\underline{r},t)\,d\underline{r}\,d\underline{v}$ is a large number of particles, and the fields in Maxwell's equations are not the true microscopic fields, but the average field in a small volume containing a large number of particles.

1.2. The linearized Vlasov equation

We split the distribution functions $\hat{f}_s(\underline{v},\underline{r},t) = n_s[f_s^o(\underline{v}) + f_s(\underline{v},\underline{r},t)]$ into an unperturbed part with homogeneous density n_s and velocity distribution $f_s^o(\underline{v})$, and a perturbation $f_s(\underline{v},\underline{r},t)$ related to a plasma wave. To simplify the notation, we will henceforth assume that f_s is nonzero only for a single species and drop the suffix s. In the presence of a homogeneous magnetic field $\underline{B}_o$, the acceleration due to the Lorentz force consists of a part $\underline{F}_o = q/m(\underline{v}\times\underline{B}_o)$, in addition to the part $\underline{F} = q/m(\underline{E}+\underline{v}\times\underline{B})$ caused by the wave fields. The unperturbed plasma is assumed to be a Vlasov equilibrium,

$$[\partial_t + \underline{v}\cdot\underline{\nabla} + \underline{F}_o\cdot\partial_{\underline{v}}]f^o = 0, \tag{4}$$

and the perturbations thus satisfy the equation

$$[\partial_t + \underline{v}\cdot\underline{\nabla} + \underline{F}_o\cdot\partial_{\underline{v}}]\, f + \underline{F}\cdot\partial_{\underline{v}} f^o = -\underline{F}\cdot\partial_{\underline{v}} f. \tag{5}$$

The right hand side of this equation is nonlinear in the amplitude of the perturbing wave, and assuming the perturbation to be small we may as a first approximation neglect this nonlinear term. We also introduce Fourier-Laplace transforms as

$$A(\underline{k},\omega) = \int_o^\infty dt e^{i\omega t} \int d\underline{r}\; e^{-\underline{k}\cdot\underline{r}} A(\underline{r},t)\ , \tag{6}$$

where $\omega = \nu + i\Gamma$ is complex, with Γ sufficiently large to ensure convergence of the integral. The linearized and transformed version of equation (5) is then written

$$g^{-1} f(\underline{v},\underline{k},\omega) + f(\underline{v},\underline{k},0) + \underline{F}(\underline{k},\omega)\cdot\partial_{\underline{v}} f^{o}(\underline{v}) = 0 \tag{7}$$

Here, the operator

$$g^{-1} = -i\omega + i\,\underline{k}\cdot\underline{v} + \underline{F}_{o}\cdot\partial_{\underline{v}} \tag{8}$$

is introduced, and the term $f(\underline{v},\underline{k},0)$ represents the initial value of the perturbation. At least formally, we can easily solve for the distribution function as

$$f(\underline{v},\underline{k},\omega) = -\,g\underline{F}\cdot\partial_{\underline{v}} f^{o}(\underline{v}) - g f(\underline{v},\underline{k},0). \tag{9}$$

The corresponding current density, calculated from equation (2b), may be written

$$\underline{j}(\underline{k},\omega) = nq\int \underline{v} f(\underline{v},\underline{k},\omega)d\underline{v} = \underline{\underline{\sigma}}(\underline{k},\omega)\cdot\underline{E}(\underline{k},\omega) + \underline{j}_{o}(\underline{k},\omega), \tag{10}$$

where we have introduced the conductivity tensor $\underline{\underline{\sigma}}(\underline{k},\omega)$ defined by

$$\underline{\underline{\sigma}}(\underline{k},\omega)\cdot\underline{E}(\underline{k},\omega) = -nq\int \underline{v} g F(\underline{k},\omega)\cdot\partial_{\underline{v}} f^{o}(\underline{v})\, d\underline{v}\ , \tag{11}$$

and the current

$$\underline{j}_{o}(\underline{k},\omega) = -nq\int \underline{v}\, g\, f(\underline{v},\underline{k},0)\, d\underline{v} \tag{12}$$

due to the initial perturbation.

1.3. The wave equation

Eliminating $\underline{B}$ from Ampere's law (1d) by means of (1a) and applying the Fourier-Laplace transform, we may write the resulting wave equation in the form

$$i\,\omega\,\varepsilon_{o}\,[\,(\underline{k}\underline{k}-k^{2}\underline{\underline{I}})c^{2}\omega^{-2} + \underline{\underline{I}}\,]\cdot E(\underline{k},\omega) - \underline{j}(\underline{k},\omega) = \tag{13}$$

$$= [\,\varepsilon_{o}\underline{E}(\underline{k},t) + i\omega^{-1}(\underline{j}(\underline{k},t) + \varepsilon_{o}\partial_{t}\underline{E}(\underline{k},t))]_{t=0}.$$

Expressing $j(\underline{k},\omega)$ as in equation (10), we collect all the initial conditions in a term

$$\underline{j}_{I}(\underline{k},\omega) = [\varepsilon_{o}(1-i\omega^{-1}\partial_{t})\,\underline{E}\,(\underline{k},t) - nq\int \underline{v}(g-i\omega^{-1})f(\underline{v},\underline{k},t)dv]_{t=0} \tag{14}$$

and rewrite the wave equation as

$$i\underline{\underline{D}}(\underline{k},\omega)\cdot\underline{E}(\underline{k},\omega) = \underline{j}_I(\underline{k},\omega), \tag{15}$$

where

$$\underline{\underline{D}}(\underline{k},\omega) = \omega\varepsilon_o\left[(\underline{k}\,\underline{k}-k^2\,\underline{\underline{I}})\frac{c^2}{\omega^2}+\underline{\underline{I}}-\frac{\underline{\underline{\sigma}}(\underline{k},\omega)}{i\,\omega\varepsilon_o}\right]. \tag{16}$$

In order to solve for the electric field in (15), we must invert the matrix $\underline{\underline{D}}$. Applying a standard method, we write the inverse matrix as

$$\underline{\underline{D}}^{-1}(\underline{k},\omega) = \frac{\underline{\underline{C}}(\underline{k},\omega)}{D(\underline{k},\omega)}, \tag{17}$$

where $D(\underline{k},\omega) = \det\underline{\underline{D}}(\underline{k},\omega)$ and $\underline{\underline{C}}(\underline{k},\omega)$ is the adjoint of $\underline{\underline{D}}(\underline{k},\omega)$, defined as the transpose of the matrix of its cofactors. We then obtain

$$\underline{E}(\underline{k},\omega) = \frac{\underline{\underline{C}}(\underline{k},\omega)\cdot\underline{j}_I(\underline{k},\omega)}{i\,D(\underline{k},\omega)}, \tag{18}$$

and the time dependence of each Fourier component is found by inverting the Laplace transform as

$$\underline{E}(\underline{k},t) = \frac{1}{2\pi}\int_{-\infty+i\Gamma}^{\infty+i\Gamma} d\omega\;\frac{\underline{\underline{C}}(\underline{k},\omega)\cdot\underline{j}_I(\underline{k},\omega)}{i\,D(\underline{k},\omega)}\,e^{-i\omega t}. \tag{19}$$

The integration in the complex ω-plane runs along the path P, as indicated in Figure 1. When carrying out the Laplace transform, we assume that Γ is chosen sufficiently large, so that all poles of the integrand in (19) lay below P. We also notice that for $t>0$, the factor $\exp(-i\omega t) = \exp(-i\nu t+\gamma t)$ makes the integrand exponentially small as $\gamma\to-\infty$. According to the roules of contour integration, we may, without crossing any poles, deform the original path into the path P' shown in Figure 1. The contribution from the horizontal part of P' can be neglected when it is pulled sufficiently far down in the complex ω-plane. From the way $\underline{\underline{C}}(\underline{k},\omega)$ and $D(\underline{k},\omega)$ were formed, we notice that any pole in $\underline{\underline{C}}(\underline{k},\omega)$ will be cancelled by a pole of at least the same order in $D(\underline{k},\omega)$. For similar reasons, there will be no contribution from the pole of $\underline{j}_I(\underline{k},\omega)$ at $\omega = 0$. Adding up the residues at the remaining singularities, the integral is evaluated in the form

$$\underline{E}(\underline{k},t) = -\sum_{\substack{\text{zeros}\\ \text{of } D(\underline{k},\omega)}}\frac{\underline{\underline{C}}(\underline{k},\omega)\cdot\underline{j}_I(\underline{k},\omega)}{\partial_\omega D(\underline{k},\omega)}e^{-i\omega t} - \sum_{\substack{\text{poles}\\ \text{of } \underline{j}_o(\underline{k},\omega)}}\frac{\underline{\underline{C}}(\underline{k},\omega)}{D(\underline{k},\omega)}\cdot\left[\text{residue of } \underline{j}_o(\underline{k},\omega)\right]e^{-i\omega t} \tag{20}$$

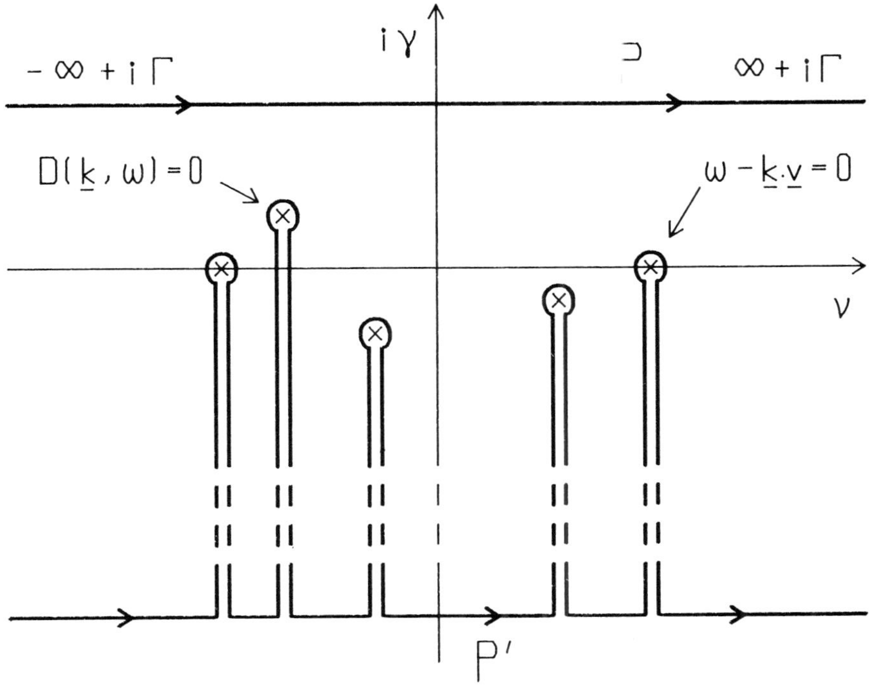

Figure 1. The original (P) and deformed (P') integration paths for inversion of the Laplace transform.

1.4. Unmagnetized plasma

Let us, as an example, assume that there is only one solution ω_k satisfying the dispersion relation $D(\underline{k},\omega_k) = 0$, and that the plasma is unmagnetized so that $g^{-1} = -i(\omega-\underline{k}\cdot\underline{v})$. From equation (12) we then have

$$\underline{j}_o(\underline{k},\omega) = nq \int \underline{v}\,\frac{f(\underline{v},\underline{k},0)}{i(\omega-\underline{k}\cdot\underline{v})}\,d\underline{v}, \tag{21}$$

and according to the formula (20) we obtain

$$\underline{E}(\underline{k},t) = -\frac{\underline{\underline{C}}(\underline{k},\omega_k)\cdot\underline{j}_I(\underline{k},\omega_k)}{\partial_\omega D(\underline{k},\omega_k)}\,e^{-i\omega_k t} + $$

$$+ inq \int \frac{\underline{\underline{C}}(\underline{k},\underline{k}\cdot\underline{v})\cdot\underline{v}}{D(\underline{k},\underline{k}\cdot\underline{v})}\,f(\underline{v},\underline{k},0)\,e^{-i\underline{k}\cdot\underline{v}t}\,d\underline{v} \tag{22}$$

The time dependence $e^{-i\omega_k t}$ of the first term is completely determined by the solution of the dispersion relation $D(\underline{k},\omega) = 0$. This term satisfies the homogeneous wave equation, and is often referred to as an eigenmode of the plasma. The second part of (22) (or (20)) is the complementary function, which is needed in order to satisfy inhomogeneous initial conditions. As t grows, the factor exp $(-i\underline{k}\cdot\underline{v}t)$ will become a rapidly oscillating function of $\underline{v}$, and if the rest of the integrand is reasonably smooth, this term will become small for sufficiently large t. In space physics, the transient response is usually of limited interest, and the complementary function can often be neglected. However, $\underline{j}_o(\underline{k},\omega)$ in equation (20) may also be taken to represent other inhomogeneous terms, such as a prescribed external current or a nonlinear current.

1.5. Velocity integrals

A problem which should be considered here is sometimes encountered when calculating velocity integrals of the type found in the conductivity tensor. In an unmagnetized plasma, where $g = i(\omega-\underline{k}\cdot\underline{v})^{-1}$, the conductivity is from equation (11) defined by

$$\underline{\underline{\sigma}}(\underline{k},\omega)\cdot\underline{E}(\underline{k},\omega) = -\, inq \int \underline{v}\, \frac{\underline{F}(\underline{k},\omega)\cdot\partial_{\underline{v}} f^{o}(v)}{\omega-\underline{k}\cdot\underline{v}}\, d\,\underline{v} \tag{23}$$

for $\omega = \nu + i\Gamma$, where Γ is a large positive constant. If we could carry out the $\underline{v}$-integration at this stage, there would be no problem, since the denominator is well defined and nonzero. However, we often need the conductivity at some frequency $\omega = \nu + i\gamma$ with $\gamma \leqq 0$, and this is given by the analytic continuation of the definition (23) to the rest of the complex ω-plane. This means, that even if we postpone the evaluation of this kind of integral until we have inserted a frequency with $\gamma \leqq 0$, we must still evaluate the integral as if ω had a positive imaginary part. Examples where this is needed are found in the second term of equation (22). Since $\underline{\underline{D}}(\underline{k},\omega)$ contains $\underline{\underline{\sigma}}(\underline{k},\omega)$ we find that $\underline{C}(\underline{k},\underline{k}\cdot\underline{v})$ as well as $D(\underline{k},\underline{k}\cdot\underline{v})$ will contain integrals of the type

$$\int \underline{v}'\, \frac{\underline{F}(\underline{k},\underline{k}\cdot\underline{v})\cdot\partial_{\underline{v}'} f^{o}(v')}{\underline{k}\cdot\underline{v}-\underline{k}\cdot\underline{v}'}\, d\,\underline{v}' \tag{24}$$

This expression is properly evaluated, if we first add a positive imaginary part to $\underline{k}\cdot\underline{v}$, carry out the integration, and then let $\underline{k}\cdot\underline{v}$ go back to its original real value.

2. ELECTROSTATIC WAVES

2.1. The dielectric function

In order to keep the algebra manageable while developing more explicit formulas from the rather formal theory presented in the previous section, we will here consider electrostatic waves. It is convenient and common practice to introduce the dielectric tensor defined by

$$\underline{\underline{\varepsilon}}(\underline{k},\omega) = \underline{\underline{I}} - \frac{\underline{\underline{\sigma}}(\underline{k},\omega)}{i\omega\varepsilon_o} , \tag{25}$$

and the dielectric function

$$\varepsilon(\underline{k},\omega) = \hat{k} \cdot \underline{\underline{\varepsilon}}(\underline{k},\omega) \cdot \hat{k} , \tag{26}$$

where the unit vector $\hat{k}$ is defined by $\underline{k} = \hat{k} k$. Equation (16) then becomes

$$\underline{\underline{D}}(\underline{k},\omega) = \omega\,\varepsilon_o \left[(\hat{k}\hat{k} - \underline{\underline{I}}) k^2 c^2/\omega^2 + \underline{\underline{\varepsilon}}(\underline{k},\omega) \right] , \tag{27}$$

and expanding the determinant we obtain

$$D = \omega\,\varepsilon_o \left\{ \varepsilon\left(\frac{kc}{\omega}\right)^4 + \left[\hat{k}\cdot\underline{\underline{\varepsilon}}\cdot\underline{\underline{\varepsilon}}\cdot\hat{k} - \varepsilon \text{ trace } \underline{\underline{\varepsilon}} \right] \left(\frac{kc}{\omega}\right)^2 + \det \underline{\underline{\varepsilon}} \right\} . \tag{28}$$

In the electrostatic approximation, which corresponds to the limit $(kc/\omega) \to \infty$, only the first term is kept and the dispersion relation is reduced to

$$\varepsilon(\underline{k},\omega) = 0. \tag{29}$$

Using equation (11), we find that the dielectric function is expressed in the form

$$\begin{aligned} \varepsilon(\underline{k},\omega) &= 1 - i\omega_p^2 \int \frac{\hat{k}\cdot\underline{v}}{\omega} g\, \hat{k}\cdot\partial_{\underline{v}} f^o(\underline{v})\, d\underline{v} = \\ &= 1 - i\frac{\omega_p^2}{k^2} \int g\, \underline{k} \cdot \partial_{\underline{v}} f^o(\underline{v})\, d\underline{v} , \end{aligned} \tag{30}$$

where $\omega_p^2 = \dfrac{nq^2}{\varepsilon_o m}$, and the last step follows from

$$\begin{aligned} 0 &= \int \underline{k}\cdot\partial_{\underline{v}} f\, d\underline{v} = \int g^{-1} g\, \underline{k}\cdot \partial_{\underline{v}} f\, d\underline{v} = \\ &= \int (-i\omega + i\underline{k}\cdot\underline{v} + \underline{F}_o\cdot\partial_{\underline{v}})\, g\, \underline{k}\cdot\partial_{\underline{v}} f\, d\underline{v} = -i \int (\omega - \underline{k}\cdot\underline{v})\, g\, \underline{k}\cdot\partial_{\underline{v}} f\, d\underline{v} \end{aligned}$$

(Use partial integration to show this.)

2.2 The g-operator

In an unmagnetized plasma, where $g = i(\omega - \underline{k}\,\underline{v})^{-1}$, we find that the dielectric function is given by the simple formula

$$\varepsilon(\underline{k},\omega) = 1 + \frac{\omega_p^2}{k^2}\int \frac{\underline{k}\cdot\partial_{\underline{v}} f^o(\underline{v})}{\omega-\underline{k}\cdot\underline{v}}\, d\underline{v}\,, \tag{31}$$

but in a magnetized plasma the algebra is more complicated. Introducing cylindrical coordinates with the z-axis along $\underline{B}_o$ in $\underline{v}$ and $\underline{k}$ - space,

$$\underline{v} = v_\perp \cos\phi\, \hat{x} + v_\perp \sin\phi\, \hat{y} + v_z \hat{z} \tag{32a}$$

$$\underline{k} = k_\perp \cos\theta\, \hat{x} + k_\perp \sin\theta\, \hat{y} + k_z \hat{z} \tag{32b}$$

we find that $\underline{F}_o \cdot \partial_{\underline{v}} = -\Omega\partial_\phi$, with $\Omega = qB_o/m$, and thus

$$g^{-1} f = [-i\omega + ik_z v_z + ik_\perp v_\perp \cos(\phi-\theta) - \Omega\partial_\phi] f. \tag{33}$$

Multiplying by the integrating factor $\exp(iA(\phi))$, where $A(\phi) = [(\omega-k_z v_z)\phi - k_\perp v_\perp \sin(\phi-\theta)]/\Omega$, we have

$$e^{iA}(i\partial_\phi A + \partial_\phi) f = \partial_\phi(e^{iA} f) = -\Omega^{-1} e^{iA} g^{-1} f\,, \tag{34}$$

which is integrated to give

$$f = -\Omega^{-1} e^{-iA(\phi)} \int_\infty^\phi e^{iA(\phi')} g^{-1} f\, d\phi'. \tag{35}$$

If we now coose $f = g\,(\ldots)$ we obtain

$$g\,(\ldots) = -\Omega^{-1} e^{-iA(\phi)} \int_\infty^\phi e^{iA(\phi')} (\ldots)\, d\phi'. \tag{36}$$

We can rewrite this formula in a more familiar form by making use of the identity

$$\exp(iz\sin\phi) = \sum_{n=-\infty}^{\infty} J_n(z)\, e^{in\phi}, \tag{37}$$

which is obtained by substituting $t = \exp i\phi$ in the generating function for ordinary Bessel functions J_n,

$$\exp\left[\frac{z}{2}(t - t^{-1})\right] = \sum_{n=-\infty}^{\infty} t^n J_n(z).$$

This allows us to express the g-operator as

$$g(\ldots) = -\Omega^{-1} \sum_{n,m} J_n J_m \int_\infty^\phi (\ldots)\, e^{i\left[\frac{\omega-k_z v_z-n\Omega}{\Omega}(\phi'-\phi) + i(n-m)(\phi'-\theta)\right]} d\phi' \tag{38}$$

where the argument of the Bessel functions is understood to be $k_\perp v_\perp/\Omega$. To me, the size of this expression seems like a strong argument in favour of the g-notation.

2.3. The electrostatic dispersion relation

We now use the formula (38) in equation (30) and integrate over the angles ϕ' and ϕ. After some algebra, we obtain the dielectric function in the very useful form

$$\varepsilon(\underline{k},\omega) = 1 + \frac{\omega_p^2}{k^2}\int \sum_n \frac{J_n^2\left(\frac{k_\perp v_\perp}{\Omega}\right)}{\omega - k_z v_z - n\Omega}\left[\frac{n\Omega}{v_\perp}\partial_{v_\perp} + k_z \partial_{v_z}\right] f^o \, d\underline{v}\,, \tag{39}$$

which is valid for any gyrotropic distribution f^o. The dispersion relation $\varepsilon(\underline{k},\omega) = 0$, with $\varepsilon(\underline{k},\omega)$ given by this formula, describes the propagation, damping or growth of electrostatic waves, and has been applied to numerous problems in space plasma physics.

2.4. Resonant wave-particle interaction

In order to investigate resonant wave-particle interactions, we assume that we have found a solution $\omega_k = \nu_k + i\gamma_k$ to the electrostatic dispersion relation. If this solution corresponds to a weakly damped or growing wave $|\gamma_k| << |\nu_k|$ we may expand $\varepsilon(\underline{k},\omega)$ in a Taylor series about ν_k,

$$\varepsilon(\underline{k},\omega_k) \approx \varepsilon(\underline{k},\nu_k) + i\gamma_k[\partial_\omega \varepsilon]_{\nu_k} \approx \varepsilon_R(\underline{k},\nu_k) + i\varepsilon_I(\underline{k},\nu_k) + i\gamma_k[\partial_\nu \varepsilon_R]_{\nu_k}\,, \tag{40}$$

where $\varepsilon(\underline{k},\nu_k) = \varepsilon_R(\underline{k},\nu_k) + i\varepsilon_I(\underline{k},\nu_k)$. Equating the imaginary part of (40) to zero, we can estimate the growth or damping rate as

$$\gamma_k \approx -\frac{\varepsilon_I(\underline{k},\nu_k)}{[\partial_\nu \varepsilon_R]_{\nu_k}}\,. \tag{41}$$

In the limit $\gamma \to 0$, the real and imaginary parts of $\varepsilon(\underline{k},\omega)$ may be separated by means of the general rule

$$\lim_{\gamma\to 0_+} \int \frac{f(x)}{x-y+i\gamma}\,dx = P\int \frac{f(x)}{x-y}\,dx - i\pi f(y)\,, \tag{42}$$

where P denotes the principal value of the integral. Notice that according to the discussion in Section 1.5, we must here assume $\gamma > 0$ even if it turns out that $\gamma_k \leqq 0$. Applying this to equation (39), we find that the imaginary part is given by

$$\varepsilon_I(\underline{k},\nu_k) \approx -\ 2\pi^2 \frac{\omega_p^2}{k^2 k_z} \sum_n \int_0^\infty v_\perp dv_\perp J_n^2\left(\frac{k_\perp v_\perp}{\Omega}\right)\left[\frac{n\Omega}{v_\perp}\partial_{v_\perp} f^o + k_z \partial_{v_z} f^o\right]_{v_z = \frac{\nu_k - n\Omega}{k_z}} \qquad (43)$$

Since γ_k is proportional to $\varepsilon_I(\underline{k},\nu_k)$, we conclude that the growth or damping is largely controlled by particles with parallel velocities satisfying $v_z = (\nu_k - n\Omega)/k_z$. The distribution f^o is usually a rapidly decreasing function of v_z, and at least when k_z is small, only a few harmonic numbers $n \approx \nu_k/\Omega$ will contribute in the summation over n.

The range of important perpendicular velocities is less well defined, but as a rule of thumb we can use that in order to obtain a reasonably large value of $J_n(k_\perp v_\perp/\Omega)$ when $n \approx \nu_k/\Omega$, the perpendicular velocity should be approximately $v_\perp \approx \nu_k/k_\perp$. Somewhat more accurate estimates can be found by considering the actual maxima of the Bessel functions.

We may of course also use these estimates to find the wavevector components $k_z \approx (\nu_k - n\Omega)/v_z$ and $k_\perp \approx \nu_k/v_\perp$, which are likely to interact strongly with particles having given velocities.

2.5. Classification of instabilities

Instabilities with a growth rate given approximately by (41), and a real frequency ν_k satisfying $\varepsilon_R(\underline{k},\nu_k) = 0$, are usually collectively called resonant (kinetic, resistive, ...) instabilities, since a group of resonant particles can be identified as responsible for the instability. If the plasma parameters are varied, a point may be reached beyond which the equation $\varepsilon_R(\underline{k},\nu) = 0$ no longer has a solution for real ν. The solutions then occur in complex conjugate pairs, and the growing solutions are called non resonant (fluid, reactive, ...) instabilities. This classification is often useful and rather unambiguous, but sometimes γ_k as given by equation (41) may be large when the solution of $\varepsilon(\underline{k},\omega) = 0$ becomes complex, and such cases may be hard to classify.

3. ENERGY DENSITY OF PLASMA WAVES

In the limit of vanishing dissipation, the energy density of a plasma wave with amplitude $\underline{E}_o$ is given by the well known formula (Landau and Lifshitz, 1960)

$$W = \frac{\varepsilon_o}{2\omega} \underline{E}_o^* \cdot \partial_\omega(\omega^2 \underline{\underline{\varepsilon}}(\omega)) \cdot \underline{E}_o . \qquad (44)$$

Here, we will generalize this expression to waves with arbitary damping or growth. Suppressing the space dependence $\sim \exp(i\underline{k}\cdot\underline{r})$ for notational convenience, we write the electric field as

$$\underline{E}(t) = \underline{E}_o e^{-i\omega_o t}, \qquad (45)$$

and assume that the frequency ω_o satisfies $D(\omega_o) = 0$. The Laplace transform of this field is $i(\omega-\omega_o)^{-1}\underline{E}_o$, and this wave satisfies the transformed homogeneous wave equation $i\underline{\underline{D}}(\omega) \cdot \underline{E}_o i(\omega-\omega_o)^{-1} = j_T$ = constant. In order to calculate the energy, we perturbe the wave by an infinitesimal external current of the form

$$\underline{j}(t) = \underline{b}(t)\underline{j}_o e^{-i\omega_o t}, \tag{46}$$

where j_o is a small constant vector, and $b(t)$ is a slowly varying function. In the presence of this perturbation, the modified electric field can be written

$$\underline{E}(t) = A(t)\underline{E}_o e^{-i\omega_o t} = (1 + a(t))\underline{E}_o e^{-i\omega_o t}, \tag{47}$$

where $a(t)$ will be slowly varying if $\underline{j}(t)$ is small. The Laplace transform of this field is $\underline{E}(\omega) = [a(\omega-\omega_o) + i(\omega-\omega_o)^{-1}]\underline{E}_o$, where $a(\omega)$ is the transform of $a(t)$. For the current, the corresponding formula is $\underline{j}(\omega) = b(\omega-\omega_o)\underline{j}_o$. The inhomogeneous wave equation $i\underline{\underline{D}}(\omega)\cdot\underline{E}(\omega) = \underline{j}_I + \underline{j}(\omega)$ may then be reduced to

$$i\underline{\underline{D}}(\omega_o+\Omega) \cdot \underline{E}_o a(\Omega) \approx i\Omega \left[\partial_\omega \underline{\underline{D}}(\omega) \right]_{\omega_o} \cdot \underline{E}_o a(\Omega) = \underline{j}_o b(\Omega), \tag{48}$$

since $a(\Omega)$ will be small unless $\Omega = \omega - \omega_o$ is small. Multiplying by $\underline{E}_o^*$ and introducing

$$\Delta = \underline{E}_o^* \cdot \left[\partial_\omega \underline{\underline{D}}(\omega) \right]_{\omega_o} \cdot \underline{E}_o, \tag{49}$$

we may invert the Laplace transform to obtain

$$a(t) = \frac{\underline{E}_o^*\cdot\underline{j}_o}{2\pi\Delta} \int \frac{b(\Omega)}{i\Omega} e^{-i\Omega t} d\Omega = -\Delta^{-1}\underline{E}_o^* \cdot \underline{j}_o \int_0^t b(t')dt'. \tag{50}$$

Using this equation and $\partial_t a(t) = -\Delta^{-1}\underline{E}_o^* \cdot \underline{j}_o b(t)$ we find

$$\partial_t |\Delta A(t)|^2 = -\Delta^* \underline{E}^*(t) \cdot \underline{j}(t) - \Delta\, \underline{E}(t) \cdot \underline{j}^*(t), \tag{51}$$

which we can rewrite as

$$\partial_t \frac{|\Delta A(t)|^2}{2\mathrm{Re}\Delta} = -\,\mathrm{Re}\,[\, \underline{E}^*(t) \cdot \underline{j}(t)\,] - \frac{\mathrm{Im}\Delta}{\mathrm{Re}\Delta}\,\mathrm{Im}\,[\, \underline{E}^*(t) \cdot \underline{j}(t)] \tag{52}$$

The first term on the right hand side represents the power dissipated by the external current, and the second term can be interpreted as the change of the dissipation in the plasma due to the small frequency shift caused by the perturbation. Clearly, the left hand side must be the change of the wave energy density, caused by the pertubing current. Expressing the energy density in a more familiar notation we have

$$W = \frac{\varepsilon_o}{2} \frac{|\omega_o^{-1}\partial_\omega(\omega^2 \underline{E}_o^* \cdot \underline{\underline{\varepsilon}}(\omega)\cdot\underline{E}_o)|^2}{\mathrm{Re}\left[\omega_o^{-1}\partial_\omega(\omega^2\underline{E}_o^* \cdot \underline{\underline{\varepsilon}}(\omega)\cdot\underline{E}_o)\right]} e^{2\gamma_o t}. \tag{53}$$

This expression is evaluated at $\omega = \omega_o$, and $\underline{E}_o$ is the wave amplitude at $t = 0$.

It may not be obvious that the form (47) of the electric field is valid in the presence of an arbitary perturbation $\underline{j}(t)$, and if the direction of $\underline{E}_o$ were changed by the external current, we could not use that $\underline{\underline{D}}(\omega_o)\cdot\underline{E}_o = 0$ in equation (48). However, without making any assumptions about the direction of $\underline{E}_o$ we may solve equation (48) as

$$\begin{aligned} \underline{E}_o a(t) &= \int \frac{\underline{\underline{C}}(\omega_o+\Omega)\cdot\underline{j}_o}{2\pi i D(\omega_o+\Omega)} b(\Omega)e^{-i\Omega t}d\Omega \\ &\approx -\left[(\partial_\omega D)^{-1}\left\{\underline{\underline{C}}\cdot\underline{J}_o\int_o^t b(t')dt' + i\partial_\omega\underline{\underline{C}}\cdot\underline{J}_o b(t)\right\}\right]_{\omega_o}. \end{aligned} \tag{54}$$

Here, the second term corresponds to a driven oscillation, which remains small as long as $\underline{j}(t)$ is small. The first term, which represents the resonant response to the driving current, may become arbitarily large after a long time. Neglecting the small second term we find $\underline{\underline{D}}(\omega_o)\cdot\underline{E}_o \sim \underline{\underline{D}}(\omega_o)\cdot\underline{\underline{C}}(\omega_o)\cdot\underline{j} = D(\omega_o)\underline{j} = 0$, which essentially is what we need to derive equation (48).

The energy density of waves in plasmas with finite dissipation has previously been discussed by for example Ginzburg (1970), Ishida (1976), and Kruscha and Kondratenko (1983).

4. WAVES IN MAXWELLIAN PLASMAS

4.1. The dielectric tensor of a Maxwellian plasma

Although plasmas in space hardly ever are in thermal equilibrium, a Maxwellian distribution function

$$f^o(\underline{v}) = (\pi^{\frac{1}{2}}V)^{-3}\exp[-(v/V)^2], \tag{55}$$

or a sum of such distributions, is often a reasonable first approximation. Inserting a Maxwellian in for example equation (11) and using the expression (38) for the g-operator, the velocity integrals may be evaluated in terms of modified Bessel functions $I_n(\lambda)$, with argument $\lambda = \frac{1}{2}(k_\perp V/\Omega)^2$, and the plasma dispersion function $Z(s_n)$, with

$s_n = (\omega - n\Omega)/(k_z V)$ (Fried and Conte, 1961). A derivation found in most textbooks on kinetic plasma theory leads to the expression (see e.g. Ichimaru, 1973; Akhiezer et al., 1975; Chen, 1984).

$$\underline{\underline{\varepsilon}}(\underline{k},\omega) = \underline{\underline{I}} + \frac{\omega_p^2}{\omega_o^2} s_o \left[2s_o \hat{z}\hat{z} + \sum_n \underline{\underline{\Pi}}_n Z(s_n) \right] \tag{56}$$

for the dielectric tensor of a magnetized Maxwellian plasma. Introducing $\Lambda_n(\lambda) = I_n(\lambda)\exp(-\lambda)$. the matrix $\underline{\underline{\Pi}}_n$ is given by

$$\underline{\underline{\Pi}}_n = \begin{bmatrix} \lambda^{-1} n^2 \Lambda_n & in\Lambda'_n & \sqrt{2/\lambda}\, ns_n \Lambda_n \\ -in\Lambda'_n & \lambda^{-1} n^2 \Lambda_n - 2\lambda\Lambda'_n & -i\sqrt{2\lambda}\, ns_n \Lambda'_n \\ \sqrt{2/\lambda}\, ns_n \Lambda_n & i\sqrt{2\lambda}\, ns_n \Lambda'_n & 2s_n^2 \Lambda_n \end{bmatrix} \tag{57}$$

This dielectric tensor provides a good starting point for numerical studies of waves in plasmas.

4.2. Numerical methods

It is sometimes possible to use equation (56) as it stands for numerical computations, although this direct approach leads to problems for some parameter values. When λ is large, the infinite sum over harmonic number n is not only slowly convergent, but it also contains terms which are much larger in magnitude than the final result, and this may cause significant truncation errors. In the limit $\lambda \to 0$, special care must be taken to treat the terms containing λ^{-1} properly. Similarly, s_n will become infinite when k_z approaches zero. As shown by Rönnmark (1982; 1983), these problems can be avoided by introducing a Padé approximant for the plasma dispersion function Z(s).

A modified Padé method was first applied to Z(s) by Martín and Gonzáles (1979), Martín et al. (1980), and Németh et al. (1981). The rational approximants they derive are extremely efficient and accurate, but in this context it is even more important that an M-pole approximant $Z_M(s)$ can be written as

$$Z_M(s) = \sum_{m=1}^{M} b_m (s - c_m)^{-1}, \tag{58}$$

where $3 \leq M \leq 10$ will be sufficient in most applications. When $Z(s_n)$ in equation (56) is replaced by an approximant of this form, we find that all the infinite sums over n may be reduced to

$$R(x,\lambda) = \lambda^{-1} \sum_{n=-\infty}^{\infty} n^2 \Lambda_n(\lambda)(x-n)^{-1} \tag{59}$$

or

$$R'(x,\lambda) = \partial_\lambda [\lambda R(x,\lambda)] = x^2 \sum_{n=-\infty}^{\infty} \Lambda'_n(\lambda)(x-n)^{-1}. \tag{60}$$

Functions closely related to R have been studied by for example Fredricks (1968), Karpman et al (1973), and Schmitt (1974). In most cases, $R(x,\lambda)$ can be evaluated by roughly the same amount of computing as $I_x(\lambda)$, by summing a power series or an asymptotic series. In terms of $R_m = R(x_m,\lambda)$ and $x_m = (\omega-k_z V c_m)\Omega^{-1}$ we can after some algebra rewrite the dielectric tensor as

$$\underline{\underline{\varepsilon}}(\underline{k},\omega) \approx \underline{\underline{I}} + \frac{\omega_p^2}{\omega\Omega} \sum_{m=1}^{M} b_m \begin{bmatrix} R_m & \frac{i}{x_m} R'_m & \frac{c_m}{x_m}\sqrt{2\lambda}\, R_m \\ -\frac{i}{x_m} R'_m & R_m - \frac{2\lambda}{x_m^2} R'_m & -i\frac{c_m}{x_m^2}\sqrt{2\lambda}\, R'_m \\ \frac{c_m}{x_m}\sqrt{2\lambda}\, R_m & i\frac{c_m}{x_m^2}\sqrt{2\lambda}\, R'_m & 2\frac{c_m^2}{x_m^2}(x_m + \lambda R_m) \end{bmatrix} \tag{61}$$

This expression is regular for all values of $\underline{k}$, but it will break down close to the poles of the Padé approximant, that is when $\omega \approx k_z V c_m$ for some m. In practice it turns out that the poles c_m are all located in the lower half of the complex s-plane, and that they will cause problems only for heavily damped waves, which usually are of limited interest.

An important advantage with equation (61) is that it is very easily translated into computer code. Three alternative methods must be used to compute R, but in the rest of the program there will be no need to use different routines for different values of the input parameters. The derivatives of $\underline{\underline{\varepsilon}}(\underline{k},\omega)$ are also rather easily calculated from equation (61).

A FORTRAN program which is based on this method is described in detail by Rönnmark (1982). This program, called WHAMP, solves the dispersion relation of Waves in Homogeneous, Anisotropic, Multicomponent Plasmas. A generalized version of (61) is used to compute $\underline{\underline{\varepsilon}}(\underline{k},\omega)$ and its derivatives. The dispersion relation $D(\underline{k},\omega) = 0$, with $\bar{D}(\underline{k},\omega)$ given by the full expression (28), is then solved by iteration in the complex ω-plane.

4.3. Dispersion surfaces

Using the program WHAMP mentioned above, André (1985) has produced a set of very useful perspective plots displaying the dispersion relation as surfaces of normalized frequency versus the components of the wavevector. The main purpose of these perspective plots is to provide our intuition with a picture of how the analytical solutions, obtained by various approximations in different limiting cases, are connected to each other.

The structure of the high frequency modes in a magnetized electron plasma is shown in Figure 2. On the vertical axis, the frequency $f = \omega/2\pi$ is normalized to the electron gyrofrequency f_{ce}, which in this case is 3.3 kHz. The wavevector components are normalized to the inverse of the gyroradius ρ_e of a 2 eV electron, and plotted on logarithmic scales. The electron distribution is Maxwellian, with a temperature of 2eV and a density of 1.4 electrons cm^{-3}. This density corresponds to a plasma frequency $f_{pe} \approx 3.22\ f_{ce}$.

Let us try to orient ourselves in Figure 2, using the labels (a-f) as a guidance. The arrow from (a) points to the well known whistler waves, which propagate along the magnetic field in a right circularly

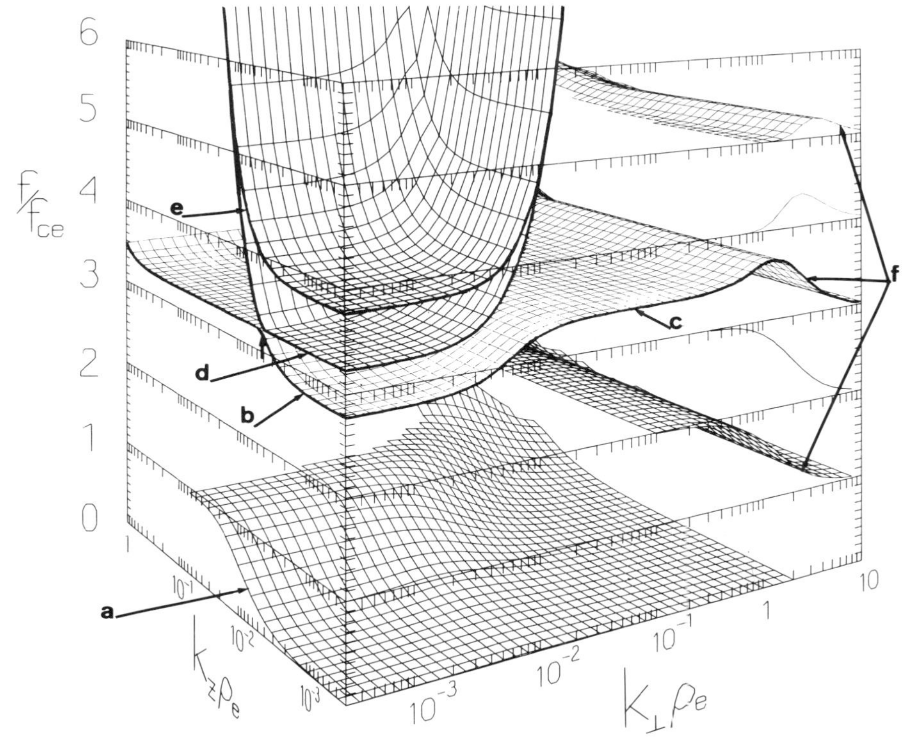

Figure 2. Perspective plot of the high frequency modes in a Maxwellian electron plasma. (André, 1985)

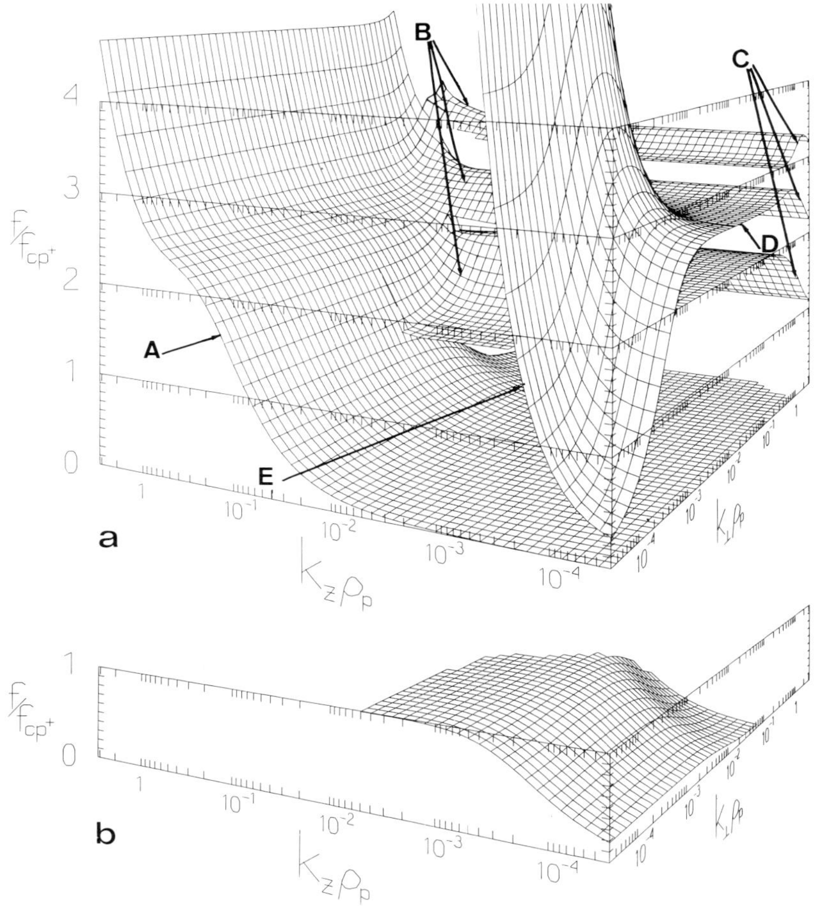

Figure 3. Perspective plot of the low frequency modes in a plasma with hot electrons and cool protons. (André, 1985)

polarized mode confined below the resonance at f_{ce}. This mode gradually becomes electrostatic as $k_\perp$ increases. On the next surface, the (b) points out left circularly polarized waves in the Z-mode, which becomes an extraordinary mode for perpendicular propagation. This surface also contains the upper hybrid plateau (c) and electrostatic Bernstein waves (f). The long wavelength electrostatic Langmuir waves (d) at the plasma frequency are on the same surface as the left-handed component of the escaping electromagnetic radiation(e).

The low frequency modes in a plasma consisting of 1.5 electrons cm^{-3} with temperature 1 keV and an equal density of 2 eV protons are shown in Figure 3. In this case, the frequency is normalized to the proton gyrofrequency $f_{cp} \approx 109$ Hz, corresponding to $f_{ce} = 200$ kHz, and ρ_p

is the gyroradius of a 2 eV proton. Since the electron temperature is much higher than the ion temperature, ion acoustic waves (A) can propagate along the magnetic field in this plasma. In Figure 3a we also recognize the low frequency end of the whistler mode (E), which connects to the compressional Alfvén wave at very low frequencies, and to the lower hybrid plateau (D) for some $k_\perp >> k_z$. At even larger $k_\perp$, we find the electrostatic ion Bernstein modes (C). The surface shown in Figure 3b exists together with the modes in Figure 3a, but it is shown separately for the sake of clarity. Electromagnetic ion cyclotron waves propagating parallel to the magnetic field are found on this surface. In the low frequency limit, these waves may also be called shear Alfvén waves.

Dispersion surfaces like these, and those presented by André (1985), can be used as an aid in the interpretation of wave observations or plasma simulations, and they may also help to resolve questions about nomenclature.

5. CONCLUDING REMARKS

In some textbooks on kinetic plasma theory, Laplace transforms are not used, and the initial value problem is not explicitly addressed. Derivations based solely on Fourier transforms are common, and the resulting formulas are often convenient, in particular for the description of plasma turbulence in the random phase approximation. Such treatments in general rely more or less implicitly on some of the results derived in Section 1. of this paper. At least I would feel uncomfortable, if I completely neglected the initial conditions without fully understanding why, and one purpose of Section 1. is to provide this understanding.

The energy concept and energy conservation laws play important roles in most branches of physics, but in the theory of plasma waves, even the definition of wave energy is not fully understood. Several important problems remain to be solved, before an equation describing for example the wave energy flow in a turbulent plasma can be derived, assuming that the waves obey linear theory but have finite growth rates.

Numerical studies of the dispersion relation, using a program like WHAMP, can in some ways fill in the gap between analytical calculations and plasma simulations. Perspective plots of dispersion surfaces can help us to understand the structure of the wave modes, and quantitative results can easily be obtained in situations which are too complicated to be approached analytically. It is also easy to check the range of validity of an analytical formula, if a general numerical method is available. Finally, any sufficiently flexible computer program, be it WHAMP or a plasma simulation code, will every now and then produce an unexpected result. Programming errors may be the most common cause of strange output, but sometimes an unforeseen numerical result indicates an interesting physical effect.

REFERENCES

André, M., 'Dispersion surfaces', J. Plasma Phys., in press, 1985.

Akhiezer, A.I., A.I. Akhiezer, R.V. Polovin, A.G. Sitenko, and K.N. Stephanov, Plasma electrodynamics, Pergamon Press, Oxford, 1975.

Chen, F.F., Plasma physics and controlled fusion, Plenum Press, New York, 1984.

Fredricks, R.W., J Plasma Phys., 2, 197, 1968.

Fried, B.D., and S.D. Conte, The plasma dispersion function, Academic Press, New York, 1961.

Ginzburg, V.L., The propagation of electromagnetic waves in plasmas, Pergamon Press, Oxford, 1970.

Ichimaru, S., Basic principles of plasma physics, Benjamin, Reading, Massachusetts, 1973.

Ishida, A., J. Phys. Soc. Japan, 41, 292, 1976.

Karpman, V.I., Ju.K. Alekhin, N.D. Borisov, and N.A. Rajabova, Astrophys. Space Sci., 22, 267, 1973.

Kruscha, K.J.G., and A.N. Kondratenko, Beitr. Plasmaphys., 23, 105, 1983.

Landau, L.D., and E.M. Lifshitz, Electrodynamics of continuous media, Pergamon Press, Oxford, 1960.

Martín, P., and M.A. Gonzáles, Phys. Fluids, 22, 1413, 1979.

Martín, P., G. Donoso, and J. Zamudio-Christi, J. Math. Phys., 21, 280, 1980.

Németh, G., Á. Ág, and Gy. Páris, J. Math. Phys., 22, 1192, 1981.

Rönnmark, K., 'WHAMP-Waves in Homogeneous, Anisotropic, Multicomponent Plasmas', KGI report 179, 1982.

Rönnmark, K., Plasma Phys., 25, 699, 1983.

Schmitt, J.P.M., J. Plasma Phys., 12, 51, 1974.

COHERENT NONLINEAR EFFECTS ON ELECTROMAGNETIC WAVE-PARTICLE INTERACTIONS

Hiroshi Matsumoto
Radio Atmospheric Science Center,
Kyoto University, Uji, Kyoto 611, Japan

ABSTRACT. In many particle simulations and space experiments, the knowledge of nonlinear characteristics of both waves and particles in plasmas is very helpful in the data analysis phase. Such knowledge is needed even more in the designing phase of appropriate diagnostics and probings in both simulations and space experiments. In this tutorial lecture, I will attempt to provide a basic introduction to the fundamental features of coherent nonlinear wave-particle and wave-wave interactions in magnetized plasmas. The present lecture covers only some of the important and basic characteristics in the coherent nonlinear interactions. The main subjects are: (1) trapping dynamics in the electrostatic wave, (2) nonlinear phase trapping and phase bunching in electromagnetic wave both in homogeneous and inhomogeneous plasmas, and (3) coherent three wave interactions.

1. INTRODUCTION

Nonlinear processes in plasmas are often classified into two categories, coherent and turbulent processes. In the coherent processes, the phases of waves and particles involved in the interaction of interest are rather clearly defined. This means that cause-and-result relations in the process are well-defined. Therefore, it is possible to explain the whole nature of the interaction by a combination of basic nonlinear elements involved in the interaction. On the contrary, phases among waves and particles in the turbulent interaction are random and not well-defined. The statistical behavior of particles and waves (ensemble average) is more important than the individual behavior of constituent elements for the turbulent interaction.

In this lecture, we will not touch the turbulent interaction problem and will restrict our discussion only to selected characteristics of the coherent nonlinear interactions which are useful in understanding and interpreting outputs and results of computer simulations of coherent nonlinear phenomena in plasmas. The selected topics are trapping and bunching of particles in both space and phase caused by a large-amplitude monochromatic or quasi-monochromatic wave, and ponderomotive force mechanism of coherent nonlinear three wave interactions.

Sections 2 and 3 are devoted to the fundamental characteristics of

Space Science Reviews **42** (1985) 429–448. 0038–6308/85.15

trapping and bunching of resonant particles. In Section 4, the basic mechanism of nonlinear coherent three wave coupling is discussed in terms of simple physical pictures of the ponderomotive force and of mode coupling.

2. FUNDAMENTAL DYNAMICS OF PARTICLE TRAPPING DUE TO LONGITUDINAL ELECTRIC FIELD

Prior to examination of phase trapping and bunching of resonant particles in the coherent transverse electromagnetic wave field, it is instructive to review a simpler case of trapping of resonant particles in the coherent longitudinal electrostatic wave field. The concept of particle trapping and associated bunching in both cases are essentially the same.

2.1 Pendulum Equation of Resonant Particles in Large-Amplitude Longitudinal Electric Field

Let us consider the case in which a large-amplitude monochromatic electrostatic wave with the following form is traveling along z axis.

$$E(z,\ t) = E_A \sin(\omega t - kz) \tag{1}$$

where E_A, ω and k are the amplitude of the wave angular frequency and wavenumber, respectively. In the wave frame which is moving with the phase velocity ω/k, the electric field is simply expressed as

$$E(z) = -E_A \sin(kz) \tag{2}$$

The equation of motion of a test particle with a mass m and a charge q is written as

$$\frac{dv}{dt} = \frac{q}{m} E(z) \quad , \quad v = \frac{dz}{dt} \tag{3}$$

which is rewritten in a form of the pendulum equation as

$$\frac{d^2\zeta}{dt^2} + \omega_T^2 \sin\zeta = 0 \tag{4}$$

where $\zeta = kz$ is a phase angle representing a relative location of the particle in one wavelength, and ω_T is the trapping frequency defined by

$$\omega_T = (\ kqE_A / \ m\ )^{1/2} \tag{5}$$

If $\zeta \simeq 2n\pi$ (n: integer), Eq.(4) reduces to the following equation of motion of a harmonic oscillator.

$$\frac{d^2\zeta}{dt^2} + \omega_t^2\zeta = 0 \tag{6}$$

This means that the trapping frequency ω_T physically determines the bounce period of particles trapped in the vicinity of the bottom of the wave potential $\phi(z)$ expressed by

$$\phi(z) = -\int E(z)dz = -\frac{E_A}{k}\cos\zeta. \tag{7}$$

Before examining the solutions of $\zeta(t)$ and $v(t)$ to the pendulum equation (4) as an explicit function of time t, let us review the relation between v and ζ in the v-ζ phase space by eliminating t. From (2) and (3),

$$\frac{dv}{d\zeta} = -\frac{\omega_T^2}{k^2}\frac{\sin\zeta}{v}. \tag{8}$$

Integration of (8) gives an equation of energy conservation of a form of

$$\frac{1}{2}mv^2 - \frac{qE_A}{k}\cos\zeta = W_o \tag{9}$$

where W_o is an integration constant which represents a total energy of the particle. The second term in the left hand side of (9) is identical to the potential energy $q\phi(\zeta)$ of the particle. The total energy W_o determines the phase space orbit of the particle. The phase space orbits for various values of W_o are shown in Fig.1. The orbits within the shaded area are of a form of the closed loops representing a bounce motion in the wave potential well. The shaded area is therefore often called " trapping region" in the phase space. The separatrix of the shaded and unshaded area is expressed by

$$\frac{1}{2}mv^2 - \frac{qE_A}{k}\cos\zeta = \frac{qE_A}{k} \tag{10}$$

Therefore the speed of the trapped particles is limited by a maximum velocity which is achieved at $\zeta = 2n\pi$ on the separatrix. The maximum value is called " trapping velocity " and is expressed by

$$V_T = 2\frac{\omega_T}{k}. \tag{11}$$

The separatrix (10) can be expressed in terms of V_T as

$$v^2 = V_T^2\cos^2(\zeta/2) \tag{12}$$

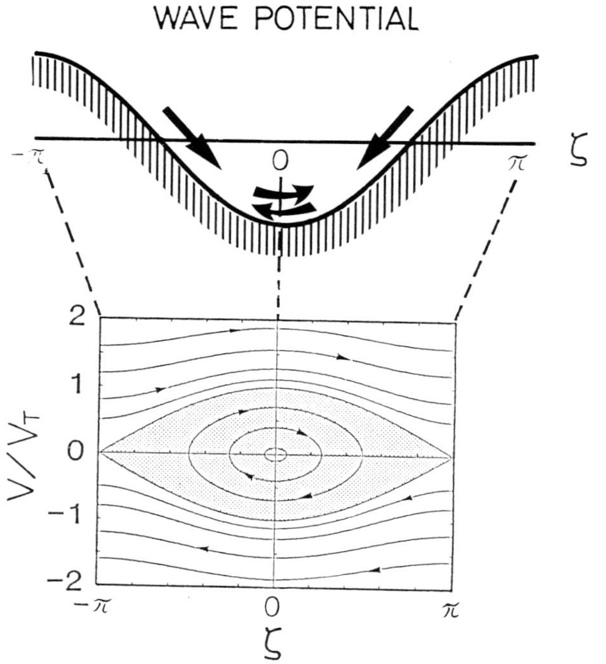

Fig.1 A concept of particle trapping in the wave potential well.

Therefore those particles with an initial velocity v_o and initial phase angle ζ_o which satisfy

$$v_o^2 \geqslant V_T^2 \cos^2 (\zeta_o/2) \tag{13}$$

are not trapped by the wave potential and travel over hill-and-valley of the potential well.

Let us now examine the solutions $\zeta(t)$ and $v(t)$ to the pendulum equation (4). Integrating (4) with respect to t after multiplying $2(d\zeta/dt)$, we have

$$\left(\frac{d\zeta}{dt}\right)^2 = \frac{2k^2}{m} W_o + 2\,\omega_T^2 \cos\zeta \tag{14}$$

which is equivalent to (9). Replacing W_o with initial v_o and ζ_o, (14) can be written in a form of

$$\left(\frac{d\zeta}{dt}\right) = \frac{4\omega_T^2}{\kappa^2}\{1 - \kappa \sin(\zeta/2)\} \tag{15}$$

where

$$\kappa^2 = \frac{V_T^2}{v_o^2 + V_T^2 \sin^2(\zeta_o/2)} \tag{16}$$

Introducing $\xi = (1/2)\zeta$, Eq.(15) reduces to

$$\left(\frac{d\xi}{dt}\right)^2 = \frac{\omega_t^2}{\kappa^2}(1 - \kappa^2 \sin^2\xi) \tag{17}$$

It is obvious from this equation that the range of possible ξ (and hence ζ)is limited if $\kappa \geqslant 1$ in such a way that $\kappa^2 \sin^2\xi \leqslant 1$. This physically means that if $\kappa \geqslant 1$ then the particle is trapped by the potential well and shows a bounce motion of $\zeta(\xi)$ within the limited range. In case of $\kappa \geqslant 1$, we need to transform the variable ξ into ξ' by a relation

$$\sin\xi' = \kappa \sin\xi \tag{18}$$

The result is

$$\left(\frac{d\xi'}{dt}\right)^2 = \omega_T^2 \left(1 - \frac{1}{\kappa^2} \sin^2\xi' \right) \tag{19}$$

which has the identical form to Eq.(17) by replacing ω_T^2 by ω_T^2/κ^2 and κ^2 by $1/\kappa^2$. Both (17) for $\kappa^2 \leqslant 1$ and (19) for $\kappa^2 \geqslant 1$ are integrable. We will show the procedure of obtaining solutions to (17) below. The solutions to (19) is obtainable by a similar procedure. Integrating (17), we have

$$\frac{\omega_T}{\kappa} t = \int_{\xi_o}^{\xi} \frac{d\xi}{\sqrt{1-\kappa^2 \sin^2\xi}}$$

$$= F(\xi, \kappa) - F(\xi_o, \kappa) \tag{20}$$

where $F(\xi, \kappa)$ is the elliptic integral of the first kind defined by

$$F(\xi, \kappa) = \int_o^{\xi} \frac{du}{\sqrt{1-\kappa^2 \sin^2 u}} \tag{21}$$

in which κ is a modulus. In order to obtain an explicit solution of ξ from (20), the following Jacobi's functions are used.

$$\begin{aligned} \operatorname{sn}(F, \kappa) &= \sin\xi, \quad \operatorname{cn}(F, \kappa) = \cos\xi \\ \operatorname{dn}(F, \kappa) &= \sqrt{1-\kappa^2\sin^2\xi} \end{aligned} \tag{22}$$

The solution to (17) is therefore expressed in the following explicit form as

$$\begin{aligned} \xi &= \sin^{-1}[\operatorname{sn}(F, \kappa)] \\ &= \sin^{-1}\left[\ \operatorname{sn}\left\{F(\xi_o, \kappa) + \frac{\omega_T}{\kappa}t, \kappa\right\}\ \right] \end{aligned} \tag{23}$$

The final solutions to (4) are thus given by

$$\zeta(t) = 2\sin^{-1}\left[\ \operatorname{sn}\left\{F\left(\frac{1}{2}\zeta_o, \kappa\right) + \frac{\omega_T}{\kappa}t, \kappa\right\}\ \right] \tag{24}$$

$$v(t) = V_T \frac{\operatorname{dn}\left[\ F\left(\frac{1}{2}\zeta_o, \kappa\right) + \frac{\omega_T}{\kappa}t, \kappa\right\}\ \right]}{\kappa} \tag{25}$$

for $\kappa \leqq 1$ and

$$\zeta(t) = 2\sin^{-1}\left[\frac{1}{\kappa}\operatorname{sn}\left\{F\left(\sin^{-1}\left[\kappa\sin\frac{1}{2}\zeta_o\right], \frac{1}{\kappa}\right) + \omega_T t, \frac{1}{\kappa}\right\}\ \right] \tag{26}$$

$$v(t) = V_T \frac{\operatorname{cn}\left[\ F\left(\sin^{-1}\left[\kappa\sin\frac{1}{2}\zeta_o\right], \frac{1}{\kappa}\right) + \omega_T t, \frac{1}{\kappa}\ \right]}{\kappa} \tag{27}$$

for $\kappa \leqq 1$. The solutions (24) and (25) with $\kappa \leqq 1$ represent the time behavior of untrapped particles, while the solutions (26) and (27) with $\kappa \geqq 1$ are for trapped particles.

2.2 Characteristic Behavior of Trapped and Untrapped Particles and Phase-Mixing

The functional forms of $\zeta(t)$ and $v(t)$ are expressed in terms of the Jacobi functions, the values of which are depicted in Fig.2. The periods of the Jacobi functions are either 4K or 2K for sn and cn or dn functions, respectively where K is the complete integral of the first kind defined by

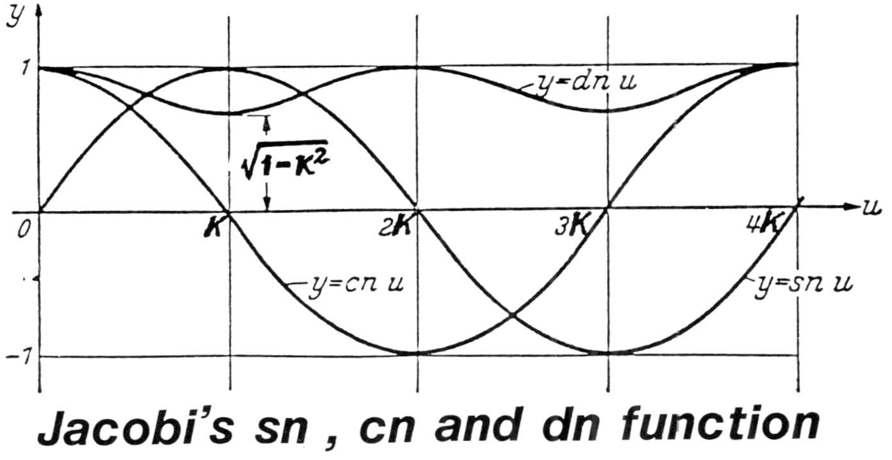

Fig.2 Functional forms of the Jacobi functions.

$$K(\kappa) = \int_0^{\pi/2} \frac{du}{\sqrt{1-\kappa^2 \sin^2 u}} \tag{28}$$

The dependence of K on the modulus κ is shown in Fig.3. As κ approaches to unity, the value of K and thus the period of the Jacobi functions become long. This physically means that the motion of the trapped particles locating in the vicinity of the separatrix in the v-ζ phase plane is much slower than that of the particles trapped near the bottom of the wave potential. Therefore the so-called phase-mixing takes place within the trapping region in the phase space. Figure 4 demonstrates how the phase-mixing takes place due to the difference of period of rotation of the trapped particles [Kadomtsev, 1976]. As time elapses, the shaded area occupied by trapped particles below the resonance velocity changes its shape as time elapses and spread in the whole trapping region as seen at time (b) and (c) due to the difference of rotation speed among particles. The lower panel shows the corresponding deformation of the velocity distribution function as a result of the phase-mixing process. In many simulaiton results of coherent wave-particle interactions the above-mentioned phase-mixing process plays a crucial role in the thermalization processes of the resonant particles.

The feedback effect of the phase trapping and subsequent phase-mixing on the wave field was first discussed by O'Neil [1965] and Al'tshul' and Karpman [1966]. The effect appears primarily as a change of the amplitude of the coherent wave which was a priori assumed to be constant in the mathematical treatment reviewed above. Typical time history of the wave amplitude as a result of the feedback of the nonlinear behavior of resonant particles is depicted in Fig.5. In this example the wave energy is absorbed by resonant particles according to the linear theory in the first half period ($t \lesssim \pi/\omega_T$) of rotation of resonant particles in the wave potential. However, as the majority of

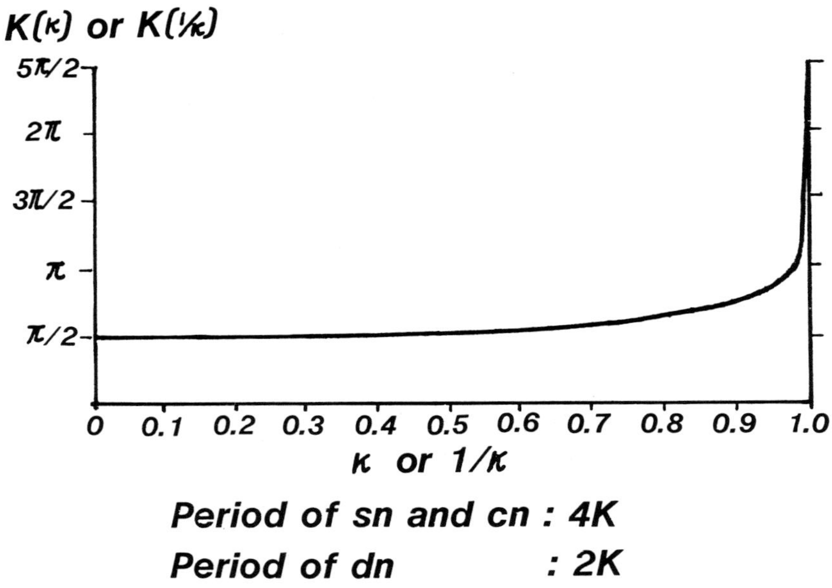

Fig.3 Period of the Jacobi functions vs. modulus κ.

the trapped particles move into the upper half plane of the trapping region, the direction of the net energy transfer is reversed. The energy is thus fed back to the wave from the particles. After repetition of similar process, the wave amplitude reaches to a steady-state. The approach to the steady-state is a result of the phase-mixing discussed in the previous paragraph.

3. PHASE TRAPPING AND BUNCHING IN THE ELECTROMAGNETIC WAVE

Suppose that a monochromatic or quasi-monochromatic electromagnetic wave with a frequency ω and a wavenumber k is propagating along the inhomogeneous magnetic field. The governing equations of motion of a particle in such wave fields are then expressed in the following form [e.g., Matsumoto and Omura, 1981].

$$\frac{dv_{\parallel}}{dt} = \Omega_w v_{\perp} \sin\zeta - \frac{v^2_{\perp}}{2\Omega_o}\frac{2\Omega_o}{2z} \tag{29}$$

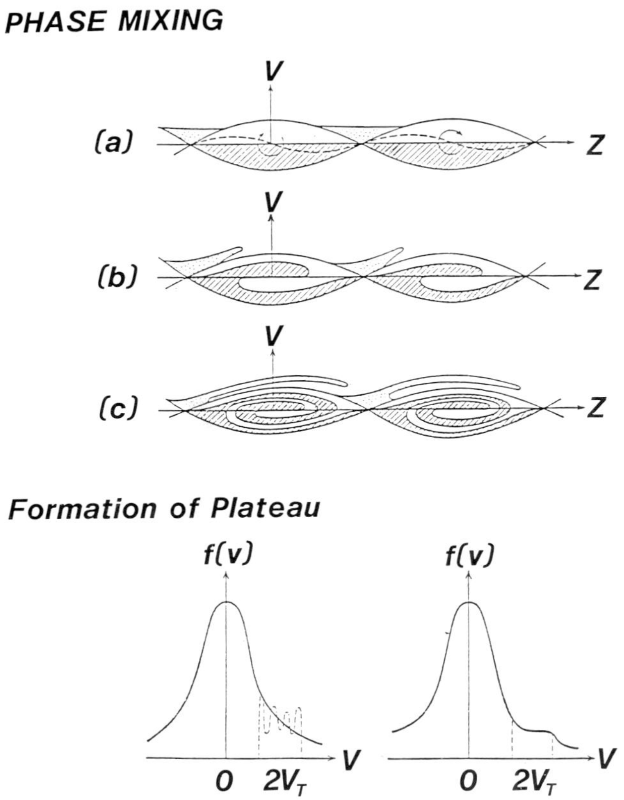

Fig.4 A schematic illustration of phase mixing and associated deformation of velocity distribution function (after Kadomtsev, 1976).

$$\frac{dv_\perp}{dt} = \Omega_w\left(\frac{\omega}{k} - v_\parallel\right)\sin\zeta + \frac{v_\parallel v_\perp}{2\Omega_o}\frac{\partial\Omega_o}{\partial z} \tag{30}$$

$$\frac{d\phi}{dt} = \frac{\Omega_w}{v_\perp}\left(\frac{\omega}{k} - v_\parallel\right)\cos\zeta - \Omega_o \tag{31}$$

where $v_\parallel$, $v_\perp$, and ϕ are the velocity components parallel and perpendicular to B_o (which is assumed to be in the z direction) and Larmour phase angle, respectively. The quantity ζ is a relative phase angle between $v_\perp$ and B_w vectors where B_w is the magnetic field component of the wave. The cyclotron frequency is expressed as $\Omega_o(z) = qB_o(z)/m$

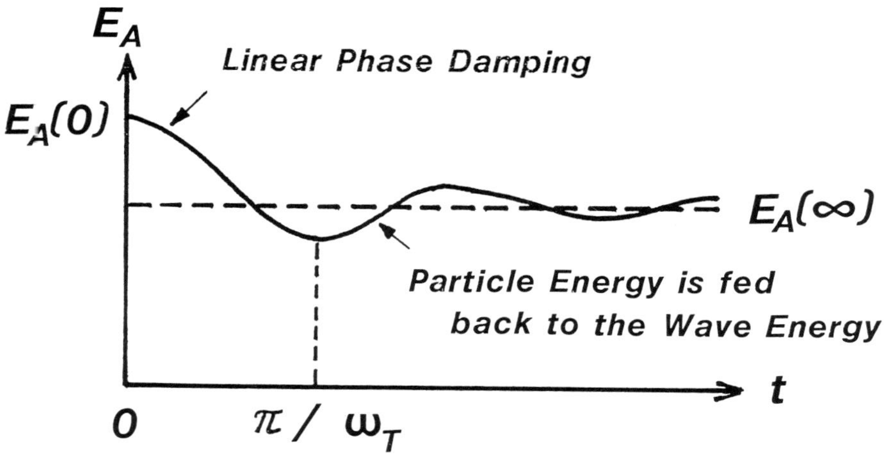

Fig.5 A concept of amplitude oscillation in the nonlinear phase.

while Ω_w is defined by $\Omega_w = qB_w/m$. The first and second terms in the r. h. s. of (29) and (30) come from the Lorentz force due to $v_\perp \times B_w$ and the mirror force, respectively.

Assuming that the wave propagates along B_0 in the whistler mode, and that the electron density is constant with distance, the basic set of equation (29)-(30) can be reduced to [Omura and Matsumoto, 1982].

$$\frac{d}{dt}[k(v_\parallel - V_R)] = \omega_T^2(\sin\zeta + R) \tag{32}$$

$$\frac{dv_\perp}{dt} = \Omega_w\left(\frac{\omega}{k} - v_\parallel\right)\sin\zeta + \frac{1}{2}\left(\frac{\omega}{k} - v_\parallel + V_R\right)\frac{v_\perp}{\Omega_o}\frac{\partial\Omega_o}{\partial z} \tag{33}$$

$$\frac{d\zeta}{dt} = k(v_\parallel - V_R) + \frac{\Omega_w}{kv_\perp}(\omega - kv_\parallel)\cos\zeta \tag{34}$$

where R is the inhomogeneity factor defined by

$$R = -\frac{1}{2\omega_T^2}\left(3V_R + \frac{kv_\perp^2}{\Omega_o}\right)\frac{\partial\Omega_o}{\partial z} \tag{35}$$

in which $V_R = (\omega + \Omega_o)/k$ is a resonance velocity and ω_T is defined by

$$\omega_T = (\ kv_\perp \Omega_w)^{1/2} \tag{36}$$

If $|\Omega_o| \sim \kappa v_\perp \sim (\Omega_o + \omega) >> \omega_T >> \Omega_w$, which is normally satisfied for the magnetospheric whistler waves, then the variation of $v_\perp$ is negligible. This means that the quantity ω_T is determined by the initial value of $v_{\perp o}$ and can be regarded as a constant. Under this assumption, the second equation (32) in the basic set (31)-(33) becomes of no concern and the second term in the r.h.s. of (33) is negligible. Thus we have

$$\frac{d}{dt}[k(v_\parallel - V_R)] = \omega_T^2(\sin\zeta + R) \tag{37}$$

$$\frac{d\zeta}{dt} = k(v_\parallel - V_R) \tag{38}$$

The set of equations (36) and (37) with R=0 is exactly the same as (3) with a replacement of $v_\parallel - V_R$ by v and ζ by $\zeta+\pi$. Therefore the relative phase angle ζ between $v_\perp$ and B_w behaves in quite a similar manner to the relative position kz of a particle in the electrostatic wave potential as discussed in detail in the previous section. The difference from the previous simple case is the additional term R in the r.h.s. of (36). The quantity R is proportional to the gradient of the external magnetic field $\partial B_o/\partial z$ and is inversely proportional to ω_T^2, i.e., the wave amplitude B_w via Ω_w.

3.2 Nature of Phase Trapping and Phase Trapping in the Homogeneous Plasma

The nonlinear behavior of the phase angle ζ in the whistler interaction in the homogeneous plasma has been extensively investigated by numerical studies without such simplification and assumptions as made in the derivation of (36) and (37) [Matsumoto, 1972; Matsumoto and Hashimoto, 1974; Gendrin, 1974]. Basically the phase behavior of resonant particles in the whistler mode wave is very similar to that of the spatial position of particles in the electrostatic electric field. This means that the phase angle ζ shows a bounce motion around $\zeta = \pi$ and thus phase-trapped. A typical example of the phase behavior as a function of time is shown in Fig.6(b). As time elapses, the initially uniformly-distributed phases are bunched around $\zeta \cong \pi$ and show a bounce motion in a very similar manner as in the case of the longitudinal spatial trapping and bunching in the electrostatic field. However, one point to which we must pay attention is that the phase bunching is possible without phase-trapping. One example is shown in Fig.6(a) where phases of resonant electrons are well-bunched but not phase-trapped. The average angle of the bunched electrons is not limited within a certain range as seen in Fig.6(b) but shows a complete rotation. The existence of such case indicates that the nonlinear phase behavior can

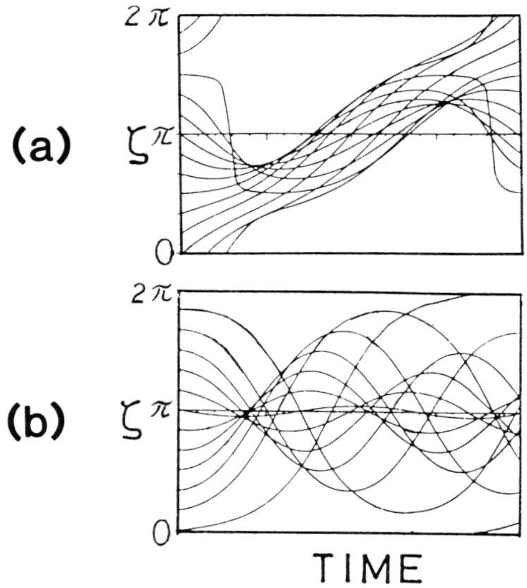

Fig.6 Examples of (a) phase bunching without phase trapping and (b) regular phase trapping by the whistler mode wave.

not be always described by the simple pendulum equation (36) and (37) if the variation of $v_\perp$ is taken into account.

3.3 Effects of Inhomogeneous Magnetic Field and NPI Amplification

The significance of the inhomogeneous magnetic field in the whistler mode interactions has been discussed by many authors [Nunn, 1971; Karpman et al., 1974, 1975; Roux and Pellat, 1978; Vomvoridis et al., 1978, 1980; Matsumoto, 1979; Matsumoto and Omura, 1981; Omura and Matsumoto, 1982; Matsumoto and Omura, 1983]. The inhomogeneous magnetic field pushes the particle along B_0 via a mirror force. The resultant effects are the shift of the center angle of the trapping region and the contraction of the trapping region in the phase space. Figure 7 shows the various shapes of the trapping region (shaded area) and associated orbits of particles in the $v_\perp$-ζ phase planes for various values of R at various locations along the geomagnetic field line in the magnetosphere. The sign of R determines the shift of the center phase angle of the trapping region. For non-zero R, not only the trapped particles in the trapping region but also untrapped particles flowing outside the trapping region contribute to the phase-bunching if the number of trapped particles is much less than the untrapped particles. One example of phase-bunching by the untrapped electrons and associated wave amplification due to the resonant current produced by the phase-bunched untrapped electrons is shown in Fig.8. The R value is approximately R = −0.5. In the middle frame of the three phase diagrams, a clear phase-bunching is seen near $\zeta \cong 270^\circ$. Correspondingly the wave amplitude starts to grow after this time at the resonance point

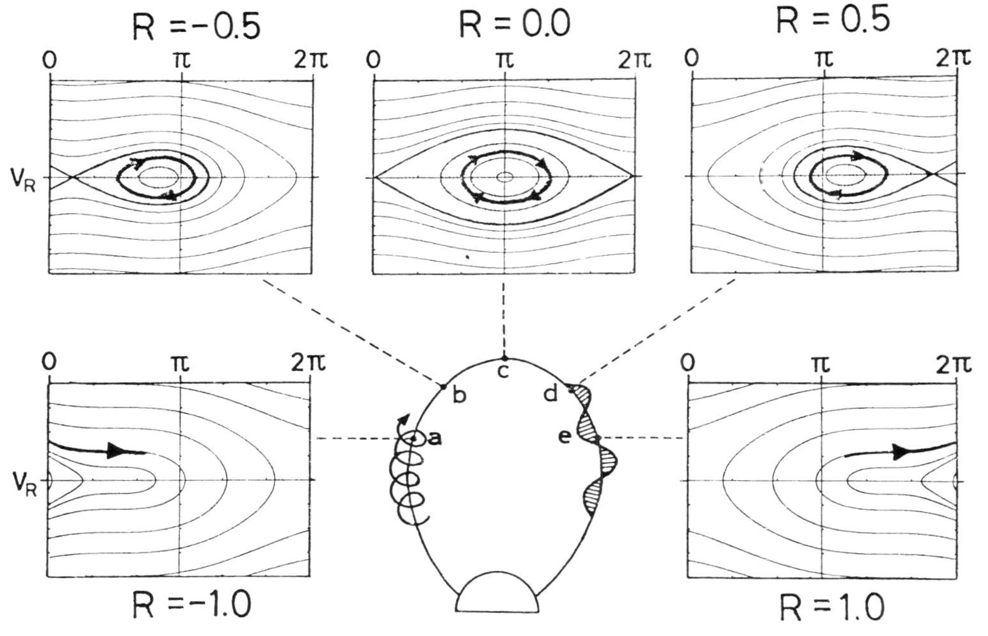

Fig.7 Phase diagrams in the inhomogeneous magnetic field for different values of R.

where $v_\parallel \cong V_R$. In this example, the population of trapped electrons is almost zero. However, if the population of trapped electrons exceeds that of the untrapped electrons under the same R, the phase bunching would occur around $\zeta \cong 150^o$ which will lead the wave damping on the contrary. It should be noted that the untrapped electrons in the medium with opposite sign of R (e.g., R = 0.5) contribute to the wave damping and trapped particles to the wave amplification [Omura and Matsumoto, 1982]. This means that the roles of trapped and untrapped particles are exchanged when the sign of R is reversed.

The trapped and untrapped resonant particles play different roles in the inhomogeneous magnetic field. Therefore the wave growth or damping in the inhomogeneous medium is strongly controlled by a relative population of trapped and untrapped particles. The mechanism of wave growth and damping on the nonlinear stage in the inhomogeneous medium is thus totally different from that in the homogeneous case. Let us consider a case where the whistler wave is unstable and grows in time. In the early phase of the wave-particle interaction, the wave grows according to the linear theory in both homogeneous and inhomogeneous medium as shown in Fig.9. In the homogeneous case, the wave reaches a saturation level and shows an amplitude oscillation by a similar mechanism as discussed in the previous section. However, in the

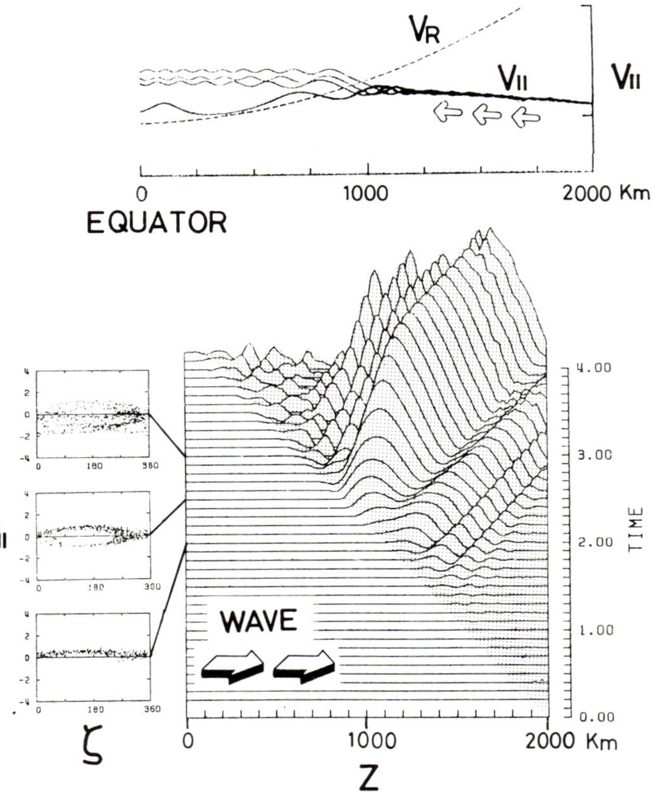

Fig.3 An example of wave amplification by untrapped resonant electrons.

inhomogeneous case, the wave grows further due to " Nonlinear plus Inhomogeneity (NPI) " effect as illustrated in Fig.9. The NPI effect is a very important mechanism fpr the strong amplification of the whistler wave in the magnetosphere [Roux and Pellat, 1978; Dowden et al., 1978; Matsumoto and Omura, 1983]. The NPI growth takes place at the off-equatorial region as illustrated in the upper panel of Fig.9 where the populations of trapped and untrapped electrons are different. The difference in number density of untrapped and trapped electrons is caused by a pumping action by the wave. Those electrons trapped at the front edge of the wave, where the population of resonant electrons is much different from that in the region near the wave tail, are phase-trapped and guided down to the tail region of the wave along the resonance velocity in the $v_{\parallel}$-z phase space. This pumping effect is due totally to the NPI effect [Matsumoto and Omura, 1981,1983].

Mathematically, the growth rate of the whistler wave in the inhomogeneous medium is expressed by

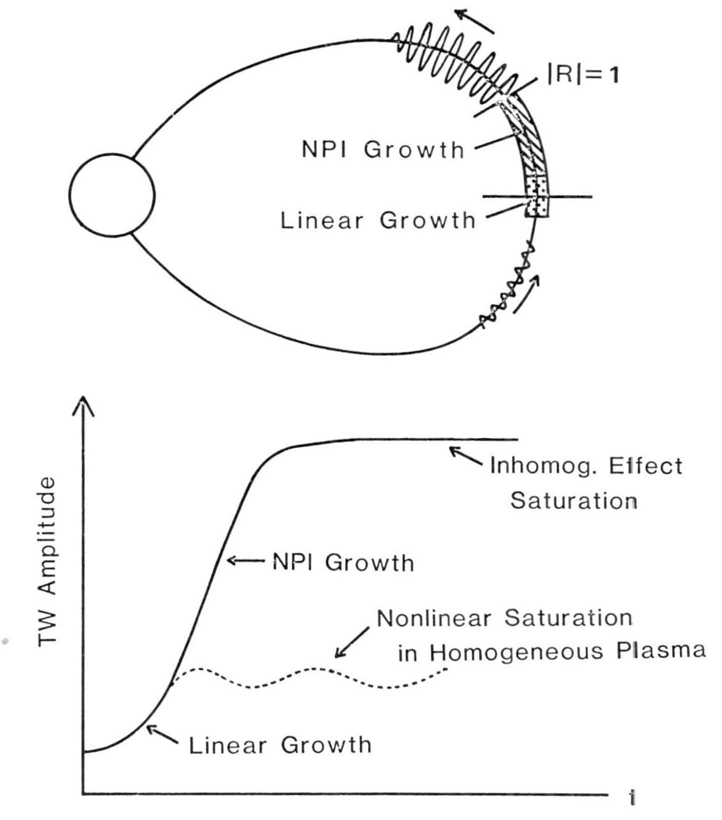

Fig.9 Region of linear and nonlinear growth of the whistler waves in the magnetosphere (upper panel) and difference of characteristics of linear and NPI growth.

$$\gamma = \gamma_L + \gamma_{NPI} \tag{39}$$

where γ_L is the linear growth rate calculated by

$$\gamma_L = \frac{\pi^2(\Omega_e - \omega)^2}{k} \int_0^\infty v_\perp^2 dv_\perp \left(\frac{\partial f}{\partial v_\perp} + \frac{kv_\perp}{\Omega_e}\frac{\partial f}{\partial v_\parallel} \right)\Bigg|_{v_\parallel = V_R} \tag{40}$$

and γ_{NPI} is the NPI effect growth rate. The NPI effect growth rate is decomposed into two parts as

$$\gamma_{NPI} = \gamma_{NPI}^{T} + \gamma_{NPI}^{UT} \tag{41}$$

where γ_{NPI}^{T} and γ_{NPI}^{UT} show the contribution from trapped and untrapped particles, respectively. The untrapped contribution γ_{NPI}^{UT} can be expressed as

$$\gamma_{NPI}^{UT} = c_1 \gamma_L \tag{42}$$

where c_1 is a positive constant. This is because the untrapped electrons are not much modified in their orbits even on the nonlinear stage. Their contribution is calculated by the velocity distribution function at a local point of interest. However the trapped particle contribution γ_{NPI}^{T} is expressed in a form of

$$\gamma_{NPI}^{T} = c_2 \gamma_L \left. \frac{f_o^F(V_R^F) - f_o(V_R)}{(V_R^F - V_R) \dfrac{\partial f_o}{\partial v_\parallel}} \right|_{v_\parallel = V_R} \tag{43}$$

where c_2 is a positive constant, V_R^F and f_o^F are the resonant velocity and unperturbed velocity distribution function at the wave front where most of the trapped particles are captured by the wave. Eq.(42) shows that γ_{NPI} is proportional to γ_L if the population difference $f_o^F(V_R^F) - f_o(V_R)$ at the wave front and a point of interest is so small that the following assumption is valid.

$$\left. \frac{\partial f_o}{\partial v_\parallel} \right|_{v_\parallel = V_R} = \frac{f_o^F(V_R^F) - f_o(V_R)}{V_R^F - V_R} \tag{44}$$

If the assumption is not valid, γ_{NPI} can have an opposite sign to that of γ_L.

4. NONLINEAR RESONANT THREE WAVE INTERACTION

In this section, we will show one specific example of nonlinear resonant three wave interaction in order to interpret a fundamental mechanism of the three-wave interaction. The example is a nonlinear excitation of electrostatic (ES) electron cyclotron harmonic wave by a strong electromagnetic O-mode wave. A schematic illustration is given in Fig.10 of the dispersion relation of the O-mode pump wave, another O-mode idler wave and the ES electron cyclotron excited wave. The dotted parallelogram shows the following well-known three-wave resonance conditions.

$$\omega_o = \omega_i + \omega_e , \quad k_o = k_i + k_e \tag{45}$$

Instead of giving a mathematical formalism of nonlinear three wave interactions [Sagdeev and Galeev, 1969 ; Chapter 4 in Davidson, 1972], we will interpret the process in terms of basic elements of the involved

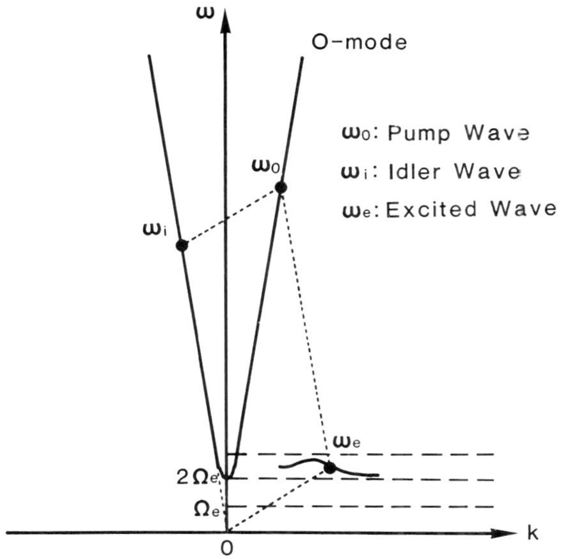

Fig.10 A schematic parallelogram for three-wave resonant interaction of the example of the present concern.

physics. In the present example, the key point is to understand how the electrostatic density modulation along the k vector (which is parallel to z axis) is caused by the purely transverse electromagnetic wave. In the linear regime, the electrons oscillate only in the transverse plane under the purely transverse wave. Therefore no density modulation along z-axis is created. In the nonlinear regime, however, the electrons are accelerated or decelerated along z-axis by a ponderomotive force produced by the strong exciting wave. The ponderomotive force is essentially the nonlinear Lorentz force $v_\perp \times B_\perp$ in the z direction. Suppose that a weak idler transverse wave component $E^i_\omega \sin(\omega_i t - k_i z)$ co-exists with the strong exciting wave $E^p_w \sin(\omega_o t - k_o z)$. The z-component of the nonlinear Lorentz force acting on electrons is given by

$$f_{NL} = -e(v_p B_i + v_i B_p)$$

$$= \frac{e^2}{m} \left[\frac{E_i B_p}{2\omega_i} \{ \sin [(\omega_o + \omega_i)t - (k_o + k_i)z] \right.$$

$$+ \sin [(\omega_o - \omega_i)t - (k_o - k_i)z] \}$$

$$+ \frac{E_p B_i}{2\omega_o} \{ \sin [(\omega_o + \omega_i)t - (k_o + k_i)z] + \sin [(\omega_o - \omega_i)t - (k_o - k_i)z] \} \qquad (46)$$

where B_p and B_i are the amplitudes of the magnetic field of the pump and idler waves, and v_i and v_p are the velocity modulations by the idler and pump wave electric field, respectively given by

$$v_p = - \frac{eE_p}{m\omega_o} \sin (\omega_o t - k_o z)$$

$$v_i = - \frac{eE_i}{m\omega_i} \sin (\omega_i t - k_i z) \quad . \qquad (47)$$

The slowly varying component of the Lorentz force f_{NL} in the z-direction is given by

$$f_{NL} = - \frac{e^2 E_p E_i}{2m \omega_o \omega_i} k_e \sin (\omega_e t - k_e z) \qquad (48)$$

where the resonance condition (44) and the relations $B_p = (k_o/\omega_o)E_p$ and $B_i = (k_i/\omega_i)E_i$ are used. Eq.(47) can be expressed in a form of $f_{NL} = -\nabla\psi_p$ where $\nabla \equiv \partial/\partial z$ and

$$\psi_p = \frac{e^2 E_p E_i}{2m \omega_o \omega_i} \cos(\omega_e t - k_e z) \qquad (49)$$

is a nonlinear ponderomotive potential. The ponderomotive potential ψ_p, which is small and negligible unless E_p is large, drives the electrons in the z direction and causes the density modulation $n_e(\omega_e, k_e)$. Now that the density modulation along z-axis with a frequency ω_e and wavenumber k_e is produced under the assumed existence of the idler transverse wave, the left task is to check how the density modulation gives a positive feedback to the assumed idler wave. The density modulation $n_e(\omega_e e, k_e)$ produces the Poisson electric field $E_e(\omega_e, k_e)$. Therefore the electrons move along z-axis with a velocity v_e —— $(eE_e/m\omega_e)\sin(\omega_e t -k_e z)$. The cross product of $v_e(\omega_e, k_e)$ and the magnetic field $B_p(\omega_o, k_o)$, which is the Lorentz force, derives the electrons oscillate in the direction parallel to the electric field of the postulated idler wave, thus produces a current in that direction. The current which includes the frequency and wave number component (ω_i, k_i) amplifies the postulated idler wave. Thus a positive feedback loop is possible to lead a parametric decay instability via three-wave interactions.

Generally the ponderomotive force determines the coupling coefficients between the element waves in the nonlinear three-wave interactions. Such microscopic analysis is very helpful in the analysis of the computer simulation of coherent wave-wave couplings.

ACKNOWLEDGEMENTS. The author would like to thank Y. Omura and T. Kimura for their discussion and K. Miwa for typing of the manuscript.

References

Al'tshul', L. M. and V. I. Karpman, On the theory of nonlinear oscillations in a collisionless plasma, Sov. Phys. JETP, 22, 361, 1966.

Davidson, r R. C., Methods in Nonlinear Plasma Theory, Academic Press, 1972.

Dowden, R. L., A. D. McKay, L. E. S. Amon, H. C. Koons, and M. H. Dazey, Linear and nonlinear amplification in the magnetosphere during a 6.6 kHz transmission, J. Geophys. Res., 83, 169, 1978.

Gendrin, R., Phase-bunching and other nonlinear processes occuring in the gyro-resonant wave-particle interactions, Astrophys. Space Sci., 28, 245, 1974.

Kadomtsev, B. B., Collective Phenomena in Plasmas, Nauka, USSR, 1976.

Karpman, V. I., Ja. N. Istomin, and D. R. Shklyar Nonlinear theory of a quasi-monochromatic whistler mode packet in inhomogeneous plasma, Plasma Phys., 16, 685, 1974.

Karpman, V. I., Ja. N. Istomin, and D. R. Shklyar Effects of nonlinear interaction of monochromatic waves with resonant particles in the inhomogeneous plasma, Physica Scripta, 11, 278, 1975.

Matsumoto, H., Theoretical studies on whistler mode wave-particle interactions in the magnetospheric plasma, Ph.D. thesis, Kyoto Univ., Kyoto, Japan, 1972.

Matsumoto, H., K. Hashimoto, and I. Kimura, Two-types of phase bunching in the whistler mode wave-particle interaction, J. Geomag. Geoelectr., 26, 365, 1974.

Matsumoto, H., and Y. Omura, Cluster and channel effect phase bunchings by whistler waves in the nonuniform geomagnetic field, J. Geophys. Res., 86, 779, 1981.

Matsumoto, H., and Y. Omura, Computer simulation studies of VLF triggered emissions: Deformation of distribution function by trapping and detrapping, Geophys. Res. Lett., 10, 607, 1983.

Nunn, D, A theory of VLF emissions, Planet. Space Sci., 19, 1141, 1971.

Omura, Y., and H. Matsumoto, Computer simulations of basic processes of coherent whistler wave-particle interactions in the magnetosphere J. Geophys. Res., 87, 3537, 1982.

O'Neil, T., Collisionless damping of nonlinear plasma oscillations, Phys. Fluids, 2255, 1965.

Sagdeev, R. Z. and A. A. Galeev, Nonlinear Plasma Theory, W. A. Benjamin, Inc., 1969.

Vomvoridis, J. L., and J. Denavit, Test particle correlation by a whistler wave in nonuniform magnetic field, Phys. Fluids, 22, 367, 1979.

Vomvoridis, J. L., and J. Denavit, Nonlinear evolution of a monochromatic whistler wave in a nonuniform magnetic field, Phys. Fluids, 23, 174, 1980.

Vomvoridis, J. L., T. L. Crystal, and J. Denavit, Theory and computer simulation of magnetospheric very low frequency emissions, J. Geophys. Res., 87, 1473, 1982.

WHISTLER MODE PROPAGATION IN THE EARTH AND PLANETARY MAGNETOSPHERES AND RAY TRACING TECHNIQUES

I. Kimura
Dept. of Electrical Engineeing II
Kyoto University, Kyoto 606
Japan

ABSTRACT. Whistler mode wave phenomena observed in the earth's environment and in the planetary magnetospheres, such as in the vicinity of the Jupiter, Venus, and Saturn are reviewed. Importance of ray tracing technique is stated in connection with the interpretation of these phenomena. A historical review of ray tracing technique is then given. Finally the most recent ray tracing technique, such as 3-D ray tracing in a non-dipolar background magnetic field model is introduced.

1. INTRODUCTION

It is well-known that the phenomenon of whistling atmospherics or whistlers is a part of the radio emissions from lightning, which is dispersed by the magnetoplasma in the ionosphere and magnetosphere. The so-called whistler mode is defined as radio waves propagating in the frequency range lower than the lower out of the electron plasma frequency and the cyclotron frequency of the medium. To this mode, the background magnetic field is essential, even though the cyclotron frequency is lower than the plasma frequency of the medium.

Characteristics of this mode, when only one positve ion species is present with electrons in the medium, are (1) right-hand polarized, (2) the refractive index being larger than unity and especially very large when the electron plasma frequency and the cyclotron frequency are much larger than the wave frequency, (3) the refractive index being greatly dependent on the direction of propagation with respect to the direction of background magnetic field. These properties cause the whistler mode propagation to have very small phase and group velocity and to propagate in a direction close to the direction of magnetic field lines, especially for frequency much less than the electron cyclotron frequency.

When more than two ion species are present with electrons, two characteristic frequencies, the cutoff and cross-over frequencies appear for each of the additional ion species. Between these two characteristic frequencies, the whistler mode is converted to left-hand polarized mode (Smith and Brice, 1964).

For each observed whistler a quantity, so-called dispersion can be measured. The dispersion is the delay time from the source divided by

Space Science Reviews **42** (1985) 449–466. 0038–6308/85.15

the square root of frequency, and it is almost constant for frequencies much lower than the electron cyclotron frequency.

The importance of observations of whistlers has been recognized from the fact that their dispersion represents an integrated information of plasma density and magnetic field along their ray paths. Therefore, if the ray paths are known, the integrated electron density can be estimated. Namely, the whistler phenomenon can be used as a means of diagnosis of the medium. Actually an electron density profile in the magnetospheric equatorial plane could be deduced and the existence of plasmapause was discovered based on the observed whistler data (Carpenter, 1963).

Since Storey (1953) gave an almost complete interpretation of whistler phenomenon by the above characteristics of the whistler propagation of low frequency radio waves in the ionsophere and magnetosphere, it was also found that there is another propagation mechanism of the whistler mode in the magnetosphere; that is field aligned duct propagation. Namely the wave can be ducted in a crest or trough of electron density along the geomagnetic field lines, just like in an optical fibre. Most of whistlers which can be observed on the ground are thought to be the ducted whistlers, whereas most of whistlers observed in the ionosphere and magnetosphere by scientific satellites are nonducted.

Out of whistler mode phenomena observed by satellites, there have been found many interesting new characteristics, which could not be found in those observed on the ground. These characteristics could have been understood by tracing their nonducted propagation paths. The technique of ray tracing has been developed since 1956 (Maeda and Kimura), when ray paths of whistlers were first calculated by a nonsophisticated graphical method. At present, we have a good computer program to trace the ray paths of whistler mode waves.

From the ray tracing, we can determine (i) ray path, (ii) instantaneous directions of the wave normal along the path, (iii) elapsed time for propagation, and (iv) doppler frequency shift as observed by a receiver on-board space vehicles.

In the present paper, several interesting whistler mode phenomena are reviewed, which required ray tracing in order to understand the characteristics. After that, the ray tracing technique and some recent new developements of ray tracing programs are explained.

On the other hand, as a result of the successful planetary exploration by spacecrafts, such as Pioneer-Venus, Voyager 1 and 2, many interesting whistler mode phenomena have been detected, with similar characteristics to those observed in the earth' environment. A brief review of these planetary observations is also given.

2. WHISTLER MODE PHENOMENA IN THE VICINITY OF THE EARTH

Few years after the appearance of Storey's paper (1953), in order to confirm actual whistler ray paths, Maeda and Kimura(1956) devised a graphical method, by which nonducted whistler ray paths were roughly traced and it was found that generally the ray paths considerably deviate from the geomagnetic field lines. The first computer program for ray tracing was made by Yabroff(1961), based on Haselgrove's differential equations for ray paths.

The first whistler phenomenon that was interpreted using ray tracing, was "Subprotonospheric (SP) whistler". This whistler was observed by rockets, the Alouette satellite and on the ground and was characterized by.a number of closely spaced components with extremely small dispersions (Carpenter et al., 1964), as if the wave component starting from the ground once reaches an altitude around 1000km, is reflected back to the ground and echoes in the same path, without traversing to the opposite hemisphere like usual whistlers. In order for the ray paths to cross the geomagnetic field line, effects of positive ions in the ionosphere must be taken into account. Therefore the ion effects were included in the ray tracing program (Kimura, 1966), by which the ray paths of SP whistlers were confirmed to be constructed in the pre-

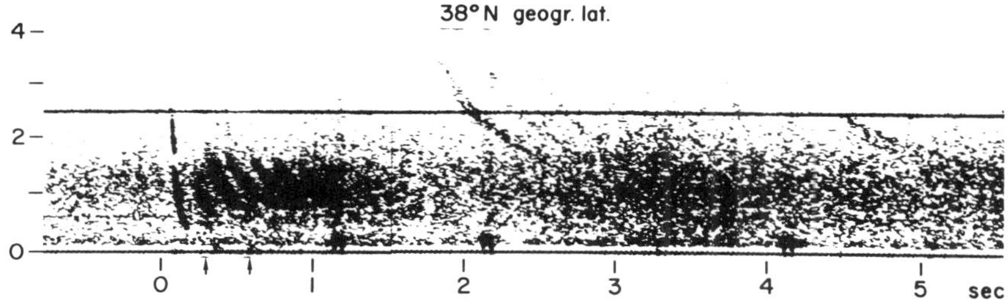

Fig. 1 SP whistlers observed by Alouette satellite (Carpenter et al., 1964)

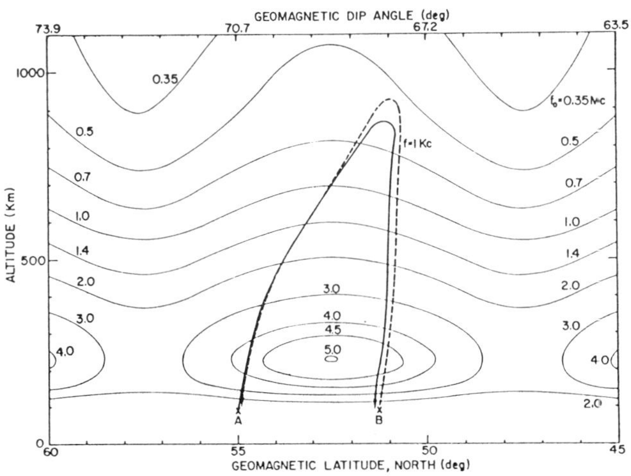

Fig. 2 Calculated ray path of SP whhistler, which can echo through almost the same paths. A sinusoidal horizontal gradient of electron density assumed (Kimura, 1966).

dicted manner. Examples of usual whistlers and SP whistlers are shown in Fig. 1 and a a result of ray tracing is shown in Fig. 2. A latitudinal density gradient was used for the result shown.

When the positive ions are included in addition to electrons, whistler mode wave can traverse across the magnetic field lines, if the wave frequency is less than the lower hybrid resonance frequency. This situation may be understood from the refractive index surface as shown in Fig. 3 and from the property that the group velocity vector is perpendicular to the refractive index surface.

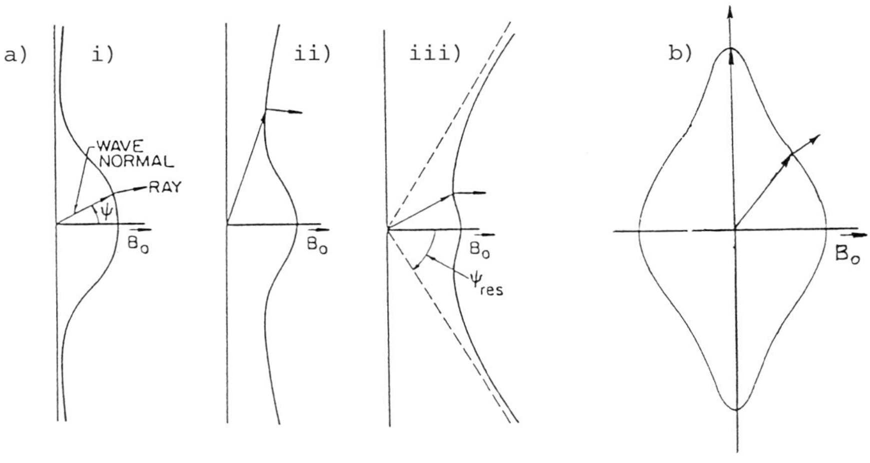

Fig. 3 (a) Refractive index surfaces for different values of f/f_H. i) $f/f_H \ll 1$, ii) $f/f_H < 0.5$, and iii) $f/f_H > 0.5$. (b) Refractive index surface for frequencies less than LHR frequency.

As a by-product of this ray tracing, it was found that magnetospherically reflected (MR) whistler mode ray paths may exist, as shown in Fig. 4. This possibility was later confirmed by satellite observations (Smith and Angerami, 1968), as shown in Fig. 5.

Nonducted propagation causes for whistlers to have many interesting characteristics, such as "walking trace(WT)" (Fig. 6 and 7), focusing of energy, large doppler shift (Walter and angerami, 1968). The walking trace and large doppler shift characteristics result from that part of the ray path over which the wave frequency is higher than the local LHR frequency and the wave normal direction is very close to the resonance cone angle, where the refractive index is very large.

LHR noise(Brice and Smith, 1965) is excited by whistlers and detected by satellites in near-earth orbits. This phenomenon was interpreted to result from the trapping of whistler energy in the valley of LHR frequency altitude profile (Smith, et al., 1966) as shown in Fig. 8.

Drifting upper cutoff phenomenon was often observed on OGO-4 satellites. This was interpreted by a frequency dependence of the ray paths that takes place when a local enhancement of electron density in the ionospheric level, such as the equatorial anomaly, exists (Kimura, 1971).

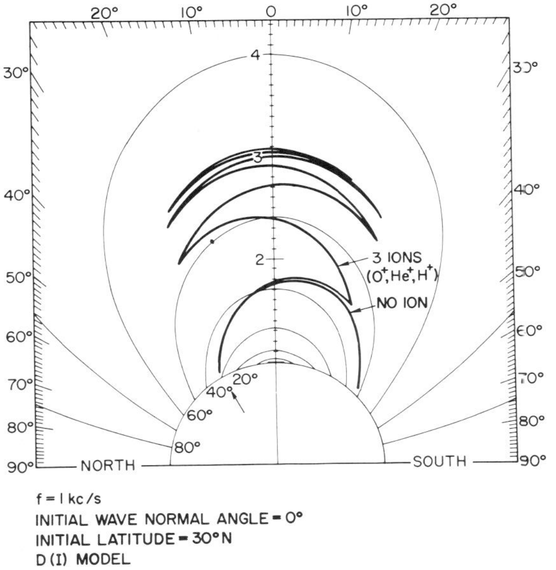

Fig. 4 A calculated ray path of a magnetospherically reflected whistler mode signal (Kimura, 1966).

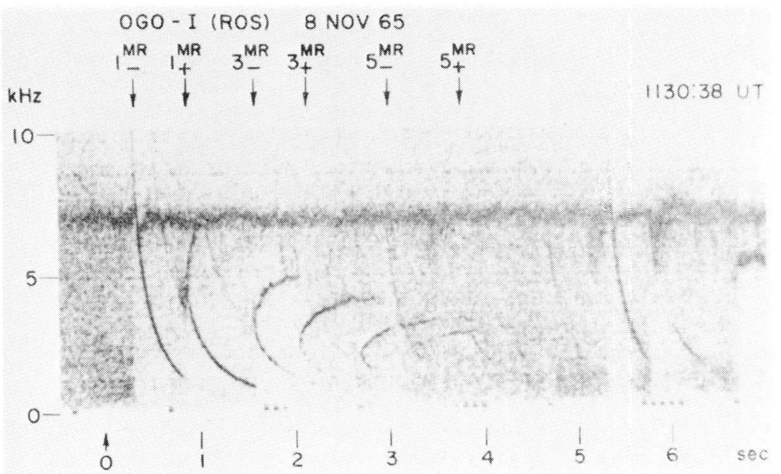

Fig. 5 MR whistlers observed by OGO-1 (Smith and Angerami, 1968)

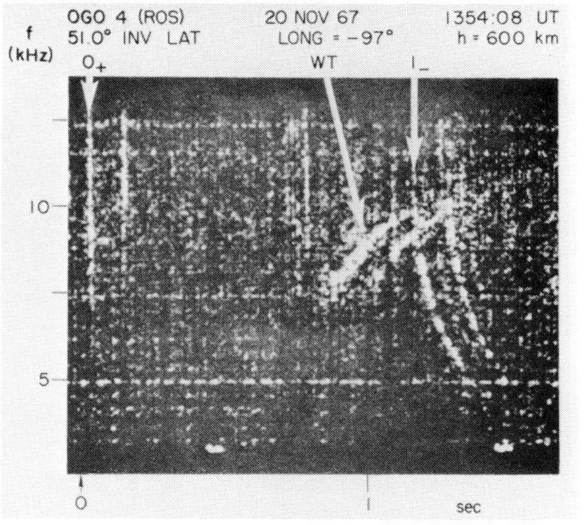

Fig. 6 Walking trace (WT) whistlers (Walter and Angerami, 1968)

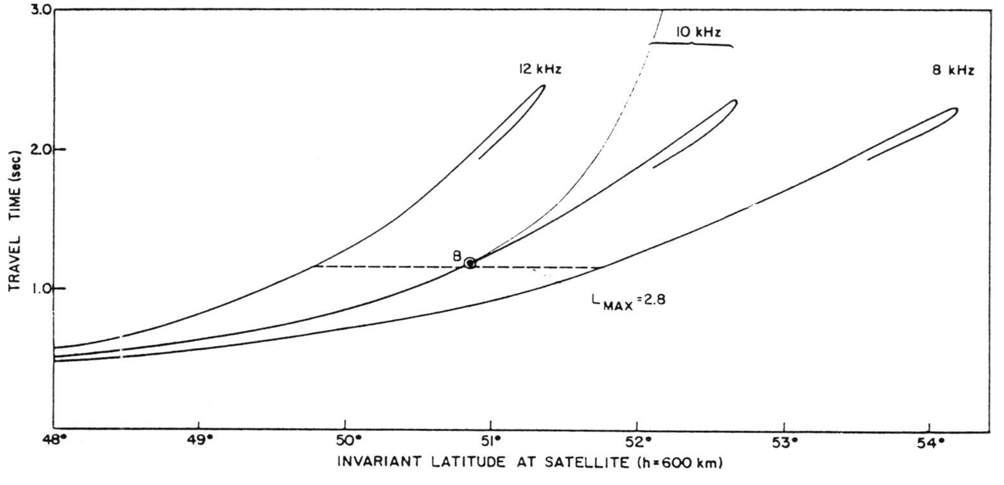

Fig. 7 Calculated travel time versus invariant latitude at satellite for different frequencies (Walter and Angerami, 1968)

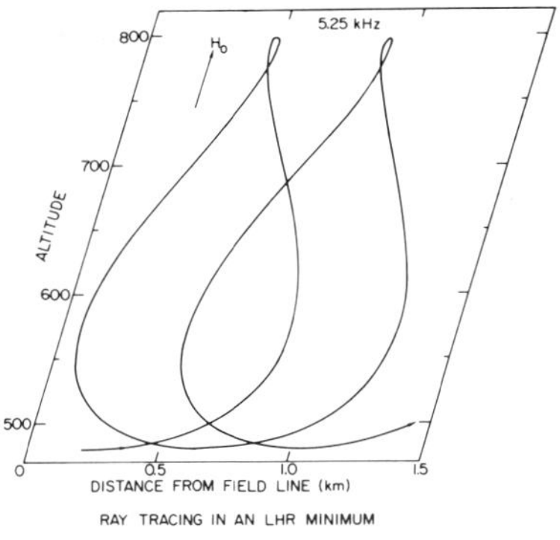

Fig. 8 Calculated ray path trapped in the valley of LHR frequency (Smith et al., 1966)

Edgar studied multipath structures of ground-based VLF transmitter signals observed on satellites, each component of the multipath signals showing different doppler shift at the satellite altitude. Such a possibility of multiple paths reaching the satellite was confirmed by ray tracing and it was found that several steps of sharp electron density decrease with altitude were required to explain the phenomenon (Edgar, 1976).

Matsuo et al.(1984) also found a similar phenomenon for Norway-Omega signals observed by ISIS-1 and 2 telemeter-tracked at Syowa station, Antarctica. In this case, there were many emissions triggered by the Omega signals. By ray tracing, we could determine the most appropriate plasma density profile in the magnetosphere, from which the wave normal direction of the triggering wave at the equatorial interaction region can be determined.

3. WHISTLERS AND WHISTLER MODE PHENOMENA IN THE PLANETARY MAGNETOSPHERE

3.1. Venus

Using what we know about whistlers and whistler mode phenomena, we can determine whether or not there exist lightning on other planets, by observing radio waves in the medium near the planets.

In December, 1978, the first in situ detection of electromagnetic radiation from Venus lightning was made by the Groza instrumentation on the Venera 11, 12 descent vehicles. The Venera wave instruments operated from an altitude of about 60 km down to the surface, and the measured low frequency noise burst were interpreted in terms of electrical discharges in the cloud layers (Ksanformaliti et al., 1979).

Just after Venera's observation, the Pioneer Venus instrument (with 100, 730 Hz, 5.4, 30 kHz band pass channels) could also detect whistler mode radiation by Venus lightning in the nightside ionosphere. These findings were confirmed by the fact that the observed impulsive radiation were detected mostly at the lowest frequency band, which was lower than the local electron cyclotron frequency, while the electron plasma frequency is much higher than electron cyclotron frequency.

3.2. Jupiter

During the Voyager 1 encounter with Jupiter, a number of discrete signals were observed with characteristics similar to whistlers generated by lightning. This fact provided the complete evidence that lightning is present in the Jovian atmosphere. All (a total of 167) of the whistlers detected were observed in a narrow range of L-shell near the inner edge of the Io plasma torus, from about 5.2 to 5.9R_j and within $\pm 10^{\circ}$ of the magnetic equator. Ray tracing was carried out in the Jovian magnetospheric model which includes the Io torus, to confirm the ray paths from the Jovian surface to the Voyager 1 and the observed dispersion characteristics (Gurnett et al., 1979; Menietti and Gurnett, 1980). All information on the characteristics of whistlers observed in the earth's environment was applied to those observed in the Jovian magnetosphere. Similarly, chorus and hiss are also observed in the Jovian magnetosphere.

3.3. Saturn

During the Voyager 1 and 2's encounters with Saturn, whistler mode emissions, similar to chorus and hiss in the earth magnetosphere were detected, whose frequencies were below the electron plasma and cyclotron frequency (Scarf et al., 1983).

4. RAY TRACING TECHNIQUE

4.1. Fundamental equations of ray tracing

In a slowly varying medium, it can be shown that the dispersion relation at an arbitrary point in the medium is the same as that of a homogeneous medium with the medium parameters of the point, according to WKB approximation. Therefore the refractive index can be defined as if the wave propagates in the homogeneous medium.

Now let us consider the ray path $\bar{r}(x,y,z)$ of a radio wave of the frequency ω propagating in the medium. Along the ray path, the next dispersion relation must be satisfied.

$$D(t, \bar{r}, \omega, \bar{k}) = \text{constant} \tag{1}$$

where t is the time elapsed and $\bar{k}$ is the propagation (or wave normal) vector. Along the ray, the right-hand side constant must be independent of the variables in the parentheses of the left hand side. From this condition, the following differential equations are derived with respect to τ which is an arbitrary parameter proportional to the ray path length (e.g. Stix, 1962).

$$d\bar{r}/d\tau = \partial D/\partial \bar{k} \tag{2}$$

$$d\bar{k}/d\tau = \partial D/\partial \bar{r} \tag{3}$$

where we have assumed that the medium parameters do not change with real time t. The physical meaning of the above ray tracing equation is as follows.

i) The ray path is constructed by a successive connection of the group velocity vector that is represented by the right hand side of eq. (2), which depends on the direction of $\bar{k}$.

ii)The spatial change rate of the wave normal direction is governed by the spatial variation of the refractive index; that is equivalent to Snell's law.

For whistler mode three dimensional ray tracing, the above equations are modified by inserting an appropriate expression of the function D, to get the following equations (see e.g. Kimura, 1966)

$$\begin{aligned} dr/dt &= (\rho_r - \mu\partial\mu/\partial\rho_r)/\mu^2 \\ d\theta/dt &= (\rho_\theta - \mu\partial\mu/\partial\rho_\theta)/r\mu^2 \\ d\phi/dt &= (\rho_\phi - \mu\partial\mu/\partial\rho_\phi)/r\mu^2\sin\theta \end{aligned} \tag{4}$$

$$\begin{aligned} d\rho_r/dt &= (\partial\mu/\partial r)/\mu + \rho_\theta d\theta/dt + \rho_\phi \sin\theta\, d\phi/dt \\ d\rho_\theta/dt &= [(\partial\mu/\partial\theta)/\mu - \rho_\theta dr/dt + r\rho_\phi \cos\theta\, d\phi/dt]/r \\ d\rho_\phi/dt &= [(\partial\mu/\partial\phi)/\mu - \rho_\phi \sin\theta\, dr/dt - r\rho_\phi \cos\theta\, d\theta/dt]/r\sin\theta \end{aligned} \tag{5}$$

where r, θ and ϕ are radius distance, colatitude and longitude respectively in the spherical coordinate system. μ is the refractive index and ρ_r, ρ_θ, and ρ_ϕ are the r, θ, and ϕ components of the refractive index vector $\bar{\rho}(\mu = |\bar{\rho}|)$. The first three eqs. (4) are derived from the eq.(2), and the second three eqs.(5) are derived from (3).

The right hand sides of the above ray path equations contain the refractive index μ and the derivatives of the refractive index with respect to spatial coordinates and to the three components of the refractive index vector. The latter derivatives can be calculated analytically, but to calculate μ and the spatial derivatives of μ, a model of plasma parameteres and a model of background magnetic field represented by functions of the spatial coordinates are required as will be explained in the next section. Once some appropriate models are given, then all quantities of the right hand side of eqs.(4) and (5) can be numerically calculated.

For numerical integration of these integral equations, an appropriate technique, such as Adam's predictor-corrector method, should be adopted. The precision of ray tracing depends on the precision of the numerical integration technique. At each step of ray tracing, μ and ρ can be determined independently, because one of the six equations ((4) and (5)) is redundant. In order to keep the two equal, the following correction is made at each step of integration;

$$\rho_{i\ corr} = \rho_i \mu / |\bar{\rho}| \tag{6}$$

Additionally, the step size of integration can be so adjusted that the relative difference between μ and ρ becomes within a pre-assigned error margin.

4.2. Refractive index and plasma parameters to be used for ray tracing

In a cold plasma, which may include several species of ions, it is well known that the refractive index μ can be calculated from the following dispersion relation which is derived from Maxwell's equation and the dielectric tensor of magnetoactive cold plasma,

$$A\mu^4 + B\mu^2 + C = 0 \tag{7}$$

where

$$\begin{aligned} A &= K_1\cos^2\psi + K_2\sin^2\psi \\ B &= -[k_3(1 + \cos^2\psi) + K_4\sin^2\psi] \\ C &= K_1K_4 \end{aligned} \tag{8}$$

$$\begin{aligned} &K_1 = 1 - \Sigma X_\alpha, \quad K_2 = 1 + \Sigma X_\alpha/(Y_\alpha{}^2-1), \\ &K_3 = K_1 \cdot K_2, \quad K_4 = K_2{}^2 - \Sigma X_\alpha Y_\alpha/(Y_\alpha{}^2-1), \\ &X_\alpha = f_{p\alpha}{}^2/f^2, \quad Y_\alpha = f_{H\alpha}/f. \end{aligned} \tag{9}$$

$f_{p\alpha}$ and $f_{H\alpha}$ are plasma and cyclotron frequency of the α-th ion or electron. ψ is the angle between the propagation vector and the earth' magnetic field, and is determined by

$$\psi = \cos^{-1}[(Y_{or}\rho_r + Y_{o\theta}\rho_\theta + Y_{o\phi}\rho_\phi)/\mu] \tag{10}$$

where Y_{or}, $Y_{o\theta}$, and $Y_{o\phi}$ are the direction cosine of the geomagnetic field vector. For a dipole field in the dipole coordinate system

$$\begin{aligned} &Y_{or} = \tan\theta/\ [\,|\tan\theta|\,(\ 1 + 0.25\tan^2\theta)^{1/2}], \\ &Y_{o\theta} = (1 - Y_{or}{}^2)^{1/2}, \quad Y_{o\phi} = 0, \end{aligned} \tag{11}$$

where θ is the geomagnetic colatitude, and i is the geomagnetic longitude. Therefore in this model the angle ψ can be easily determined by these equations.

On the other hand, as the plasma density profile in the earth magnetosphere, so-called the diffusive equilibrium model is widely used, which is based on the physical process that the thermal diffusion of the all constituents of charged particles must be balanced with the gravity force on the particles along each magnetic field line (Angerami and Thomas, 1964). In this model, the electron density N_{de} and ion density N_i are given by

$$N_{de} = N_o\ [\Sigma_i\ \eta_i\ \exp(-z/H_i)]^{1/2} \tag{12}$$

$$N_i = N_{de}\ \eta_i\ \exp(-z/H_i)/[\Sigma_i\eta_i\ \exp(-z/H_i)] \tag{13}$$

where η_i and H_i are is the relative constituent and the scale height of the i-th ion, and N_o is the electron density at a reference altitude r_o.

$$z = r_o(r-r_o)/r$$

is the geopotential height. The scale height is determined by

$$H_i = kT_i/m_i g(r_o),$$

where T_i is the ion temperature, m_i is the ion mass and $g(r_o)$ is the gravitational acceleration at r_o.

This distribution can be applied to an altitude range higher than the peak altitude of the F_2 peak altitude. Below this level, an appropriate function N_L which decreases with decreasing altitudes, may be multiplied to this function. The electron density at the reference level can be a function of latitudes and longitudes. In order to simulate the presence of the plasmapause, an analytical function was devised (e.g. Aikyo and Ondoh, 1971; Inan and Bell, 1977) which smoothly connects the diffusive equilibrium model and the collisionless model that is considered as suitable outside the plasmapause (Eviatar et al., 1964).

4.3. Effect of plasma temperature on ray tracing

So far we have neglected plasma temperature, by adopting the cold plasma treatment. The effect of plasma temperature on the refractive index is, however, not negligible for wave normal angles near and greater than the resonance cone angle.

In a hot plasma, the exact dispersion equation was derived by e.g. Stix (1962). However, that is very complex and not analytic, so that its use in the above ray tracing is not easy. In order to estimate the temperature effect, we now adopt a warm-plasma approximation (Hashimoto et al., 1977) based on Allis et al. (1970). In this approximation, the following assumptions are made; (i) neglection of collisions, (ii) Maxwellian distribution of hot electrons, and (iii) moderate electron temperature, where the thermal velocity of electrons being much smaller than the wave phase velocity. Under these assumptions the dielectric tensor can be approximated by

$$K = K^{(0)} + \tau K^{(1)} \tag{14}$$

where $K^{(0)}$ and $K^{(1)}$ are the cold-plasma dielectric tensor and the warm plasma correction, respectively (see e.g. Hashimoto et al., 1977).

The dispersion equation derived from the above equation is the first-order expansion in τ of the hot-plasma dispersion and is a cubic equation in n^2. This equation gives only the real part of the refractive index, i.e. the imaginary part and thus the damping effect cannot be calculated.

Fig. 9 shows that the refractive index can be well approximated by this warm plasma approximation as far as the real part is concerned at least up to a temperature of 10^4 K. It is evident from the figure that the refractive index deviates much in warm and hot plasma from those in the cold plasma at least for the wave normal angle greater than the

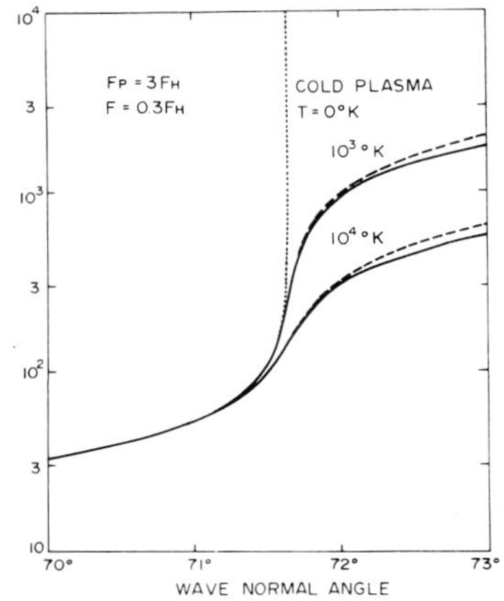

Fig. 9 Temperature effect on refractive index versus wave normal angle for $f/f_H = 0.3$ and $f_p/f_H = 3$.

resonance cone angle. Therefore this warm plasma approximation for the refractive index can be used for ray tracing with acceptable accuracy. For example, at 10^3 K, the ray paths of signal transmited from the ground thus calculated are found to be almost the same as those calculated by the cold plasma treatment. At 10^4K, the ray tracing process stops at its wave normal angle, close to the resonance cone angle, where the warm plasma dispersion relation does not provide any self consistent real solution. Before this point the ray path is still almost the same as those by the cold plasma treatment. Beyond that point where real part of refractive index begins to deviate from that of cold plasma treatment, the exact hot plasma treatment tells us that a damping effect suddenly becomes dominant. Therefore beyond that region the tracing of ray becomes meaningless.

In conclusion, for a plasma temperature less than several thousand K, and for ray paths starting at the ionosphere, the cold plasma treatment is the sufficient approach. However, if we consider ray paths starting from the deep magnetosphere, such as chorus propagation, the warm plasma treatment can provide the ray path until the damping term becomes dominant.

5. RAY TRACING IN A NON-DIPOLAR GEOMAGNETIC FIELD

5.1. Non-dipolar geomagnetic field model

In the dipole model, a geomagnetic field line passing through a point(r_o, θ_o) is represented by the following simple functin of (r, θ).

$$r = r_o \sin^2\theta/\sin^2\theta_o \qquad (15)$$

On the other hand, a more general geomagnetic field and field lines can be determined in the geographical coordinated system (r, Θ, Φ) from the follolwing geomagnetic potential V

$$V = a \sum_n (a/r)^{n+1} \sum_m [g_n^m \cos m\Phi + h_n^m \sin m\Phi]\, P_n^m(\Theta) \qquad (16)$$

where a is the mean earth radius, $P_n^m(\Theta)$ are the associated Legendre spherical harmonic functins of order m and degree n, and g_n^m and h_n^m are the Gauss coefficients. A small distortion of the earth from a sphere is neglected here. Any contribution to eq. (16) from external current sources such as the ring current, is also neglected although the effect is not always negligible during a high solar activity period.

The r, Θ, Φ components of the field (B_r, B_Θ, B_Φ) are given by $B = -\nabla V$, or

$$\begin{aligned} B_r &= -\partial V/\partial r \\ B_\Theta &= -(1/r)\partial V/\partial\Theta \\ B_\Phi &= -(1/r \sin\Theta)\partial V/\partial\Phi \end{aligned} \qquad (17)$$

The Gauss coefficients are adjusted to fit the observed geomagnetic field. In one such geomagnetic field model, called IGRF (Internatinal Geomagnetic Reference Field) model, the coefficients g_n^m and h_n^m are

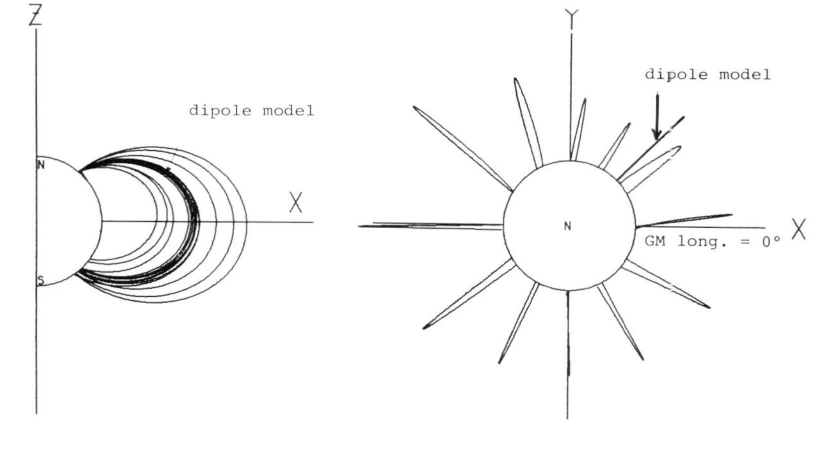

Fig. 10 Magnetic field lines based on IGRF 1980 model, starting at a geomagnetic latitude of 50° (in detail, see text).

known to be n = m = 10 (see e.g. Peddie, 1982).

By integrating the direction cosines of the geomagnetic field vector, field lines can be traced. Fig. 10 illustrates a longitudinal dependence of the geomagnetic field lines starting at geomagnetic latitude

50° on the earth's surface, which are calculated by using the Gauss coefficients of IGRF 1980. (a) shows the projection of the all traced field lines on one geomagnetic meridian plane. A dipole field line is also shown for comparison. (b) is the projection of the field lines on the geomagnetic equatorial plane, as it is viewed from the geomagnetic north pole. The difference of the field lines from the dipole model is evident. Especially longitudinal deviation becomes great between geomagnetic longitudes of 0° and $+90^{\circ}$, due to Brazilian anomaly.

In calculating the right hand side of (1), the derivatives of B_r, B_Θ, B_Φ with respect to r, Θ, Φ are required, so that the 1st and 2nd derivatives of $\underline{V}$ with respect to r, Θ, Φ must be calculated. The Θ derivatives of P_n^m which appear in the 2nd derivatives of V can be analytically calculated (see e.g. Cain et al., 1967).

5.2. Processes in calculating ray paths under the non-dipolar model

All necessary quantities associated with the geomagnetic field can be calculated at any point in the course of ray tracing, by giving the geographical coordinates and assuming the IGRF model. However, in calculating the local plasma density, we need the electron density (N_o in (12)), ion density, and their temperatures at the foot (reference altitude) of the corresponding field line, because the plasma density represented by (12) is the density profile along a geomagnetic field line.

If the plasmapause effect is taken into account, the L value is also needed at each step of the ray path. In the following, the L value of a field line is defined as the ratio of the longest radial distance of each field line to the earth radius.

In the course of ray tracing, the field line foot point and L value are required at every step. Practically, the following procedure is used, in order to reduce the computational time: a necessary number of field line tracings are made in advance before the ray tracing process, starting from the earth°s surface. On each field line, several mesh points are distributed, at which necessary information of the field line is registered. In the course of ray tracing the information necessary for determining plasma densities is obtained by interpolation from the mesh points (Kimura et al., 1985).

In order to determine the mesh points, field line tracing is made prior to ray tracing, in an adopted geomagnetic field model, such as IGRF. Field lines traced from the earth°s surface every 1° in geomagnetic latitude and every 5° in geomagnetic longitude suffice. We draw latitudinal planes (rigorously curved surfaces) at every 5° of geomagnetic latitude. All crossing points of the above mentioned field lines and the above latitudinal surfaces become the mesh points. At all mesh points along a field line, the parameters, i.e. geomagnetic latitude (θ_o) and longitude (ϕ_o) of the field line at the reference radial distance r_o, and L value of the field line are common and are, therefore, commonly registered. The geographical coordinates of each mesh point are written in the computer memory, together with other necessary information.

5.3. Examples of ray paths in the IGRF model

Some examples of ray paths calculated in the IGRF model (Kimura et al., 1985) are shown in Fig. 11 (a), (b). In this case, an electron density model without latitudinal dependence of N_o in eq.(12) is adopted for ray tracing, so that the interpolation of parameters using the mesh points was not necessary. Namely (a) of the figure shows the projection on a meridian plane of the ray paths of 5 KHz starting from an altitude of 300 km at 45° geomagnetic latitude and at every 30° of geomagnetic longitudes. (b) is the projection on the geomagnetic equatorial plane as viewed from the north pole. In (a) and (b), a ray path in the dipole

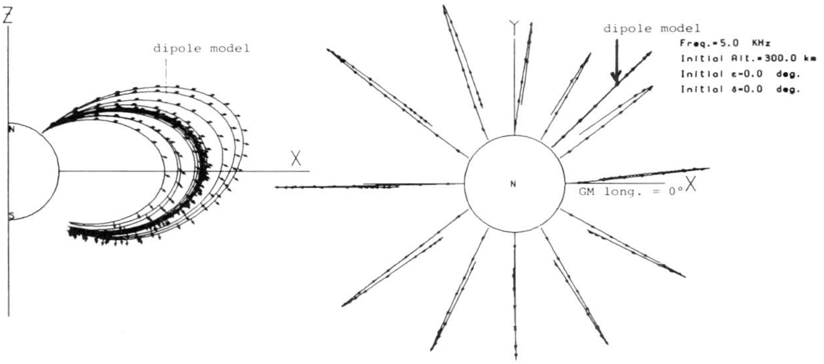

Fig. 11 Ray paths calculated in the IGRF model, for a frequency of 5 kHz (in detail, see text).

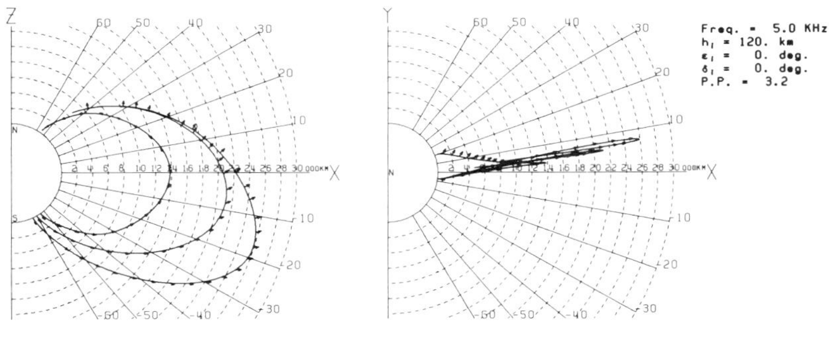

Fig. 12 Ray paths of 5 kHz signals starting from the Siple geomagnetic longitude (7.64°) at geomagnetic latitude of 55, 60, and 65°S.

model is also shown for comparison.

Fig. 12(a) illustrates three ray paths of a 5 kHz signal starting from geomagnetic latitudes 55°, 60° and 65° near the Siple Station (64.78°S) geomagnetic longitude (7.64°W) under a plasmaspheric model with the plasmapause at L = 3.2. In this figure, all ray paths are projected on the Siple geomagnetic meridian plane. Fig. 12(b) shows the same ray paths projected on the geomagnetic equatorial plane seen from the north pole. As is evident from this figure, the ray paths deviate eastwards by about 15°, even though the initial wave normal direction at the bottom of the ionosphere is assumed as vertical. This eastward deviation appears to be consistent with the results of the EXOS-B (Jikiken) satellite observation (Kimura et al., 1983), in which there was a tendency for the Siple signal to be observed more frequently and with stronger intensity in the east side of Siple meridian than the west side.

6. CONCLUSIONS

We have presented a review of whistler mode wave phenomena in the earth and planetary magnetospheres, whose ray paths were important in interpreting their properties. Effects of ions and plasma temperature were also discussed. Other topics concerned with plasma instabilities have not been discussed, since these will be covered by other tutors.

3-D ray tracing in nondipolar models will be an interesting new item, because ample data of plasma wave phenomena in the earth and planetary magnetosphere have been obtained by space vehicles, so that we can investigate, in more detail, any effect of actual magnetic field lines on their ray paths.

The technique of ray tracing is not always concerned with only whistler mode propagation. Those techniques described here can be applied to e.g. AKR and electrostatic wave phenomena, that have been observed in the terrestrial and planetary magnetospheres.

Acknowledgements: The author thanks Dr. U. S. Inan for his careful reading of this manuscript and for his valuable comments.

7. REFERENCES

Aikyo, K. and T. Ondoh, 'Propagation of nonducted VLF waves in the vici- vicinity of athe plasmapause', J. Raio Res. Labs, 18, 153-181, 1971.

Allis, W. P., S. J. Buchsbaum, and A. Bers, Waves in anisotropic plasmas, M.I.T. Press, Cambridge, 1963

Angerami, J. J. and J. O. Thomas, 'The distribution of ions and electrons in the earth's exosphere', J. Geophys. Res., 69, 4537, 1964

Brice, N. M. and R. L. Smith, 'Lower hybrid resonance emissions', J. Geophys. Res., 70(1), 71, 1965.

Carpenter, D. L., 'Whistler evidence of a "knee" in the magnetospheric ionization density profiles', J. Geophys. Res., 68(6), 1675, 1963.

Carpenter, D. L., N. Dunckel, and J. F. Walkup, 'A new, very low frequ-

ency phenomenon: whistler trapped below the protonosphere', J. Geophysical Res., 69, 5009, 1964.
Edgar, B. C., 'The upper- and lower-frequency cutoffs of magnetospherically reflected whistlers', J. Geophys. Res., 81, 205, 1976.
Edgar, B. C., 'Theory of VLF doppler signatures and their relation to magnetospheric density structure', J. Geophys. Res., 81, 3327, 1976.
Eviater, A., A. M. Lenchek and S. F. Singer, 'Distributin of density in an ion-exosphere of a non rotating planet', Phys. Fluid, 7, 1775, 1964.
Gurnett, D. A, R. R. Shaw, R. R. Anderson, and W. S. Kurth and F. L. Scarf, 'Whistlers observed by Voyager 1: Detection of lightning on Jupiter', Geophys. Res. Lett, 6, 511, 1979.
Hashimoto, K., I. Kimura and H. Kumagai, 'Estimation of electron temperature by VLF waves propagating in directions near the resonance cone', Planet. Space Sci., 25, 871, 1977.
Inan, U. S. and T. F. Bell, 'The plasmapause as a VLF wave guide', J. Geophys. Res., 82, 2819, 1977.
Kimura, I., 'Effects of ions on whistler mode ray tracing', Radio Sci., 1, 269, 1966.
Kimura, I., 'Drifting whistler cut-off phenomena --striations-- observed by POGO satellites', Space Res., XI, 1331, 1971.
Kimura, I., H. Matsumoto, T. Mukai, K. Hashimoto, T. F. Bell, U. S. Inan, R. A. Helliwell, and J. P. Katsufrakis, 'EXOS-B/Siple Station VLF wave-particle interaction experiments: 1. General description and wave-particle correlations', J. Geophys. Res., 88(A1), 282, 1983.
Kimura, I., T. Matsuo, M. Tsuda, and K. Yamauchi, '3-D ray tracing of whistler mode waves in a non-diplar magnetosphere', contributed to J. Geomag. Geoelectr., 1985.
Ksanfomaliti, L. V, N. M. Vasil°chikov, O. F. Ganpantserova, E. V. Petrova, A. O. Sowvorov, G. F. Filippov, O. V. Vablonskaya, L. V. Yabrova, 'Electrical dischanges in the atmosphere of Venus', Pisma Astron. Zh., 5, 229, 1979
Maeda, K. and I. Kimura, 'A theoretical investigation on the propagation path of the whistling atmospherics', Rept. Ionos. Res. Japan, 10, 105, 1956.
Matsuo, T., I. kimura and H. Yamagishi, 'ISIS-I and ISIS-II observation of emissions triggered by doppler shifted Norway Ommega signals', Mem. Natl. Inst. Polar Res. in press, 1985.
Menietti, J. D. and D. A. Gurnett, 'Whistelr propagation in the Jovian magnetosphere', Geophys. Res. Lett, 7, 49, 1980.
Scarabucci, R. R., 'Satellite observations of equatorial phenomena and defocusing of VLF electromagnetic waves', J. Geophys. Res., 75(1), 69, 1970.
Scarf, F. L., W. W. Taylor, C. T. Russell and L. H. Brace, 'Lightning on Venus: Orbiter detection of whistler signals', J. Geophys. Res., 85, A13, 8158, 1980.
Scarf, F. L., D. A. Gurnett and W. S. Kurth, 'Measurements of plasma wave spectra in Jupiter°s magnetosphere', J. Geophys. Res., 86, A10, 8181, 1981.
Scarf, F. L., D. A. Gurnett, W. S. Kurth and R. L. Poynter, ' Voyager plasma wave measurements at Saturn', J. Geophys. Res., 88, A11, 8971,

1983.
Smith, R. L. and N. M. Brice, 'Propagation in multicomponent plasmas', J. Geophys. Res., 69, 5029, 1964.
Smith, R. L., I. Kimura, J. Vigneron, and J. Katsufrakis, 'Lower hybrid resonance noise and a new ionospheric duct', J. Geophys. Res., 71(7), 1925, 1966.
Smith, R. L. and J. J. Angerami, 'Magnetospheric properties deduced from OGO 1 observations of ducted and nonducted whistlers', J. Geophys. Res., 73, 1, 1968
Stix, T. H., The theory of plasma waves, McGraw-Hill Book Co. Inc, 1962.
Storey, L. R. O, 'An investigation of whistling atmospherics', Phil. Trans. Roy. Soc. Lond., A246, 113, 1953.
Walter, F. and J. J. Angerami, 'Nonducted mode of VLF propagation between conjugate hemispheres: Observations on OGO's 2 and 4 of the 'walking trace' whistler and of doppler shifts in fixed frequency transmissions', J. Geophys. Res., 74(26), 6352, 1969.
Yabroff, I., 'Computation of whistler ray paths', J. Res. NBS 65D, 485, 1961.

COUPLING OF MACROSCOPIC AND SMALL SCALE PHENOMENA

C. T. Dum
Max Planck Institut für Physik und Astrophysik
Institut für extraterrestrische Physik
D 8046 Garching
West Germany

Abstract. Collisionless microscopic phenomena such as anomalous resistivity, particle acceleration and heat conduction have been successfully treated by particle simulations. Such simulations are usually restricted to volume elements and time scales that are small compared to global scales or even the space and time steps in macroscopic codes. Despite the recent code advances and increases in computing power, it remains necessary to determine the effect of macroscopic dynamics on small scale phenomena and vice versa. The sensitivity of microscopic simulation results to macroscopic boundary conditions is demonstrated. Macroscopic codes, on the other hand, are examined for their dependence on microscopic details. The consequences for the design and analysis of simulation experiments in space physics are discussed. Combining macroscopic and microscopic aspects in a single simulation, despite the usual disparity of scales, will remain a challenging problem.

I. INTRODUCTION

Fluid dynamic phenomena or the electrodynamics of a solid depend on material properties which arise from microscopic processes. These processes are uniquely determined by the macroscopic state. A similar situation can arise in collision dominated plasmas. Although the Coulomb force has a long range compared to interparticle forces in a neutral gas, collisions correspond to the short wavelength part of the fluctuation spectrum which, quite independently of phenomena on a longer scale, is determined by densities and temperatures of the plasma constituents. These fluid variables thus uniquely determine transport coefficients which enter the transport relations for the macroscopic state variables. Macroscopic and microscopic processes in a collision dominated plasma are in essence decoupled.

The plasma we encounter in interplanetary space or in the magnetosphere, however, is nearly collisionless. Nevertheless, the most widely used description of this plasma is magnetohydrodynamics which strictly speaking, applies to collision dominated conducting fluids. It is thus implied that collective processes (connected with plasma turbulence) assume the role of collisions, in providing anomalous transport coefficients. Nonlinear theories for the saturation of microinstabilities are supposed to establish fluctuation levels and hence effective collision frequencies which are specified by fluid variables of the plasma.

Sagdeev's formula

$$\eta = \frac{m}{ne^2}\nu^* = \frac{4\pi}{\omega_{pe}} A \frac{u}{v_e}\frac{T_e}{T_i} \tag{1}$$

for anomalous resistivity is the best known example of this kind [Galeev and Sagdeev, 1979]. In this relation n is the density, u is the relative drift velocity between electrons and ions having

Space Science Reviews **42** (1985) 467–484. 0038–6308/85.15

temperatures T_e andT_i, respectively, $v_e = \sqrt{T_e/m}$ is the electron thermal velocity, ω_{pe} is the plasma frequency, and A is a constant of $O(10^{-2})$. The physics behind this relation is that the relative drift u, corresponding to the current $j = -neu$, excites ion acoustic turbulence which is saturated by nonlinear Landau damping.

A relative drift between the bulk of the electrons and the ions arises also in the presence of an electron heat flux, although there may be no net current. Ion acoustic turbulence in this case is expected to lead to anomalous heat conduction. One can also imagine shear driven microscopic instabilities which result in anomalous viscosity etc. Particle simulations are the prime tool for verifying these predictions. They nearly always reveal a rather complex dynamic behavior in which microscopic and macroscopic processes are strongly coupled, rather than the quasi-static microscopic behavior that is implied by the existence of simple transport relations. The reasons for this behavior and the consequences for space plasma simulation are analyzed in this paper.

II. SIMULATIONS OF ANOMALOUS RESISTIVITY

II.1 Homogeneous Systems

It appears that the simplest configuration, i.e. a homogeneous system with periodic boundary conditions is adequate for the simulation of anomalous resistivity, because relevant microscopic scale lengths, although larger than the Debye length, are much smaller than typical macroscopic scales. This configuration has been used in many particle simulations, starting with the work by Buneman [1959]. He considered a pure initial value problem in which the two stream instability develops from a configuration consisting of Maxwellian electrons and ions with a relative drift velocity much larger than the thermal velocities. (Limitations of computing power actually made it necessary to speed up this development by also initiating the linearly most unstable wave mode.) The electric field energy rises rapidly above the thermal level, according to the linear growth rate. Self-consistent particle motion in the wave field (sloshing) implies an increase in the kinetic energies relative to the particle mean drift velocities and a decrease in the relative electron-ion drift. The instability is reactive. Energy and momentum transfer are thus reversibly related to wave growth. Some randomization of (one dimensional) electron and ion sheets was apparent after several wave e-foldings, typically after about 100 ω_{pe}^{-1}, but this turbulent phase could not be studied extensively by Buneman, due to computational limitations. As no energy was supplied to the system, the "thermal energy," along with the energy in the electric field fluctuations, must have been extracted from the kinetic energy of drift motion. The braking of this drift thus is often ascribed to an effective collision frequency which is of the order of the linear growth rate. The concept of an anomalous collision frequency, strictly speaking, applies more properly to a phase of the two stream instability in which irreversible scattering of (resonant) electrons dominates.

The most conventional definition of resistivity relates the current, $j = -neu$, to the applied electric field, $E = \eta j$. The limitation of the current for electric fields larger than Dreicer's runaway field, which is characteristic of electrons subject to Coulomb collisions, is then attributed to anomalous resistivity. Accordingly, particle simulations have also been carried out for homogeneous systems with an imposed time independent electric field. They reveal a dynamic behavior that is quite different from the pure initial value problem. The drift velocity continues to increase on the average linearly with time, slowed down only somewhat by repeated bursts of wave growth. As the thermal energy also increases at the same rate, by virtue of energy conservation laws, this rate corresponds to a fraction of the free streaming rate [Boris et al., 1970]. Drift and electron thermal velocity evolve on average such that $u \approx v_e$. The fact that power $\mathbf{E} \cdot \mathbf{j}$ is continuously supplied, without any compensating losses, apparently forces the system into this behavior, see Fig. 1.

Dum and Chodura [1979] argued that frequently macroscopic conditions are more realistically represented by a constant current source rather than by imposing a constant voltage. In their simulations the current is kept constant by adjusting the applied electric field at each time step.

A very different picture of the two stream instability emerges, compared to the case of constant applied electric field. The system is now free to return to a stable state with no dissipation, and hence $E = 0$. The way this takes place was rather surprising. It was expected that strong electron heating would gradually change the underlying instability mode from the Buneman regime $u \gg v_e$ to the ion sound regime with $u \ll v_e$, but $T_e \gg T_i$. The simulations of Dum and Chodura [1979] showed that this transition, instead, was effected rather abruptly by some nonlinear wave decay process. The Buneman regime appears as an intense but relatively short transient which, however, was important in creating the conditions for the much longer lasting ion sound phase. The ion sound phase could not have been directly entered for our initial conditions, which correspond to the common case $T_{eo} \approx T_{io}$.

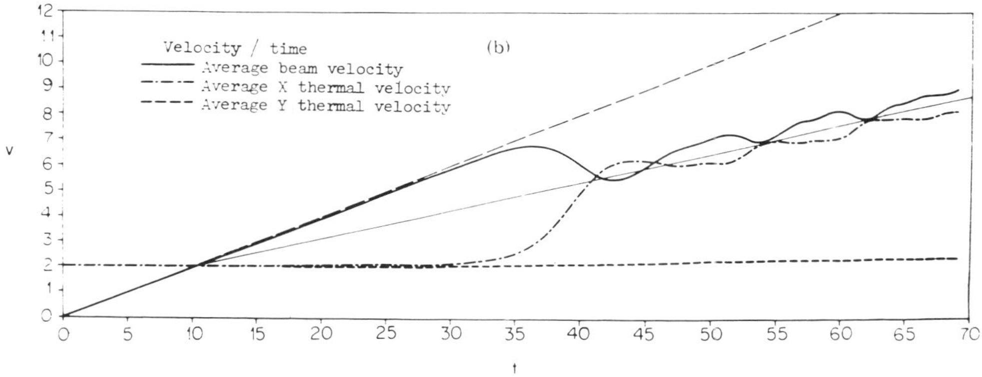

Fig. 1 Evolution of the two stream instability for constant applied electric field. Electron drift and thermal velocity increase such that on average $u \approx v_e$. Energy conservation implies then that acceleration is reduced to fraction of the free acceleration rate [Boris et al., 1970].

The ion acoustic phase of the two stream instability has been extensively investigated in separate simulations. Biskamp and Chodura [1971] presented a set of runs corresponding to different plasma parameters and orientations of magnetic field and simulation plane. Because the instability involves resonant electrons which are scattered nearly elastically (just as in electron-ion collisions) it is essential to have a multidimensional simulation, which allows for oblique wave modes. One dimensional simulations in which wave modes are aligned with the drift velocity, not too surprisingly, show rapid stabilization by quasilinear plateau formation. Biskamp and Chodura [1971] concluded from their 2 D runs that the ion acoustic instability is saturated by ion trapping. An anomalous resistivity formula in terms of plasma parameters would result in this case, albeit different from (1) which is based on saturation of wave growth by nonlinear Landau damping on the ions. Particle trapping is frequently invoked as a saturation mechanism, although the usual arguments apply only to one dimensional particle motion in a single coherent wave mode electric field. Such a situation may be created in a (poor man's) 1 D simulation, but usually does not arise in real life.

Dum et al.[1974] made an earnest effort to determine the wave saturation mechanism from a set of appropriately designed simulation runs. If there is to be any steady state anomalous resistivity, i.e. momentum transfer from the electrons to waves, saturation of wave growth must be due to ions which can pick up this momentum. In order to allow concentration on such a mechanism, a weak magnetic field perpendicular to the simulation plane was applied which maintains a gyrotropic, albeit non-Maxwellian electron distribution. Due to its drift relative to the ions, this distribution always contributes to wave growth. If the electron cyclotron frequency satisfies $\Omega_{ce} \ll \omega_{pe}$, the linear dispersion relation is not changed from the B=0 case. A careful analysis of the runs demonstrated that quenching of wave growth is due to quasilinear effects. Nonlinear effects such as particle trapping or mode coupling play no significant role. Initially nonresonant diffusion establishes an enhanced tail of the ion distribution function which leads to sufficiently strong *linear* Landau damping. The positive contribution to the growth rate by the electrons also diminuishes by flattening of the electron distribution and the decrease in the ratio u/v_e resulting from continued electron heating. These contributions to the growth rate were directly determined from the slopes of the observed distribution functions. They show the approach to marginal stability, $\gamma = \gamma_e + \gamma_i \approx 0$, followed by a slow decay, $\gamma < 0$, in accord with the observed wave spectrum and the heating rates (see Sec.III).

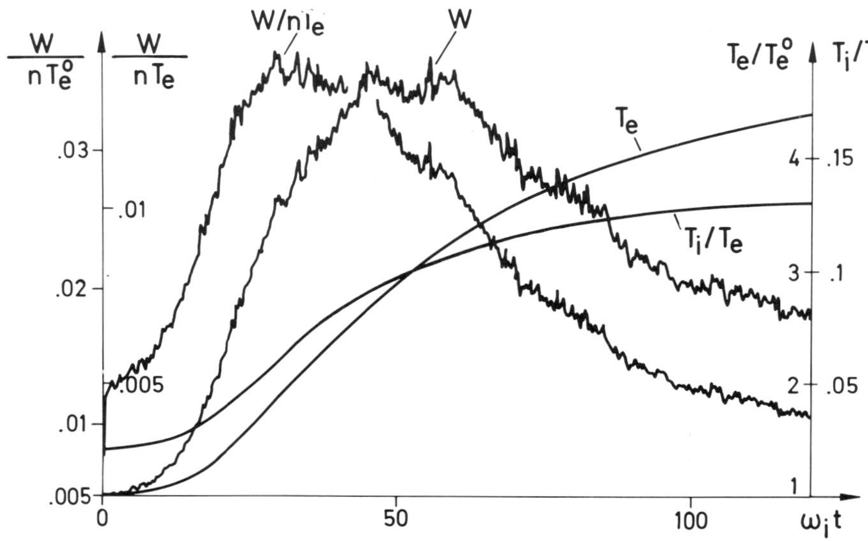

Fig. 2 Wave energy W, fluctuation level W/nT_e, electron temperature T_e, and ion to electron temperature ratio T_i/T_e for a 2D simulation of ion acoustic turbulence. The ratio of relative electron-ion drift to electron thermal velocity is kept constant, $u/v_e = 0.75$, by the applied electric field, in order to allow a better check for any nonlinear wave saturation mechanisms. Quenching of wave growth is actually due to ion tail formation and *linear* Landau damping [Dum et al., 1974].

A necessary pre-condition for the applicability of a resistivity formula like (1), is that the wave spectrum becomes quasi-steady with respect to the fluid parameters. We have therefore imposed the condition $u/v_e = const$, in another set of runs. The temperature ratio T_e/T_i automatically approaches then a constant value, which for the ion acoustic regime is much larger than unity. The only condition for the validity of the resistivity formula (1) that is not met, is that the shapes of the distribution functions remain Maxwellian. They, instead, evolve self-consistently such that the instability is again quenched by quasilinear effects, in much the same way as for the case of constant drift. Claims of having observed nonlinear wave saturation, frequently, have arisen from simulation runs that were too short, to also observe the decay phase. Certainly, nonlinear wave saturation is possible, but often it may happen not to be the most spectacular and dominant effect. The theory leading to the resistivity formula (1) is physically sound, but as the simulations of Dum et al.[1974] have shown, it is at least very difficult, if not impossible, to enter its regime of validity. The maximum fluctuation level W/nT_e arises not from nonlinear saturation, but simply corresponds to the time at which electron heating overtakes the growth of electric field energy,

$$\frac{d \ln W/nT_e}{dt} = \frac{d \ln W}{dt} - \frac{d \ln T_e}{dt} = 0. \tag{2}$$

The general trend of the fluctuation level represented by (1), i.e. the increase with increasing drift and electron to ion temperature ratio is likely to be shared by a number of plausible saturation mechanisms, thus cannot be taken as a proof that nonlinear Landau damping is operative. If the applied electric and magnetic fields are parallel, electrons develop a runaway tail, which compared to the case of perpendicular magnetic field, adds still more dynamic features to anomalous resistivity. Observed maximum fluctuation levels are lower in this case [Biskamp and Chodura, 1971; Lindman, 1984]. The reason is that for the same current, the effective drift velocity for instability, which in essence corresponds to the drift of low energy electrons, is reduced in order to compensate for the runaway current. A most fascinating development occurs if the experiment is started in the Buneman regime of drifts large compared to the electron thermal velocity. A weak electron tail develops first at velocities intermediate between electron and ion mean velocities. It is due to nearly elastic scattering off the unstable Buneman wave spectrum. After the transition to the ion acoustic regime, electrons become nearly isotropic, except for a new tail that develops now in the direction of electron acceleration [Dum and Chodura, 1979]. A variety of other experiments, which may further elucidate the microscropic mechanisms connected with anomalous resistivity, could still be designed within the simple homogeneous simulation model.

A prominent role in simulation experiments is always attributed to studies showing the scaling with respect to different physical parameters. It is, of course, important to first reduce physical parameters to the set of dimensionless parameters which actually enters the numerical experiments. Aside from parameters that characterize numerical features such as space and time steps and the collisionality of simulation macro-particles, this set includes Vlasov parameters such as the mass ratio and the ratio Ω_{ce}/ω_{pe} which measures the magnetic field strength. Some of the parameters, like the ratios of temperatures, drift velocity to electron thermal velocity, etc. are usually only characteristic of the initial state, because these ratios evolve in the course of the experiment. This fact alone, aside from the evolution of microscropic details that is evident in all such simulations, usually dwarts the common hope of checking the validity of resistivity laws of the type (1), simply from the observed scaling. The fluctuation level usually reaches only a temporary maximum which moreover, depends on the macroscopic driving conditions, such as the wave form of the applied electric field. Actually the most outstanding conclusion from many simulation experiments is, in my view, that quasilinear theory carries us a lot further then had been anticipated from the familiar one dimensional bump on tail like models. The bump on tail model, on the other hand, has developed into one of the most interesting nonlinear theories, which applies to strong beam driven Langmuir turbulence [Goldman, 1984].

Particle simulations of the kind we have described can be quite expensive. As one usually needs a set of experiments, corresponding to different choices of plasma and numerical parameters, this may even nowadays, take many hours of computing time. For simulations of ion acoustic turbulence one should have a spatial grid consisting of at least 128 × 128 grid points, in order to resolve a sufficiently large number wave modes, especially at late times in the run. In order to keep the thermal noise background tolerable, one may then require several hundred thousand simulation particles. Accuracy requirements and stability limits of conventional (explicit) particle codes limit time and space steps to $O(\omega_{pe}{}^{-1})$ and $O(\lambda_{De})$ (Debye length), respectively. Typical parameters, e.g. in the simulations of Dum et al. [1974] are $\Delta x = 0.4 - 1\lambda_{De}$ and $\Delta t = 0.33\omega_{pe}{}^{-1}$. About 200000 simulation particles of each species were generally used. In order to observe ion tail formation and to enter the decay phase of ion acoustic turbulence, one requires $t > 100\omega_{pi}{}^{-1}$ or > 12000 time steps, for an ion to electron mass ratio of 1600! Much smaller mass ratios result in an unrealistic enhancement of ion tails, but speed up computations. Such simulations may still be useful in investigating other aspects of the problem.

If only low frequency phenomena are of interest, implicit particle codes nowadays allow a much more economic choice of space and time steps. Winske et al. [1985] e.g. have used an implicit fully electromagnetic particle code [Brackbill and Forslund, 1982] to investigate phenomena near the lower hybrid frequency, $\omega_{LH} = \omega_{pi}/\sqrt{1 + (\omega_{pe}/\Omega_{ce})^2}$, which are excited by currents or ion-ion drifts across the magnetic field. The optimal choice of parameters, naturally, is quite different from simulations of ion acoustic turbulence. If $\omega_{pe} \gg \Omega_{ce}$, the latter regime extends to much higher frequencies, $\omega = O(\omega_{pi})$, and wave numbers $k\lambda_{De} = O(1)$. It cannot be resolved by the choice $\Delta x = 0.75c/\omega_{pe} = 23.7\lambda_{De}$, and $\omega_{pe}\Delta t = 5$ of Winske et al. [1985], but was eliminated already from the outset by also choosing an initial electron to ion temperature ratio of 1/1.5, compared to typically 50/1 in the ion acoustic runs. The fully electromagnetic runs were carried to $\omega_{LH}t \approx 30$ or $\omega_{pi}t \approx 300$, for $\omega_{pe}/\Omega_{ce} = 10$. The mass ratio, however, in this particular run was only $M/m = 100$, and only 65536 particles of each species were used on a 64×64 mesh, which resulted in a rather high noise level. Because the chosen set of plasma parameters is quite different from the cases considered by Dum et al. [1974], microscopic details like wave spectra and the shape of the distribution functions, obviously, are also different. But from a more macroscopic viewpoint one sees in both cases, like in many other simulations, a growth of the wave spectrum, starting from the thermal fluctuation level, which initially is in good accord with linear instability theory. Wave growth is accompanied by a rise in electron and ion temperatures. Wave growth then slows down, and eventually is followed by a slow decline of the wave energy.

II.2 SHOCKS AND RECONNECTION

For a particle simulation to be successful, it generally must be specifically designed to provide information on a particular theoretical problem, such as a mechanism for wave saturation. It then usually does not also provide at the same time a realistic representation of the global system in which this mechanism may be operative (see Dum [1985] for a detailed discussion on the design and analysis of space plasma simulations). Limitations of computing power are an obvious reason for this restriction, but it is often also desirable to purposely reduce the level of complexity, compared to actual observations, in order to facilitate the analysis.

The particle simulations on anomalous resistivity that we have just discussed are a good example. Both the ion acoustic instability and the lower hybrid instability are deemed important in providing a dissipation mechanism for perpendicular shocks. With the recent advances in computing power and (implicit) codes it is now possible to self-consistently include ion acoustic turbulence in a two dimensional particle simulation of perpendicular shocks or quasi-perpendicular shocks, although a better resolution of the microscopic mechanism for anomalous resistivity, certainly, can be achieved in a simulation that is specifically designed for this purpose. Simulations of quasi-perpendicular shocks, however, as a byproduct have also provided some rare information on heat flux driven instabilities [Forslund et al., 1984]. Depending on whether the cross-field current is in the shock simulation plane or points out of this plane, dissipation by this current can be included or eliminated, allowing an evaluation of its effect, compared to dissipation by ion reflection. The current driven lower hybrid instability, which we mentioned above, on the other hand, is virtually excluded from two dimensional simulations of perpendicular shocks, because these, obviously, must contain the direction of inhomogeneity (shock normal), thus cannot also contain the perpendicular directions of current and magnetic field as required for this instability (a high β version of the modified two-stream instability).

The same problem arises for two dimensional simulations of reconnection where again, the current is perpendicular to the plane of the simulation, (x,z), in the usual coordinates for the magnetotail. Anomalous resistivity which is thought to be important in this process, is thus again excluded from a self-consistent treatment. Tanaka and Sato [1981], therefore have carried out a separate particle simulation in a plane which contains the directions of the current (y) and the direction (z) of magnetic field reversal and density gradient, but not the direction (x) of the magnetic field. Periodicity is used only in y. The induction electric field in the reconnection configuration is modelled by a constant applied electric field E_{oy}. It leads to repeated bursts of lower hybrid wave growth and anomalous resistivity, rather than the single growth and decay phase which is observed in the absence of this field. The repetition period appears to be governed by the time required for particles to $E \times B$ drift across the neutral sheet. It would be interesting, therefore, to extend the pure initial value calculations, $E_{oy} = 0$, of Brackbill et al. [1984], which use an implicit code and thus allow for a system size that is an order of magnitude larger, along with realistic mass ratios, to this boundary value problem.

We can see already from the relatively simple examples given here that the results of particle simulations may strongly depend on the boundary conditions imposed by the macroscopic system, but that on the other hand it is often not numerically feasible to extend the calculation to this system. One solution of this dilemma is to incorporate conclusions from particle simulations into fluid like macroscopic models, and to use boundary conditions in particle simulations that approximate macroscopic conditions. [Dum, 1985].

III. QUASILINEAR THEORY AND FLUID EQUATIONS

Design and analysis of simulation experiments have much in common with the corresponding procedures for physical experiments. The result may be something totally unexpected, or something that should have been known, but was not realized by the experimenter, nor anyone else. Most often results are not this spectacular. It happens then not too infrequently that different people draw different conclusions from essentially the same runs. A rather painstaking analysis of such runs is thus required for obtaining useful results.

We pointed out already that a major result of particle simulations is an unexpectedly large range for the validity of quasilinear theory. We first want to explore the consequences for the simple homogeneous system used in the simulation of anomalous resistity, before turning to transport effects in inhomogeneous systems.

III.1 Quasilinear Effects in Anomalous Resistivity

Anomalous resistivity is due to momentum transfer from electrons to waves and then ultimately to the ions. The simulations of ion acoustic turbulence definitively show that the electron wave interaction is quasilinear, that is the (space) averaged electron distribution evolves according to the quasilinear diffusion equation and makes a contribution to wave growth that is in accord with linear theory. We mentioned above that this statement is easiest to verify if gyrotropy of the electron distribution is maintained by a weak magnetic field perpendicular to the simulation plane. In general quasilinear theory requires by itself a rather elaborate simulation. The linear dispersion relation must be evaluated in terms of the distribution functions,

$$\epsilon(\mathbf{k},\omega) = 1 + \sum_j \frac{4\pi e_j^2}{m_j k^2} \int d\mathbf{v} \frac{1}{\omega - \mathbf{k}\cdot\mathbf{v}} \mathbf{k}\cdot\frac{\partial f_j}{\partial \mathbf{v}} = 1 + \sum_j \epsilon_j = 0, \tag{3}$$

and then solved for mode frequency and growth rate, $\omega = \omega_{\mathbf{k}} + i\gamma_{\mathbf{k}}$. The energy in the wave modes may then be advanced, using

$$\frac{\partial W_{\mathbf{k}}}{\partial t} = 2\gamma_{\mathbf{k}} W_{\mathbf{k}}. \tag{4}$$

The resulting spectrum is used to compute a diffusion tensor

$$\mathbf{D}_j(\mathbf{v}) = \frac{8\pi e_j^2}{m_j^2} \int d\mathbf{k} W_{\mathbf{k}} \frac{\gamma_{\mathbf{k}}}{(\omega_{\mathbf{k}} - \mathbf{k}\cdot\mathbf{v})^2 + \gamma_{\mathbf{k}}^2} \hat{\mathbf{k}}\hat{\mathbf{k}}. \tag{5}$$

Finally, the distribution functions can be advanced using

$$\frac{\partial f_j}{\partial t} + \frac{e_j}{m_j}\mathbf{E} \cdot \frac{\partial f_j}{\partial \mathbf{v}} = \frac{\partial}{\partial \mathbf{v}} \cdot \mathbf{D} \cdot \frac{\partial f_j}{\partial \mathbf{v}}. \tag{6}$$

The complete set (3-6), written for a homogeneous system with an applied electric field but no magnetic field, represents a formidable numerical problem. Chodura et al. [1971] obtained a 1 D solution, starting with equal electron and ion temperatures and a drift that was already close to marginal stability. The results are in essential agreement with the corresponding 1 D particle simulations (Fig. 1). Electron drift and thermal velocity increase rapidly, but oscillate about $u \approx v_e$, with corresponding bursts of wave growth. The electron distribution develops a plateau, but little happens to the ions.

Field and Fried [1966] were aware of the importance of multidimensional effects for the ion acoustic regime, but for numerical reasons, were forced to make a number of approximations. The critical discussion of these approximations still makes recommended reading, especially since the crudest of these approximations are among the most popular ones today. We are referring in particular to the ***moment approach*** which circumvents the difficult solution of the diffusion equation (6) by assuming simply that distributions remain Maxwellian or bi-Maxwellian. These distributions are uniquely specified by their lowest order velocity moments. The evolution of these moments is governed by the usual fluid equations, which, however, contain now transfer terms for momentum and energy due to turbulence. The equations are obtained by taking velocity moments of the kinetic for the (ensemble) averaged distribution function. The rate of momentum transfer from fluctuations is given by [e.g. Dum, 1978 a]

$$\mathbf{R}_j = \langle \delta \mathbf{E} \delta \sigma_j \rangle + \left\langle \frac{1}{c} \delta \mathbf{J}_j \times \delta \mathbf{B} \right\rangle = 2\mathrm{Im} \int d\mathbf{k} W_{\mathbf{k}} \epsilon_j (\mathbf{k}, \omega_{\mathbf{k}} + i\gamma_{\mathbf{k}})\mathbf{k}, \tag{7}$$

where the first relation is general and a rigorous consequence of the ensemble averaged kinetic equation, and only the second relation assumes that fluctuations are electrostatic, with the perturbations of charge density and current quasi-linearly related to the fluctuation of the electric field. This relation is also obtained by taking the velocity moment of (6). For the rate of energy transfer one obtains, similarly,

$$K_j = \langle \delta \mathbf{E} \cdot \delta \mathbf{J}_j \rangle = 2\mathrm{Im} \int d\mathbf{k} \epsilon_j (\mathbf{k}, \omega_{\mathbf{k}} + i\gamma_{\mathbf{k}})(\omega_{\mathbf{k}} + i\gamma_{\mathbf{k}}). \tag{8}$$

The heating rate is given by $Q_j = K_j - \mathbf{R}_j \cdot \mathbf{u}_j$. It is equally straightforward, although more tedious, to derive transfer rates for other moments or for the general electromagnetic case [e.g. Winske et al., 1985]. Just as the dielectric constants (2), these rates, however, contain the average distribution function f_j, which also enters the left hand side of the moment equations in terms of higher order moments such as heat flux etc. As is well known from classical transport theory, a closed set of equations cannot be derived in this way, unless one assumes the shape of the distribution function. In classical transport theory these are nearly Maxwellian, but corrections are important. These corrections determine not only heat flux and viscosity, but also cause a substantial modification of resistivity. These facts are generally ignored for turbulent transport, although there is no mechanism in this case that would even approximately maintain Maxwellian distributions. The transfer rates also depend on the shape of the wave spectrum. Field and Fried [1966] evaluated the integrals (7-8) by a saddle point method, which in essence retains only the fasted growing wave mode.

We may consider the following set of model equations, in order to gain a rough understanding of the general dynamical behavior [Dum, 1981].

$$\frac{du}{dt} = -\frac{e}{m}E - \nu^*(u^* - u_w), \tag{9}$$

$$\frac{d\ln W}{dt} = 2\sigma_1\frac{u^* - u_c}{v_e} - 2\sigma_2\frac{W}{nT_e} - \sigma_3\frac{d\ln T_e}{dt}, \tag{10}$$

$$\frac{d\ln T_e}{dt} = \frac{2}{3}\left(\frac{u - u_w}{v_e}\right)\nu^*\left(\frac{u^* - u_w}{v_e}\right), \tag{11}$$

$$\nu^* = \rho\frac{W}{nT_e} \quad , \tag{12}$$

$$dT_i \approx \frac{u_w}{u - u_w}dT_e. \tag{13}$$

In these relations u^* is the effective drift velocity for instability, which for skewed non-Mawellian distributions may differ from the mean drift u. The mean phase velocity u_w replaces the ion mean velocity of the conventional friction relation. The critical drift velocity for instability, u_c, can be substantially larger than u_w, because of ion Landau damping. Linear wave growth is described by σ_1 and σ_2 describes nonlinear wave damping. Adiabatic effects may be included by σ_3. Equ. (12) expresses the linear relation between fluctuation level and anomalous collision frequency. The last relation for the relative ion heating rate is valid near marginal stability, c.f. (7-8). More generally, the energy extracted from the electrons at rate $-K_e$ goes into ion heating and wave growth. A set of this structure can be derived by taking moments of the kinetic equations [Dum, 1981]. The various coefficients represent spectral averages and depend also on the shape of the distribution functions. They are thus, in reality, also time dependent and the set (9-13), strictly speaking, is not closed. The rate of momentum transfer from ion acoustic turbulence to electrons e.g. takes the form

$$\mathbf{R}_e = -(2\pi)^{1/2}\left\langle a_{-3,\hat{\mathbf{k}}}\frac{1}{k\lambda_{De}}\hat{\mathbf{k}}\hat{\mathbf{k}}\cdot(\mathbf{u}^*_{\hat{\mathbf{k}}} - \frac{\omega_{\mathbf{k}}}{k}\hat{\mathbf{k}})\right\rangle\frac{W}{nT_e}nm\omega_{pe} = -nm\underline{\underline{\nu}}^*\cdot(\mathbf{u}^* - \mathbf{u_w}), \tag{14}$$

where the spectral average is defined by (7)and $a_{-3,\hat{\mathbf{k}}}$ is a form factor (=1, for a Maxwellian) which characterizes the slope of the low energy electron distribution. For an anisotropic distribution this form factor and also the effective drift velocity depend on the direction $\hat{\mathbf{k}}$ of wave propagation [Dum, 1978 b]. With some guidance on this kinetic behavior, obtained e.g. from particle simulations, equations of the type (9-13) can still be useful, although the effort of a detailed numerical solution probably is not warranted.

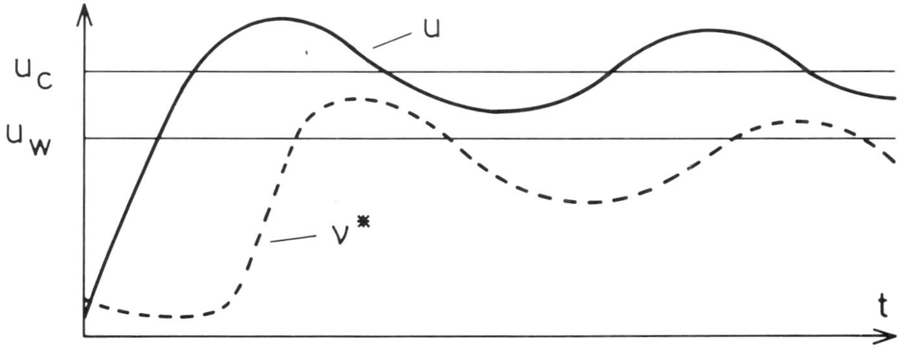

Fig. 3 Schematic representation of relaxation oscillations about marginal stability. Free acceleration by an applied electric is stopped once the drift exceeds the instability threshold u_c and turbulence reaches a sufficiently high level. Anomalous resistivity tries then to reduce the drift to the mean phase velocity $u_W < u_c$ and oscillations of drift and effective collision frequency (fluctuation level) result.

Field and Fried [1966] found little heating for the duration of their runs. In this case we can reduce the set (9-13) to coupled equations for for the linear growth rate, $\gamma = \sigma_1(u^* - u_w)/v_e$, and the effective collision frequency. They take the form

$$\frac{d\gamma}{dt} = \hat{E} - \nu^*(\gamma + \Delta\gamma), \tag{15}$$

$$\frac{d\nu^*}{dt} = 2\nu^*(\gamma - \hat{\sigma}_2\nu^*), \tag{16}$$

where it was assumed for simplicity that u^* is equal (or proportional) to u, and the normalizations $\hat{E} = -\sigma_1(eE/mv_e)$, $\hat{\sigma}_2 = \sigma_2/\rho$ have been introduced. The term $\Delta\gamma = \sigma_1(u_c - u_w)/v_e$ decribes the excess electron drift that is required for linear instability in the presence of ion Landau damping. The behavior of the drift and fluctuation level found by Field and Fried [1966] is obtained for $\Delta\gamma = 0$, $\hat{\nu}_2 = 0$, and constant $\hat{E}$. The essential result for their runs with $T_e/T_i \gg 1$ was that linear acceleration by the electric field is stopped once ion acoustic turbulence develops to a sufficiently high level. The drift is then reduced to a critical velocity of the order of the ion sound velocity, $c_s = \sqrt{T_e/M}$, which corresponds to marginal stability. If $\Delta\gamma > 0$, the drift no longer monotonically approaches u_c, after a single overshoot, but undergoes *relaxation oscillations*. These oscillations about marginal stability, $\gamma \approx 0$, are also reflected in the fluctuation level and in the effective collision frequency, see Fig. 3.

An important point to observe from (15-16) is that this behavior will qualitatively stay the same if there is also some nonlinear damping, $\hat{\sigma}_2 > 0$. The fixpoint in the (γ, ν^*) phase space, which corresponds to the stationary state, is shifted somewhat in this case from the values $\gamma_0 = 0$ and $\nu_0^* = \hat{E}/\Delta\gamma$ for $\sigma_2 = 0$. It may be obtained by setting $d\gamma/dt = d\nu^*/dt = 0$ in (15-16). Weak nonlinear effects such as nonlinear ion Landau damping thus are unlikely to be discernable in the observed macroscopic behavior. Relaxation oscillations about a state of marginal stability, with bursts of wave growth, are very common in collisionless plasmas. The model (15-16) shows such oscillations, but clearly is too simple to represent the actual macroscopic and microscopic processes which shape this behavior.

Continuous electron heating plays a dominant role if the drift $u = const.$ The growth rate is reduced by the increase in v_e. The fluctuation level , and hence ν^* and E, reach a maximum when wave growth is overtaken by electron heating, (2). Quenching of the instability begins at the latest, just as for the pure initial value problem with no applied electric field, when the characteristic phase velocity $u_w \approx c_s$ begins to exceed the drift velocity u. Ion heating at the rate (13), also leads to some additional increase in the critical drift velocity $u_c \gtrsim u_w$. It is clear that electron heat losses, which are excluded in periodic simulations, will significantly affect fluctuation levels. They may also cause relaxation oscillations. Heat flux driven ion sound turbulence in the solar wind, indeed, consists of short individual bursts, which are related to relaxation oscillations about marginal stability. They are associated , however, with very minute changes in the *shapes* of electron and ion distribution functions, rather than with changes in the moments of these distributions [Dum, 1983]. The usual moment approach even fails in representing heat flux. It can represent a current, but Field and Fried [1966] found already that the bi-Maxwellian model for the electron distribution is too rigid in coupling drift and wave growth. Their improved model therefore determines at least the electron velocity distribution parallel to the applied field, while still assuming a Maxwellian distribution in the perpendicular direction. The macroscopic behavior of drift and fluctuation level was qualitatively similar to the moment approximation, but the wave spectrum became more and more oblique as distortions of the electron distribution reduced the growth rate for modes along the drift direction. At the end of the run, modes propagating antiparallel to the drift actually had the largest growth rate. The need for a complete solution for the electron distribution function was indicated by their results. Dum et al. [1974] showed from their simulation results that even if changes in the electron contribution to growth rate can be reduced to a minimum, nonresonant diffusion creates high energy ion tails which ultimately lead to quenching of the ion sound instability by linear Landau damping.

With some caution, and using information provided by particle simulations, relations of the form (9-13) can be applied to infer some of the dynamics in this process [Dum, 1978 a]. The important observation from the particle simulation in this case is that almost all of the momentum and energy extracted from the drifting electrons is delivered by a resonant interaction to a strongly anisotropic high energy ion tail which extends only in the direction of the electron drift. As a very approximate model for this tail one can thus use a half-Maxwellian. Momentum and energy conservation laws are then sufficient to determine the tail density and temperature from the electron transfer rates. For a constant drift u one finds that the tail temperature stays relatively constant and of order of the electron temperature, in agreement with the simulations. The ion heating rate (13) which refers to the total energy delivered to the ions, thus corresponds in effect to the rate of increase in the relative number n_h/n of tail particles. The usual application of the moment approach, using Maxwellian ions or a distribution consisting of two separate Maxwellian components, naturally, would give constant particle densities and interpretate the energy increase in essence as a temperature increase. The usual definition of temperature in simulations, i.e. kinetic energy relative to the mean velocity, would not reflect this difference, but a very different picture of the effect of ion Landau damping on wave growth would result in this case. Ion tail formation is more effective in quenching the instability, succeeding even if the ratio u/v_e is kept constant. Damping by this process is also supported by a shift of the wave spectrum towards larger wave numbers and hence smaller phase velocities. A more detailed description of these dynamic processes requires a solution of the quasilinear kinetic equations (3-6). Improved numerical techniques (finite element methods) allow such solutions [Appert et al., 1976]. These techniques have been applied to the problem of anomalous resistivity by ion sound turbulence [Appert and Vaclavik, 1981]. The results support the conclusions of Dum et al. [1974].

We tried to demonstrate that a combination of particle simulations, solution of quasilinear equations, and the moment approach proves to be most fruitful in understanding the microscopic physics. Particle simulation, certainly, is the most basic description, with the least assumptions, but is comparatively expensive and not as flexible in the choice of parameters as the other methods. Relatively large drifts or accelerating electric fields must be chosen in order to get wave growth within a reasonable computing time and to overcome noise problems. Note that electric fields $eE/mv_e\omega_{pe} \approx \nu^* u/\omega_{pe}v_e = 0(10^{-3})$ must be considered large in the context of e.g. auroral observations, but essentially free acceleration would then typically consume the bulk of the computing time in particle simulations, before the process depicted in Fig. 3 is even entered. If particle simulation shows the absence of a dominant nonlinear process, quasilinear theory not only can, but should come into play. It offers the possibility to isolate any nonlinear effects and also allows experimentation with different elements of the quasilinear interaction such as resonant and nonresonant diffusion etc., in order to assess the relative importance of these effects. In contrast to particle simulations, it is also straightforward to extend two dimensional calculations to a fully three dimensional problem, by making use of axisymmetry, e.g. for a current along the the magnetic field [Ishihara and Hirose, 1983]. Quasilinear simulations need not resolve wave frequencies, but only growth rates and the evolution of average distribution functions, thus may allow a larger time step. Still, the solution of the kinetic equations for the average particle distribution and the wave spectra is a rather elaborate numerical problem which apparently has not been carried to time or space intervals that are significant on a macroscopic scale. Fluid type equations offer this possibility, if one can, as in classical transport theory, proceed beyond the simple moment approximation with prescribed (Maxwellian) distribution functions. The simple moment approach can serve for orientation purposes, especially, if some of the relevant microscopic features are already known from particle simulations.

III.2 Anomalous Transport Processes

The chief motivation for the study of plasma turbulence is its effect on transport properties. Just as in classical transport theory one would like to arrive at a closed set of equations which relates the macroscopic variables of the plasma. For classical transport the possibility of such a description arises from the existence of a universal relaxation process, which causes any distribution function to relax towards a Maxwellian. This distribution and the fluctuation spectrum causing the relaxtion are uniquely specified by the fluid variables density, mean velocity and temperature. In a collisionless turbulent plasma, as we have seen, this is generally not the case. The coupled evolution of the distribution functions and the wave spectrum can be observed in particle simulations, or can to be determined from a set of coupled kinetic equations, once we know that the wave particle interaction is quasilinear. For a realistic description of the macroscopic configuration it still appears necessary to use fluid type equations which, however, should incorporate at least the dominant effects found in separate microscopic studies. This task must be carried out separately for each underlying microinstability. Owing to the complexity of the problem, one can hardly expect to find a useful general theory for anomalous transport.

The conclusions reached from particle simulations of ion acoustic turbulence made it possible to develop an anomalous transport theory for electrons which in completeness approaches in some ways classical transport theory [Dum, 1978 a, b]. The development of this theory, in fact, is based on the observation that scattering of electrons by ion acoustic turbulence is in many ways similar to scattering by electron-ion collisions. We can consider ion acoustic turbulence as the strongly enhanced long wavelength extension of the fluctuation spectrum responsible for electron-ion collisions. Because typical electron velocities are much larger than the phase velocities of ion acoustic waves, $\omega/kv_e \ll 1$, the quasilinear resonance condition

$$\omega - \mathbf{k}\cdot\mathbf{v} = 0, \tag{17}$$

is satisfied for the bulk of the electron distribution by the approximation $\mathbf{k}\cdot\mathbf{v} \approx 0$, which implies nearly elastic scattering, and hence a tendency towards isotropization of the distribution. Residual inelastic scattering tends to flatten the distribution function, resulting in the absence of other forces in a distribution of the form

$$F(v) = n(C/v_o^3)\exp[-(v/v_o)^5], \tag{18}$$

where $C = 0.27$ is a normalization to unit density and the usual thermal velocity (second moment) is given by $v_e = 0.47v_0$. The rapid approach to this distribution from the usual initial Maxwellian can be observed in particle simulations and is in excellent agreement with a Green's function solution of the corresponding 2D quasilinear diffusion equation [Dum, 1978 a]. The normalization of (18) is different in the 2D case, the shape of the distribution, however, results from the v^{-3} speed dependence of the effective scattering frequency, which is the same in 2D and 3D. The change in the distribution function invalidates the usual moment approach with a Maxwellian distribution. Both the wave dispersion relation and the transfer rates (7-8) are significantly changed. The form factor a_{-3}, which enters both the growth rate and the anomalous resistivity relation (14), e.g. is reduced from unity to 0.44 in 3D and 0.28 in 2D. The theory at this point is applicable to current flow across the magnetic field, as it occurs in perpendicular shocks. Weak anisotropies in the electron distribution, however, are responsible for changes in resistivity associated with parallel current flow, and most importantly, for heat flux. Heat flux is even important in perpendicular shocks in carrying energy away from the localized anomalous dissipation region. The moment approach to shock simulation [Manheimer and Boris, 1972; Liewer and Krall, 1973] excludes heat conduction by imposing Maxwellian or bi-Maxwellian distributions. The fact that anomalous resistivity is dynamically related to macroscopic shock evolution could nevertheless be demonstrated by these models. The actual spatial or temporal structure of a resistive shock transition, changes, however, drastically if electron transport processes and related modifications of wave growth are properly accounted for Dum [1978 b], as could be demonstrated by a comparison with observations [Söldner et al., 1977] and in modified hybrid simulations with fluid electrons subject to these new anomalous transport relations [Chodura et al., 1977]. The transport relations are derived by splitting the diffusion equation into a system of equations for the isotropic part of the distribution, $F(v)$, and its anisotropic part $\hat{f}(\mathbf{v})$. Dominance of elastic scattering by ion sound turbulence allows then to determine $\hat{f}$ as a functional of F which is linear in the perturbing gradients, drifts or electric field. Insertion of $\hat{f}$ into the equation for F, closes this equation. Solution of this reduced diffusion equation can be avoided if the perturbing forces are weak enough in comparison to *inelastic* scattering which will then establish the distribution (18). Transport relations can now be derived by taking moments of the complete distribution $f = F + \hat{f}$. Although this procedure is similar to classical transport theory, even the resultant *structure* of transport relations , and not only the size of the transport coefficients, differs from classical relations with Maxwellian F. The theory does not extend to ion transport and still requires the knowledge of the wave spectrum. Fortunately, because of the nature of the wave resonance (17), only a few characteristics (moments) of this spectrum, such as the total wave energy and the mean phase velocity, c.f. (14), are required. It is admitted that the theory is not simple, but there is no apparent shortcut to obtaining these results. Simplifications appropriate to a particular macroscopic configuration are possible, however, which facilitate a practical application, as has been demonstrated by the application to perpendicular shocks. The correct interrelationship between the various transport processes and wave growth proved to the most important feature. Hopefully, transport theories related to other instabilities also can be further developed, a posteriori, from the understanding gained in microscopic simulations.

IV. Electrodynamic and Mechanical Coupling

Particle simulations are primarily concerned with microscopic processes as such and do not extend to real macroscopic configurations occuring in space or the laboratory. The most conceivable direct extension of these calculations by current advances in computing power to a more realistic space plasma configuration may involve *magnetospheric transition regions* such as the bow shock, magnetopause, etc. The choice of appropriate upstream and downstream boundary conditions is a delicate problem [Dum, 1985], which for bigger systems, hopefully, becomes less severe. It is not really conceivable at this time, however, that microscopic simulations can be directly extended to the entire coupled solar wind-magnetosphere-ionosphere system. The coupling by the solar wind is mainly *mechanical*, i.e. through particle and heat flux from the solar corona. Propagation of hydromagnetic waves is thought to play some role in solar wind acceleration near the corona, but the chief driving force, clearly, is electron heat flux from the hot corona which is converted into kinetic energy of expansion. The simplest description of this process is given by the one fluid equation for steady state radial flow,

$$r^{-2}\frac{d}{dr}r^2\rho u\left[\frac{u^2}{2}+\frac{5}{2}\frac{p}{\rho}-\frac{GM_o}{r}\right] = -r^{-2}\frac{d}{dr}r^2 q, \tag{19}$$

where ρ is the mass density, p is the pressure, u is the solar wind speed, the last term in the bracket is the gravitational potential and q is the heat flux. Unfortunately, the solar wind becomes essentially collisionless at not too large heliocentric distances. A total breakdown of classical heat conduction may occur at a distance of a few solar radii, and much effort has been spent in trying to find the right "fudge factor" in the classical heat flux relation, $q = -\kappa\nabla T_e$, which would reduce the large discrepancies between observations and numerical fluid models. These efforts have not met much success, however. The other extreme assumption, collisionless, steady state flow (exospheric theory) also fails badly, in particular, by predicting extremely anisotropic distribution functions. The detailed in situ observations that are now available indicate that a combination of turbulent scattering and of some remnants of exospheric expansion controls electron heat flux [Dum, 1983]. No complete theory, nor a simulation, exists for this challenging transport problem. A somewhat similar problem exists in the magnetosphere-ionosphere interaction. Exospheric theory is applied, in particular, to particle flux and electric potential drops along magnetic flux tubes in the auroral zone, when it is known at the same time that these flux tubes contain layers of intense turbulence and that auroral processes are anything else but stationary. There is also detailed observational evidence for

electrodynamic coupling by shear Alfvén waves and associated field aligned currents. The problem is to tie the macroscopic electrodynamic description with a kinetic description of microscopic processes. The latter is clearly needed to describe such important phenomena as particle acceleration, but cannot, according to the evidence from particle simulations that we have presented above, be treated independently of the electrodynamic energy supply mechanism in this very extended system.

The vast difference in microscopic and macroscopic scales, although it is a notorious problem in simulations, must also allow for some significant simplifications. If we are not interested in the details of wave growth and particle distributions, it would appear that one can safely integrate over these microscopic features. Finite difference schemes for fluid equations, in effect, amount to such an integration, as is directly seen for schemes originating from the conservation form

$$\frac{\partial \mathbf{U}}{\partial t} + \nabla \cdot \mathbf{F} = \mathbf{S}. \tag{20}$$

The finite difference scheme is obtained by integrating over a finite time interval and a volume element, using some interpolation scheme for values between grid points,

$$\frac{\mathbf{U}_m^{n+1} - \mathbf{U}_m^n}{\Delta t} = \frac{\partial \hat{\mathbf{U}}}{\partial t} = \frac{1}{\Delta t \Delta V}\left[-\int_{\Delta t} dt \oint_S d\mathbf{s} \cdot \mathbf{F} + \int_{\Delta t} dt \int_{\Delta V} dV \mathbf{S}\right]. \tag{21}$$

The flux $\mathbf{F}$ thus is time averaged over the time interval Δt and integrated over the surface of the volume element surrounding the spatial grid point m. The source term $\mathbf{S}$ is both space and space averaged. If the averaged intensity $\hat{\mathbf{U}}$ is defined in the same way, its partial time derivative at midtime corresponds then to the time difference operator in (21). A more general ***filter*** for phenomena on scales smaller than the time and space steps could also be used. The time average on the r.h.s. of (21) has to be approximated by an expression in the state variables at previous times (explicit scheme), or including the new time t_{n+1} (implicit scheme), e.g. $\hat{\mathbf{S}} = (1-\theta)\mathbf{S}^n + \theta \mathbf{S}^{n+1}$, where $1 \geq \theta \geq 0$.

In the case of the energy conservation relation (19), we need only the average heat flux through the surfaces of a spherical shell, which if controlled by turbulent scattering, should depend only on the average fluctuation level in the volume element. We may integrate in particular over the short individual bursts of wave growth. The special difficulty in the case of solar wind heat flux is that a substantial contribution is made by energetic electrons which are nearly unaffected by scattering from ion acoustic turbulence or Coulomb collisions, and thus sample fluctuations and ambient fields over large radial distances. A local transport relation in terms of gradients no longer holds for these energetic electrons [Dum, 1983].

The problem with energetic particles, naturally, is less severe for the lower order moments of the distribution function, such as the field aligned current in electrodynamic coupling, although the acceleration of runaway particles by the field aligned electric field is of great interest by itself. The observational evidence for the auroral zone suggests that we do not postulate a stationary state, but treat such a state as a limiting case of a more general evolution [Lysak and Dum, 1983; Dum and Lysak, 1985]. Let us see now, how microscopic processes enter into this global problem. For the time and length scales we consider, Maxwell's equations can be reduced to the following set,

$$\frac{\partial B_y}{\partial t} = -c\left[\frac{\partial E_x}{\partial z} - \frac{\partial E_z}{\partial x}\right], \tag{22}$$

$$j_x = -\frac{c}{4\pi}\frac{\partial B_y}{\partial z}; \qquad j_z = \frac{c}{4\pi}\frac{\partial B_y}{\partial x}. \tag{23a, b}$$

Neglect of the y dependence correponds to the east-west alignment of auroral structures. The quasineutrality approximation is consistent with the neglect of the displacement current, $\nabla \cdot \mathbf{j} = 0$. Particle kinetics is needed to determine the relation between currents and electric fields. For frequencies small compared to the ion cyclotron frequency and perpendicular length scales large compared to the ion Larmor radius, the dominant motion perpendicular to the ambient magnetic field B_o, which is in the z direction, is the $E \times B_o$ drift of electrons and ions. Ion inertia, however gives rise to the polarization current

$$j_x = \frac{1}{4\pi}\left(\frac{c}{v_A}\right)^2 \frac{\partial E_x}{\partial t}, \tag{24}$$

where $v_A = \sqrt{B_o^2/4\pi n M}$ is the Alfvén velocity. This relation can be used to advance E_x in time. The usual magnetohydrodynamic shear Alfvén waves, with dispersion relation $\omega = k_z v_A$, are obtained by combining (22) with $E_z \equiv 0$, (23 a) and (24). Obliquely propagating waves, $k_x \gg k_z$ in our case, are associated with a field aligned current j_z which is carried by electron motion along the magnetic field, neglecting small terms of the order of the electron to ion mass ratio. The equation of motion is used to determine the parallel electric field,

$$E_z = \frac{m}{ne^2}\frac{\partial j_z}{\partial t} - \frac{1}{ne}\frac{\partial p_{ez}}{\partial z} + \frac{1}{ne}R_{ez}. \tag{25}$$

If the bulk of the electrons has velocities small compared to the Alfvén velocity, we can neglect the pressure term. This approximation implies that electrons respond locally to the electric field of the passing Alfvén wave, similar to electrons in a metallic conductor. The quasi-static approximation, which balances electric field and pressure term, corresponds to a nonlocal response which is appropriate to energetic electrons with $v_z \gg v_A$. Using (23 b), the electron inertia term can be converted to a diffusion like term on the l.h.s. of (22) which introduces the c/ω_{pe} electron inertia scale for the x dependence. If, however, the current density, or more exactly, the relative electron -ion drift $u = -j_z/ne$, is sufficiently high , microinstabilities will develop and anomalous momentum transfer R_{ez} is expected to dominate in (25). As we have already seen, this introduces a complicated dynamical behavior, but we can use the arguments corresponding to (21) in order to assess the macroscopic effect. Using Stokes's theorem, the curl operation in (22) is converted to a spatial average of the electric fields along the circumference of a (x, z) cell in the numerical grid. The typical grid spacing in the simulations of Lysak and Dum [1983] corresponds to $\Delta z = 600\text{km} \approx 2.10^5\lambda_{De}$, and $\Delta x = 0.42\text{km} \approx 140\lambda_{De} \approx 0.4(c/\omega_{pe})$. An even larger z spacing may be appropriate if one is interested in effects of the actual dipolar configuration of auroral flux tubes [Dum and Lysak , 1985]. In any case, it is seen that anomalous momentum transfer is averaged over distances which exceed the length of particle simulation boxes by several orders of magnitude. Spatial structures seen in such simulations, including the occurrence of narrow potential spikes or double layers, thus are not directly relevant in the electrodynamic problem. The latter, however, provides the appropriate boundary conditions, for the microscopic simulation. Using again the arguments corresponding to (21), we see that the time integration of (22) prescribes a current (23 b) that is constant during the time step or has a very slow linear rise, depending on whether an explicit or implicit scheme is used for the time integration. Resolution of the diffusion like behavior in x actually requires a scheme that is implicit in x, but not necessarily in z [Lysak and Dum, 1983]. Explicit calculation of the z dependence restricts then the time step to say, $(v_A\Delta t/\Delta z) = 0.2$ (CFL condition), giving a time step correponding to $\omega_{pe}\Delta t = 1375$ or $\Omega_{ci}\Delta t = 2.8$, which is about the total time covered by a typical particle simulation. (In this example, the electron velocity is $v_e = 750\text{km/sec}$ and the Alfvén velocity is $v_A = 26160\text{km/sec}$.) We see that details of the instability growth and decay phases are also averaged out by the time integration in the MHD like code. The conclusion is then that a relatively simple anomalous resistivity law which represents the main features of the microscopic dissipation process, should suffice in macroscopic codes. Such laws may be derived by the methods discussed in Sec. III. They should account for the fact, that instability sets in once a critical drift velocity is exceeded, and that the anomalous drag force on the electrons increases with the drift in some way. In this sense, "laminar"double layers (in reality they are always accompanied by turbulence) can also be included in resistivity model, although the microscopic mechanism in this case, involves also the convective and pressure terms of the electron equation of motion. We also do not require that the energy taken out of the wave motion is fully dissipated, but on the contrary, expect that a substantial fraction of this energy goes into acceleration of runaway electrons [Dum, 1981]. The resistivity model in the the calculations of Lysak and Dum [1983] and Dum and Lysak [1985] for definiteness, is derived from a threshold-like saturation mechanism (resonance broadening) for electrostatic ion cyclotron turbulence [Dum, 1981]. Because electrons are magnetized for this instability, the drag force (not so much the fluctuation level) depends strongly on the slope of the parallel velocity distribution, and thus on quasilinear flattening and compensating effects by particle influx. It is possible to assess these uncertainties in some way by varying the parameters in the anomalous resistivity law in a sensible manner. It turns out that the outcome of the macroscopic simulation, although drastically affected by the introduction of a nonclassical electron drag term, is quite insensitive to its detailed form. The macroscopic configuration responds to an increase in the efficiency of energy dissipation by decreasing the energy supply. Similar feedback mechanisms occur also in other configurations, e.g. resistive shocks, but in our case is connected with enhanced wave reflection from the dissipation regions, which communicates these changes to the magnetospheric dynamo.

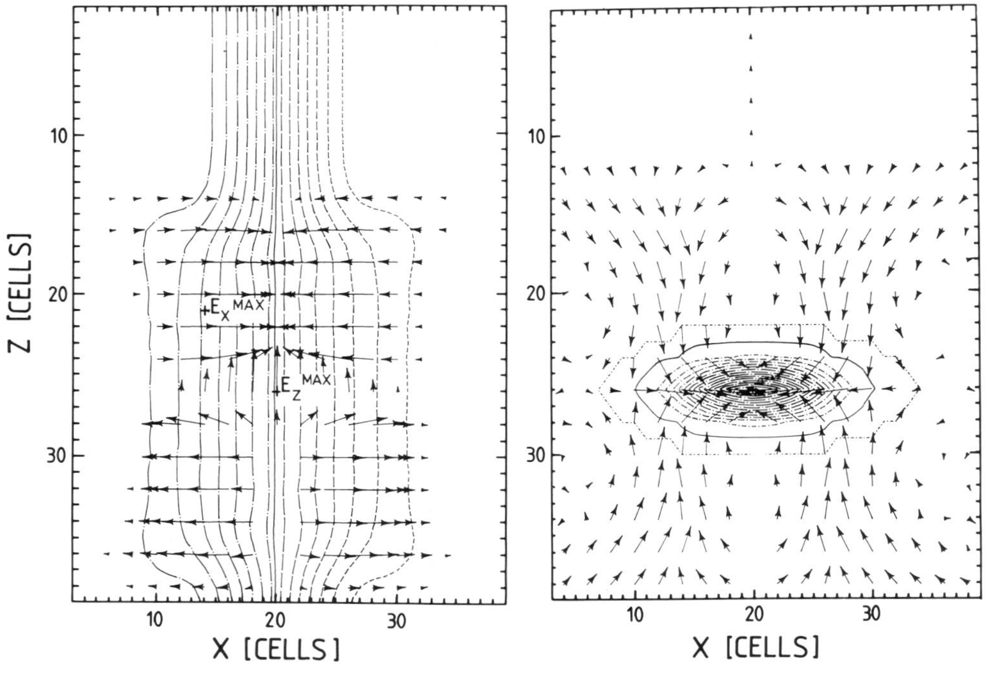

Fig. 4 Broadening of a current channel by a sudden and localized increase in anomalous resistivity. Stream lines of the current (contours of the magnetic field perturbation) are shown on the left, along with the electric field vector (E_x, $1400E_z$). The Poynting flux corresponding to Alfvén waves emitted by the dissipation zone is shown on the right. The perturbation has just reached the ionosphere, which acts as a reflector. It is eventually also communicated to the magnetospheric dynamo [Dum and Christiansen, 1985].

If the drag force increases suddenly and locally in a narrow channel (in x), the current responds by flowing around this obstacle, and this fact is again communicated by Alfvén waves to the magnetospheric dynamo and the ionosphere, see Fig. 4 and [Dum and Christiansen, 1985]. Microscopic simulations should also take notice of such changes, by choosing current and electric field profiles in accord with these macroscopic results.

V. CONCLUSION

We have demonstrated that microscopic and macroscopic processes in a collisionless plasma are in general strongly and dynamically coupled. Quasilinear effects play the dominant role in this process. Dimensionality restrictions in particle simulations and the large disparity of microscopic and macroscopic scales mostly do not allow a straightforward extension of the simulation to a realistic configuration. A cautious step by step approach which incorporates key features obtained from a separate microscopic study into a macroscopic model, and vice versa, works well in many cases, as was shown for phenomena connected with anomalous resistivity. Phenomena which crucially depend on microscopic features of particle influx and outflow from a plasma volume, such as anomalous heat flux, or double layer formation, present a more severe problem. Present day simulations of such open systems are dominated by boundary effects which makes an extrapolation to space plasmas rather difficult [see Sigov, 1982; Dum, 1985, for a detailed discussion]. For double layers, in particular, more effort is still required in creating "softer" boundaries for particle flow and in incorporating electrodynamic coupling.

REFERENCES

Appert, K.,T.M. Tran, and J. Vaclavik, 'Finite Element Approximation for the Wave-Particle Interaction in Weakly Turbulent Plasmas', *Computer Phys. Commun.*, **12**, 135, 1976.

Appert, K. and J. Vaclavik, 'Tail Formation by Nonresonant Interaction of Ions with Ion-Acoustic Turbulence', *Plasma Phys.*, **23**, 763, 1981.

Biskamp, D., and R. Chodura, 'Computer Simulation of Anomalous Resistance', *in Plasma Physics and Controlled Fusion Research*, II, p. 265, IAEA, Vienna, 1971.

Boris, J. P., J. M. Dawson, and J. H. Orens , 'Computations of Anomalous Resistivity', *Phys. Rev. Letters*, **25**, 706, 1970.

Brackbill, J. U., and D. W. Forslund, 'An Implicit Method for Electromagnetic Plasma Simulation in Two Dimensions', *J. Comput. Phys.*, **46**, 271, 1982.

Brackbill, J. U., D. W. Forslund, K. B. Quest, and D. Winske , 'Nonlinear Evolution of the Lower-Hybrid Drift Instability', *Phys. Fluids*, **27**, 2682, 1984.

Buneman, O. , 'Dissipation of Currents in Ionized Media', *Phys. Rev.*, **113**, 1959.

Chodura, R., G. Bardotti, and F. Engelmann, 'Numerical Investigation of the Anomalous Resistivity due to Two-Stream Instability', *Plasma Phys.*, **13**, 1099, 1971.

Chodura, R., C. T. Dum, F. Söldner, and K.-H. Steuer, 'Anomalous Transport in Shock Produced Plasmas', *8th Europ. Conference on Controlled Fusion and Plasma Physics*, I, p. 93, Prague, 1977.

Dum, C. T., 'Simulation Models for Space Plasmas and Boundary Conditions as a Key to their Design and Analysis', in *Computer Simulation of Space Plasmas*, edited by H. Matsumoto and T. Sato, p., Terra Scientific Publishing Company, Tokyo, to appear, 1985.

Dum, C. T., 'Electrostatic Waves and Anomalous Transport in the Solar Wind', *Proc. Solar Wind V, NASA Conf. Publ.*, **2280**, 369,1983.

Dum, C. T., 'Anomalous Resistivity and Plasma Dynamics', in *Physics of Auroral Arc Formation, Geophys. Monogr. Ser.*, **25**, edited by S.-I. Akasofu and J. R. Kan, p. 408, AGU, Washington, D. C., 1981.

Dum, C. T., 'Anomalous Heating by Ion Sound Turbulence', *Phys. Fluids*, **21**, 945, 1978 a.

Dum, C. T., 'Anomalous Electron Transport Equations for Ion Sound and related Turbulent Spectra', *Phys. Fluids*, **21**, 956, 1978 b.

Dum, C. T., and R. Chodura , 'Anomalous Transition from Buneman to Ion Sound instability', in *Wave Instabilities in Space Plasmas*, edited by P. J. Palmadesso and K. Papadopoulos, p. 135, Reidel, 1979.

Dum, C. T., R. Chodura, and D. Biskamp , 'Turbulent Heating and Quenching of the Ion Sound Instability', *Phys. Rev. Letters*, **32**, 1231, 1974.

Dum, C.T., and P. J. Christiansen, 'Electrodynamic Coupling and its Relation to Microscopic Dissipation', *Symposium on the Physics of Magnetosphere-Ionosphere Coupling*, COSPAR, Graz, 1984, and to published, 1985.

Dum, C. T., and R. L. Lysak , 'Electrodynamics of Auroral Magnetic Flux Tubes', *J. Geophys. Res.*, **90**, to be published.

Field, E. C., and B. D. Fried, 'Solution of the Kinetic Equation for an Unstable Plasma in an Electric Field', *Phys. Fluids*, **7**, 1937, 1966.

Forslund, D. W., K. Quest, J. U. Brackbill, and K. Lee , 'Collisionless Dissipation in Quasi-Perpendicular Shocks', *J. Geophys. Res.*, **89**, 2142, 1984.

Galeev, A. A., and R. Z. Sagdeev, 'Nonlinear Plasma Theory', *Rev.of Plasma Physics*, **7**, 1, Consultants Bureau, New York , 1979.

Goldman M. V., 'Strong Turbulence of Plasma Waves', *Rev. Mod. Phys.*, **56**, 709, 1984.

Ishihara, O., and A. Hirose, 'Quasilinear Evolution of Current Driven Ion-Acoustic Turbulence in a Magnetic Field', *Phys. Rev. Letters*, **50**, 1783, 1983.

Liewer, P. C.,and N. A. Krall, 'Self-consistent Approach to Anomalous Resistivity, applied to Theta Pinch Experiments', *Phys. Fluids*, **16**, 1953, 1973.

Lindman, E. L., 'Numerical Simulation of Ion Acoustic Turbulence', *J. Statistical Physics*, submitted, 1984.

Lysak, R. L., and C. T. Dum, 'Dynamics of Magnetosphere-Ionosphere Coupling Including Anomalous Transport', *J. Geophys. Res.*, **88**, 365, 1983.

Manheimer, W. M., and J. Boris, 'Selfconsistent Theory of a Collisionless Resistive Shock', *Phys. Rev. Letters*, **28**, 659, 1972.

Sigov Yu. S., 'Computer Simulation of Plasma Turbulence in Open Systems', *Physica Scripta*, **T2/2**, 367, 1982.

Söldner, F., C.T. Dum, and K.-H. Steuer, 'Electron and Ion Heating in a High-VoltageBelt Pinch', *Phys. Rev. Letters*, **39**, 194, 1977.

Tanaka, M., and T. Sato, 'Multiple Excitation of Lower-Hybrid Drift Waves in the Neutral Sheet', *Phys. Rev. Letters*, **47**, 714, 1981.

Winske, D., M. Tanaka, C. S. Wu, and K. Quest, 'Plasma Heating at Collisionless Shocks due to the Kinetic Cross -Field Streaming Instability', *J. Geophys. Res.*, **90**, to appear, 1985.

GLOBAL ENERGY REGULATION IN THE SOLAR WIND-MAGNETOSPHERE-IONOSPHERE SYSTEM

Tetsuya Sato
Institute for Fusion Theory
Hiroshima University
Hiroshima 730
Japan

ABSTRACT. Some basic concepts which are essential in the understanding of global energy regulation in the solar wind-magnetosphere-ionosphere system are introduced. The importance of line-tying concept is particularly emphasized in connection with the solar wind enetry, energy release in the magnetopshere and energy dissipation in the ionosphere.

1. INTRODUCTION

In contrast to the magnetic fusion plasma, the primary interest in the space plasma lies in its non-confinement or its evolutional nature. The principal element characterizing this nature is the solar wind. The ever-lasting and ever-changing solar wind does not allow the magnetosphere to stay in the same state. Not only the configuration but also the energy contents of the magnetosphere are continuously changed subject to the ever-changing solar wind condition.

Another element governing the energy regulation in the magnetosphere is the ionosphere. The existence of the ionosphere as an adjacent neighbor of the magnetosphere makes the magnetospheric evolution interesting in the following two senses. The first is that globally speaking, the ionosphere acts as a continuous energy absorber of the magnetospheric energy. In spite of its small volume compared with the magnetosphere, the enormous number of background neutral particles can absorb the magnetospheric free energy sufficiently that the heat capacity of the ionosphere is considered to be infinite in a practical sense. The other point is that the ionosphere acts as a partial reflector for relatively rapid or local variations. Because of this, not only as a simple passive energy absorber, but the ionosphere can also actively couple the magnetosphere via Alfven waves, hence, it becomes possible that a feedback electrodynamic coupling plays a decisive role in the magnetospheric evolution [see, Sato, 1982].

There is an additional key element that governs the magnetospheric energy regulation, that is, the fact that the solar wind is practically

Space Science Reviews **42** (1985) 485–498. 0038–6308/85.15

an infinite energy source. Because of this fact, the magnetospheric configuration, hence, the internal energy state, can significantly be changed before the ionosphere can absorb the whole excess energy given by the solar wind. Speaking in terms of ionospheric line-tying, this implies that line-tying cannot be perfect. Owing to this imperfect line-tying, the solar wind momentum can enter deep inside the magnetosphere and the energy can be stored in terms of the convection energy, which suffers a continuous dissipation by the line-tying current in the ionosphere. When the solar wind enters into the magnetosphere as a result of dayside reconnection, the dayside magnetic flux is transported into the nightside, thereby the tail current being enhanced. Subsequently, spectacular energy release phenonemon of the stored tail current energy is observed in the magnetosphere-ionosphere system.

2. EQUATIONS GOVERNING GLOBAL ENERGY REGULATION

The most convenient and useful way of describing the global energy regulation in the magnetosphere is to rely on the magnetohydrodynamics (MHD). The basic equations are given by

$$\partial\rho/\partial t + \nabla\cdot\rho\vec{v} = 0 \tag{1}$$

$$\rho d\vec{v}/dt = \vec{J}x\vec{B} - \nabla p + \nu\nabla^2\vec{v} \tag{2}$$

$$\partial\vec{B}/\partial t = -\nabla x\vec{E} \tag{3}$$

$$dp/dt = -\gamma p\nabla\cdot\vec{v} + (\gamma-1)\eta J^2 \tag{4}$$

$$\vec{E} = -\vec{v}x\vec{B} + \eta\vec{J} \tag{5}$$

$$\vec{J} = (\mu_o)^{-1}\nabla x\vec{B} \tag{6}$$

These equations consists of six macro-variables, namely, plasma density ρ, static pressure p, bulk velocity $\vec{v}$, magnetic field $\vec{B}$, electric field $\vec{E}$ and electric current $\vec{J}$, and three plasma parameters, namely, elecrical resistivity η, inter-particle viscosity ν and adiabatic constant γ. The heat conductivity may play some role in the energy regulation. In the present problem, however, this term may well be neglected. In the MHD treatment there is no self-consistent way to obtain the transport coefficients due to particle discreteness such as ν and η, so that we must assume them to be given.

In order to visualize the global energy transport and conversion, we shall transform Eqs.(1)-(6) into the following forms:

$$\frac{\partial K}{\partial t} = -\int \frac{1}{2}\rho v^2\vec{v}\cdot d\vec{S} + \int \vec{J}x\vec{B}\cdot\vec{v}dV - \int \nabla p\ \vec{v}dV \tag{7}$$

$$\frac{\partial W}{\partial t} = -\int \frac{\vec{E}x\vec{B}}{\mu_o}\cdot d\vec{S} - \int \vec{J}x\vec{B}\cdot\vec{v}dV - \int \eta J^2 dV \tag{8}$$

$$\frac{\partial T}{\partial t} = -\int \frac{\gamma}{\gamma-1} p\vec{v}\cdot d\vec{S} + \int \nabla p\cdot\vec{v}dV + \int \eta J^2 dV \tag{9}$$

with

$$K = \int \frac{1}{2}\rho v^2 dV \tag{10}$$

$$W = \int \frac{B^2}{2\mu_o} dV \tag{11}$$

$$T = \int \frac{p}{\gamma-1} dV \tag{12}$$

where K is the total bulk flow energy, W is the total magnetic energy and T is the total thermal energy in the system.

Eqs.(7)-(9) state that there are two different agents leading to the change in the energy state of the system. One is the external agent which supplies the external energy into the system or absorbs the internal energy. The other is the internal one which converts one energy state into another within the system. The first three terms on the right hand side of Eqs.(7)-(9) represent the energy exchange with the external agent and the other six terms represent the internal energy conversion. Fig. 1 shows the diagram of such external and internal energy exchanges.

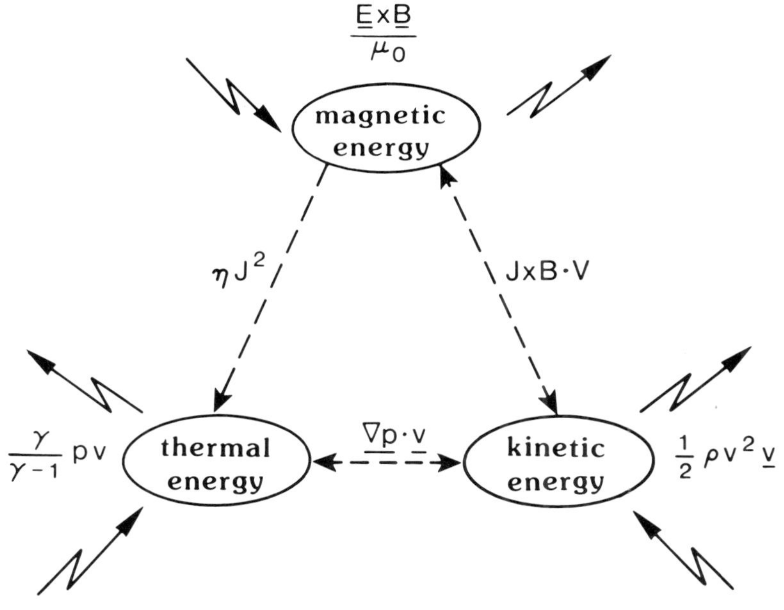

Figure 1. Diagram showing the macroscopic energy exchange with the external energy agents and the internal energy conversion.

3. BASIC CONCEPTS OF ENERGY REGULATION

3.1 Energy Conversion

In the simulation study of the global energy regulation we must first define the size of the system we wish to study and the boundaries enclosing it. Then, the energy (magnetic, bulk flow and thermal) contents in the system and the states of the plasmas in contact with the system through the boundaries must be defined. As can be seen from Eqs.(7)-(9), the energy entry from the boundaries is governed by the incoming plasma flow perpendicular to the boundaries. Therefore, the boundary conditions must be such that the bulk flow speeds across the boundaries can definitely be determined by either external states or the combination of the external and internal conditions.

The incoming plasma flow deforms the magnetic configuration in the system, consequently, the energy distribution as well as the total energy contents are changed. Of particular importance in this change of internal energy state is the change of the current distribution. This is because the free magnetic energy of the system that could be released is stored in terms of the electric current. Depending on the initial magnetic field configuration and the incoming flow pattern, usually there appear particular regions where electric currents are preferably concentrated. In such current-concentrated regions the free magnetic energy preferentially dissipates into the thermal energy due to the resistivity, classical or anomalous. More importnatly, the local current concentration is accompanied by the generation of anti-parallel field components. These two aspects lead to a conclusion that in such regions magnetic reconnection could be triggered as the currrent sufficiently develops. Such reconnection is called "driven reconnection". Once reconnection is triggered, free magnetic energy is converted to the bulk flow energy as well as the thermal energy. When this happens in a time scale shorter than the time scale of the configurational change of the magnetic field subject to the incoming plasma flow, a rapid energy release phenomenon is observed. Magnetospheric substorms and solar flares would be typical examples of such rapid energy releases.

Of particular interest in the energy conversion is the conversion from the magnetic energy to the flow (kinetic) energy and/or the thermal energy. There are other conversions, namely, from flow to thermal or from thermal to flow energy. The former process is the compression and the other is the expansion. While compression often plays an essential role in the energy build-up phase when an anti-parallel field configuration is created by an enhanced diamagnetic current, expansion can sometimes contribute to parallel acceleration of bulk plasma, particularly, when a slow shock is formed as a result of strong driven reconnection.

3.2 Energy Output

Here we define the energy output (loss) as the outgoing energy through the boundaries of the system under study. As is shown in Eqs.(7)-(9), the system energy can be lost in terms of flow energy, magnetic energy and/or thermal energy. However, it would be more physical to classify it in a different form. We divide the energy loss into the parallel loss and the perpendicular loss. Speaking in terms of the MHD waves, the carrier of the perpendicular loss is the fast wave, while that of the parallel loss is either the shear Alfven wave or the slow wave.

The perpendicular loss is the so-called convective loss. Globally speaking, this cannot be an intrinsic energy loss, since the plasma is frozen in the magnetic field in a collisionless space plasma. When we look at a fixed local system, however, the system energy is lost by this convective outflow through the boundaries. For example, when a free boundary condition is employed in a simulation model, the outflow through the free boundary is this convective loss.

The parallel loss consists of the loss associated with the field-current and the loss asocaited with the parallel plasma outflow. Since the loss due to parallel flow is essentially a convective loss, we do not comment any further on this part but let us see the loss due to field-aligned currents. Alfven waves are accompanied by field-aligned currents and inertia (polarization) currents which make closure currents. In a dissipation-less medium the phase between the polarization current $\vec{J}_p$ and the polorization field $\vec{E}_p$ differs 90 degrees so that there is no net energy loss or gain associated with Alfven waves. However, if Alfven waves are partially absorbed through the boundaries, the phase between the polarization current and field deviates from 90 degrees. In this case, $\mathrm{Re} \int \vec{J}_p \cdot \vec{E}_p dV < 0$, so that the system energy is lost in net. This loss is equivalent to the Poynting outflux through boundaries.

Let us consider the parallel loss due to Alfven waves in a more detailed way. We suppose that the field lines concerned are immersed in a resistive medium outside the system. Then, there are two essential cases that Affven waves carry out the energy. The first one is the case where twin vortices perpendicular to the magnetic field are excited in the system, as is shown in Fig. 2a. The second one is the

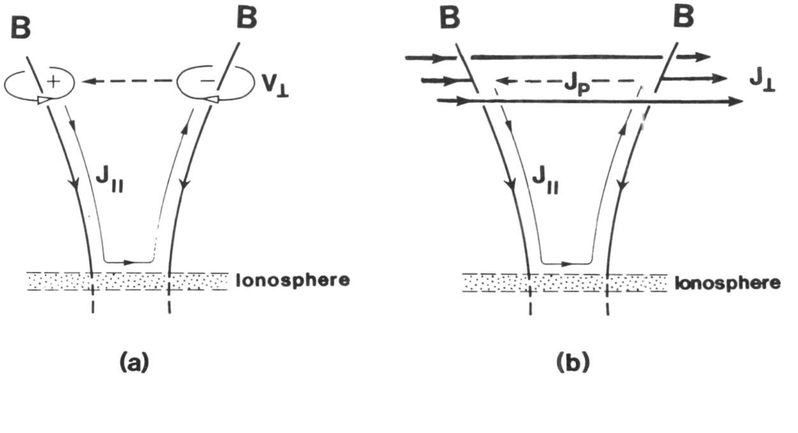

Figure 2. Field-aligned dissipation of perpendicular bulk flow energy (a) and of perpendicular current energy (b).

case where the closure current perpendicular to the magnetic field (usually the diamagnetic current) is locally interrupted by whatever mechanism it is, as is shown in Fig. 2b. In the twin-vortex case, it is readily known from Eq.(5)

$$\Omega = -Q/\varepsilon_o B \tag{13}$$

where Ω is the vorticity given by $\Omega=\vec{B}\ \nabla x\vec{v}/B$, Q is the space charge and ε_o is the dielectric constant in the vacuum. In this case the essential part is represented by the discharge of space charges stored in a condenser C (magnetosphere) into a resistance R (ionosphere), which can be modeled by an electric circuit shown in Fig. 3a. Namely,

$$dQ/dt = -(CR)^{-1}Q \tag{14}$$

This equation implies that the space charge disappers in the time scale of τ=CR. In other words, the electric field disappears in this lime scale, hence, the plasma flow is stopped. This is nothing but line-tying short-ciurcuit. Thus, we can conclude that unless space charges (electric field) are supplied externally by whatever mechanism, the plasma motion cannot survive for a longer time than CR.

In the second case where the perpendicular current is locally interrupted, the governing dynamics is different essentially from the previous case. In the previous case the basic governing equation is the equation of motion, namely, $\rho d\vec{v}/dt=\vec{j}x\vec{B}$, while in the present case the governing equation is the induction equation, namely, $\partial\vec{B}/\partial t=-\nabla x\vec{E}$. In other words, the first case corresponds to the loss of the bulk flow energy, while the second to the loss of the magnetic energy.

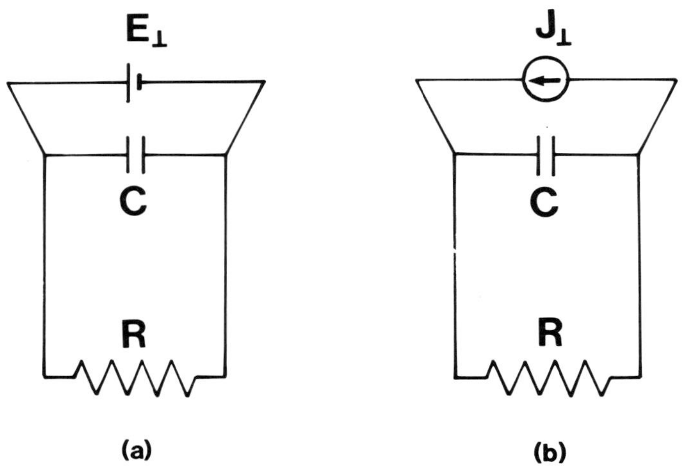

Fig. 3 An equivalent circuit of Fig. 2a (a) and of Fig. 2b (b). Field-aligned inductance is ignored in this circuit.

As is well known, the magnetic energy is stored in terms of the diamagnetic current which constitutes a closure (divergence free) current in a plane perpendicular to the magnetic field. Therefore, it is not so easy to divert the diamagnetic current to the field-aligned current as in the first case. However, once the loop current is locally blocked and a field-aligned current is generated by some mechanism, the magnetic energy is dissipated by the line-tying ionospheric current. Whatever the blocking mechanism is, it must be associated with the generation of polarization charges due to local polarization current opposing the original diamagnetic current. Thus, the electric circuit can be represented by Fig. 3b. The principal difference between Fig. 3a and Fig. 3b is that the condenser in Fig. 3a is initially charged, while that in Fig. 3b is initially empty but is charged as the blocking starts.

In the magnetosphere-ionosphere system the line-tying time is given by

$$\tau_{LT} = L/\langle\mu_o V_A^2\rangle\Sigma_P \tag{15}$$

where <...> implies the average over the field line length L, V_A is the Alfven speed and Σ_P is the height-integrated Pedersen conductivity. When we compare the line-tying time with the Alfven transit time between the magnetosphere and the ionosphere, $\tau_{AT}=L/V_A$, we obtain

$$\tau_{LT}/\tau_{AT} = (\mu_o V_A \Sigma_P)^{-1} \tag{16}$$

where μ_o is the vacuum permeability. Recalling that the $\mu_o V_A$ is the magnetospheric characteristic impedance and Σ_P^{-1} is the ionospheric impedance, we see that this ratio is the impedance ratio. On the earth this imdedance ratio is of the order of unity. Therefore, the line-tying time is comparable with the Alfven transit time, which is of the order of one minute. This indicates that line-tying effect is relatively fast and strong.

For the field-aligned current system to live long, therefore, the space charge or the plasma flow must be supplied from the external source more swiftly than the ionosphere absorbs it. When the plasma flow is introduced into the magnetosphere by dayside reconnection, the solar wind directly drives the reconnected field lines. Since the solar wind dynamo is supposed to be much stronger than the ionospheric drag, line-tying cannot stop the plasma motion however strong it is. In this sense it can be said that dayside reconnectiion is a "hard" process, or a forced voltage process. In the case where the plasma flow is induced by a viscous interaction at the magnetopause, on the other hand, it is not certain whether line-tying can be strong enough to stop the flow or not. The charging rate (momentum transfer rate) is proportional to the viscosity ν. For the twin-convection due to the viscous interaction to be maintained persistently, the viscosity must be larger than a certain value so that the charging rate can exceed the discharging rate. The value is estimated by the following inequality.

$$\tau_{vis} = \nu/w^2 < \tau_{LT} \tag{17}$$

where w is the width of the viscous layer. In contrast to the dayside reconnection process, the viscous interaction can be said a "soft" process.

As we have seen here, it is important to understand that the line-tying effect is by itself strong but that depending on the energy source driving the plasma flow in the magnetosphere, it can on some occations play a crucial role in energy regulation but not on others.

3.3 Energy Input

When we restrict ourselves to the simulation study, we must clearly separate the input boundries from the output boundaries. Of course this must be done based on a deep physical consideration. Once this separation is done, then we must define the input boundary conditions in such a way that the six macro-variables appearing in Eqs.(1)-(6) are definitely determined at each time. To this end the states of the plasma coming through an input boundary must be clarified. In this respect, it is fundamental to choose the simulation system in such a way that no significant energy feedback occurs from inside to outside [see, Sato, 1985].

4. A KEY PROCESS OF ENERGY CONVERSION - MAGNETIC RECONNECTION

In the previous section, basic concepts in dealing with the energy flow process are described. In this section we shall briefly touch on physical processes leading to powerful energy conversion.

As we already mentioned, there are three macroscopic energy states, namely, magnetic, thermal and flow energies. There are three agents dealing with energy conversion among them. They are the Ampare force, pressure-gradient force and the ohmic heating, as can be seen in Fig. 1. In the magnetosphere, the ohmic heating may be neglected as an energy conversion process, because of the high Raynolds number. Note, however, that this does not imply the neglect of an auxiliary role of the resistivity in energy conversion, namely, triggering of reconnection.

In order to have an efficient energy conversion, therefore, it is invoked that either the perpendicular current is intensified or a sharp pressure gradient is produced. The best way to intensify the current and to increase the pressure gradient is to converge the plasma flow induced by an external energy agent into as narrow a layer as possible. Formation of an intensified current layer causes two effects. One is the generation of local anti-parallel magnetic field and the other is the local enhancement of ohmic dissipation. These two facts strongly suggest that driven reconnection could be stimulated there. Simulation results indicate that oblique slow shocks can be created as a result of driven reconnection if fast mode compression of the current layer occurs effectively [Sato, 1979]. Remembering that slow shocks are characterized by the peaked current layer and the sharp pressure gradient, we come to an important conclusion, namely, that the creation of slow shocks would be the most efficient way to cause a powerful energy conversion. If the magnetospheric substorm is really an unusually rapid energy conversion phenomenon, slow shocks are very likely to be involved. The strong Ampare force and the pressure gradient associated with slow shocks can cause rapid acceleration, perpendicular and parallel, of bulk plasmas. Besides, if slow shocks are three-dimensionally formed, it is quite likely that the diamagnetic current is blocked due to their strong nonlinearity, whereby field-aligned currents are generated. 3D simulations indeed confirmed the generation of field-aligned currents by blocking the diamagnetic current [Sato et al., 1984].

5. ENERGY TRANSPORT PROCESSES

5.1 Magnetic Reconnection

We have seen that strong plasma acceleration occurs when slow shocks are created by driven reconnection. However, it is not always the case that slow shocks are created by driven reconnection. When we examine simulation results, we find that slow shocks are formed only when the neutral sheet is strongly compressed by fast magnetosonic waves. On the dayside, for example, there is a strong asymmetry on both sides of the magnetopause and the magnetic pressure of the solar wind can be neglected compared with the dynamic pressure. In such a case the change of the magnetic field direction in the solar wind does not lead to additional compression. Thus, it is rather unlikely that slow shocks are formed in association with dayside reconnection.

Even though slow shocks are not created, magnetic reconnection plays an equally important role in energy regulation in the magnetosphere, that is, the magnetic flux transport. Once reconnection occurs on the dayside magnetopause, the solar wind momentum is introduced into the magnetosphere and the line-tying ionospheric current is almost instantaneously induced, thus part of the solar wind energy being dissipated in the ionosphere. There is a much more important role of dayside reconnection than the momentum transport, however. The role is to replace the dayside (potential) magnetic flux to the nightside magnetosphere. Globally speaking, this replacement of the magnetic flux does not change much the total magnetic energy in the whole magnetosphere. However, this replacement is importnat in the sense that the dayside potential magnetic energy is changed to the free magnetic energy on the nightside. Because of this fact, free magnetic energy is increased in the tail region, which awaits to be released and returned to the original dayside potential field by return plasma flow resulting from some energy release mechanism.

Let us here take some space to comment on "driven" and "unloading" controversy of substorms [see, Hones, 1984]. As we have discussed above, the energy introduced directly by the solar wind is the flow energy. As we have discussed in section 2, the line-tying time of the flow energy is one minute or so. In addition, the introduced solar wind acts to transport the dayside magnetic flux to the nightside. Correspondingly, there should be two energy dissipation processes. One is the energy dissipation associated with the line-tying ionospheric current which almost instantaneously reacts to the solar wind condition.

The other one is the release of the dayside magnetic energy replaced to the nighside. Substorms, if any, should be the latter one. In view of its quick (one minute) response to the solar wind, the line-tying dissipation corresponds to so-called "driven" part, while the release of the replaced magnetic energy corresponds to the "unloading" part, since it awaits some special conversion process for its dissipation. If the special conversion process occurs in a relatively rapid time scale, we may call it a substorm. In this sense, substorms should be unloading phenomena of the magnetic energy transported from the dayside the the nightside by the solar wind. As we have discussed in the previous section, driven rconnection is the most likely unloading agent. We here remind the reader that the word "driven" used for reconnection and that for the substrom model have completely different meanings.

5.2 Viscous Transport

It is quite natural that at the contact surface of two plasmas with different natures, plasma instabilities, macroscopic or microscopic, take place to smoothen the difference. The solar wind plasma and the magnetospheric plasma which are in contact at the magnetopause have considerably different plasma natures, e.g., in temperature, density, magnetic field intensity and bulk flow speed. Among others, the most likely instability in transporting the solar wind momentum into the magnetosphere would be the Kelvin-Helmholtz instability, since the bulk flow energy of the solar wind is directly involved in it. The instability can create anomalous eddy viscosity which transports the momentum to the magnetosphere [Miura, 1984].

An important aspect of the viscous transport in relation to the energy regulation is that the electric field driving the magnetospheric plasma is not directly connected to the solar wind dynamo as is so for the dayside reconnection case. Because of the indirect drive through such a soft conversion process as anomalous viscosity, ionospheric line-tying cannot be disregarded. Unless the momentum transfer time, namely, the space charge supply time, is faster than the discharge time to the ionosphere (namely, the line-tying time), field lines are tied to the ionosphere and no global convection would be established in the magnetosphere. Conversely speaking, we can estimate the viscosity required for having a global convection from the line-tying time. As we already discussed, if $\tau_{vis}<\tau_{LT}$, namely,

$$\nu_{an} > w^2/\tau_{LT} \tag{18}$$

is satisfied, it becomes possible that a global convection is induced in the magnetosphere by the viscous interaction between the solar wind and the magetospheric plasma. Let us suppose that w=100km and τ_{LT}=100sec, then we obtain that

$$\nu_{an}>10^8 m^2/sec \tag{19}$$

is the necessary condition for having a viscosity-induced convection.

Note that this value is consistent with the value obtained from the energy consideration [Axford, 1964]. Speaking in terms of the electric circuit, if Eq.(19) is satisfied, the induced convection provides a constant voltage source and a steady field-aligned current is maintained. The current intensity is proportional to $\Sigma_P E$ and the dissipated energy is proportional to $\Sigma_P E^2$ where E is the induced electric field. Note that this is proportional to the square of the solar wind speed. Incidentally, the dissipated energy assocaited with the flow induced by dayside reconnection is also expressed by the same relation, since this case, as well, corresponds to a constant voltage circuit.

6. MAGNETOSPHERE-IONOSPHERE COUPLING

Some basic concepts of the magnetosphere-ionosphere coupling are already mentiioned. But, since this is essential in the understanding of global energy regulation, we shall summarize in this section.

The exchange of informations between the magnetosphere and the ionosphere is done by the field-aligned current. Let us therefore start with relating the field-aligned current to the magnetospheric macro-variables and also to the ionospheric variables.

The field-aligned current of magnetospheric origin $J_{\parallel}$ is given by the following equation [Hasegawa and Sato, 1979; Vasyliunas, 1984].

$$B \frac{\partial}{\partial s} \left(\frac{J_{\parallel}}{B}\right) = - \rho \frac{d}{dt} \left(\frac{\Omega_{\parallel}}{B}\right) + \frac{2}{B} \vec{J}_{\perp} \cdot \nabla B \qquad (20)$$

where only two dominanat terms are retained in this equation; Ω is the parallel vorticity (vorticity due to the perpendicular convection) and $\vec{J}_{\perp}$ is the perpendicular current.

From the divergence-free condition of the total current at the ionosphere, the field-aligned current at the ionospheric height is connected to the height-integrated ionospheric current $\vec{I}$ by

$$J_{\parallel} = -\nabla \cdot \vec{I} \qquad (21)$$

The height-integrated ionospheric current is given by

$$\vec{I} = \Sigma_P \vec{E}_I - \Sigma_H \vec{E} \times \vec{B}_I / B_I \qquad (22)$$

where Σ_H is the height-integrated Hall conductivity.

Coupling these equations with supplement of the magnetospheric equations Eqs.(1)-(6), we can solve the magnetosphere-ionosphere coupling [see, Sato, 1982]. In this lecture, however, we do not go into details but just touch on the field-aligned current.

As we have seen in the previous section, the energy introduced by the direct interaction with the solar wind, whether it is due to dayside reconnection or viscous interaction, takes the form of convection, or electric field. In terms of the electric circuit, this energy is equivalent to the energy stored in the magnetospheric condenser as space charges. On the other hand, the magnetic flux transported from the dayside region to the nighside region by dayside reconnection is stored in terms of the electric current.

The discharge of the space charge stored in the magnetospheric condenser is represented by the first term on the right hand side of Eq.(20). The second term on the right hand side represents the field-aligned current related to the magnetospheric current. Importantly, however, there is no time derivative or integral included in this term. This indicates that there is no direct causality between the magnetospheric current and the field-aligned current; in other words, the field-aligned current cannot directly result from the magnetospheric current. The magnetospheric current is flowing in a closed circuit disconnected from the ionospheric circuit in a normal state. In order to connect this circuit to the ionospheric circuit, therefore, some nonlinear process is required. As 3D simulations have shown, three dimensional slow shocks can be a potential candidate for connecting the magnetospheric current to the field-aligned current [Sato, 1982; Sato, et al., 1984]. In this case, the connected circuit may be represented by a constant current circuit, as shown in Fig. 3b.

7. CAN FIELD-ALIGNED CURRENTS ORIGINATE FROM THE IONOSPHERE

In the previous section the argument was done assuming field-aligned currents to originate from the magnetosphere, namely, Eq.(20). As we already discussed in Section 3, however, the ionospheric impedance is comparable to the magnetospheric impedance. Therefore, physically it is quite likely that field-aligned currents originate from the ionosphere. Since it casts a stir on the magnetosphere-ionosphere coupling, let us consider this problem in this separate section.

Certainly, it is possible that field-alaigned currents originate from the ionosphere. As Eq.(21) states, it apppears that field-aligned currents originate from the ionosphere, for example, when the ionospheric conductivity is inhomogeneous. More than a dozen years ago, such ideas were prevailing [see, for example, Coroniti and Kennel, 1972;Bostrom, 1974] and seem to have revived recently to explain an imperfect Cowling blockage [Kan et al., 1984]. However, one must recognize one key point missing in such ideas. As is clear in Eq.(21), there is no time derivative included on the right hand side. As we discussed in the previous section, therefore, it cannot be a cause of a field-aligned current unless some special mechanism is introduced in the ionosphere. This can be explained in an alternative way. The magnetospheric impedance is in general purely reactive, not resistive. This implies that the magnetospheric resistance is infinite, so that no dc field-aligned current can originate from the ionopshere, irrespective of whether field lines are closed or open. More specifically, field-aligned currents cannot be driven by the inhomogeneity of the ionospheric conductivity, either Hall or Pedersen, unless a time varying ionospheric conductivity in the rest frame of the field lines of interest exists. If and only if a time varying conductivity exists, an ac field-aligned current can originate from the ionosphere. In such a case one must solve the ionophsere-magnetosphere coupling self-consistently, since the magnetospheric impedance is comparable to the ionospheric impedance. A self-consistent treatment on this problem is given in a literature [Sato, 1978, 1982], which shows the existence of a positive feedback coupling. Incidentally, when we consider the field-aligned current of magnetospheric origin, we do not necessarily have to be so nervous about the feedback treatment. This is because usually the field-aligned current of magnetospheric origin is connected to an energy source with a huge capacity such as the solar wind.

8 SUMMARY

Some basic concepts, which are essential but not necessarily well understood in the understading of the global energy regulation in the solar wind-magnetosphere-ionosphere system, are explained. Emphasized particularly are the line-tying concept and the association of field-aligned currents with the magnetospheric and ionospheric dynamics. The importance of careful consideration of the cause and effect relationship based on sound physics is also emphasized. Depending on the way the solar wind enters into the magnetosphere, the subsequent dynamical response of the magnetosphere differs largely and depending on how the magnetosphere responds, the subsequent energy release also differs significantly.

Development of computers and facilities has enabled us to comprehend the dynamics quantitatively and three dimensionally. It is true, on the other hand, that simulation people tend to put everthing in computers without having any physical elaboration before compuation. If such tendencies prevail, we would soon find ourselves walking far behind other sciences. Critical discussions based on physics, not on semantics, are highly required.

Acknowledgements
This work is supported by the grant-in-aid from the Ministry of Education, Science and Culture in Japan. This work is also supported by NASA STTP grant NAGW-78 and NSF grant ATM 83-11102 in the United States.

REFERENCES

Axford, W.I., Viscous interaction between the solar wind and the earth s magnetosphere, Planet. Space Sci., 12, 45, 1964.
Bostrom, R., Ionosphere-magnetosphere interaction, in Magnetospheric Physics, edited by B.M.McCormac, D. Reidel Pub. Co., 45, 1974.
Coroniti, F.V. and C.F. Kennel, Polarization of the auroral electrojet, J. Geophys. Res., 77, 2835, 1972.
Hasegawa, A. and T. Sato, Generation of field-aligned current during substorm, in Dynamics of the Magnetosphere, edited by S.-I. Akasofu, D. Reidel Pub. Co., 529, 1979.
Hones, E.W., Jr., Magnetic Reconnection, AGU Geophys. Mono. 30, 1984.
Kan, J.R., R.L. Williams and S.-I. Akasofu, A mechanism for the westward travelling surge during substorms, J. Geophys. Res., 89, 2211, 1984.
Miura, A., Anomalous transport by magnetohydrodynamic Kelvin-Helmholtz instabilities in the solar wind-magnetosphere interaction,

Geophys. Res., 89, 801, 1984.

Sato, T., A theory of quite auroral arcs, J. Geophys. Res., 83, 1042, 1978.

Sato, T., Strong plasma acceleration by slow shocks resulting from magnetic reconnection, J. Geophys. Res., 84, 7177, 1979.

Sato, T., Chapt.4 Auroral physics, in Magnetospheric Plasma Physics, edited by A. Nishida, Center for Academic Pub. Japan annd D. Reidel Pub. Co., 197, 1982.

Sato, T., Principles of magnetohydrodynamic simulation in space plasmas in Computer Simulation of Space Plasmas, edited by H. Matsumoto and T. Sato, Tera Sci. Pub. Co. and D. Reidel Pub. Co., 133, 1985.

Sato, T., R.J. Walker and M. Ashour-Abdalla, Driven magnetic reconnecction in three dimensions: Energy conversion and field-aligned current generation, J. Geophys. Res., 89, 9761, 1984.

Vasiliunas, V.M., Fundamentals in current description, in Magnetospheric currents, edited by T.A. Potemra, AGU Geophys. Mono. 28, 63, 1984.

AURORAL ARC FORMATION: KINETIC AND MHD EFFECTS

C. K. Goertz
Dept. of Physics and Astronomy
University of Iowa
Iowa City, IA 52242

ABSTRACT. Recent observations by the Dynamics Explorer satellites indicate that auroral field lines are not equipotentials. The field-aligned "conductivity" is the same for up and downgoing currents and of the order of several 10^{-8} to several 10^{-9} $\Omega^{-1}m^{-2}$. This is not in agreement with models based on the resistance provided by the magnetic mirror force alone. It is suggested that the resistivity is due to damping of kinetic shear-Alfven wave packets. This damping may be nonlinear in nature or Landau damping. We briefly review the properties of kinetic Alfven waves.

1. INTRODUCTION

It is well known that discrete auroral arcs are associated with upward-directed field-aligned currents and downward precipitating energetic electrons ($E_e \gtrsim 1$ keV). The fact that there is often a peak in the energy spectrum of precipitating electrons, that their loss-cone is enlarged and that the distribution is field aligned suggest that these electrons are accelerated through a field-aligned potential drop $V_{\parallel}$. The dissipation of electromagnetic energy and its transformation into particle heating or acceleration is given by the product of $V_{\parallel}$ and the field-aligned current density, $j_{\parallel}$. Various mechanisms for maintaining a stationary parallel electric field have been reviewed by Shawhan et al. (1978), Goertz (1979), Kan and Lee (1981) and Chiu et al. (1981). For a stationary potential drop Lyons (1980), Chiu and Cornwall (1980) and Fridman and Lemaire (1980) have derived a linear relation between the upgoing current density $j_{\parallel}$ and the potential drop $V_{\parallel}$ which holds over a fairly large range of $V_{\parallel}$. Their theory emphasizes the role of the magnetic mirror force which can provide a resistance for upgoing currents carried by precipitating electrons. However, for downgoing currents a different "conductivity" (ratio of $V_{\parallel}/j_{\parallel}$) is expected. Several authors have argued that this theory cannot account for all observations (see, e.g., Goertz, 1984 and references therein). And recently Weimer et al. (1985) and Goertz

Space Science Reviews **42** (1985) 499–513. 0038–6308/85.15

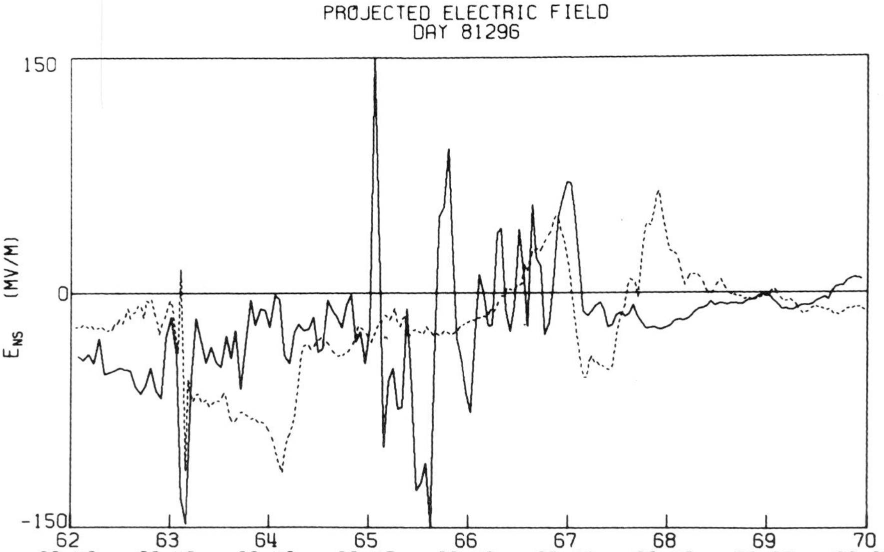

INV LAT	62	63	64	65	66	67	68	69	70
UT.DE-1	03:49	03:48	03:46	03:45	03:43	03:42	03:40	03:39	03:37
UT.DE-2	04:02	04:01	04:01	04:01	04:00	04:00	04:00	04:00	03:59
ΔT	+00:12	+00:13	+00:14	+00:15	+00:17	+00:18	+00:19	+00:20	+00:22
ALT.DE-1(KM)	8173	8436	8707	8987	9277	9576	9885	10205	10536
ALT.DE-2(KM)	896	899	903	906	909	912	915	917	921
MLT.DE-1(HR)	20.41	20.39	20.36	20.34	20.31	20.29	20.26	20.23	20.20
MLT.DE-2(HR)	20.00	19.96	19.92	19.87	19.83	19.78	19.72	19.66	19.59

Figure 1. Projected electric fields measured by the high altitude satellite DE-1 (solid line) and the low altitude satellite DE-2 (dashed line) on day 296 (October 23), 1981. Positive electric fields correspond to southward pointing fields. This figure is identical to Figure 3 of Weimer et al. (1985).

and Weimer (1985) have shown that the same conductivity applies to both up and downgoing currents.

In this paper we will describe some of these observations which require a somewhat different model for auroral arc formation than the one described by the authors mentioned above. We will argue that a proper understanding of the field-aligned current-voltage relation requires an evolutionary model. This does not necessarily mean that observed arcs cannot be considered as a stationary phenomenon. However, there are different possible stationary states. To understand which one is actually realized in nature requires consideration of the nonstationary evolution of field-aligned currents and electric fields. We are only at the beginning of such a theory and its development is much less advanced than the stationary theory. The theoretical part of this paper is closely related to the papers by Lysak and Kan in this volume.

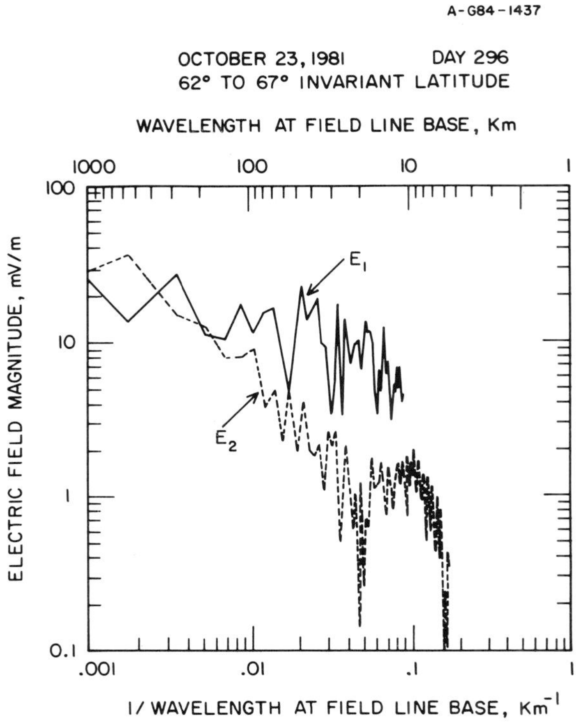

Figure 2. Electric field spectra from day 296 (October 23), 1981. These spectra are obtained from the electric field data between 62° and 67° invariant latitude. The solid line is for the DE-1 (high altitude) data. This figure is Figure 4 of Weimer et al. (1985).

II. RECENT DYNAMICS EXPLORER OBSERVATIONS

Weimer et al. (1985) have compared electric fields by the two Dynamics Explorer (DE-1 and DE-2) satellites when they were magnetically conjugate, i.e., on the same magnetic field. The high altitude (DE-1) and low altitude (DE-2) electric fields were "projected" to a common altitude (the base of the ionosphere) by a mapping scheme which accounts for the magnetic field convergence and assumes that the parallel electric field between the two satellites is zero. They also assume that the time variations seen by the two spacecraft are solely due to their motion through a stationary structure, i.e., $d/dt = \vec{v}\cdot\nabla$. It is then convenient to compare the spatial Fourier transforms of the electric fields rather than the fields themselves.

In Figure 1 (their Figure 3) the actual electric field components in the N-S direction are shown. Clearly the high altitude observations show much larger fluctuations of the electric field, i.e., a much richer spatial variation, than the low altitude observations.

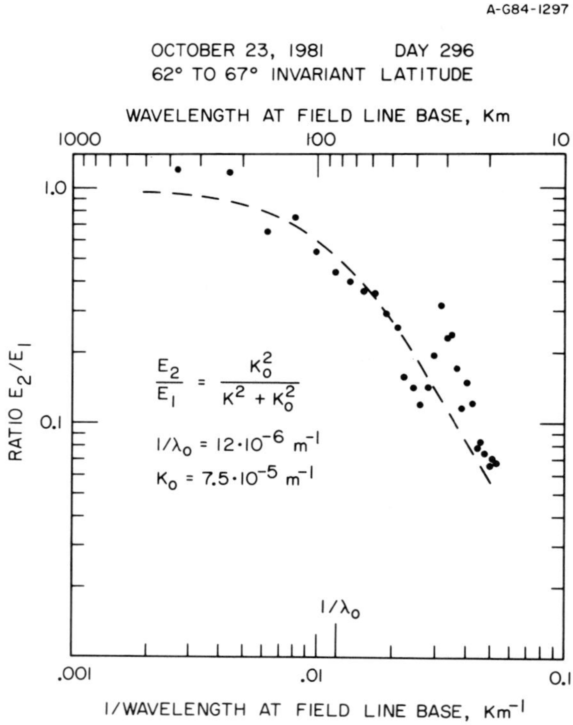

Figure 3. The ratio of the Fourier amplitudes of the electric field spectra from Figure 2 as a function of wavenumber divided by 2π ($k/2\pi = 1/\lambda$). The dashed line is the theoretical prediction of Eq. (9) with k_o chosen to give the best visual fit. This figure is Figure 17 of Weimer et al. (1985).

The Fourier spectra are shown in Figure 2 (their Figure 4). The ratio of Fourier amplitudes versus wavenumber is shown in Figure 3 (their Figure 17). This ratio decreases with increasing wavenumber. At small wavenumbers (or large spatial wavelengths) the two fields agree, as expected. However, small-scale variations with ground wavelengths of less than 100 km are much stronger at high altitudes than close to the ionosphere.

This could be due either to the existence of a field-aligned potential drop ($V_{\parallel}$) between the two spacecraft or due to real-time variations ($\partial/\partial t \neq 0$). We have considered the possibility that the fluctuations are due to Alfven waves but have found the data to be inconsistent with that hypothesis. They are, however, in excellent agreement with the static theory to be described in the next section. They thus provide a means for determining the parallel conductivity of auroral field lines.

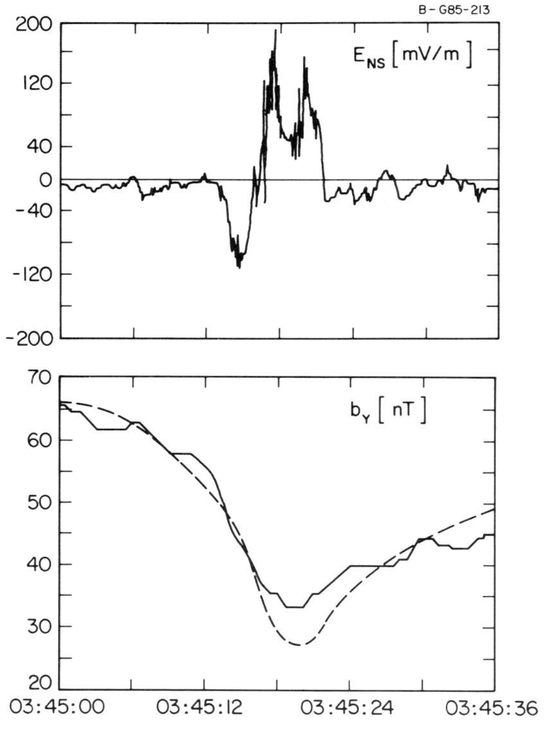

Figure 4. Top panel: The north-south electric field measured by the high altitude DE-1 satellite on day 296 (October 23), 1981. The signature is typical of so-called electrostatic shocks. Note that the width of the structure is only about 30 km which maps to about 8 km on the ground. Bottom panel: The magnetic field perturbation b_y (solid line) and the second integral of the electric field (dashed line). Note that the magnetic field indicates an upward field-aligned current and a downward current.

Small-scale electric field variations are associated with variations of the east-west magnetic field (b_y) which indicate the presence of field-aligned currents. An example is shown in Figure 4. The electric field signature is typical of so-called "electrostatic shocks": E_x changes sign and is very small outside the structure. The magnetic field changes in a step-like fashion. The direction of the field-aligned current in this case is out of the ionosphere before 3:45.18 and into it after that time. Other examples (data not shown) have only one sign of current coming either out of or going into the ionosphere. We will indicate in the next section how these data can also be used to infer the parallel conductivity.

III. THEORY

We consider an electric field in the meridianal plane $E_x(L,s)$ which varies along the magnetic field (s) and across the field lines labelled by their L parameter ($r = L\, R_e \sin^2\theta$). It is convenient to work in this L-s space for purposes of mapping. The variations of E_x and the associated east-west magnetic field (b_y) are taken to be due to the motion of the spacecraft.

$$\frac{d}{dt} = \bar{v}\cdot\nabla = v_L \frac{\partial}{\partial L} + v_s \frac{\partial}{\partial s} + v_y \frac{\partial}{\partial y} \tag{1}$$

We assume that the variation in the azimuthal direction is negligible ($\partial/\partial y \approx 0$). We also assume that the variations along the field lines are much smaller than those across the field ($v_L\, \partial/\partial_L \gg v_s\, \partial/\partial_s$). The electric field is the gradient of a potential which is written in terms of its Fourier spectrum.

$$\Phi(s,L) = \int \tilde{\Phi}(s,k_L) e^{ik_L L}\, dk_L \tag{2}$$

The potential on the ground is

$$\Phi(o,L) = \Phi(s,L) - V_\parallel(L) \quad . \tag{3}$$

The relation holds for the Fourier amplitudes also

$$\tilde{\Phi}(o,k_L) = \tilde{\Phi}(s,k_L) - \tilde{V}_\parallel(k_L) \quad . \tag{4}$$

The parallel potential $V_\parallel$ consists of two parts. The potential drop in the ionosphere which scales with the ratio of the Pedersen to parallel conductivity of the ionosphere (Spreiter and Briggs, 1961) can be shown to be very small. The potential drop above the ionosphere is related to the parallel current density by the relation

$$j_\parallel = -a\, V_\parallel \quad . \tag{5}$$

Although several authors mentioned above have derived this relation from theoretical arguments, it is not clear whether it is valid for all values of $V_\parallel$. We will use the data to test this relation.

For static conditions the currents must close in the ionosphere by horizontal currents. For a uniform ionosphere, i.e., $\nabla\Sigma_P = \nabla\Sigma_H = 0$ (where Σ_P and Σ_H are the height-integrated Pedersen and Hall conductances, respectively), the Fourier amplitude of the parallel current density is

$$\tilde{j}_\parallel(s,k_L) = \tilde{j}_\parallel(o,k_L) = -k_L^2\, \Sigma_p \tilde{\Phi}(o,k_L) \left(\frac{dL}{dx_i}\right)^2 \tag{6}$$

where x_i is the north-south coordinate on the ground ($dx_i = R_e\, d\theta$). The constancy of $\tilde{j}_\parallel(k_L)$ along the field line follows from current continuity and the absence of perpendicular currents in the magnetosphere.

Combining these equations we find

$$\tilde{V}_\parallel(k_L) = \frac{k_L^2}{k_L^2 + k_{oL}^2}\, \tilde{\Phi}(s,k_L) \tag{7a}$$

where

$$k_{oL}^2 = \frac{a}{\Sigma_p}(dL/dx_i)^2 \quad . \tag{8a}$$

Weimer et al. (1985) have written this relation in terms of $k = k_L\, dL/dx_i$.

$$\tilde{V}_\parallel(k) = \frac{k^2}{k^2+k_o^2}\, \tilde{\Phi}(s,k) \tag{7b}$$

$$k_o^2 = \frac{a}{\Sigma_P} \tag{8}$$

The ratio of the electric field amplitudes at high altitudes ($\tilde{E}(k,s)$) and at low altitudes ($\tilde{E}(k,o)$) is given by

$$\tilde{E}(k,o) = \frac{k_o^2}{k^2 + k_o^2}\, \tilde{E}(k,s) \quad . \tag{9}$$

This relation is superimposed on the data as the dashed line in Figure 3. The fit is remarkably good. We conclude that the linear relation (5) is a valid approximation with the conductivity independent of the sign and magnitude of $V_\parallel$. The field-aligned "conductivity" in this case was calculated by Weimer et al. to be $a = 3.2 \times 10^{-8}\ \Omega^{-1}m^{-2}$.

Equation (9) can be explained by a very simple model. The total current in a current sheet of width L_x is $I_\parallel = -aL_x V_\parallel$. Thus, the parallel resistance is $R_\parallel = 1/aL_x$. In the ionosphere we have $I_\perp = \Sigma_P E_x = \Sigma_P V_\perp/L_x$ and the ionospheric resistance is $R_\perp = L_x/\Sigma_P$. A potential applied in the magnetosphere will then be distributed along the field and in the ionosphere with the ratio of $V_\parallel/V_\perp$(ionosphere) $= R_\parallel/R_\perp = \Sigma_P/aL_x^2$. Thus small-scale structures $L_x^2 \ll \Sigma_P/a$ have most of the applied potential along the field line. The current is then determined by the parallel conductivity and rather independent of the ionospheric conductance.

This suggests that by looking at the very small-scale structure ($k_L \gg k_{oL}$) the parallel conductivity (a) can be determined from the high altitude observations alone. Using equations (7a) and (5) we get for $k_L^2 \gg k_{oL}^2$

$$\tilde{j}_{\parallel}(s,k_L) \simeq a\,\tilde{\Phi}(s,k_L) \tag{10}$$

or in real space

$$j_{\parallel}(s,x) \simeq a\,\Phi(s,x) \quad . \tag{11}$$

Since $j_{\parallel} = \mu_o\,\partial b_y/\partial x$ and $\Phi = -\int E_x d_x$ we have

$$\frac{\partial^2 b_y}{\partial x^2} = -\frac{a}{\mu_o} E_x(x,s) \quad . \tag{12}$$

In Figure 4 we have superimposed the second integral of the observed electric field ($\int\int E(x'')dx''dx'$) on the magnetic field variation. We see that, indeed, the agreement is very good. We have twice integrated E_x rather than twice differentiated b_y because integration smoothes out fluctuations whereas differentiation would have amplified them and made a comparison more difficult.

The value of a for which the best fit is obtained is $a = 4 \times 10^{-8}\ \Omega^{-1}m^{-2}$. This is very similar to the value obtained by Weimer et al. (1985) from the data shown in Figure 3. We have looked at several other examples of "electrostatic shocks" in the DE data and found good agreement with the theory for up and downgoing currents. The value of a always seems to be in the range of several 10^{-9} to several 10^{-8} $\Omega^{-1}m^{-2}$ (Goertz and Weimer, 1985).

Clearly, the magnetic mirror force cannot impede or provide a resistance for a current into the ionosphere carried by upgoing electrons. Thus, we conclude that the finite parallel conductivity must be due to some other effect, representing a real resistivity or dissipation of electromagnetic energy between the high altitude DE-2 spacecraft and the ionosphere. We also note that outside the auroral zone Weimer et al. (1985) find that electric fields at all wavenumbers for which data are available map without attenuation from high to low altitudes. (Structures with very large values of k cannot be resolved by the instruments.) This indicates that k_o outside the auroral zone is much larger and suggests that the field-aligned conductivity is greater outside the auroral zone.

IV. EVOLUTION OF FIELD-ALIGNED CURRENTS

Two explanations for the existence of a finite parallel conductivity have been published. In the model of Chiu and Cornwall (1980) the resistance is provided for by the upward magnetic mirror force on downgoing electrons. It is evident that this mechanism cannot provide a resistance for downgoing currents carried by upward going electrons. However, upward going ionospheric electrons must overcome the ambipolar electric field and linear current-voltage relation has also been derived by Chiu and Cornwall (1980) for this case. But they show that the "conductivity" for downgoing currents is, in general, larger than that for upgoing currents. No such distinction seems to be evident in the data.

Lysak and Dum (1983) have derived a finite conductivity (or resistivity) from considerations of current-driven instabilities. They argue that when the field-aligned current exceeds a certain threshold, electrostatic ion cyclotron waves become unstable which provide a resistance. The resistance (1/a) is zero for $j_{\parallel} < j_c$. Although it may be difficult to detect such a threshold effect in the data, it appears that the linear relation (5) holds quite well for all values of $V_{\parallel}$ and $j_{\parallel}$.

We would like to point out another possibility which is based on similar considerations as those of Lysak and Dum. We must consider the temporal evolution of the auroral current. The magnetosphere acts as a generator for the perpendicular electric field and/or magnetic field perturbation. The generator physics is not well understood at the present time. It was shown by Goertz and Boswell (1979) that the perturbation propagates along the field line as a shear Alfven wave. Since the generator has a finite extent perpendicular to the magnetic field line, the perturbation must contain obliquely propagating shear Alfven waves ($k_{\perp} \neq 0$). If $k_{\perp}$ is not too large ($k_{\perp}c/\omega_{pe} \leqslant 1$) the wave is guided along the field line, i.e., the group velocity has only a very small perpendicular component. Any perturbation must be considered as a shear Alfven wave packet propagating along the field line. The packet consists of a superposition of obliquely propagating waves. The dispersion relation for low frequency MHD waves ($\omega < \Omega_i$) propagating in the x-z plane with the background field along the z-axis is

$$\begin{vmatrix} K_{xx} - n_z^2 & 0 & n_x n_z \\ 0 & K_{yy} - n^2 & K_{yz} \\ n_x n_z & K_{zy} & K_{zz} - n_x^2 \end{vmatrix} \vec{E} = 0 \qquad (13)$$

where

$$\vec{n} = \vec{k}\, c/\omega$$

$$K_{xx} = c^2/V_A{}^2$$

$$K_{yz} = -K_{zy} = \frac{k_x}{k_z}\frac{\omega_{pi}^2}{\omega\Omega_i}(\alpha_{oe}F_{oe} - \alpha_{oi}F_{oi}) \tag{14}$$

$$K_{yy} = 1 + \sum_s \frac{\omega_{ps}^2}{\Omega_s^2}(1 + i\frac{k_x^2}{k_z^2}(1 + i\,\alpha_{os}^2 F_{os})/\alpha_{os}^2)$$

$$K_{zz} = 1 + 2\sum_s \frac{\omega_{ps}^2}{\omega^2}\alpha_{os}^2(1 + i\,\alpha_{os}F_{os}) \quad .$$

The summation is over ions (i) and electrons (e) whose plasma and gyrofrequencies are ω_{ps} and Ω_s, respectively. The plasma dispersion function is given by

$$F_{os}(\alpha_{os}) = \pi^{1/2}e^{-\alpha_{os}{}^2} + 2i\, e^{-\alpha_{os}{}^2}\int_0^{\alpha_{os}} e^{t^2}dt$$

$$\alpha_{os} = \frac{\omega}{k_z V_{ths}} \tag{15}$$

where the thermal velocity is $V_{ths} = \sqrt{kT_s/m_s}$.

For $\beta \ll 1$ the K_{yz} term can be neglected and the fast mode ($K_{yy} - n^2 = 0$) decouples from the shear ($K_{xx} - n_z{}^2 = 0$) and slow modes ($K_{zz} - n_x{}^2 = 0$). However, for oblique propagation the shear Alfven and the slow mode are coupled

$$(K_{xx} - n_z{}^2)(K_{zz} - n_x{}^2) = n_x{}^2\, n_z{}^2 \quad . \tag{16}$$

The dispersion relation for a cold plasma ($\beta \ll m_e/m_i$) is

$$\frac{k_z^2}{\omega^2 V_A^2} = 1 + \frac{c^2 k_x^2}{\omega_{pe}^2} \quad . \tag{17a}$$

For a warm plasma ($m_e/m_i \ll \beta \ll 1$) it reads

$$\frac{\omega^2}{k_z^2 V_A^2} = 1 + k_x^2 R_{gi}^2 \left(\frac{3}{4} + \frac{T_e}{T_i}\right) \tag{17b}$$

where $V_A = B/\sqrt{\mu_o n m_i}$ is the Alfven speed and R_{gi} is the ion gyroradius. These obliquely propagating shear Alfven waves have been called kinetic Alfven waves by Hasegawa (see Hasegawa and Uberoi (1982)).

They carry a perpendicular electric field E_x and a parallel electric field E_z in phase with E_x

$$\frac{E_z}{E_x} = \left(\frac{\omega^2 V_A^2}{k_z^2} - 1\right) \frac{k_z}{k_x}$$

$$= - \frac{c^2 \; k_x k_z / \omega_p^2}{1 + c^2 k_x^2/\omega_p^2} \quad \text{for} \quad \beta << \frac{m_e}{m_i} \tag{18}$$

$$= + \; k_x k_z R_{gi}^2 \left(\frac{3}{4} + \frac{T_e}{T_i}\right) \quad \text{for} \quad \frac{m_e}{m_i} << \beta << 1 \quad .$$

The existence of a parallel electric field can be understood by the following simple argument. An Alfven wave contains a perpendicular polarization current $j_x \; \alpha \; \partial E_x/\partial t$. This is a fluid or MHD effect. For a finite perpendicular wavenumber, k_x, this current has a finite divergence $k_x j_x \neq 0$. However, for small frequencies the divergence of the current must be zero. Since by definition $k_y = 0$ the polarization current must close by field-aligned currents ($k_z j_z = -k_x j_x$). A field-aligned current can only be due to a field-aligned motion of electrons relative to the ions. Thus, electrons must be accelerated by a parallel electric field. This is a kinetic effect--hence the name kinetic Alfven wave. We also see that $\nabla \cdot E = i(k_x E_x + k_z E_z) \neq 0$. Thus the kinetic Alfven wave carries a charge density. Hasegawa and Mima (1976) have shown that solitary kinetic Alfven waves can exist in a warm plasma ($\beta >> m_e/m_i$). Recently, Goertz (1984) has suggested that the so-called electrostatic shocks can be described in terms of such solitary waves in a cold plasma ($\beta << m_e/m_i$). However, Alfven waves are reflected at the ionosphere (Goertz and Boswell, 1979) and the fields above the ionosphere should consist of the superposition of downgoing Alfven wave packets and reflected waves as described by Lysak and Dum (1983). Without damping of the waves in the magnetosphere, a steady state will be reached only after many tens of reflections (Goertz and Boswell, 1979). The data seem to indicate that a steady state is reached much more quickly and that energy dissipation in the magnetosphere is important. For that reason Lysak and Dum (1983) have introduced a phenomenological damping rate or effective collision frequency.

The field-aligned current carried by the kinetic Alfven wave can be calculated from Ampere's and Faraday's laws

$$j_{\parallel} = -i \, \frac{1}{\mu_o} \, \frac{\omega}{k_z V_A^2} \, k_x E_x \tag{19}$$

Note that this is a capacitive current and does not dissipate energy. Lysak and Dum argue that when $j_{\parallel}$ becomes large the current excites electrostatic ion cyclotron waves and the waves are damped nonlinearly. However, even small currents can be damped when $\beta \approx O(m_e/m_i)$. In that case $V_A \sim V_{the}$ and the kinetic Alfven wave is damped by Landau resonant electrons.

In both cases the dispersion relation becomes complex. Lysak and Carlson (1981) and Lysak and Dum (1983) write for a cold plasma ($\beta \ll m_e/m_i$)

$$\frac{k_z^2 V_A^2}{\omega^2} = 1 + \frac{k_x^2 c^2}{\omega_{pe}^2} \left(1 + i \frac{\nu}{\omega}\right) \tag{20}$$

where ν is an effective collision frequency for the momentum exchange between electrons and ions. The nonlinear collision frequency

$$\nu = \nu_1 + \nu_2 \left\{ \frac{|j_z/n_e| - u_c}{u_c} \right\}^2 \tag{21}$$

is derived from electrostatic ion cyclotron turbulence saturated by resonance broadening. The critical velocity for the excitation of ion cyclotron waves is u_c.

For $\beta \lesssim m_e/m_i$ a reasonable approximation to the collisionless dispersion relation (16) can be written as

$$\frac{k_z^2 V_A^2}{\omega^2} \approx 1 + \frac{c^2 k_x^2}{\omega_{pe}^2} \left[1 + i\, 2\alpha_{oe}^3 \sqrt{\pi}\, e^{-\alpha_{oe}^2}\right] \tag{22}$$

with $\alpha_{oe} = \omega/k_z V_{the}$. This is similar in form to the dispersion relation of Lysak and Carlson (1981). Landau damping is significant only for $\alpha_{oe} \sim O(1)$. Since for large-scale structures ($k_x^2 c^2/\omega_{pe}^2 \ll 1$) $\omega/k_z \approx V_A$ Landau damping will be limited to regions where $\beta \sim O(m_e/m_i)$. Along auroral field lines the Alfven speed varies from 40,000 km/s at 1.4 R_e to 8000 km/s at 5 R_e (Lysak, 1985). Thus, β is usually smaller than m_e/m_i and $\alpha_{oe} \ll 1$. Electrons with energies between 400 eV and 10 keV are in resonance. Since the bulk of the electrons has smaller energies Landau damping should not be important. However, small-scale structures ($k_x^2 c^2/\omega_{pe}^2 > 1$) propagate along the magnetic field at a phase speed smaller than V_A and the Landau resonance energy will be smaller. The properties of such Alfven waves with $k_x^2 c^2/\omega_{pe}^2 \geqslant 1$ and $\beta \sim O(m_e/m_i)$ (smaller but not much smaller) have not been investigated. But these arguments suggest that if damping is due to Landau resonance, the parallel conductivity should depend on k_x (a should decrease with increasing k_x). It is not clear that the data are compatible with this prediction.

We conclude this section by noting that none of the models advanced seems to provide a completely satisfactory explanation for the observed parallel conductivity (a). The magnetic mirror force model predicts a dependence of a on the sign of $V_\parallel$. The anomalous resistivity predicts a dependence of a on the current density, $j_\parallel$. The Landau damping model predicts a dependence of a on the perpendicular wavenumber k_x.

V. DISCUSSION

The numerical simulations by Lysak and Dum (1983) and Lysak (1985) of Alfven waves carrying a supercritical current density have shown that quasi-steady structures of perpendicular and parallel electric fields can develop in a few Alfven travel times. As we have mentioned above, their model requires sufficiently large current densities or electric field amplitudes to generate the electrostatic ion-cyclotron waves. It is not clear that the phenomenological form for the collision frequency (Eq. 21) is a good representation of the physics. It seems that a particle simulation is needed to investigate the microphysics of damping of Alfven waves in more detail. A completely realistic simulation would be too difficult and expensive. The following consideration may be important for making progress on this problem.

Over most of the auroral field line the current carried by the Alfven waves is subcritical and Landau damping is negligible. The cold plasma approximation ($\beta \ll m_e/m_i$) should yield an adequate description of the field-aligned currents, perpendicular and parallel electric fields. The region where $j_\parallel > j_{\parallel c}$ or where $\alpha_{oe} \sim O(1)$ is thus limited in extent along the field line. In this region kinetic effects are important and an electrostatic, two-dimensional particle simulation should be used. Since the magnetic field perturbation of the Alfven wave are small ($\Delta b/B_o \ll 1$) the magnetic field can be assumed constant and equal to the background field ($\vec{B}_o = B_o\hat{z}$). The coupling between these two regions (the MHD region where Alfven waves propagate and the kinetic region where electrostatic effects are important) is provided by the boundary conditions. The current into the simulation region is given by the Alfven relation (Eq. 19)

$$j_\parallel = \pm \frac{1}{\mu_o V_A} \frac{\partial E_x}{\partial x} (x,\ t) \Bigg|_{z = \begin{smallmatrix}0\\L\end{smallmatrix}} \quad . \tag{23}$$

The length of the system along the z-direction is L. The ± sign refers to waves propagating antiparallel and parallel to the background field.

The electric field is calculated self-consistently from Poisson's equation using periodic boundary conditions in the x-direction and $E_z(z = 0,L) = 0$. The perpendicular electric fields $E_x(x,t)$ on the top ($z = 0$) and bottom ($z = L$) boundary of the system act as generators of upward and downward going Alfven waves, respectively. These in turn change the current into the system according to Eq. (23). In this way the evolution of the electric fields in the simulation region will affect the particle distribution of particles injected into the system which in turn are used to calculate the electric field from Poisson's equation.

We hope to report the results of such a simulation in the near future.

ACKNOWLEDGEMENTS

Part of this work was done in collaboration with Dr. D. R. Weimer. Helpful discussions with Dr. D. Gurnett and Mr. T. Whelan are gratefully acknowledged. This work was supported by NASA grants NSG-7632, NGL-16-001-043, NAGW-386, and NSF grant ATM-811126.

REFERENCES

Chiu, Y. T., and J. M. Cornwall, 'Electrostatic model of a quiet auroral arc', J. Geophys. Res., 85, 543, 1980.

Chiu, Y. T., A. L. Newman, and J. M. Cornwall, 'On the structures and mappings of auroral electrostatic potentials', J. Geophys. Res., 86, 10029, 1981.

Fridman, M., and J. Lemaire, 'Relationship between auroral electron fluxes and field-aligned potential difference', J. Geophys. Res., 85, 664, 1980.

Goertz, C. K., 'Double layers and electrostatic shocks in space', Rev. Geophys. and Space Phys., 17, 418, 1979.

Goertz, C. K., 'Kinetic Alfven waves on auroral field lines', Planet. Space Sci., 32, 1387, 1984.

Goertz, C. K., and R. W. Boswell, 'Magnetosphere-ionosphere coupling', J. Geophys. Res., 84, 7239, 1979.

Goertz, C. K., and D. R. Weimer, 'Conductivity of auroral field lines', in preparation.

Hasegawa, A., and K. Mima, 'Exact solitary Alfven waves', Phys. Rev. Lett., 37, 690, 1976.

Hasegawa, A., and C. Uberoi, The Alfven Wave, DOE Critical Review Series, DOE/TIC-11197, 1982.

Kan, J. R., and L. C. Lee, 'Theory of imperfect magnetosphere-ionosphere coupling', Geophys. Res. Lett., 7, 633, 1980.

Lyons, L. R., 'Generation of large-scale regions of auroral currents, electric potentials, and precipitation by the divergence of the convection electric field', J. Geophys. Res., 85, 17, 1980.

Lysak, R. L., 'Auroral electrodynamics with current and voltage generators', J. Geophys. Res., submitted, 1985.

Lysak, R. L., and C. W. Carlson, 'Effect of microscopic turbulence on magnetosphere-ionosphere coupling', Geophys. Res. Lett., 8, 269, 1981.

Lysak, R. L., and C. T. Dum, 'Dynamics of magnetosphere-ionosphere coupling including turbulent transport', J. Geophys. Res., 88, 365, 1983.

Shawhan, S. D., C.-G. Fälthammar, and L. P. Block, 'Cn the nature of large auroral zone electric fields at 1-R_E altitude', J. Geophys. Res., 83, 1049, 1978.
Weimer, D. R., C. K. Goertz, D. A. Gurnett, N. C. Maynard, and J. L. Burch, 'Auroral zone electric fields from DE-1 and -2 at magnetic conjunction', J. Geophys. Res., in press, 1985.

CONSEQUENCES OF HYDROMAGNETIC WAVES ON MAGNETOSPHERIC PARTICLE DYNAMICS

R. Gendrin
Centre de Recherches en Physique de l'Environnement
Centre National d'Etudes de Télécommunications
38-40 rue du Général Leclerc
92131 Issy-les-Moulineaux, France

ABSTRACT. Since long ULF waves are known to play a dominant role in the dynamics of energetic protons. Recent observations have shown that they also contribute to the heating and/or acceleration of both heavy ions and electrons. This review will deal with all the aspects of wave particle interactions that involve electromagnetic ULF waves in the equatorial region of the magnetosphere. We will consider successively Pc-3-4-5 pulsations, magnetosonic waves and ion cyclotron waves. In the latter case, for which both the experimental data and the theoretical interpretation are more advanced, we will discuss the linear, quasi-linear and nonlinear aspects of the interaction. Results of recent numerical simulations of this type of interaction will also be reported. A final section will be devoted to ULF waves observed in the vicinity of the Io torus.

1. INTRODUCTION

In recent years a lot of experimental data, mainly obtained at geostationary L-values, has allowed to study in detail magnetospheric processes involving ultra low frequency (ULF) waves. This is the result of many technical improvements in spacecraft instrumentation: increase of sensitivity and time resolution of wave and particle detectors and, as far as waves are concerned, possibility of measuring both the electric and magnetic components; existence of sophisticated composition experiments; full coverage of the plasma energy range from low (thermal) energy to high energy. It is also the consequence of coordinated data analysis involving one or two spacecraft or one spacecraft and ground stations.

The processes by which ULF waves are generated are basically understood and well documented by a large set of experimental data. These waves are known to have important effects on particle dynamics but the exact role which they play in the exchange of energy between the different magnetospheric particle populations is not yet fully established.

The purpose of this paper is to give an overview of phenomena involving ULF waves as observed in the equatorial magnetosphere, the processes which occur at auroral altitudes being covered by other articles in this book. In the ULF range, five different phenomena can be distinguished (Table 1). Except for the turbulence which is associated with localized current structures and which is the subject of other reviews, these phenomena will be covered in the three following sections. A

Space Science Reviews **42** (1985) 515–539. 0038–6308/85.15

Table 1. Different types of ULF pulsations observed in the equatorial magnetosphere

Type and frequency range	Possible origin	Consequences	Section
Long period pulsations 3-50 mHz (Pc-3,4,5)	- drift mirror instability - coupling of KH waves with shear Alfven waves	- Joule heating of the ionosphere - diffusion of energetic particles	2
Turbulence associated with localized current structures 10-1 000 mHz (Pi-1,2)	- current driven instability - kinetic Alfven waves	- diffusion of protons and acceleration of electrons	-
Ion cyclotron waves 0.1-2 Hz (Pc-1,2)	- anisotropic proton distributions in multicomponent plasmas	- heating of heavy ions	4
Magnetosonic waves 1-10 Hz	- ring distributions - steep spatial gradients in the hot particle distribution	- precipitation of energetic ions	3
ELF turbulence 10-500 Hz	- trapping and detrapping of thermal electrons	- generation and precipitation of suprathermal electrons	4

final section will be devoted to the role of ULF phenomena in the Jovian magnetosphere since the characteristics of this medium are somewhat different from those which prevail in the Earth's magnetosphere and since the present lack of detailed and systematic measurements leaves the door open to more speculative conclusions.

2. LONG PERIOD PULSATIONS

2.1. Experimental improvements

Progress has been made in the field of long period pulsations (LPP) because of many technical improvements: more complete sets of experiments on a given spacecraft, use of multi-spacecraft observations, conjunction of ground and spacecraft observations.

To the first category belong the measurements of electric fields in association with magnetic fields, the reliable measurement of cold plasma density and plasma composition, the measurement of detailed particle distribution functions with a good time resolution. All these quantities have been used simultaneously to study the structure of the waves and their association with particle dynamics. Plasma data which are essential to compute the Alfven velocity, V_a have also served to check the agreement between observed and theoretical periods (section 2.2).

The second category of improvements has helped to separate the spatial and temporal evolution of pulsations. From such measurements, mainly obtained with ISEE-1 and -2 (Singer et al., 1982) but also with these two spacecrafts and GEOS-1 (Hughes and Grard, 1984) or with ATS-6, SMS-1 and SMS-2 (Takahashi et al., 1984) the size of the resonating region has been determined and the variation of the wave characteristics with magnetic latitude has been studied. Note that in the past multi-spacecraft experiments has allowed to determine the azimuthal wavelength of the pulsations (Hughes et al., 1978).

Finally the conjunction between measurements made onboard geostationary spacecraft and at the ground have been helpful for determining the characteristics of the waves (e.g. Tonegawa et al., 1984; Wedeken et al., 1984). This is especially true for electric field intercomparison: with STARE it is possible to study the spatio-temporal characteristics of the drift velocity associated with the wave electric field existing at ionospheric altitudes (e.g. Villain, 1982). As an example, figure 1 clearly shows the poloidal nature of the plasma oscillation associated with a wave detected onboard GEOS-2 (Walker et al., 1982).

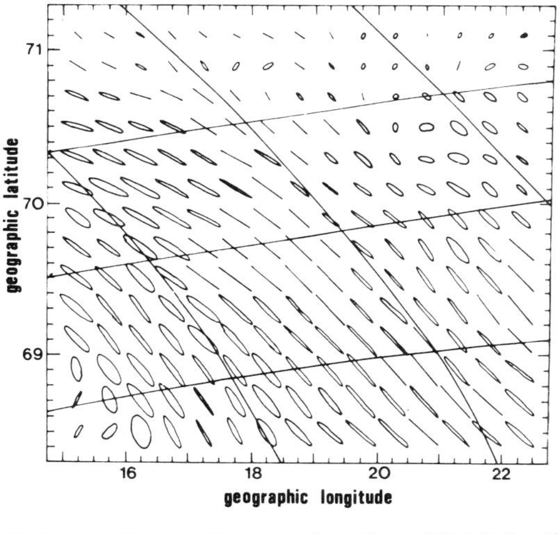

Figure 1. Polarization ellipses in the STARE field of view for the Fourier component of the drift velocity with period 213 s. The grid lines map to lines of constant azimuth and radial distance in the equatorial plane. The mode is clearly poloïdal (Walker et al., 1982).

The two problems which have to be addressed when studying long period pulsations are their three-dimensional structure and their origin (see the review by Southwood and Hughes, 1983). Recent data concerning these two problems are briefly presented below.

2.2. Structure of LPP's:

As far as the structure is concerned the first step is to determine the parity $n = 2\lambda/\mathscr{L}$ of the mode, λ being the wave length and $\mathscr{L}$ the length of the field line. At the equator odd (respectively even) modes present a node for the magnetic (respectively electric) field and an antinode for the electric (respectively magnetic) field. The fundamental (i.e. the first harmonic) which is an odd mode has been clearly observed (figure 2) on the electric antennas of GEOS-2 in the absence, from a statistical point of view, of magnetic fluctuations (Junginger et al., 1983, 1984). However odd modes have been also clearly detected on the ATS-6 magnetic antennas (Takahashi et al., 1984; Tonegawa et al., 1984). At higher latitudes GEOS-1 and ISEE-1 and -2 data, both electric and magnetic, show predominently the existence of second harmonics (Singer et al., 1982; Hughes and Grard, 1984). When two modes are simultaneously present, as observed for instance by Tonegawa et al. (1984), and by Junginger and Baumjohann (1984), their relative intensity should depend on the latitude of the observation (Takahashi and McPherron, 1982). For a given mode the amplitude radio $E/b\,V_a$ or the phase relationship between b and E as a function of latitude are two characteristics which can be used to deduce the parity of the mode (Singer et al., 1982; Hughes and Grard, 1984).

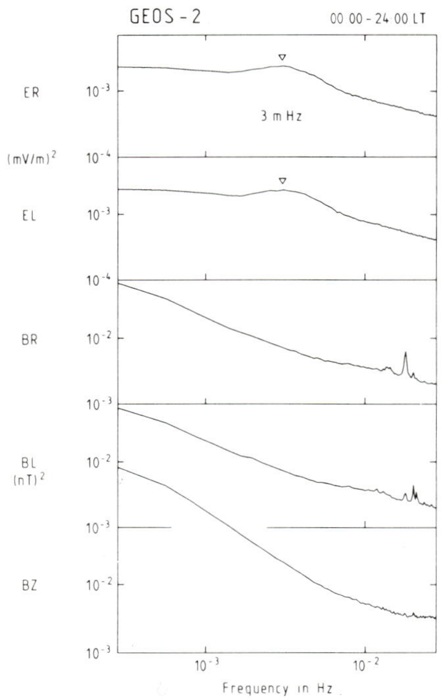

Figure 2. Mean frequency spectra of the electric and magnetic fields measured onboard GEOS-2 within a 186-day period. Note the presence of a peak in the E-field spectra which is not seen on the magnetic field. Note also that the compressional component B_z is larger than the two circular components B_R and B_L (Junginger et al., 1984).

One important aspect of these studies is the comparison between the measured wave frequency and the frequency which is obtained by using the measured plasma parameters and by applying Cummings et al's (1969) theoretical results. Two difficulties are encountered when doing such a comparison. First plasma composition data are not always available so that a 'reasonable' mass density has to be assumed in order to compute the theoretical wave period. Moreover when plasma composition data are available they do not cover the energy range below 10 eV (the assumption that the plasma below 10 eV is pure hydrogen is generally made). The second difficulty lies in the fact that the frequency which is measured is not always the frequency of the 'resonant' field line, as the excited region may have a radial extension larger than 1 R_e (Singer et al., 1982).

Notwithstanding these difficulties interesting comparisons between observed and theoretical values have been obtained. These comparisons are meaningful when one has to deal with second or higher harmonics since the corresponding periods do not depend on the nature (poloidal or toroidal) of the mode. An example of the result thus obtained is given in figure 3 (Hughes and Grard, 1984). But when dealing with the fundamental mode, one must know the wave polarization since the periods may differ by more than 40 %. Indeed in an event when a long period (T = 230 s) toroidal wave (fundamental) was observed simultaneously with a shorter period (T = 95 s) poloidal second harmonic, Junginger and Baumjohann (1984) had difficulties to reconcile the two observations with a unique model.

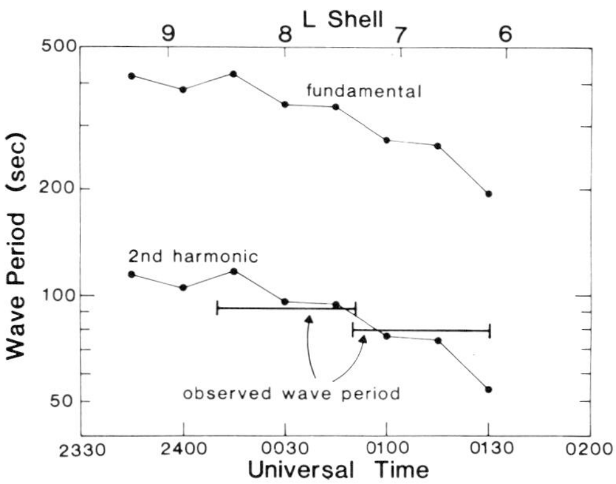

Figure 3. Comparison between observed and computed eigenperiods of an electric field pulsation detected onboard ISEE-1. The magnetic latitude of the spacecraft was smaller than 10° and the local time was around 02.30 (Hughes and Grard, 1984).

2.3. Origin of LPP's:

Strangely enough the answer to the question: what is the origin of LPP's? depends on the type of measurements which one analyses. Synoptic electric or magnetic field measurements seem to confirm the 'SCH' theory of Southwood (1974), Chen and Hasegawa (1974) which involves a Kelvin-Helmoltz (KH) instability generated at the magnetopause, whereas detailed analysis of particle fluxes measured during

LPP's seem to support the theory of local generation by either a drift mirror instability (e.g. Hasegawa, 1969) or a bounce resonance mechanism (e.g. Southwood, 1981). Both mechanisms are probably at work in conditions which still need to be specified.

By using their GEOS-2 high-quality electric field data Junginger et al. (1983) were able to draw maps of the polarization vector of the drift velocity as a function of local time. A statistical study then showed that the wave was left-handed in the forenoon sector and right-handed in the afternoon in agreement with the SCH theory, provided that GEOS-2 be situated earthward of the resonance field line, this point being confirmed by the observed value of the oscillation period ($T \sim 230$ s) which is higher than the average value ($T \sim 190$ s) observed for $L = 6.6$ (Samson and Rostoker, 1972). Tonegawa et al. (1984) also support the SCH mechanism, but with a wide band spectra in the excitation region, in order to interpret the simultaneous occurrence of the fundamental and second harmonic modes. The same hypothesis is made by Takahashi et al. (1984) who, by using data from three geosynchronous satellites ATS-6, SMS-1, and SMS-2, have observed up to five harmonics simultaneously though the fundamentals had different frequencies on each spacecraft, due to the different values of the plasma parameters as a function of local time.

On the other hand the observations made by Kremser et al. (1981) according to which one category of LPP's is associated with butterfly distributions for electrons ($dJ_e/d\alpha \gtrless 0$ for $\alpha \gtrless 90°$) seem to support the drift mirror instability generation. A quantitative analysis of the April 5, 1979 events by Wedeken et al. (1984) leads to the same conclusion, which is also supported by Walker et al. (1982) who made a detailed analysis of the coupling between a drift mirror wave and a standing Alfven wave. However this theory is confirmed only for those LPP's for which the electron fluxes are out-of-phase with ion fluxes (the 'out-of-phase' events as defined by Kremser et al., 1981). The other category of oscillations for which ion and electron fluxes are 'in-phase' and for which electrons have a pancake distribution ($dJ_e/d\alpha \gtrless 0$ for $\alpha \lessgtr 90°$) are not interpreted. Another controversy has been raised as wether or not the ion flux modulation is maximum for ions with 90° pitch-angles (Walker et al., 1982) or for ions with other pitch-angles (e.g. Su et al., 1980; Lin and Parks, 1982).

The bounce resonance mechanism has also been proposed or the basis of qualitative or semi-quantitative arguments (Singer et al., 1982; Walker et al., 1983; Hughes and Grard, 1984; Junginger and Baumjohann, 1984). A model of global compressional oscillations of the magnetosphere has also been presented for explaining ~8 mn oscillations observed onboard ISEE-1 (Kivelson et al., 1984).

2.4. Conclusion

The power which is dissipated in the ionosphere by pulsations is rather low as compared with the power dissipated by Joule heating due to ionospheric currents: 1 to 2 GW (Junginger et al., 1984) as compared to 10-100 GW (Greenwald and Walker, 1980). Consequently these pulsations do not play a fundamental role in the energy budget of the magnetosphere. However their importance with respect to the radial diffusion of energetic particles is well known. They may also have importance on the pitch-angle distributions of drifting particles.

As far as simulation is concerned it is doubtful that one could simulate such a complex phenomenon: it involves three dimensions and a very inhomogeneous medium. Besides one must introduce not only a full plasma composition (hot and cold, protons and heavy ions) but also specific pitch-angle distributions for each energetic species. External sources might also be introduced.

3. MAGNETOSONIC WAVES

3.1. Wave properties

GEOS has been the first spacecraft on which a sensitive three-component ULF antenna has been operating in a wave-form mode in the whole frequency range from ~0.1 Hz to 10 Hz (S-300 experimenters, 1979). Since at the geostationary orbit the proton gyrofrequency, F_H is of the order of 1 Hz it has been possible to study electromagnetic waves both below and above F_H. Below F_H waves have been observed whose polarization is mainly perpendicular to the DC magnetic field $\underline{B}_o$ and whose dispersion characteristics are modified by the presence of heavy ions. These are the ion cyclotron waves (ICW) which will be discussed in the next section. Above F_H waves have been observed which are mainly polarized along $\underline{B}_o$ and which consequently have large perpendicular wave numbers, $k_\perp$. These electromagnetic waves propagate in the magnetosonic mode.

Their spectral features are much typical. They consist of different quasi-monochromatic emissions which are harmonically related (figure 4). At the

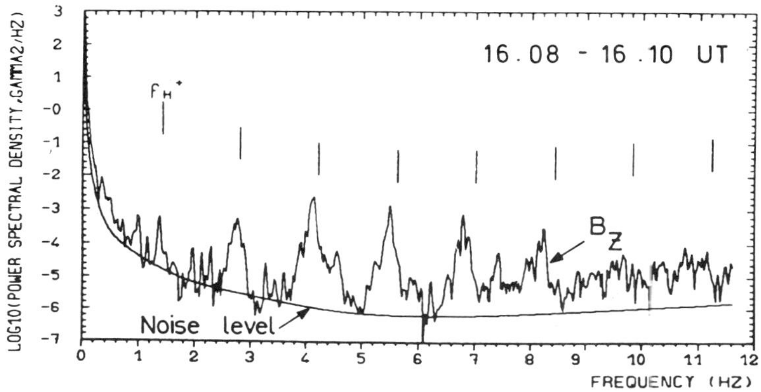

Figure 4. Magnetic power spectrum of a magnetosonic waves detected onboard GEOS-2. The z component is parallel to the local DC magnetic field (Perraut et al., 1982).

beginning of the event the frequency of the fundamental is often equal to the local proton gyrofrequency and it increases with time. Sometimes the fundamental is missing. Sometimes also the signal splits into two or three different emissions with independant harmonic structures. Linear combinations of the different frequencies may also exist. Examples of these different types of emissions are given in the papers published by the GEOS experimenters (e.g. Perraut et al., 1978, 1982; Perraut, 1982; Korth et al., 1983).

3.2. Associated particle dynamics

Attempts have been made to correlate these emissions with some characteristic features of the energetic particle populations. Two correlations have been found.

First it has been observed that during some magnetosonic wave events the energetic proton distribution shows a structure in which there is a flux maximum both at a certain energy (of the order of 10 keV) and at 90° pitch-angle (Perraut et al., 1982). Such a distribution can be roughly modeled by a ring of energetic protons to which an isotropic cold proton distribution is added:

$$f = \frac{n_r}{2\pi v_\perp} \delta(v_{/\!/}) \, \delta(v_\perp - U_\perp) + \frac{N_c}{2\pi v_\perp} \delta(v_{/\!/}) \, \delta(v_\perp) \tag{1}$$

where n_r and N_c are the ring and cold plasma densities, and where U is the ring velocity. Depending on the ratio $U_\perp/V_a$ and on the ratio $\rho = n_r/N_c$ the growth rates are positive ($U_\perp/V_a > 1$ is a necessary condition for small values of ρ) or negative. The growth rate maximizes where the cold plasma dispersion curve ($\omega = k_\perp V_a$) intersects the ringlike solutions ($\omega = n\Omega_H$). The growth rate increases with n until some value, which may explain why sometimes only higher harmonics are observed (figure 5). The frequency increase which occurs during the emission is probably due to the inward diffusion of the ringlike distribution, the wave propagating outward to the spacecraft. This would be consistent with the fact that the emission starts at the local proton gyrofrequency, provided that it could be demonstrated that such waves can propagate only toward regions of lower gyrofrequencies. Ray tracing of magnetosonic waves in an inhomogeneous medium still needs to be done.

Korth et al. (1983, 1984) have also observed magnetosonic waves (associated with wide band impulsive noise) in conjunction with displacements of energetic proton boundaries. The radial gradient in the energetic proton distribution is determined by the large East-West assymmetry observed during one spin period. Its characteristic length, L is of the order of a few gyroradii, ρ_i for $\sim$20 keV protons. In such a situation a compressional Alfven wave becomes coupled with a drift wave (e.g. Hasegawa, 1975, chapter 3). The instability exists provided that the inequality

$$n_h/N_c < (\beta/2)\,(L/\rho_i)^2 \tag{2}$$

where n_h is the energetic proton density and $\beta = \mu_o n_h m_p v_p^2/B_o^2$ is the 'β' of the hot plasma. Note that (2) is equivalent to

$$L\,\delta_{ic} < \rho_i^2 \tag{3}$$

where $\delta_{ic} = c/\omega_{pi}$ is the inertial length associated with the cold ion plasma density. From (3), one sees that the instability condition does not depend explicitely on the hot proton concentration but only on the hot proton temperature and spatial gradient.

Korth et al. (1983, 1984) have tried to interpret some of the events which they observed in the frame of this theory but they did not make a comparison

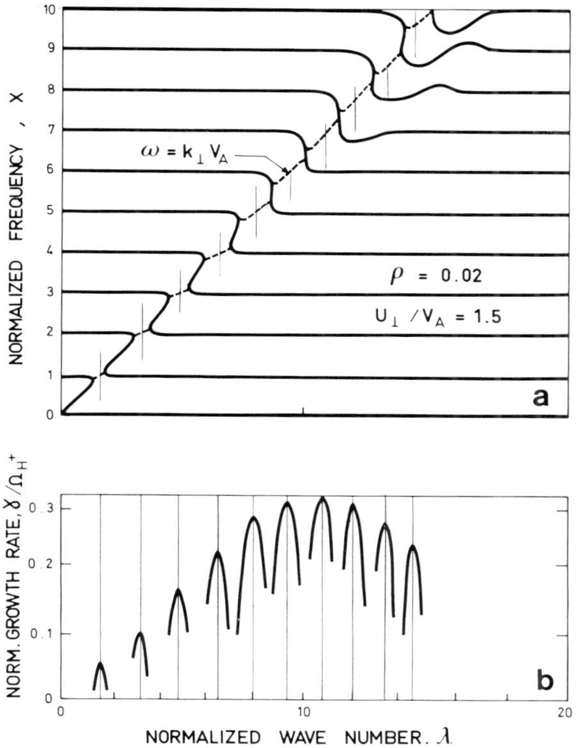

Figure 5. Real and imaginary parts of the dispersion curves for waves propagating in a direction perpendicular to the DC magnetic field in the presence of a ring distribution of hot protons. $\lambda = k_\perp U_\perp / \Omega_H$ where $U_\perp$ is the central velocity of the ring. ρ is the ratio between the ring and cold plasma densities (Perraut et al., 1982).

between the observed and theoretical spectra. Hasegawa's theory does not involve emission of F_H and its harmonics. Moreover for the two events which were studied there is no pitch-angle data for energies below 27 keV whereas in the first theory maximas are observed in the energy and pitch-angle distributions around 10 keV. It may well be that ring distributions around 10 keV are associated with the sharp spatial gradients observed above 27 keV. More experimental data are needed before the origin of these multi-harmonic emissions can be ascertained.

3.3. Conclusion

Magnetosonic waves are often observed in the equatorial magnetosphere, though their intensity is usually not large ($\leqslant$ 0.3 nT). Since they are generated near the proton gyrofrequency they could resonate with the thermal plasma, provided they have a small but finite parallel wave number. Therefore they could be at the origin of the heating of the thermal plasma of ionospheric origin to a few eV as is often observed (Décréau et al., 1982).

The diurnal occurrence of these waves has a shape which is similar to the diurnal occurrence of proton fluxes precipitating in the auroral zone and whose average energy lies in the 10 keV range (e.g. Galperin et al., 1976 and Perraut et al., 1982, figure 10). Equatorial interaction between these protons and magnetosonic waves may be one of the mechanisms by which these protons are precipitated. The numerical simulation of such an interaction could well be undertaken since ring distributions are easy to model (see f.i. Ashour-Abdalla et al., 1981 and Ashour-Abdalla and Okuda, 1984 who have studied a similar problem at low altitudes for electrons and electrostatic ion cyclotron waves). The simulation would give information about the time constants with which energetic protons are precipitated into the loss cone and with which cold particles are heated to temperatures higher than the ionospheric one.

Finally it must be noted that in this same frequency range multi-harmonic electrostatic ion cyclotron waves have been also observed (e.g. Perraut et al., 1978, figure 11), the harmonic being related with the He^+ gyrofrequency. A detailed theoretical study of these waves in multi-component plasma has been done by Barbosa (1982).

4. ION CYCLOTRON WAVES

4.1. Review of basic properties

At frequencies smaller than F_H large amplitude waves have been observed which are mainly polarized in the plane perpendicular to $\underline{B}_o$. Their frequency spectra often consists of two narrow bandwidth emissions separated by a gap which is situated between the He^+ gyrofrequency, F_{He} and a frequency $f_{co} < F_{He}$. This frequency has been identified as the cut-off frequency for the left-hand mode in a plasma containing a certain fraction of thermal He^+ ions. Indeed these waves are observed when thermal or suprathermal He^+ ions are present (e.g. Mauk et al., 1981; Young et al., 1981; Roux et al., 1982). Their propagation characteristics are well documented from both the experimental and theoretical stand points (Young et al., 1981; Perraut, 1982; Rauch and Roux, 1982). Their consequences with respect to the simultaneously observed heating of the He^+ ions to suprathermal energies are the subject of sections 4.3 to 4.6.

The presence of a large amount of O^+ ions may lead to similar effects at lower frequencies (see section 4.7).

4.2. Linear amplification

The mechanism by which these waves are generated is a cyclotron resonance between hot and anisotropic protons and left-handed (LH) ULF waves. A quantitative verification of this mechanism has been done by Roux et al. (1982), who introduced the energetic proton distribution function measured at the time of ICW emissions and who computed the theoretical convective growth rates for different events. For a given event this growth rate maximizes at the two frequencies (one below F_{He} and one above f_{co}) for which the wave spectra, for LH waves, also presents a maximum (figure 6). In one case, for which the theoretical computation gave a unique frequency range in which the growth rate was large, the observed spectra contained also a unique band of emission (their figure 9). Amplification is reinforced for waves generated at frequencies larger than the

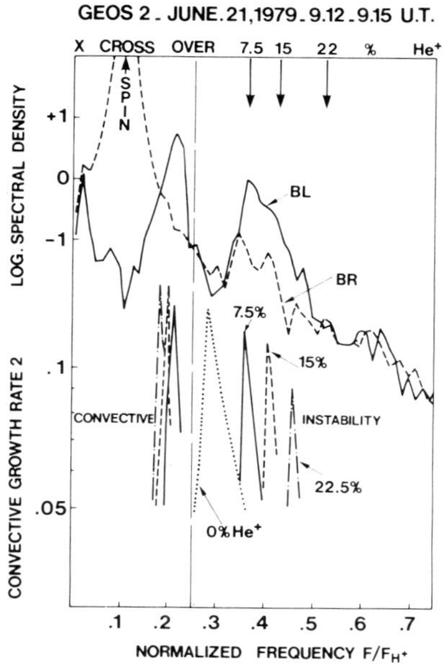

Figure 6. Comparison between the observed wave spectra and the integrated growth rate computed from the measured energetic proton distribution function. The agreement is good if the cold He^+ concentration is assumed to be equal to 7.5 % (Roux et al., 1982).

equatorial cross-over frequency since these waves are reflected at some distance from the equator because of the presence of He^+ ions (Rauch and Roux, 1982). As a consequence these waves undergo an enhanced convective amplification as they travel back and forth between the He^+-induced mirrors.

The variation of the spatial and temporal growth rates of ICW's in multi-component plasmas has been the subject of many theoretical papers for homogeneous media (e.g. Gomberoff and Cuperman, 1982; Gomberoff and Neira, 1983; Gendrin et al., 1984) or inhomogeneous media (Mauk, 1982a). In particular Gendrin et al. (1984) have shown that when the cold plasma was assumed to have a finite temperature waves could not be generated at frequencies just below F_{He}, thus preventing the generated waves to interact with the bulk of the He^+ distribution. For example figure 7 shows that, when the He^+ temperature is equal to 1.5 eV, waves have positive growth rates only for frequencies which resonate with He^+ ions that have a parallel energy that is larger than 3.5 eV (respectively 6.5 eV) when the He^+ concentration is 5 % (respectively 20 %).

But this effect cannot explain the width of the frequency gap which the emissions present below F_{He} and which, according to the synoptic studies independently performed by Fraser (1982) and Mauk (1982a), is equal to 0.05 F_H. Indeed in order that no emission be generated above 0.20 F_{H^+} the He^+ temperature must be larger than ~20 eV (respectively 150 eV) for a He^+ concentration equal to 20 % (respectively 5 %). This can be seen on figure 7.

Mauk (1982a, 1983) has extended the cold plasma linear analysis to inhomogeneous media by integrating the growth rate along lines of force between ± 15°

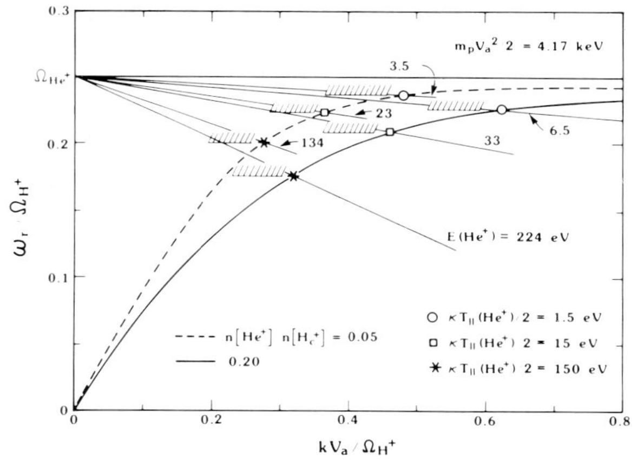

Figure 7. Variation of the marginal frequency and of the marginal parallel velocity of He^+ ions for three He^+ temperatures and two He^+ concentrations. The hatched areas correspond to damped waves. The marginal parallel velocity is always larger than the parallel temperature (Gendrin et al., 1984).

from the equator. He showed that even by taking inhomogeneity effects into account, the width of the gap below F_{He} cannot be explained. Consequently nonlinear effects must be invoked to explain both the gap and the He^+ heating (section 4.5).

4.3. Quasi linear diffusion

When waves are generated in the vicinity of the He^+ gyrofrequency ($\omega - \Omega_{He} \ll \Omega_{He}$), He^+ ions with a small parallel velocity can resonate with the wave since the gyroresonance equation, for $\underline{k} = \underline{k}_{/\!/}$

$$\omega - kv_{/\!/} = \Omega_{He} \tag{4}$$

can be verified for small values of $v_{/\!/}$ (i.e. $v_{/\!/} \ll V_a$ since $v_{/\!/} \ll \Omega_{He}/k$ which is of the order of V_a in most part of the frequency range). The quasi-linear (QL) diffusion in velocity space which is associated with this resonance has been studied by Gendrin and Roux (1980) who computed the shape of the diffusion curves and who showed that this process may lead to a large increase of He^+ kinetic energy, mainly in the direction perpendicular to $\underline{B}_o$. This He^+ energy increase is taken at the expense of the energy of the wave whose amplitude is itself maintened at the expense of the hot proton energy. One has in principle an efficient energy transfer mechanism between different particle species in a collisionless plasma (Gendrin, 1983b).

However, as explained in the previous section, this process may not be as efficient as expected since the waves cannot be generated at frequencies that involve the bulk of the He^+ distribution, and in fact they are not observed in the very vicinity of F_{He}. This problem was already raised by Gendrin (1983a). Notwithstanding this difficulty it is interesting to compute the maximum energy which can be reached by He^+ ions who have a sufficient initial parallel velocity to "catch" the frequency band where waves are generated. The computation is done as follows.

One assumes for simplification that the wave phase velocity is equal to the Alfven velocity, V_a, an approximation which is valid, at least for low values of the He^+ concentration, in most part of the frequency range where waves can be generated. Figure 8 gives a schematic representation of the interaction with the low frequency (LF) or the high frequency (HF) branch of the dispersion curve in both the ω, k plane and the $v_\perp$, $v_{//}$ plane. Waves are assumed to be generated between ω_{min} and ω_{max}. With our assumption, the resonant velocity is given by:

$$v_{//} = (1 - \Omega_{He}/\omega)\, V_a \tag{5}$$

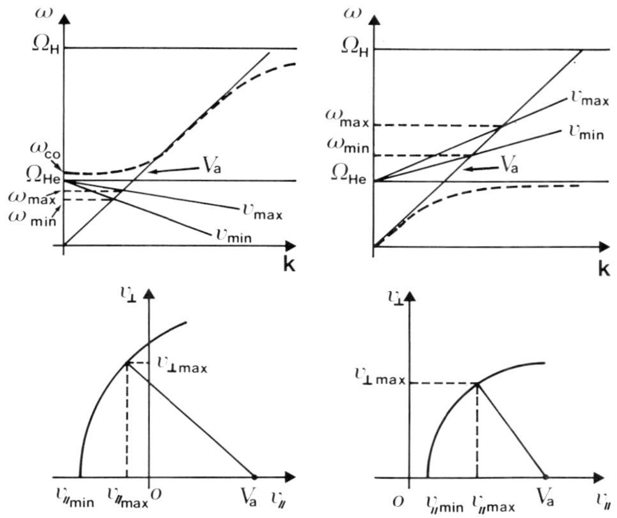

Figure 8. Quasi-linear diffusion of He^+ ions interacting with an ICW. The frequency range in which waves are assumed to be generated by hot protons is defined by ω_{min}, ω_{max}. Left: interaction in the LF range. Right: interaction in the HF range. The dashed lines represent the true dispersion curve which is not much different in the frequency range of interest from a straight line.

Assuming that the initial perpendicular velocity is zero, one easily gets, from the examination of the lower panels of figure 8:

$$v_{\perp max}^2 + (V_a - v_{//max})^2 = (V_a - v_{//min})^2 \tag{6}$$

Since $2E_{max}/m = v_{\perp max}^2 + v_{//max}^2$, one obtains

$$\frac{2E_{max}}{mV_a^2} = 1 + \left(\frac{\Omega_{He}}{\omega_{min}}\right)^2 - 2\left(\frac{\Omega_{He}}{\omega_{max}}\right) \tag{7}$$

which allows to compute the maximum energy which a He^+ ion can gain, knowing the frequency range of the emission. As an illustration, we may assume that ω_{min} and ω_{max} correspond to the frequency for which the linear growth rate is important (e.g. figure 2 in Gendrin et al., 1984). One gets the following results:

LF : $\omega_{min} = 2\Omega_{He}/5$, $\omega_{max} = 4\Omega_{He}/5$ $E_{max} \sim mV_a^2$

HF : $\omega_{min} = 6\Omega_{He}/5$, $\omega_{max} = 8\Omega_{He}/5$ $E_{max} \sim 0.25\ mV_a^2$

which show that heating is more efficient in the LF branch than in the HF branch. However, because of the 'mirror effect' which may occur in the HF branch, waves can be more intense in this frequency range and the time that is necessary to reach E_{max} would be shorter.

Note that the energy gain $\Delta E = E_{max} - E_{min}$ is then given by the simple expression:

$$\Delta E = mV_a . \Delta v_{//} \tag{8}$$

where $\Delta v_{//} = v_{//max} - v_{//min}$.

4.4. Associated generation of suprathermal electrons

By comparing the simultaneously observed ULF and ELF GEOS data Cornilleau-Wehrlin (1981) has shown that ELF emissions in the $\sim 0.1 - 0.5$ kHz range are often observed in conjunction with He^+-associated ICW's. These ELF emissions are electrostatic and their amplitude is modulated at the ICW frequency. By looking at the GEOS suprathermal plasma analyzer Norris et al. (1983) were able to demonstrate that during these modulated ELF emissions the suprathermal plasma distribution had a bump in the $\sim 10-50$ eV range and that this bump was mainly observed by the detector looking in the direction parallel to $\underline{B}_o$.

This effect has been interpreted theoretically by Roux et al. (1984) who showed that electrons can be trapped in the electric field of the wave. This is possible because some ICW's have a non negligible electric component parallel to $\underline{B}_o$ when the 'mirror effect' is operating (i.e. when the wave frequency is larger than the equatorial cross-over frequency). In such a case electrons can be trapped in the wave potential. Because of the increase of the Alfven velocity with latitude, the trapped electrons gain energy. The process works until the detrapping force ejects the electrons from the wave. This occurs after a few trapping lengths. The final energy which the electrons can reach is of the order of 20 eV for realistic values of the parallel electric field of the wave ($E \sim 10 \mu V.m^{-1}$). This is true for not too large plasma densities ($n_e \sim 20\ cm^{-3}$). When the cold plasma density is larger the Alfven velocity is smaller as is the electron final energy. It is worth noting in this respect that ULF-associated ELF emissions are not observed when the plasma density is larger than 25 cm^{-3} (Cornilleau-Wehrlin, 1981). After being detrapped the suprathermal electron beam is supposed to generate the ELF emission, but the detailed theory of this emission has not yet been worked out.

4.5. Non resonant acceleration

The energization of He^+ ions up to suprathermal energies in conjunction with the existence of ICW's is a well demonstrated experimental fact (e.g. Young et al., 1981). But in some occasions a modulation of 90° pitch-angle He^+ ions has been also observed, at a frequency equal to (or in the vicinity of) the wave frequency (Young et al., 1981; Mauk et al., 1981). In the case of GECS data (for which the energy resolution is poor) the hypothesis according to which this modulation is due to the linear plasma displacement which is associated with the magnetic field fluctuation of the wave cannot be ruled out (Mauk, private communication). But in the case of ATS-6 data, for which the energy resolution is good (~9 %) periodic impulses are clearly seen on suprathermal energy channels (e.g. Mauk et al., 1981, figure 4).

Mauk (1982b, 1983) has proposed that these wave-associated pulses could be the result of non linear phase bunching. Phase bunching (as opposed to phase trapping) occurs for non resonant particles with small initial velocities. In that case the electric force acting on the particle is smaller than the Laplace force so that there is almost no change in the parallel velocity and that the continuing action of the electric field can give rise to high perpendicular velocities. In phase bunching the phase of the particle positions along their gyration motion (with respect to the electric field) tends to be concentrated in a narrow range of angles as the interaction proceeds. Mauk (1982b, 1983) has made a computer simulation of this process (figure 9). It shows that after 10 He^+ gyroperiods the perpendicular energy of He^+ ions has been multiplied by 25. In principle there is no reason why this energy increase should not continue as time goes on (see below). The dimensionless parameters which have been used in this example are $EP = q\,\delta E\, r_o/E_o$ and $BP = \delta B/B_o$ where δE and δB are the electric and magnetic wave components, r_o and $E_o = m\, v_{\perp o}^2/2$ are the He^+ initial gyroradius and perpendicular energy. It is easy to show that

$$v_{\perp o}/v_\varphi = 2\, BP/EP \tag{9}$$

where v_φ is the wave phase velocity which is equal in this example to $0.2\,V_a$ since the wave frequency is equal to $0.24\,F_{He}$ ($v_\varphi = V_a\,(1-\omega/\Omega_{He})^{1/2}$).

However Mauk's (1982b) computations were done by neglecting the ion parallel displacement. Such an assumption is not allowed. Even when the He^+ initial parallel velocity, $v_{/\!/ o}$ is small with respect to v_φ, as soon as $v_\perp$ will increase, the $v_\perp \times B_o$ force will modify $v_{/\!/}$ and the phase relationship between $v_\perp$ and δE will be modified. Berchem and Gendrin (1984) has reformulated the problem and they have solved it both analytically and numerically. They have shown that this effect puts an upper limit to the maximum perpendicular velocity which the He^+ ion can reach (as opposed to an infinite increase) and that there is an optimum frequency for which the process is the most efficient ($\omega \sim 0.23\,\Omega_{He}$). The maximum velocity is given by

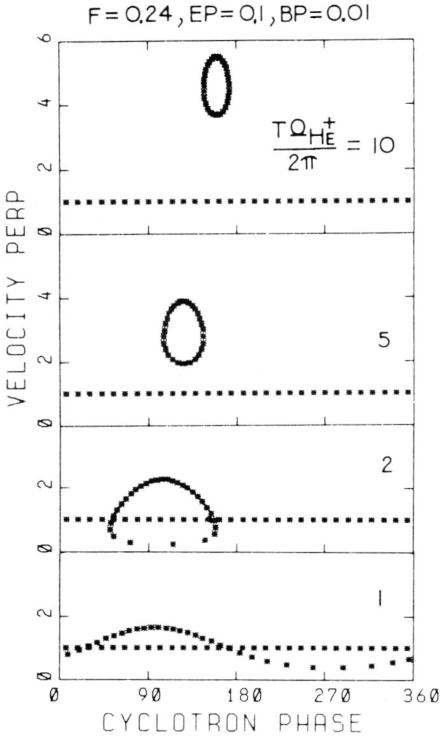

Figure 9. Non linear phase bunching of 32 He^+ ions in an ICW with a frequency $f = 0.24\ F_H$. See text for the definition of the parameters EP and BP (Mauk, 1982b).

$$v_{\perp max} \sim \frac{2\omega_1 v_\varphi}{\Omega_{He} - \omega} \qquad \text{when} \left(\frac{\Omega_{He} - \omega}{3\omega}\right)^3 \gg \left(\frac{\omega_1}{2\omega}\right)^2 \tag{10}$$

$$v_{\perp max} \sim 2\left(\frac{\omega_1}{\omega}\right)^{1/3} v_\varphi \qquad \text{when} \left(\frac{\Omega_{He} - \omega}{3\omega}\right)^3 \ll \left(\frac{\omega_1}{2\omega}\right)^2 \tag{11}$$

in which $\omega_1 = q\ \delta B/m_{He}$ is a measure of the wave field intensity. The first formula, which is valid for low values of δB and/or for frequencies far from the He^+ gyrofrequency corresponds to the linear phase ($v_{\perp max} \propto \delta B$). The second formula is valid for large values of δB and/or for frequencies in the vicinity of F_{He}. It corresponds to the truly non linear phase of the interaction ($v_{\perp max} \propto \delta B^{1/3}$).

4.6. Numerical simulation

Two attempts have been made to simulate the generation of ICW's in a multi-component plasma. Both attempts are based upon the use of an hybrid electromagnetic code in which all ions (H^+ cold, He^+ cold, H^+ hot) are treated by the particle-in-cell technique whereas electrons are treated like an inertialess fluid.

The first attempt (Tanaka and Goodrich, 1984) is mainly directed towards the study of such effects in the foreshock region of the magnetopause. The average energy of the hot protons is low ($\beta = (n_h/n_c)\ \kappa T/m V_a^2 \sim 0.1$) and their anisotropy is large ($T_\perp/T_{/\!/} \sim 20\text{-}60$), although in one run $T_\perp/T_{/\!/}$ was chosen equal to 3. Four types of heavy ions have been considered: He^+, He^{++}, D^+, T^+.

Because of the large value of the anisotropy and of the low value of the hot proton parallel velocity with respect to V_a the initial growth rate maximizes in the HF branch of the dispersion curve ($\omega/\Omega_H \sim 0.9$) and it is rather large ($\gamma/\Omega_H \sim 0.6$). Therefore the linear phase lasts only a few proton gyroperiod. After a time $t \sim 50\ \Omega_H^{-1}$ the hot proton anisotropy has decreased drastically ($T_\perp/T_{/\!/} \sim 2$), the Helium temperature has been greatly enhanced ($T_\perp/T_o \sim 20$), mainly in the perpendicular direction ($T_\perp/T_{/\!/} \sim 4$), and the central emission frequency has decreased ($\omega/\Omega_H \sim 0.4$). In the above formulas T_o is the average proton temperature

$$T_o = (n_c T_c + n_h T_h)/(n_c + n_h) \tag{12}$$

where $n_h/n_c = 0.1$ and $T_h/T_c \sim 2T_h/T_o \sim 10$. One interesting result of this study, apart from the demonstration of the existence of phase bunching, is that the final heavy ion temperature depends on the charge number, Z and on the mass, m_i, through the empirical formula

$$(T_\perp/T_o)_i \sim 40\ Z^2\ (m_p/m_i) \tag{13}$$

Since an important heating is observed in the simulation though the wave frequency is far from the heavy ion gyrofrequency, the authors conclude that a non resonant mechanism is at work.

The second attempt has been done by Omura et al. (1984) who have chosen plasma parameters that better correspond to magnetospheric conditions. Specific numerical values (instead of dimensionless ones) were selected. They are characteristic of dayside geostationary altitude. For cold species $\kappa T_\perp = \kappa T_{/\!/} = 1.7$ eV, $n(H_c^+) = 10\ cm^{-3}$, $n(He^+) = 2\ cm^{-3}$. For hot protons $\kappa T_\perp = 2\kappa T_{/\!/} = 34$ keV, $n(H_h^+) = 1.1\ cm^{-3}$. A large number of superparticles were introduced: 64 per grid point for each cold species, 512 per grid point for hot protons in order to reduce initial thermal fluctuations. There were 64 grid points in the system.

Because of the large initial parallel energy of the hot protons ($V_{/\!/}/V_a \sim 1.8$) the initial growth rate maximizes in the LF branch of the dispersion curve. Its values is small with respect to $\Omega_H (\gamma/\Omega_H \sim 0.02)$ since $n_h/n_c \sim 0.1$ is small. Consequently the linear growth rate can be observed on a rather long period of time ($\Omega_H t \sim 200$). Saturation occurs around $t \sim 300\ \Omega_H^{-1}$ and the wave field intensity at saturation, b is of the order of $0.06\ B_o$.

Before saturation the progressive increase of $v_\perp$ induces a variation of $v_{/\!/}$ through the $\underline{v}_\perp \times \underline{B}_o$ force. Since $v_\perp$ oscillates the associated oscillations in $v_{/\!/}$

leads to a spatial merging of particles which were initially phase-bunched. This corresponds to true heating. After a time $t \sim 600\,\Omega_H^{-1}$ the particle distribution characteristics do not vary much. Their final values at $t = 1\,200\,\Omega_H^{-1}$ are for H_c^+: $\kappa T_{/\!/} \sim 20$ eV, $\kappa T_\perp \sim 10$ eV; for He^+: $\kappa T_{/\!/} \sim 80$ eV, $\kappa T_\perp \sim 150$ eV; for H_h^+: $\kappa T_{/\!/} \sim 24$ keV, $\kappa T_\perp \sim 30$ keV. The loss of hot proton energy is responsible for the heating of cold species and for the energy which is contained in the wave field). The increase in hot proton parallel energy is responsible for the decrease of the wave frequency (from $\omega = 0.16\,\Omega_H$ at the beginning to $\omega/\Omega_H \sim 0.1$ at the end). Table 2 gives an evaluation of the energy budget obtained during this run. The slight difference between energy loss and energy gain can be attributed to numerical errors.

Table 2. Energy budget of an H^+-ICW-He^+ interaction

(Numerical values are taken from Omura et al's (1984) simulation results)
Total energy per species : $n\kappa\,(T_\perp + T_{/\!/}/2)$

HOT PROTONS		
$E_i = 34 + 8.5 = 42.5$ keV	$\Delta E = 0.7$ keV	$n\Delta E = 770$ eV.cm^{-3}
$E_f = 29.8 + 12 = 41.8$ keV	$n = 1.1$ cm^{-3}	
energy loss : 770 eV.cm^{-3}		
COLD PROTONS		
$E_i = 1.7 + 0.8 = 2.5$ eV	$\Delta E = 17.5$ eV	$n\Delta E = 175$ eV.cm^{-3}
$E_f = 10 + 10 = 20$ eV	$n = 10$ cm^{-3}	
COLD HELIUM		
$E_i = 1.7 + 0.8 = 2.5$ eV	$\Delta E = 187.5$ eV	$n\Delta E = 375$ eV.cm^{-3}
$E_f = 150 + 40 = 190$ eV	$n = 2$ cm^{-3}	
WAVE ENERGY		
$E_i = 0$	$b_f \sim 9$ nT	$E_f = 202$ eV.cm^{-3}
energy gain : 752 eV.cm^{-3}		
estimated computational error at the end of the run : ~ 50 eV.cm^{-3}		

It would be interesting to use such codes for studying the values of maximum He^+ energy, final wave amplitude, characteristic times, etc... for different initial plasma conditions. The detailed study of the plasma bunching process (see section 4.5) is less obvious since in these simulations waves with opposite wave numbers are simultaneously present.

4.7. Oxygen cyclotron waves

There is no reason why similar effects involving O^+ ions should not occur. Indeed Fraser and McPherron (1982) have reported that out of 78 ULF events detected onboard ATS-6 during a four-month period, 19 showed evidence of an associated O^+ spectral slot. Fraser (1982) computed the theoretical dispersion curve for a mixture of H^+, He^+ and O^+ whose relative concentrations where chosen to match the oxygen and helium spectral slot frequencies (his figure 9). The numbers he arrived at seem realistic ($n(O^+)/n_c = 0.03$; $n(He^+)/n_c = 0.05$).

But whereas the presence of the cold He^+ ions with non negligible concentrations (~ 5-20 %) is ascertained, that of thermal O^+ ions with a concentration larger than ~1 % is not. In fact this presence is difficult to explain and no model presently exists which includes O^+ transport (Young, 1983). On the contrary energetic O^+ ions are often present in large concentration (e.g. Young et al., 1982). If one assumes that they have some positive anisotropy they can contribute to wave growth for frequencies below the O^+ gyrofrequency. Therefore they could be at the origin of some Pc-2 waves. The possibility of such a process has been demonstrated by Inhester et al. (1984) for one event during which two ULF emissions were observed onboard GEOS-2: one above F_{He} and one below F_O. Simultaneous energy spectra and ion composition measurements show that the Pc-2 emission starts at the time at which energetic O^+ ions are seen. From the measurements the following plasma parameters were obtained: before the emission $n(H^+) = 2\ cm^{-3}$, $T(H^+) = 20$ keV, $n(O^+) = 0$; at the beginning of the emission $n(H^+) = 0.6\ cm^{-3}$, $T(H^+) = 20$ keV, $n(O^+) = 0.8\ cm^{-3}$, $T(O^+) = 6$ keV. By assuming an anisotropy for both species equal to 2 and a cold plasma density of the order of $4\ cm^{-3}$ the authors were able to compute growth rates which maximize at approximately the two observed frequencies.

There seems to be a difference between He^+-associated and O^+-associated ICW's. In the first case the cold heavy ion which is in sufficient concentration plays an important role in the generation mechanism either because of the lowering of the wave group velocity or because of the mirror-induced propagation effect (section 4.2). As a consequence of the emission the heavy ion is strongly energized. In the second case there are not enough cold heavy ions to produce the same effect. But the presence of an energetic component (whose origin is to be found elsewhere) may induce an emission at a frequency smaller than the O^+ gyrofrequency. The direct emission by hot protons in this very low frequency range is unlikely to occur because of the excessively large energy which is needed.

5. JOVIAN ULF EMISSIONS

5.1. Propagation characteristics

The Jovian magnetosphere does present a unique situation for studying other mechanisms that involve ULF waves. It is well known that the extended Io plasma torus ($6 < L < 15\ R_J$) is magnetically conjugated with regions of intense auroral activity. Simultaneously the torus is a region where energetic ions exhibit a dramatic decrease in phase space density, which could be accounted for only by strong pitch-angle diffusion which in turn can account for the observed intensity of auroral emissions (e.g. Gehrels and Stone, 1983).

Voyager-1 plasma wave observations have shown the existence of strong emissions ($E^2/\Delta f \sim 10^{-6} (V.m^{-1})^2.Hz^{-1}$) at frequencies below the equatorial proton gyrofrequency (~25 Hz) around L = 6 (Thorne and Scarf, 1984). However these high values are observed only during strong events and the average power at these frequencies is one or two orders of magnitude lower (Gurnett and Goertz, 1983;

Thorne and Scarf, 1984).

Since the Io torus is characterized by a large latitudinal variation of the ion composition along the field lines propagation of ULF waves is more complex there than in the Earth's vicinity. At low latitudes (< 10°) heavy ions (S^+, S^{++}, O^+, O^{++}) are the dominant species and the proton concentration is smaller than 10 %. At higher latitudes H^+ becomes the dominant ion (e.g. Bagenal and Sullivan, 1981; Tokar et al., 1982). Consequently the shape of the dispersion curve for ULF waves is much dependent on latitude. This is illustrated in figure 10 where the

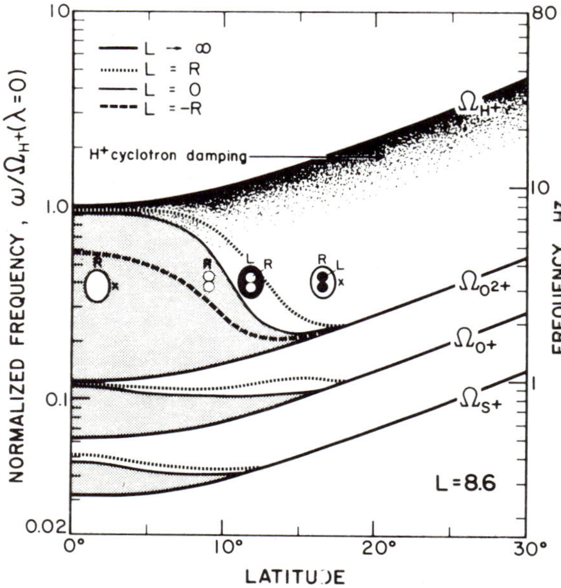

Figure 10. Changes in the ICW characteristic frequencies as a function of latitude along a field line in the outer Io torus (Thorne and Moses, 1983).

characteristic frequencies are displayed as a function of latitude along a field line (Thorne and Mose, 1983). These frequencies are the gyroresonances ($L \rightarrow \infty$), the cut-offs ($L = 0$), the cross-over ($L = R$) and the bi-ion hybrid resonance ($L = -R$). The hatched regions in figure 10 are those where LH waves cannot propagate. Therefore any wave which is observed in the equatorial region and the frequency of which is above the heavy ion gyrofrequency and below the associated cut-off frequency must be an RH wave. The same wave could be observed at higher latitudes but its polarization would be reversed.

5.2. Generation mechanisms

These propagation characteristics which are specific of the Io torus region has led to two different interpretations concerning the origin of the ULF waves that have been observed onboard Voyager-1.

According to one hypothesis (Gurnett and Goertz, 1983) the waves are generated in the equatorial region in the R-mode. The generation mechanism is either a normal gyroresonant interaction with energetic electrons that must be very energetic in order to generate such low frequencies or an anomalous resonance with positive ions. In the first case electrons are loosing energy and pitch-angle and thus they are precipitated. In the second case the ions must be energetic ($v_{/\!/} \gtrsim 3 V_a$). Their angular distribution must be such that $T_{/\!/} > T_{\perp}$ and their pitch-angle increases during the interaction (e.g. Gendrin, 1981), which means that they cannot precipitate.

When propagating to higher latitudes the wave is converted into a L-H cyclotron wave which can interact with either protons or heavier ions. But in both cases it can be shown that the pitch-angle of the interacting particles either increases or does not decrease significantly (Gendrin and Roux, 1980; Gendrin, 1981). Apparently it is impossible to explain with such a mechanism the precipitation of energetic protons or heavy ions.

The second hypothesis is exactly opposite to the previous one. In this hypothesis waves are assumed to be generated off the equator in the LH mode (Thorne and Moses, 1983). The generation mechanism is a normal gyroresonance interaction with protons or heavy ions that leads to a pitch-angle decrease and therefore to precipitation. Recently Thorne and Moses (1984) made a quantitative analysis of the associated amplification by including the most important resonances ($m = 0, \pm 1, \pm 2$ in the resonance equation $\omega - k_{/\!/} v_{/\!/} = m \Omega_i$) that exist when the wave k vector is not strictly parallel to $\underline{B}_o$. They showed that L-mode generation off the equator is more efficient than R-mode generation at the equator.

An argument in favor of this interpretation is that when propagating toward the equator the wave (whose polarization has been reversed) can again interact with heavy ions. Whatever the anisotropy of these ions is (provided that $T_{/\!/}$ is not very large with respect to $T_{\perp}$) their pitch-angle will decrease, thus leading to heavy ion precipitation (see Gendrin, 1981, figure 3b).

More measurements of the electromagnetic wave field (principally of its magnetic component) and of particle distribution and composition are needed at different latitudes in the vicinity of the Io torus in order to ascertain one or the other hypotheses. Besides a synoptic study of ULF turbulence in the Jovian inner magnetosphere would be of interest to ascertain the role that stochastic diffusion could play in the acceleration of pick up heavy ions which are observed with magnetic moments of the order of 10^4-10^5 MeV/(nucl. G) (Barbosa et al., 1984).

6. CONCLUSION

The present review demonstrates the complexity of phenemena involving ULF waves in planetary magnetosphere. Important progress has been made in the interpretation of these phenomena. But more measurements are still needed in both the Earth's and Jupiter's environments. Important problems related to wave generation, quasi-linear diffusion or non-linear effects are still to be theoretically studied. Some of these problems can be attacked with the help of numerical simulation techniques.

ACKNOWLEDGEMENTS. The help of Maryse Le Floch and Michelle Pennec for carefully preparing this camera-ready copy is gratefully acknowledged.

REFERENCES

Ashour-Abdalla, M., H. Okuda, and C. Z. Cheng, 'Acceleration of heavy ions on auroral field lines', Geophys. Res. Lett., 8, 795-798, 1981.

Ashour-Abdalla, M., and H. Okuda, 'Turbulent heating of heavy ions on auroral field lines', J. Geophys. Res., 89, 2235-2250, 1984.

Bagenal, F., and J. D. Sullivan, 'Direct plasma measurements in the Io torus and inner magnetosphere of Jupiter', J. Geophys. Res., 86, 8447-8466, 1981.

Barbosa, D. D., 'Multi-ion resonances in finite temperature plasma', Astrophys. J., 254, 376-390, 1982.

Barbosa, D. D., A. Eviatar, and G. L. Siscoe, 'On the acceleration of energetic ions in Jupiter's magnetosphere', J. Geophys. Res., 89, 3789-3800, 1984.

Berchem, J., and R. Gendrin, 'Non resonant interaction of heavy ions with electromagnetic ion cyclotron waves', submitted to J. Geophys. Res., 1984.

Chen, L., and A. Hasegawa, 'A theory of long-period magnetic pulsations, 1, Steady state excitation of field line resonance', J. Geophys. Res., 79, 1024-1032, 1974.

Cornilleau-Wehrlin, N., 'A new ULF-modulated electrostatic wave detected in the extremely low frequency range onboard GEOS', J. Geophys. Res., 86, 1365-1373, 1981.

Cummings, W. D., R. J. O'Sullivan, and P. J. Coleman, Jr., 'Standing Alfven waves in the magnetosphere', J. Geophys. Res., 74, 778-793, 1969.

Décréau, P. M. E., C. Béghin, and M. Parrot, 'Global characteristics of the cold plasma in the equatorial plasmapause region as deduced from the GEOS 1 mutual impedance probe', J. Geophys. Res., 87, 695-712, 1982.

Fraser, B. J., 'Pc 1-2 observations of heavy ion effects by synchronous satellite ATS-6', Planet. Space Sci., 30, 1229-1238, 1982.

Fraser, B. J., and R. L. McPherron, 'Pc 1-2 magnetic pulsation spectra and heavy ion effects at synchronous orbit: ATS 6 results', J. Geophys. Res., 87, 4560-4566, 1982.

Galperin, Y. I., N. V. Jorjio, R. A. Kovrazhkin, F. Cambou, J. A. Sauvaud, and J. Crasnier, 'On the origin of auroral protons at the dayside auroral oval', Ann. Geohys., 32, 117-129, 1976.

Gehrels, N., and E. C. Stone, 'Energetic oxygen and sulfur ions in the Jovian magnetosphere and their contribution to the auroral excitation', J. Geophys. Res., 88, 5537-5550, 1983.

Gendrin, R., 'General relationships between wave amplification and particle diffusion in a magnetoplasma', Rev. Geophys. Space Phys., 19, 171-184, 1981.

Gendrin, R., 'Effects of heavy ions on microplasma physics in the magnetosphere', in High Latitude Plasma Physics, B. Hultqvist and T. Hagfors eds., Plenum Press, New-York, pp. 415-436, 1983a.

Gendrin, R., 'Wave particle interactions as an energy transfer mechanism between different particle species', Space Sci. Rev., 34, 271-287, 1983b.

Gendrin, R., and A. Roux, 'Energization of helium ions by proton-induced electromagnetic waves', J. Geophys. Res., 85, 4577-4586, 1980.

Gendrin, R., M. Ashour-Abdalla, Y. Omura, and K. Quest, 'Linear analysis of ion cyclotron interaction in a multicomponent plasma', J. Geophys. Res., 89, 9119-9124, 1984.

Gomberoff, L., and S. Cuperman, 'Combined effect of cold H^+ and He^+ ions on the proton cyclotron electromagnetic instability', J. Geophys. Res., 87, 95-100, 1982.

Gomberoff, L., and R. Neira, 'Convective growth rate of ion cyclotron waves in a H^+-He^+ and H^+-He^+-O^+ plasma', J. Geophys. Res., 88, 2170-2174, 1983.
Greenwald, R. A., and A. D. M. Walker, 'Energetics of long period resonant hydromagnetic waves', Geophys. Res. Lett., 7, 745-748, 1980.
Gurnett, D. A., and C. K. Goertz, 'Ion cyclotron waves in the Io plasma torus: polarization reversal of whistler mode noise', Geophys. Res., 10, 587-590, 1983.
Hasegawa, A., 'Drift mirror instability in the magnetosphere', Phys. Fluids, 12, 2642-2650, 1969.
Hughes, W. J., R. L. McPherron, and J. N. Barfield, 'Geomagnetic pulsations observed simultaneously on three geostationary satellites', J. Geophys. Res., 83, 1109-1116, 1978.
Hughes, W. J., and R. J. L. Grard, 'A second harmonic geomagnetic field line resonance at the inner edge of the plasma sheet: GEOS 1, ISEE 1, and ISEE 2 observations', J. Geophys. Res., 89, 2755-2764, 1984.
Inhester, B., U. Wedeken, A. Korth, S. Perraut, and M. Stockholm, 'Ground satellite coordinated study of the April 5, 1979 events: observation of O^+ cyclotron waves', J. Geophys., 55, 134-141, 1984.
Junginger, H., O. H. Bauer, G. Haerendel, F. Melzner, B. Higel, and E. Amata, 'Plasma drift measurements with the electron beam experiment on GEOS-2 during long period pulsations on April 7, 1979', Geophys. Res. Lett., 10, 667-670, 1983.
Junginger, H., and W. Baumjohann, 'Resonant harmonic Alfven waves in the magnetosphere: a case study', J. Geophys. Res., 89, 10757-10762, 1984.
Junginger, H., G. Geiger, G. Haerendel, F. Melzner, E. Amata, and B. Higel, 'A statistical study of dayside magnetospheric electric field fluctuations with periods between 150 and 600 s', J. Geophys. Res., 89, 5495-5505, 1984.
Kivelson, M. G., J. Etcheto, and J. G. Trotignon, 'Global compressional oscillations of the terrestrial magnetosphere : the evidence and a model', J. Geophys. Res., 89, 9851-9856, 1984.
Korth, A., G. Kremser, A. Roux, S. Perraut, J. A. Sauvaud, J. M. Bosqued, A. Pedersen, and B. Aparicio, 'Drift boundaries and ULF wave generation near noon at geostationary orbit', Geophys. Res. Lett., 10, 639-642, 1983.
Korth, A., G. Kremser, S. Perraut, and A. Roux, 'Interaction of particles with ion cyclotron waves and magnetosonic waves. Observations from GEOS 1 and GEOS 2', Planet. Space Sci., 32, 1393-1406, 1984.
Kremser, G., A. Korth, J. A. Fejer, B. Wilken, A. V. Gurevich, E. Amata, 'Observations of quasi-periodic flux variations of energetic ions and electrons associated with Pc 5 geomagnetic pulsations', J. Geophys. Res., 86, 3345-3356, 1981.
Lin, C. S., and G. K. Parks, 'Modulation of energetic particle fluxes by a mixed mode of transverse and compressional waves', J. Geophys. Res., 87, 5102-5108, 1982.
Mauk, B. H., 'Helium resonance and dispersion effects on geostationary Alfven/ion cyclotron waves', J. Geophys. Res., 87, 9107-9119, 1982a.
Mauk, B. H., 'Electromagnetic wave energization of heavy ions by the electric 'phase-bunching' process', Geophys. Res. Lett., 9, 1163-1166, 1982b.
Mauk, B. H., 'Frequency gap formation in electromagnetic cyclotron wave distributions', Geophys. Res. Lett., 8, 635-638, 1983.
Mauk, B. H., C. E. McIlwain, and R. L. McPherron, 'Helium cyclotron resonance within the Earth's magnetosphere', Geophys. Res. Lett., 8, 103-106, 1981.

Norris, A. J., J. F. E. Johnson, J. J. Sojka, G. L. Wrenn, N. Cornilleau-Wehrlin, S. Perraut, and A. Roux, 'Experimental evidence for the acceleration of thermal electrons by ion cyclotron waves in the magnetosphere', J. Geophys. Res., 88, 889-898, 1983.

Omura, Y., M. Ashour-Abdally, K. Quest, and R. Gendrin, 'Heating of thermal Helium in the equatorial magnetosphere: a simulation study', submitted to J. Geophys. Res., 1984.

Perraut, S., 'Wave-particle interactions in the ULF range: GEOS-1 and -2 results', Planet. Space Sci., 30, 1219-1227, 1982.

Perraut, S., R. Gendrin, P. Robert, A. Roux, C. de Villedary, and D. Jones, 'ULF waves observed with magnetic and electric sensors on GEOS-1', Space Sci. Rev., 22, 347-369, 1978.

Perraut, S., A. Roux, P. Robert, R. Gendrin, J. A. Sauvaud, J. M. Bosqued, G. Kremser, and A. Korth, 'A systematic study of ULF waves above F_{H^+} from GEOS 1 and 2 measurements and their relationships with proton ring distributions', J. Geophys. Res., 87, 6219-6236, 1982.

Perraut, S., R. Gendrin, A. Roux, and C. de Villedary, 'Ion cyclotron waves: direct comparison between ground-based measurements and observations in the source region', J. Geophys. Res., 89, 195-202, 1984.

Rauch, J. L., and A. Roux, 'Ray tracing of ULF waves in a multicomponent magnetospheric plasma: consequences for the generation mechanism of ion cyclotron waves', J. Geophys. Res., 87, 8191-8198, 1982.

Roux, A., S. Perraut, J. L. Rauch, C. de Villedary, G. Kremser, A. Korth, and D. T. Young, 'Wave-particle interactions near Ω_{He^+} observed onboard GEOS 1 and 2, 1. Generation of ion cyclotron waves and heating of He^+ ions', J. Geophys. Res., 87, 8174-8190, 1982.

Roux, A., N. Cornilleau-Wehrlin, and J. L. Rauch, 'Acceleration of thermal electrons by ICW's propagating in a multicomponent magnetospheric plasma', J. Geophys. Res., 89, 2267-2273, 1984.

Samson, J. C., and G. Rostoker, 'Latitude-dependent characteristics of high-latitude Pc 4 and Pc 5 micropulsations', J. Geophys. Res., 77, 6133-6144, 1972.

Singer, H. J., W. J. Hughes, and C. T. Russell, 'Standing hydromagnetic waves observed by ISEE 1 and 2: Radial extent and harmonic', J. Geophys. Res., 87, 3519-3529, 1982.

Southwood, D. J., 'Some features of field line resonances in the magnetosphere', Planet. Space Sci., 22, 483-491, 1974.

Southwood, D. J., 'Low frequency pulsations generation by energetic particle', in ULF Pulsations in the Magnetosphere, D. J. Southwood ed., D. Reidel, pp. 75-88, 1981.

Southwood, D. J., and W. J. Hughes, 'Theory of hydromagnetic waves in the magnetosphere', Space Sci. Rev., 35, 301-366, 1983.

Su, S.-Y., R. L. McPherron, A. Konradi, and T. A. Friz, 'Observations of ULF oscillations in the ion fluxes at small pitch-angle with ATS', J. Geophys. Res., 85, 515-522, 1980.

S-300 Experimenters, 'Measurements of electric and magnetic wave fields and of cold plasma parameters onboard GEOS-1, Preliminary results', Planet. Space Sci., 27, 317-339, 1979.

Takahashi, K., and R. L. McPherron, 'Harmonic structure of Pc 3-4 pulsations', J. Geophys. Res., 87, 1504-1516, 1982.

Takahashi, K., R. L. McPherron, and W. J. Hughes, 'Multispacecraft observations of the harmonic structure of Pc 3-4 magnetic pulsations', J. Geophys. Res., 89, 6758-6774, 1984.

Tanaka, M., and C. C. Goodrich, 'Simulations of heavy ion heating by electromagnetic ion cyclotron waves induced by proton temperature anisotropy', submitted to J. Geophys. Res., 1984.

Thorne, R. M., and J. Moses, 'Electromagnetic ion-cyclotron instability in the multi-ion Jovian magnetosphere', Geophys. Res. Lett., 10, 631-634, 1983.

Thorne, R. M., and J. Moses, 'Resonant instability near the two-ion cross-over frequency in the Io plasma torus', submitted to J. Geophys. Res., 1984.

Thorne, R. M., and F. L. Scarf, 'Voyager 1 evidence for ion-cyclotron instability in the vicinity of the Io plasma torus', Geophys. Res. Lett., 11, 263-266, 1984.

Tokar, R. L., D. A. Gurnett, F. Bagenal, and R. R. Shaw, 'Light ion concentrations in Jupiter's inner magnetosphere', J. Geophys. Res., 87, 2241-2245, 1982.

Tonegawa, Y., H. Fukunishi, T. Hirasawa, R. L. McPherron, T. Sakurai, and Y. Kato, 'Spectral characteristics of Pc 3 and Pc 4/5 magnetic pulsation bands observed near L = 6', J. Geophys. Res., 89, 9720-9730, 1984.

Villain, J. P., 'Characteristics of Pc 5 micropulsations as determined with the STARE experiment', J. Geophys. Res., 87, 129-137, 1982.

Walker, A. D. M., R. A. Greenwald, A. Korth, and G. Kremser, 'STARE and GEOS 2 observations of a storm time Pc 5 ULF pulsation', J. Geophys. Res., 87, 9135-9146, 1982.

Walker, A. D. M., H. Junginger, and O. H. Bauer, 'GEOS 2 plasma drift density measurements associated with a storm time Pc 5 pulsation', Geophys. Res. Lett., 10, 757-760, 1983.

Wedeken, U., B. Inhester, A. Korth, K.-H. Glassmeier, R. Gendrin, L. J. Lanzerotti, H. Gough, C. A. Green, E. Amata, A. Pedersen, and G. Rostoker, 'Ground-satellite coordinated study of the April 5, 1979 events: flux variations of energetic particles and associated magnetic pulsations', J. Geophys., 55, 120-133, 1984.

Young, D. T., 'Heavy ion plasma in the outer magnetosphere', J. Geophys., 52, 167-175, 1983.

Young, D. T., S. Perraut, A. Roux, C. de Villedary, R. Gendrin, A. Korth, G. Kremser, and D. Jones, 'Wave-particle interactions near Ω_{He^+} observed on GEOS 1 and 2, 1. Propagation of ion cyclotron waves in He^+-rich plasma', J. Geophys. Res., 86, 6755-6772, 1981.

Young, D. T., H. Balsiger, and J. Geiss, 'Correlations of magnetospheric ion composition with geomagnetic and solar activity', J. Geophys. Res., 87, 9077-9096, 1982.

Spectral Computation of Tandem Mirror Equilibria with Finite Larmor Radius Effects

B.McNamara, L.D. Pearlstein
Lawrence Livermore National Laboratory,
P.O. Box 5511,
Livermore,
California, 94550.

ABSTRACT. A direct iterative method of solving for Tandem equilibria by moving magnetic field lines in a manner to satisfy the linearized equilibrium equations converges much more rapidly than standard relaxation techniques, typically in under a fifty iterations. At the highest β's the number of iterations increase, but is still far less than other methods. In quadrupole tandem mirror equilibrium, octupole and higher distortions of the flux surfaces are important which forces us to abandon finite differences in the angle-like flux variable and resort to a spectral decomposition to solve the equilibrium equations. We display equilibria at the high β expected for MFTF-B and show how Finite Larmor Radius (FLR) effects strongly suppress these azimuthal distortions.

1. INTRODUCTION

This paper on tandem mirror equilibria was presented at the school to outline recent developments in equilibrium theory and to show the need for a spectral representation of these numerical solutions. To put the work in context we start with a brief overview of the theory of plasma equilibrium and comment on its relevance for the magnetosphere.

On average the solar wind may be regarded as being in a non-stationary (flowing) equilibrium up to the earth's bow-shock. Behind this shock the plasma flow speeds are much less than the Alfven speed and we have a static equilibrium maintained by slow diffusive processes. Towards the magneto-tail and beyond,the plasma flow must match again with the solar wind and there is no reason to expect a stationary state as the tail may flap. The space and time scales of features in the magnetosphere vary greatly and one would not expect a single model to encompass all these possibilities.

Space Science Reviews **42** (1985) 541–551. 0038–6308/85.15

Research in Magnetic Fusion has been directed towards creating plasma equilibria, mainly static, and each of the models which have been developed[1] have a region of application in space plasma studies .

At the microscopic level equilibrium is established if the particle distributions are functions only of constants of the motion in the self-consistent electric and magnetic fields. At high β the magnetic fields may be determined from the guiding centre fluid equations, $J \times B = \nabla \cdot P, \nabla \times B = J, \nabla \cdot B = 0$. The Tandem Mirror uses axial electric fields for axial plasma confinement but these do not contribute to the total pressure balance [2]. Similar electric fields exist in the magnetosphere to confine electrons in flux tubes which reach the earth's atmosphere. Electric fields perpendicular to the magnetic field cause the plasma to drift at velocity $E \times B/\mathrm{B}^2$ which must therefore be a rotation to have a confined equilibrium. Some work has been done on rapidly rotating equilibria and their dynamical stability [3].

Static axisymetric equilibria with scalar pressure satisfy the Grad-Shafranov equation for the poloidal flux, Ψ

$$\triangle^* \Psi = -p'(\Psi) - r^2 F F'(\Psi)$$

The two apparently free functions, p and F, should be determined from the integral constants of the MHD equations, entropy and magnetic helicity in each flux tube [4]. The same considerations would apply to the magnetosphere as it is compressed or dilated by adiabatic changes in the solar wind. Other changes in an equilibrium induced by plasma transport and heating are described by coupling the equilibrium equations on the Alfven time scales to the transport equations for the integral constants of the ideal equations. This is standard practice for axisymetric devices like the Tokamak and is being extended to 3D configurations like the Stellarator [5].

In three dimensions the guiding center fluid equations can be arranged to describe plasma magnetisations due to Larmor motion of the particles, drift currents, and currents driven by external sources [6]. This gives an algorithm for finding equilibria from a set of field-line integrals and the solution of a 3D Poisson equation. The connection with particle distributions is made through the concept of omnigenity: When the plasma-field configuration is sufficiently symmetric all the particles on a given field line will drift on the same drift-surface, ψ. This means that the longitudinal invariant, J, depends only on ψ, the particle energy, H, and the magnetic moment, μ, so the particle distribution is $f = f(H, \mu, \psi)$ and the pressure tensor becomes $\underline{\mathbf{P}} = \underline{\mathbf{P}}(\mathrm{B}, \psi)$. The magnetic configuration will determine the drift surfaces for most particles in a fusion device - the magnetic surfaces in a tokamak. Perpendicular electric fields, arising from the more rapid transport of the non-omnigenic particles, will tend to "anneal" the drift surfaces of the these particles towards the average surface.

Avoiding the details of the underlying physics of an equilibrium, the steepest descents method [7] is simple to implement for any 3D system and moves each fluid element in the direction of the net force on the element. Pressure balance is rapidly established, but the slow, shear-Alfven waves take 1000's of iterations to eradicate. The method also avoids solving a 3D elliptic equation if there is no vacuum region in the problem.

The rest of this paper discusses a direct method for finding 3D tensor-pressure equilibria in the paraxial approximation, which reduces the problem to 2D X 1D and a sequence of 2D elliptic equations to be solved.

2. TANDEM MIRROR EQUILIBRIA

In general, tandem mirrors are characterized by the geometric description "long-thin". By this we mean that field lines are nearly parallel to a straight magnetic axis. Consequently curvatures are weak, perpendicular components of the magnetic field are small compared with the main $B\hat{z}$ field, radial scale lengths are short compared with axial scale lengths and we have as an expansion parameter($\lambda = R/L$). Ordering with respect to this smallness parameter is generally called the paraxial or long-thin aproximation and a consistent equilibrium can then be described through $O(\lambda^2)$. A key advantage is that we need only solve a 2D elliptic equation for one scalar function on each z-plane and never have to invert a 3D operator. In this limit we also find that FLR effects, because they exist throughout the central cell, are competitive with these weak MHD forces and thus contribute to both equilibrium and stability of quadrupole tandem mirrors [8]. These terms enter as an additive correction to the main $B\hat{z}$ field of the same order as the curvature.

The best representation of the plasma boundary is achieved by using the flux coordinates for the magnetic field, $B = \nabla\psi \times \nabla\theta$, as the independent variables. Then, to take advantage of the standard symetry of quadrupole tandem mirrors we need only compute in an octant since $x(\psi, \theta, z) = y(\psi, \pi/2 - \theta, -z)$, where (x, y, z) are the coordinates of a field line.

Now, there are many ways to solve these nonlinear equations. We have found that a direct iterative procedure [9], which closely follows the β-expanded analytic solution to the equilibrium [10], converges rapidly, accurately and robustly. This method is to be contrasted with the commonly used relaxation techniques for 3-dimensional problems which take advantage of the property that removing energy from the system forces the equations to relax to equilibrium [11]. An important limitation of these techniques is that the onset and saturation of the mean octupole distortion of the flux surfaces [12] evolves very slowly. In our method this appears imediately from a constraint and would lead to the analytic equilibrium in a single step at low β.

Another property of the numerical procedure is that all equations are spectrally decomposed in the angle-like flux variable. Generally, it is found that the greater accuracy of such procedures outweighs the added computational overhead. In our case the strong ellipticity and high β's ,which generate substantial angular distortions in the flux surfaces, and 4th derivatives introduced by FLR, force us to use the spectral method. We are then able to use the additional information available to better evaluate the convergence properties of a numerical equilibrium.

The first section displays the equations ,in the form used, with further comment on their physical meaning and how they are solved. The next section expands on some of the numerical techniques which were necessary. We end with a comparison of MFTF-B equilibria with a range of values of FLR at the nominal operating point. Without FLR they would violate the local flute-stability criteria over a portion of the plasma. Not only is this a stability violation in the MHD limit, it also delineates a breakdown in the numerical procedure and the existence of a likely bifurcation point in the equilibrium[10]. It should be emphasized that FLR moderates this constraint. With minimal FLR values the octupole distortions of the center-cell are considerable, whereas these distortions are almost totally eliminated with FLR values of the expected magnitude. A further point that should be made, is that in these comparisons we have used the "special equilibrium", $\omega_{*i} = -2\omega_{E\times B}$.

3. LONG-THIN EQUILIBRIUM EQUATIONS WITH FLR.

The equations determining equilibrium, in the MHD limit, have been derived independently by Newcomb[2] and Strauss[11]. A paraxial expansion in λ was used to separate the space and time scales in the problem. In zeroth order in λ they obtain perpendicular pressure balance, to first order they have parallel pressure balance, and second order gives the parallel-current equation. It is only at this level that FLR contributes [13]. In general the plasma pressures, densities, and potentials could be functions of both flux variables and $B = |B|$. Here we limit ourselves to isorrhopic equilibria that are separable functions of ψ and B

$$\underline{\mathbf{P}} = \omega(\psi)\hat{\underline{\mathbf{P}}}(\mathrm{B}) \tag{1}$$

where ω is a radial profile weighting and $\hat{\underline{\mathbf{P}}}$ must always satisfy the local mirror-mode and firehose restrictions. These MHD equilibria will be valid on the drift time scales if they are omnigenous [6] , that is, all particles on a ψ-surface drift only on that surface, regardless of pitch angle. Paraxial tandem mirrors are omnigenous to leading order because the $E \times B$ drift and, for $\beta \gg \lambda^2$, the ∇B drift dominate the curvature drifts.

The plasma satisfies the radial pressure balance equation, which is $O(\lambda^0)$,

$$p_\perp + \frac{B^2}{2} = \frac{1}{2}B^2_{vac}(z) \tag{2}$$

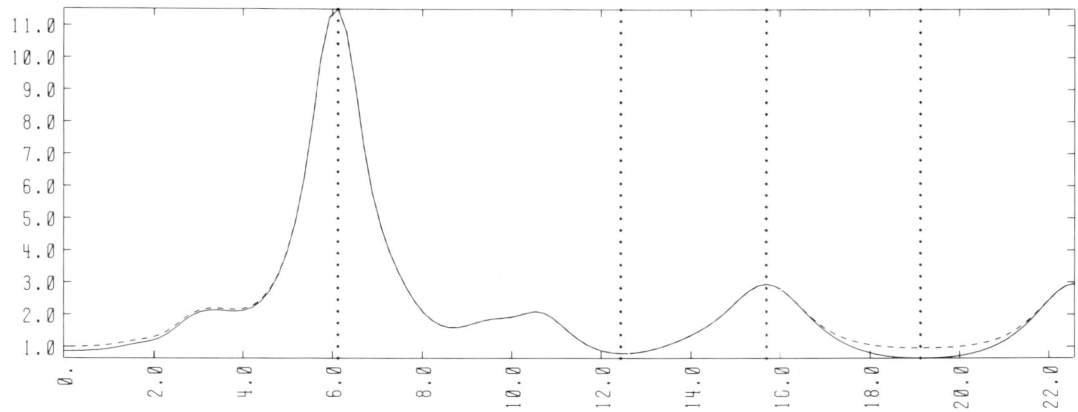

Figure 1: Axial magnetic field profiles in MFTF-B. These are B_{vac} (dotted), and $B(\psi = 0, z)$ at $\beta_{center-cell} = .25$, $\beta_{plug} = .55$. The axis is stretched to put fewer points in the center-cell. Cardinal planes are indicated for later plots.

where B_{vac} is the axial vacuum magnetic field. This is readily solved for $B(\psi, z)$ which is shown in Fig. (1). The next time scale, $O(\lambda)$, is eliminated by the relation, $\partial(p_{\parallel}/B)/\partial B = -p_{\perp}/B^2$. To determine the flux line geometry, $x_{\perp}(\psi, \theta, z)$, we first must satisfy the Jacobian relation

$$\frac{1}{B(\psi, z)} = [x, y] \equiv x_{\psi} y_{\theta} - x_{\theta} y_{\psi} \tag{3}$$

Since there are two unknowns (x, y) in the above equation, we need an additional relation. If the system were axisymetric this would be the extra condition. With quadrupole symetry we need the additional relation, the parallel current equation. To obtain this last condition we could solve $J \times B = \nabla \cdot \underline{P}$ to $O(\lambda^2)$. However because the FLR terms contribute additively[14] it is most convenient to derive this relation from the energy principle[2,13]. We construct the potential energy

$$W = \frac{1}{2} \int d\psi d\theta \frac{dz}{B} \left(Q x_z^2 + Y x_\theta^2 \right) \tag{4}$$

where $Q = B^2 + p_{\perp} - p_{\parallel}$ and $Y = -\rho\,(\omega_{*i} + \omega_{E\times B})\,(\omega_{E\times B} + \omega_{\nabla B})$ are the MHD and FLR stresses and both must be positive for well posed equilibrium. This simple form of Y is for an isothermal Maxwellian plasma.

The first variation of W, subject to the Jacobian constraint, yields the final equation to be solved, though in fact we use the line-integrated form

$$H \frac{Q}{B^2} \equiv \frac{iQ}{B^2} - \int_z^L \frac{dz}{B} \left(\frac{\partial Q}{\partial \psi} I\!O - \frac{\partial Y}{\partial \psi} I\!O_F - \frac{Y}{B^2} \frac{\partial i_F}{\partial \theta} \right) + \frac{1}{2} F(\psi, \theta) = 0 \tag{5}$$

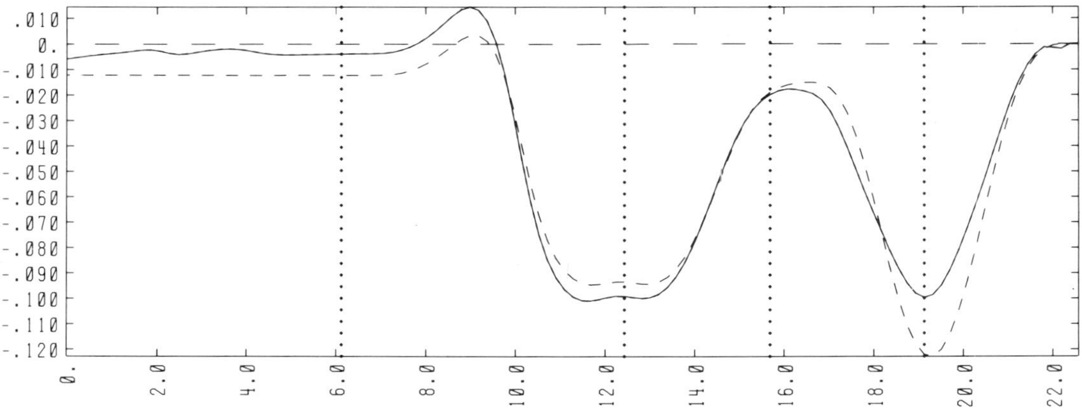

Figure 2: Parallel Current in MFTF-B at Equilibrium. The dotted line is the parallel current to lowest order in β.

The geodesic curvature is $\kappa = x_\theta x_{zz} + y_\theta y_{zz}$, the parallel current per unit flux is $i = B\left([x_z, x] + [y_z, y]\right)$, $i_F = B\left([x_\theta, x] + [y_\theta, y]\right)$ and $\kappa_F = (x_{\theta\theta} x_\theta + y_{\theta\theta} y_\theta)$. The integration constant ,$F(\psi, \theta)$, vanishes when equilibrium is reached. However, during the iteration as equilibrium is approached, F must be chosen to preserve the symetry and is equal to the total integral from -L to +L in the above equation. The equilibrium current is shown in Fig. (2).

To solve we note that, provided (3) is satisfied initially, we need only move field lines incompressibly to preserve this Jacobian constraint. This is accomplished with the equations of motion

$$\dot{x} = B[x, U] \qquad \dot{y} = B[y, U] \tag{6}$$

and the initial conditions

$$\begin{aligned} x &= \sqrt{2\int d\psi 2\psi (B_{vac}(z)/B(\psi, z))}\sigma(z)\cos\theta \\ y &= \sqrt{2\int d\psi 2\psi (B_{vac}(z)/B(\psi, z))}\tau(z)\sin\theta \end{aligned} \tag{7}$$

The vacuum field is prescribed by $B_{vac}(z) = 1/(\sigma(z)\tau(z))$ with the symetry $\sigma(z) = \tau(-z)$ and $\nabla \cdot B = 0$ determines the perpendicular components of the magnetic field, $B_\perp = Bx_z$.

All that remains is to determine the stream function U by substituting the linearized solution of the equations of motion, $\delta x = B[x, U\delta t]$, into (5). This produces an equation of the form

$$\nabla_\perp^2 \left((BU)_z / B\right) = H \tag{8}$$

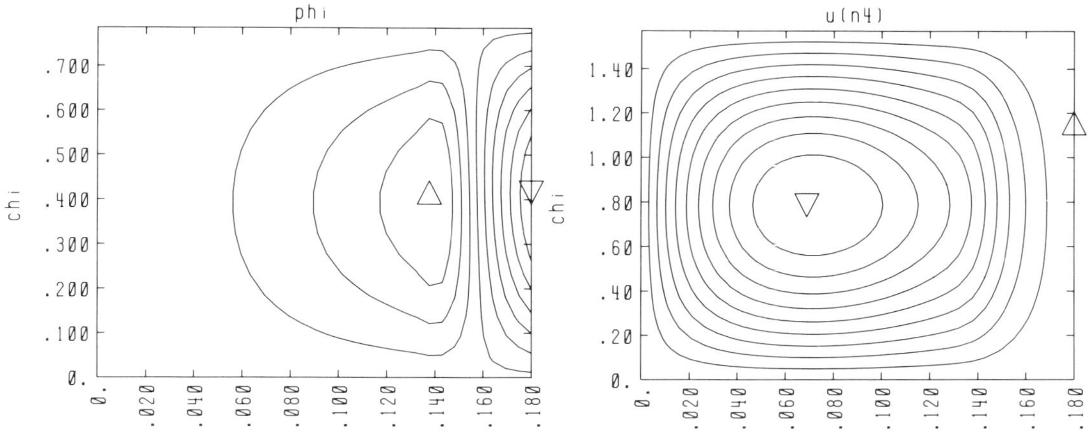

Figure 3: Field Line Flows to Equilibrium. Shown in flux coordinates, ψ, θ. (a) The only flow in the midplane is from Φ. (b) Flow in the Plug from U

plus some unimportant terms. Strictly speaking this is the solution for $\delta t = 1.0$. In practice we only move field lines part of the way with this value of U. This is necessary for accuracy, so as to preserve the Jacobian, and for numerical stability. The equation can be solved for $(BU)_z$ on each z-plane independently, subject to the boundary condition that $(BU)_z = 0$ at a lateral boundary. The initial shape of this boundary specifies the vacuum field. We integrate the solution, $(BU)_z$, to obtain the stream function U

$$BU = \int_0^z (BU)_z \, dz + \Phi(\psi, \theta) \tag{9}$$

The function of integration, Φ, is determined from the constraint that $F = 0$, which forces the parallel current to vanish at the ends of the plasma. Here too we linearize by substituting Φ into the equations of motion and then into F such that $F + (\delta F/\delta\Phi) \cdot \Phi = 0$. This is a partial differential equation for Φ which is second order in θ in the MHD limit and fourth order when the FLR terms are included. It is at this stage that the $\sin 4m\theta, (m = 1, 2...)$, distortion of the flux surface develops. These distortions at high β and the 4th-order derivatives make a finite difference representation impractical. The flows, U and Φ, which carry the field lines to equilibrium are shown in Fig. (3). The integral in (9) makes no contribution at the mid-plane where the motion is determined by Φ.

The transformed constraint has the form

$$A_2 \cdot \frac{\partial^2 \tilde{\Phi}}{\partial \psi^2} + A_1 \cdot \frac{\partial \tilde{\Phi}}{\partial \psi} + A_0 \cdot \tilde{\Phi} = \tilde{F}, \tag{10}$$

where $\tilde{F}$ is the Fourier representation of F and the A's are matrices which vanish outside the plasma. The matrix A_2 is proportional to the FLR stress, Y. One boundary condition is $\Phi = 0$ at $\psi = 0$ and, with FLR, we have the further condition at the plasma boundary that the solution be well behaved since A_2 vanishes more rapidly than the other coefficients. Beyond the plasma boundary any Φ would suffice since motion generated by Φ only corresponds to a relabelling of the field lines without altering the equilibrium. However, it is necessary to assure numerical smoothness across the boundary. This is achieved by continuing the equation into the vacuum by dividing out the common factor $d\omega/d\psi$. As $\hat{\underline{\mathbf{P}}}(B)$ and $\hat{Y}$ are defined in the vacuum, we have a smooth continuation. An open boundary condition is then applied at the distant wall.

The matrix, A_0, contains the MHD flute-stability integral on each field line and is singular if these are zero anywhere. Local violation of the flute stability criterion in the equilibrium field is the demarcation between what we call low and high β. The FLR contributions allow us to find stable equilibria beyond this limit. However, when the FLR terms are too large the iterations fail to converge. This may be due to the fact that z-derivative are treated implicitly and the FLR θ-derivatives explicitly in Eq.(5), but Stupakov [15] has raised the possibility of nearby equilibria with short scale θ-structures.

4. NUMERICAL METHODS

All equations are solved directly as a spectral decomposition in θ and finite differenced in (ψ, z) with 2nd-order accuracy everywhere. The ψ and z coordinates can be stretched to place mesh points where they are needed and a typical mesh is $(35 \times 10 \times 120)$. The dominant ψ-dependence in G is extracted from the Fourier components by

$$U(\psi, \theta) = \sum_{m=1}^{M} U_m \psi^m \sin(2m\theta) \tag{11}$$

All of this leads to the inversion of large block-tridiagonal matrices for each equation on each z-plane. All pressure profiles are typically chosen to vanish cubically in B and ψ. The pressure balance equations, (2), are solved on a fine z-mesh and interpolated to the coarser 3D-mesh on which we compute. Second derivatives in z are eliminated by an integration by parts.

As we have outlined, equilibrium is attained when both H and F are driven to zero. The iterative procedure to do this normally treats both processes on an equal level. However there are cases when it is preferable to iterate more conservatively until $F \approx 0$ for each iteration on H. This guarantees that the parallel current vanishes at the ends for each "outer" iteration.

We satisfy several convergence criteria for an equilibrium: The residuals of the current balance equation, (5), are checked locally and globally. The Jacobian, (3), is conserved to .1% even at high β. The most sensitive tests are provided by iter-

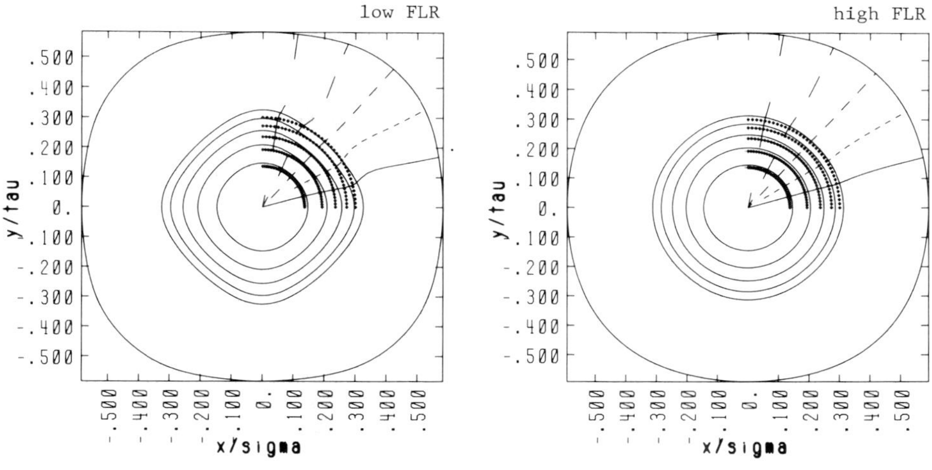

Figure 4: MFTF-B Flux Surfaces in the center cell with Low and Design-point FLR. Plasma surfaces (left) have large octupole distortion at low FLR. Heavy dots indicate the flux surfaces in the vacuum field. Also shown are θ-lines at equilibrium.

ating until the volume averages of each Fourier harmonic of the motion change by less than 1%. Measures of the kinetic energy of the flows have proved useless as a convergence test as this can drop many orders of magnitude when the equilibrium still has 1% errors.

5. MFTF-B EQUILIBRIA ABOVE THE MHD FLUTE LIMIT.

In the design of MFTF-B the TEBASCO [10] β-expanded equilibrium code was used to minimize the parallel current in the center-cell. This is still small at high-β, as shown in Fig(2). The ψ and θ flux surfaces for a case with 1/3 of the FLR effect expected at the design point is shown in Fig.(4). The outermost surface is where boundary conditions were applied and is far from the plasma. Its square shape arises from the continuation of Φ into the vacuum and has no significance, as noted earlier, if the wall is far enough away. We see considerable distortion of the flux surfaces from their vacuum values. At the expected FLR, this distortion is almost completely eliminated.

6. CONCLUDING REMARKS

The method we have described will not apply directly to a fully 3D equilibrium but it does shows promising directions for developing fast and accurate methods. The full equations are to be cast in Lagrangian form in the coordinates $\alpha^i = (\psi, \theta, s)$ through the covariant representation in terms of the basis vectors $\partial X/\partial\alpha^i$. The equations are to be linearised and solved for the direct displacement towards equilibrium. The various terms can be ordered in the parameter, λ, and the higher order parts treated explicitly in the iterations. The lower order parts are those identified here. Any vacuum solution will have to be calculated from Laplace's equation with suitable boundary conditions.

Attempts to simulate the earth's magnetosphere have problems with resolution at the atmosphere which could be solved naturally by a multipole expansion and spectral representation. The integral constraint on the parallel current essentially determines the lateral shift of the earth's magnetic field by the solar wind at each iteration. Finite Larmor radius effects may well be important in some regions of the magnetosphere where the plasma is hot and the curvature weak.

Insofar as the magnetosphere may be divided into separate regions of flow and static equilibrium, the methods of this paper may have some direct application. Even for simulation in the large these methods might be adapted to damp shear Alfven motions more rapidly.

7. ACKNOWLEDGEMENTS
In a short paper it is impossible to give credit to all the contributors in the field so the references we have given are our selection of current topics and understanding in magnetic fusion. We gratefully acknowledge many discussions with Bruce Cohen, Larry Hall, Tom Kaiser,Bill Nevins, Bill Newcomb and Hank Strauss

Work performed under the auspices of the U.S. Department of Energy by Lawrence Livermmore National Laboratories under contract number W-7405-ENG-48.

Bibliography

[1] B.McNamara, Methods in Computational Physics, Vol. 16,

[2] W.A. Newcomb. J.Plasma Phys. 26,529 (1981) (1976). Academic Press.

[3] T.Hayashi, D.D.Meyerhofer, F.J.Wysocki, M.Yamada, T.Sato, Bull. Am. Phys. Soc., 29,1202 (1984)

[4] R.N.Byrne, H.H.Klein, J.Comp. Phys. 26,352,(1978)

[5] P.N.Hu, H.Grad, Bull. Am.Phys. Soc.29,1324,(1984)

[6] L.S. Hall, B.McNamara. Phys. Fluids, 18,552 (1975).

[7] R.A.Chodura, A.Schluter, J. Comp. Phys. 41,68(1981)

[8] W.A. Newcomb Ann. Phys.(N.Y.) 150,172(1983).

[9] B. McNamara. "Rapidly convergent algorithms for 3D Tandem and Stellarator Equilibria". US-Japan Workshop on 3D MHD, ORNL Conf-840370 (1984) and UCRL-90437.

[10] L.D. Pearlstein, T.B.Kaiser, W.A.Newcomb. Phys. Fluids, 24,1326 (1981).

[11] H.R. Strauss. Nuclear Fusion 22,893 (1982)

[12] G.V. Stupakov. Sov. J. Plasma Phys. 5,486 (1979)

[13] W.A. Newcomb Ann. Phys.(N.Y.) 81,231(1973).

[14] L.D. Pearlstein and J.P. Freidberg. Phys. Fluids, 21,1218 (1978).

[15] G.V. Stupakov. Lecture at ,"Twenty Years of Plasma Physics" ,Trieste (1984)

ISSS-2 PARTICIPANTS AND CONTRIBUTORS

NAME	ADDRESS
Kazuhiro Akimoto	Astronomy Program University of Maryland College Park, MD 20742
Charles H. Aldrich	Los Alamos National Laboratory MS E531 Los Alamos, NM 87545
John Ambrosiano	Berkeley Scholars P.O. Box 852 Springfield, VA 22150
Maha Ashour-Abdalla	Institute of Geophysics & Planetary Physics University of California at Los Angeles Los Angeles, CA 90024
Kile Baker	Applied Physics Laboratory Johns Hopkins University Laurel, MD 20707
A.R. Barakat	CASS Utah State University Logan, UT 84322
Lee Frost Bargatze	Department of Earth & Space Sciences University of California at Los Angeles Los Angeles, CA 90024
Chris W. Barnes	Los Alamos National Laboratory MS E531 Los Alamos, NM 87545
Jean Berchem	Institute of Geophysics & Planetary Physics University of California at Los Angeles Los Angeles, CA 90024
Joachim Birn	Los Alamos National Laboratory MS D438 Los Alamos, NM 87545

Manfred H. Boehm	Space Sciences Laboratory University of California at Berkeley Berkeley, Ca 94720
David L. Book	Naval Research Laboratory Washington D.C 20375
Jeremiah Brackbill	Los Alamos National Laboratory MS E531 Los Alamos, NM 87545
Stephen H. Brecht	Berkeley Research Associates P.O. Box 241 Berkeley, CA 94701
Oscar Buneman	STAR Laboratory Stanford University Stanford, CA 94305
Peter J. Cargill	Astronomy Program University of Maryland College Park, MD 20742
Curt Carlson	STAR Laboratory Stanford University Stanford, CA 94305
Robert L. Carovillano	Department of Physics Boston College Chestnut Hill, MA 02167
Tom Chang	M.I.T. Center for Space Research Massachusetts Institute of Technology Cambridge, MA 02139
Kimball Clark	Department of Physics & Astronomy University of Iowa Iowa City, Iowa 52242
Paul J. Coleman, Jr.	Inst. of Geophysics & Planetary Physics University of California at Los Angeles Los Angeles, CA 90024
John W.D. Connolly	NSF Office of Advanced Scientific Computing Washington, D.C. 20550
F.V. Coroniti	Department of Physics/Astronomy Department of Astronomy University of California at Los Angeles Los Angeles, CA 90024

John D. Craven
Department of Physics & Astronomy
University of Iowa
Iowa City, Iowa 52242

Geoffrey B. Crew
M.I.T. Center for Space Research
Massachusetts Institute of Technology
Cambridge, MA 02139

Patrick A. Crowley
Defense Nuclear Agency
Washington D.C. 20305

John M. Dawson
Department of Physics
University of California at Los Angeles
Los Angeles, CA 90024

Dwight T. Decker
Space Data Analysis Laboratory
Boston College
Chestnut Hill, MA 02167

Robert B. Decker
Applied Physics Laboratory
Johns Hopkins University
Laurel, MD 20707

Viktor K. Decyk
Physics Department
University of California at Los Angeles
Los Angeles, CA 90024

Jacques Denavit
Lawrence Livermore National Laboratory
P.O. Box 5508, L-472
Livermore, CA 94550

Delia Donatelli
Air Force Geophysics Laboratory
Hanscom AFB, MA 01731

Christian T. Dum
Max-Planck-Inst. f. Physik und Astrophysik
Institut f. extraterrestrische Physik
D8046 Garching
FEDERAL REPUBLIC OF GERMANY

Pat Edmiston
Department of Physics
University of California at Los Angeles
Los Angeles, CA 90024

Richard Elphic
Inst. of Geophysics & Planetary Physics
University of California at Los Angeles
Los Angeles, CA 90024

Gary M. Erickson
Department of Space Physics & Astronomy
Rice University
P.O. Box 1892
Houston, TX 77251

Carl-Gunne Falthammar
Department of Plasma Physics
Royal Institute of Technology
10044 Stockholm, 70
SWEDEN

Joseph F. Fennell
The Aerospace Corporation
P.O. Box 92957
Los Angeles, CA 90009

David W. Forslund
Los Alamos National Laboratory
MS E531
Los Alamos, NM 87545

L.A. Frank
Department of Physics & Astronomy
University of Iowa
Iowa City, IA 52242

B.J. Fraser
Physics Department
Newcastle University
N.S.W. 2308
AUSTRALIA

Alex Friedman
Lawrence Livermore National Laboratory
P.O. Box 808, L-477
Livermore, CA 94550

Supriya Ganguli
Code 4780, SAIC
Naval Research Laboratory
Washington, D.C. 20375

G. Ganguli
Code 4780, SAIC
Naval Research Lab
Washington, D.C. 20375

S. Peter Gary
Los Alamos National Laboratory
ESS-8, MS D438
Los Alamos, NM 87545

Roger Gendrin
CNET/CRPE
38 Avenue du general Leclerc
92131 Issy-les-Moulineaux
FRANCE

Keith Goetz
School of Physics & Astronomy
University of Minnesota
116 Church St. SE
Minneapolis, MN 55455

C.K. Goertz
Department of Physics & Astronomy
University of Iowa
Iowa City, IA 52242

Martin V. Goldman	Department of Astrophysical, Planetary & Atmospheric Sciences University of Colorado Box 391 Boulder, CO 80309
Charles Goodrich	Astronomy Program University of Maryland College Park, MD 20742
James Lauer Green	NASA/MSFC Huntsville, AL 35812
Tohru Hada	Inst. of Geophysics & Planetary Physics Department of Physics University of California at Los Angeles Los Angeles, CA 90024
Kozo Hashimoto	Department of Electrical Engineering Kyoto University Kyoto 606, JAPAN
Lin-Ni Hau	Department of Space Physics & Astronomy Rice University Houston, TX 77251
Bonnie Hausman	NOAA/NESDIC, E1GC2 325 Broadway Boulder, CO 80309
John F. Hawley	Theoretical Astrophysics California Institute of Technology Pasadena, CA 91125
Michael Heinemann	Air Force Geophysics Laboratory AFGL/PHK 32 Daniel Street Hanscom AFB, MA 01731
Dennis Hewett	Lawrence Livermore National Laboratory P.O. Box 5508, L-472 Livermore, CA 94550
Edward W. Hones, Jr.	Los Alamos National Laboratory MS D438 Los Alamos, NM 87545
Naoko Hori	Geophysics Research Laboratory University of Tokyo Hongo Bunkyo-ku, Tokyo 113 JAPAN

Ritoku Horiuchi	IFT Hiroshima University Higashisenda-machi Naka-ku Hiroshima 730 JAPAN
J.L. Horwitz	Department of Physics The University of Alabama in Huntsville Huntsville, AL 35899
Masahiro Hoshino	Institute of Space & Astronautical Science 4-6-1 Komaba, Meguro Tokyo 153, JAPAN
Cheryl Huang	Department of Physics & Astronomy University of Iowa Iowa City, IA 52242
Mary K. Hudson	Physics & Astronomy Department Dartmouth College Hanover, NH 03755
Bengt Hultqvist	Kiruna Geophysical Institute P.O. Box 704 S-98127 Kiruna, SWEDEN
Keizo Inagaki	Radio Atmospheric Science Center Kyoto University Uji, Kyoto 611 JAPAN
Umran S. Inan	STAR Laboratory Stanford University Stanford, CA 94305
Christer Juren	Kiruna Geophysical Institute P.O. Box 704 S-98127 Kiruna, SWEDEN
Joseph R. Kan	Geophysical Institute University of Alaska Fairbanks, AL 99701
Iwane Kimura	Department of Electrical Engineering Kyoto University Sakyoku, Kyoto 606 JAPAN

Margaret Kivelson	Inst. of Geophysics & Planetary Physics University of Californa at Los Angeles Los Angeles, CA 90024
Kenichi Komori	Radio Atmospheric Science Center Kyoto University Uji, Kyoto 611 JAPAN
Kanya Kusano	IFT Hiroshima University Higashisenda-machi Naka-ku, Hiroshima 730 JAPAN
Shu T. Lai	Air Force Geophysics Lab/PHK Hanscom AFB Bedford, MA 01731
A.B. Langdon	Lawrence Livermore National Laboratory P.O. Box 5508, L-472 Livermore, CA 94550
William K. Levedahl	Space Science Laboratory University of California at Berkeley Berkeley, CA 94720
Nai Guo Lin	Department of Physics University of Minnesota Minneapolis, MN 55455
Jon Linker	Inst. of Geophysics & Planetary Physics University of California at Los Angeles Los Angeles, CA 90024
Willis Livesey	Inst. of Geophysics & Planetary Physics University of California at Los Angeles Los Angeles, CA 90024
Stefano Livi	Max Planck Institut fur Aeronomie Postfach 20 D3411 Katlenburg Lindau-3 FEDERAL REPUBLIC OF GERMANY
Tony Lui	Applied Physics Laboratory Johns Hopkins University Johns Hopkins Road Laurel, MD 20707
John G. Lyon	Naval Research Laboratory Washington, D.C. 20375

Robert L. Lysak	University of Minnesota Tate Laboratory of Physics 116 Church St. SE Minneapolis, MN 55455
Shinobu Machida	Department of Physics & Astronomy University of Iowa Iowa City, IA 52242
Mark Mandt	Geophysics Institute University of Alaska Fairbanks, AL 99701
George Mantjoukis	Dept. of Space Physics & Astronomy Rice University P.O. Box 1892 Houston, TX 77251
Eckart Marsch	Max Planck Institut fur Aeronomie Postfach 20 D 3411 Katlenburg-Lindau FEDERAL REPUBLIC OF GERMANY
Richard F. Martin, Jr.	NOAA Space Environment Laboratory R/E/SE 325 S. Broadway Boulder, CO 80303
Hiroshi Matsumoto	Radio Atmospheric Science Center Kyoto University Uji, Kyoto 611 JAPAN
Michael McCarthy	Physics Dept. FM-15 University of Washington Seattle, WA 98195
Brendan McNamara	Lawrence Livermore National Laboratory P.O. Box 5511, L630 Livermore, CA 94550
Mary Mellott	Department of Physics & Astronomy University of Iowa Iowa City, IA 52242
John D. Mihalov	NASA 245-3, NASA-Ames Research Center Moffett Field, CA 94035

Kyoung W. Min
Princeton University
C-Site, PPL
Princeton, NJ 08544

Akira Miura
Geophysics Research Laboratory
University of Tokyo
Tokyo 113
JAPAN

Forrest Mozer
Space Science Laboratory
University of California
Berkeley, CA 94720

Laurent Muschietti
Department of APAS
University of Colorado
Campus Box 391
Boulder, CO 80309

Tadas Nakamura
Geophysics Reserach Laboratory
University of Tokyo
Bunkyo-ku
Tokyo 113,
JAPAN

Alice Newman
The Aerospace Corporation
Space Sciences Laboratory
Box 92957
Los Angeles, CA 90009

J. David Nichols
Center for Astrophysics & Space Science
University of California at San Diego
La Jolla, CA 92093

Ken-Ichi Nishikawa
Department of Physics & Astronomy
University of Iowa
Iowa City, IA 52242

Tatsuzo Obayashi
ISAS
Meguroku-Komaba
Tokyo 153
JAPAN

Tatsuki Ogino
Institute of Geophysics & Planetary Physics
University of California at Los Angeles
Los Angeles, CA 90024

Shin-ichi Ohtani
G.R.L.
Hongo 7-3-1
Bunkyo-ku Tokyo, 113
JAPAN

Hideo Okuda	Plasma Physics Laboratory Princeton University Princeton, NJ 08544
Nojan Omidi	I.P.S.T University of Maryland College Park, MD 20740
Yoshiharu Omura	Radio Atmospheric Science Center Kyoto University Uji, Kyoto 611 JAPAN
Peter Palmadesso	4700IP Department of the Navy Naval Research Laboratory Washington D.C. 20550
Michael J. Pangia	Department of Physics FM-15 University of Washington Seattle, WA 98195
George K. Parks	Department of Physics FM-15 University of Washington Seattle, WA 98195
Dennis S. Peacock	NSF Atmospheric Science Section 1800 G.Street NW Washington D.C. 20550
John S. Perko	Space Science Center Demeritt Hall University of New Hampshire Durham, NH 03824
John L. Phillips	Inst. of Geophysics & Planetary Physics University of California at Los Angeles Los Angeles, CA 90024
Victor J. Pizzo	High Altitude Observatory P.O. Box 3000 Boulder, CO 80307
Thomas A. Potemra	Applied Physics Laboratory Johns Hopkins University Johns Hopkins Road Laurel, MD 20707

Channon Price	Geophysical Institute University of Alaska Fairbanks, Alaska 99701
Philip Pritchett	Department of Physics University of California Los Angeles, CA 90024
Zu-Yin Pu	Dept. of Geophysics Peking University Beijing, CHINA
Kevin B. Quest	Los Alamos National Laboratory MS E531 Los Alamos, NM 87545
C.E. Rasmussen	Utah State University CASS UMC-34 Logan, Utah 84322
Lee A. Reinleitner	NASA/ES53 Space Science Laboratory Marshall Space Flight Center Huntsville, Alabama 35812
John M. Retterer	Boston College Space Data Analysis Laboratory Chestnut Hill, MA 02167
Robert L. Richard	Inst. of Geophysics & Planetary Physics University of California at Los Angeles Los Angeles, CA 90024
Kjell Rönnmark	Kiruna Geophysical Institute University of Umea S-90187 Umea SWEDEN
Ilan Roth	Space Sciences Laboratory University of California at Berkeley Berkeley, CA 94720
Tetsuya Sato	IFT Hiroshima University Higashisenda-machi Nakaku Hiroshima 730 JAPAN
Herbert H. Sauer	NOAA-Space Environment Laboratory R/E/SE 325 S. Broadway Boulder, CO 80303

Frederick L. Scarf	TRW One Space Park Redondo Beach, CA 90278
David Schriver	Inst. of Geophysics & Planetary Physics University of California at Los Angeles Los Angeles, CA 90024
R.W. Schunk	CASS UMC-34 Utah State University Logan, Utah 84322
Yoichi Serizawa	IFT Hiroshima University Higashisenda-machi Nakaku, Hiroshima 730 JAPAN
Stanley Shawhan	Code EES NASA Headquarters Washington D.C. 20546
Nagendra Singh	CASS Utah State University Logan, Utah 84322
Larry Smarr	Department of Physics University of Illinois at Urbana Urbana, IL 61801
Yan Song	University of Minnesota Department of Physics Minneapolis, MN 55455
Robert Spiro	Department of Space Physics Rice University Box 1892 Houston, TX 77251
Walther N. Spjeldvik	Physics Dept. Boston College Chestnut Hill, MA 02167
Elmer K. Stover	Los Alamos National Laboratory MS E531 Los Alamos, NM 87545
Robert Strangeway	Inst. of Geophysics & Planetary Physics University of California at Los Angeles Los Angeles, CA 90024

Daniel W. Swift	Geophysical Institute University of Alaska Fairbanks, Alaska 99701
Richard Sydora	University of Texas Department of Physics Austin, TX 79712
Kazue Takahashi	Los Alamos National Laboratory ESS-8, MS D438 Los Alamos, NM 87545
Hiroshi Tanaka	Radio Atmospheric Science Center Kyoto University Uji, Kyoto 611 JAPAN
Motohiko Tanaka	Institute for Fusion Theory Hiroshima University Hiroshima 730, JAPAN
Michael Temerin	Space Sciences Laboratory University of California at Berkeley Berkeley, CA 94720
T. Terasawa	The Inst. of Space & Astron. Sciences 6-1, Komaba 4-chome Meguro-ku, Tokyo 153 JAPAN
Richard M. Thorne	Department of Atmospheric Sciences University of California at Los Angeles Los Angeles, CA 90024
Catherine L. Thornton	Jet Propulsion Laboratory 4800 Oak Grove Drive, MS 264-748 Pasadena, CA 91109
Robert L. Tokar	Los Alamos National Laboratory ESS-8, MS D438 Los Alamos, NM 87545
Masayuki Ugai	Ehime University Matsuyama 790 JAPAN
J. Manuel Urrutia	Plasma Physics Laboratory University of California at Los Angeles Los Angeles, CA 90024

Hannes Voigt
Space Physics Department
Rice University
P.O. Box 1892
Houston, TX 77251

R.J. Walker
Inst. of Geophysics & Planetary Physics
University of California at Los Angeles
Los Angeles, CA 90024

Haruichi Washimi
The Reseach Institute for Atmospherics
Nagoya University
Honohara, Toyokawa 442
JAPAN

Kunihiko Watanabe
Inst. of Geophysics & Planetary Physics
University of California at Los Angeles
Los Angeles, CA 90024

Daniel Weimer
Department of Physics & Astronomy
University of Iowa
Iowa City, IA 52242

Terry Whelan
Department of Physics & Astronomy
University of Iowa
Iowa City, IA 52242

Donald J. Williams
Applied Physics Laboratory
Johns Hopkins University
Johns Hopkins Road
Laurel, MD 20707

Dan Winske
Los Alamos National Laboratory
MS E531
Los Alamos, NM 87545

Daniel Winterhalter
Inst. of Geophysics & Planetary Physics
University of California at Los Angeles
Los Angeles, CA 90024

Michael Wiskerchen
Stanford University
STARLAB/SEL
Stanford, CA 94305

Toru Yamada
Radio Atmospheric Science Center
Kyoto University
Uji, Kyoto 611
JAPAN

Andrew Zachary	Astronomy Department University of California at Berkeley 601 Campbell Hall Berkeley, CA 94720
Steven T. Zalesak	Naval Research Laboratory Code 4780 Washington, D.C. 20375
Xiaoming Zhu	Inst. of Geophysics & Planetary Physics University of California at Los Angeles Los Angeles, CA 90024
Wolfgang Zwingmann	Theoretische Physik IV Ruhr-Universität Bochum 4630 Bochum 1 FEDERAL REPUBLIC OF GERMANY

ISSS-2 SUPPORTING ORGANIZATIONS

International Union of Radio Science, Commission H

United States
- Air Force Geophysics Laboratory
- California Space Institute, University of California, San Diego
- Los Alamos National Laboratory
- National Aeronautics and Space Administration
- National Science Foundation
- Office of Naval Research
- Universities of Space Research Associates
- University of California, Los Angeles

Japan
- Fujitsu Ltd.
- Hitachi Ltd.
- IBM Japan Ltd., Corporation
- Mitsubishi Electric Corporation
- NEC Corporation
- Nissan Motor Co., Ltd.
- Toray Science Foundation
- Toshiba Corporation
- Toyota Motor Corporation

Second International School for Space Simulations
February 4 – 15, 1985
Kauai BeachBoy Hotel
Kapaa, Kauai, Hawaii

Organizing Committee
- M. Ashour-Abdalla
- P. J. Coleman, Jr.
- R. Gendrin
- H. Matsumoto
- T. Obayashi
- H. Okuda
- T. Sato

Program Committee
- P. M. Banks
- T. J. Birmingham
- J. M. Dawson
- M. K. Hudson
- B. Hultqvist
- I. Kimura
- J. M. Kindel
- A. Nishida
- H. Oya
- P. J. Palmadesso
- K. Papadopoulos
- D. Peacock
- P. L. Pritchett
- R. C. Sagalyn
- M. Schulz
- S. D. Shawhan
- T. Tamao

Poster Previewers
- C.-G. Falthammar
- S. P. Gary
- H. Matsumoto
- P. J. Palmadesso

Closing Colloquium
- R. L. Carovillano
- C. T. Dum
- J. F. Fennell
- E. W. Greenstadt
- M. G. Kivelson
- H. Matsumoto
- G. K. Parks

SUBJECT INDEX

Acceleration region 216
Active magnetospheric particle 386
Active relaxation sounder 284
Adaptive gridding 163
Adiabatic invariants 376, 377, 378
Ampere's law 12
Amplitude-frequency-time spectrograms 257
Angular momentum 6
Anomalous resistance 400, 401, 402, 403, 404, 405
Anomalous resistivity 467, 468, 511
Anomalous transport processes 477
Antenna 115
Area weighting 12
Assembly language 10, 148
ATS-6 360
Aurora 295
Auroral acceleration region 313
Auroral activity 228
Auroral and polar cap regions 337
Auroral arc formation 499
Auroral electrons 316
Auroral emissions 306
Auroral ionosphere 313
Auroral kilometric radiation 22, 258

Beam 54
Beam of ions 60
Beam velocity 284
Bi-ion frequency 366
Birkeland currents 295, 298, 299, 300
Boundary conditions 8, 13, 108, 131, 140, 142, 143, 170, 174, 180, 488
Boundary layer model 225
Boundary layers 289
Bow shock 54, 55, 59, 313
Broadband electrostatic noise 221
Bunching 436
Buneman instability 468
Butterfly distribution 520

Cartesian case 8
Cerenkov processes 11
CFL condition 19
Characteristic time scales 68
Charge composition explorer 386
Charge weighting 10
Code optimization 146
Cold component 277
Cold plasma 525
Collisionless shocks 53
Computer inactivity 149
Computer libraries 148
Computers 187
Conductive transport 154, 158
Conical distributions 284
Convection patterns 303
Coronal holes 212
Coulomb gauge 12
Counter streaming electrons 346
Courant condition 11
Courant limit 12
Cray 146, 148
Crossover frequency 366
Cross-tail electric field 223
Current distribution 488
Current-driven instabilities 507
Cusp electron precipitation 343
Cyclotron maser instability 23, 26

Darwin 30, 32, 55, 67
Darwin codes 36, 79
Database 114
Data management 9, 108
Dayside precipitation 343
Debye length 11
Decomposition of vectors 34
Dessler-Parker relation 381
Diagnostic methods 195
Diagnostics 110, 113

Diffuse aurora 337, 338
precipitation 339, 399, 400, 405, 406, 407
Dimensionality 170
Direction of propagation 119, 265
Direct method 15, 71
Discrete arc precipitation 339
Discrete grid 5
Discretisation 104
Disks 150
Displacement current 11, 31
Distribution functions 5, 131, 137, 139
Distribution of trapped particles 381
Documentation 149
Donor cell differencing 158, 160, 163
Double layers 313
Downward electron beams 346
Downward flowing ion beams 350
Drift velocity 520
Driven reconnection 488
Dynamic explorer observations 501

Earth's magnetosphere 257
Eigenmodes 116
Electric dipole antenna 261
Electric field corrections 93
Electric fields 313
Electrodynamic coupling 479, 480
Electron beam 53
Electron cyclotron maser instability 22
Electron density 204, 279, 280, 282, 284, 288
Electron distribution 23, 24
functions 339
Electron fluxes 520
Electron gyrofrequency 259
Electron heating 476
Electron plasma frequency 259
Electron precipitation 337, 338, 344
Electron temperatures 217
Electron trapping 528
Electrons as a fluid 37
Electromagnetic codes 11, 18
Electromagnetic direct implicit method 78
Electromagnetic energy 25
Electromagnetic ion beam instability 57, 59
Electromagnetic microinstabilities 41
Electromagnetic model 47
Electromagnetic simulation 17, 23
Electromagnetic waves 258
Electrostatic fields 31
Electrostatic Poynting vector 121
Electrostatic shocks 313
Electrostatic waves 258, 416-420
Energetic electrons 15
Energetic protons 522
Energy absorption $\vec{F}\cdot\vec{V}$ 119
Energy and momentum 121
Energy conversion 488
mechanisms 261
Energy density of plasma waves 420-422
Energy exchanges 487
Energy flow vector 123
Energy input 492
Energy loss 489
Energy output 488
Energy regulation 488
Energy transfer 526
Equation of motion 13, 69
Exospheric theory 479
Explicit particle code 10
Explicit time differencing of particle 69
Extraction process 284

Faraday's law 12
Fast Fourier transform 13
Field-aligned currents 219, 295, 399, 400, 404, 408, 495, 496, 506, 507, 509
Field-aligned electron distributions 346
Field-aligned ions 284
Field-aligned potential 282
Finite difference methods 158
Finite-Larmor-radius effects 41
Finite size particle 56
Flaring model 230
Flow of kinetic energy 124
Fluid equations 85, 473
Fluid models 153
Flux-corrected transport 177
Flux tubes 216, 217

Fortran optimization 147
Fourier transform 18
Funnel-shaped auroral hiss 264
Future 187

Gaussian distribution 135, 136
Gaussian elimination 13
Geomagnetic fields 214
Geomagnetic storms 375
Geomagnetic tail 318
GEOS-1 361
GEOS-2 361
Geospace 257
Global momentum conservation 129
Global simulations 169, 170
Graphical output 149
Grid mesh 105
Gudonov scheme 175
Guiding center approximation 376
  Drift approximations 41
Gyroresonance and streaming instabilities 261

Hall effect 296
Hamiltonian method 53, 54, 58, 60, 62
Hand coding 148
Harang discontinuity region 301
Heavy ion species in the thermal plasma 359
He^+ concentrations 369
He^+ cyclotron frequency 362
High latitude boundary of the plasma sheet 313
High Mach number shocks 61
High order schemes 177
Hot ion density 284
Hot or streaming plasma 258
Hot plasma 277
Hybrid 79
  Codes 53, 54, 55, 57, 62, 85
  Method 14
  Models 59
  PIC-fluid 67
  Schemes 177, 178
Hydromagnetic whistlers 358

ICW 524, 525
  Linear amplification 524
IMF 171, 172, 174
  Orientation 302
Implicit algorithm 91
Implicit electromagnetic codes, applications of 81
Implicit electromagnetic fields 79
Implicit particle codes 472
  Limitations of 76
Implicit particle simulations 84
Implicit susceptibility 72, 73
Implicit techniques 14
Implicit time differencing of the particle 69
Implicit time integration 67
Incidental radiation from power transmission systems 257
Index of refraction 259
Inelastic scattering 479
Inhomogeneous magnetic field 440, 441
Injected particles 131, 134
Injection 132
Intense laser light 15
Intentional man-made transmitters 257
Interferogram 116
Intermediate speed memory device 9
Internal tension flow 125
Interplanetary magnetic field 344
Interpolation 107
  Formula 7
  Techniques 12
'Inverted-V' precipitation 339, 343
Io plasma torus 368
Ion acoustic instability 469
Ion and electron drift speeds 217
Ion beam 62, 349
Ion bulk flows 215
Ion bulk velocities 215, 216
Ion composition 221
Ion conic distributions 317
Ion conics 348, 349
Ion cyclotron frequencies 244
Ion cyclotron instabilities 57
Ion cyclotron resonance 282
Ion cyclotron wave propagation 365
Ion cyclotron waves 316, 360, 521
Ion density 281

Ion distribution 213
Ion fluxes 520
Ion ring velocity 54
Ion tail formation 477
Ion temperatures 215, 217
Ion velocity distributions 213
Ionosphere 212, 485
Ionospheric currents 303
Ionospheric current system 172
Ionospheric ion flow 347
Ionospheric waveguide propagation 371
Irrotational vectors 33
ISEE-1 wideband receiver 263
Isotropic cold protons 522
Isotropy 284

Jovian ULF emissions 533
Jupiter observations 243
Jupiter tail 245

Kinetic plasma behavior 67

Landau damping 11
Langmuir solitons 313
Lax-Wendroff differencing 158, 160
Lax-Wendroff scheme 175
Leapfrog scheme 19, 20
Leap-frog time integration scheme 177
Left-hand polarized transverse wave 362
Lightning 455-456
Lightning discharges 257
Linear polarization 363
Linear wideband receiver 263
Line-tying 490, 491
Linked lists 10
Lion roars 258
Liouville equation 4
LLP 520
LLP origin 519
LLP structure 518
Local wavenumber 119
Long period pulsations 516
Long time-scale behavior 67
Low frequency correlator 261
Low frequency wave phenomena 53
  Upstream of shocks 53
Lower hybrid drift instability 36
Lower hybrid resonance 244
Low β plasma 47
L wave stop band 367

Macroparticle 9
Magnetic disturbances 298
Magnetic field induced electron transport 15
Magnetic field lines 228, 277
Magnetic loop antenna 261
Magnetic power spectrum 521
Magnetic reconnection 492, 493
Magnetic shock waves 15
Magnetic soliton 132, 133, 143
Magnetic storms 287, 295
Magnetic substorms 218
Magnetic reconnection 57
Magnetic topology 212
Magnetohydrodynamics 486
Magnetoinductive 30
Magnetopause 54, 58, 213
  Boundary layers 214
  Crossing 290
Magnetosheath 212, 213, 214
Magnetosonic waves 521, 522, 523
Magnetosphere 54, 313, 485
Magnetosphere-ionosphere coupling 495
  Interaction 479
Magnetospheric currents 172
Magnetospheric plasma entry 216
Magnetospheric simulations 171
Magnetotail 172, 211, 212, 213, 218, 219, 229
  Acceleration 217
  Lobes 227, 228
Marginal stability 470
Maxwells' equation 4, 17
Maxwell stress tensor 128
Measurement techniques 278
Mechanical coupling 479
Memory banks 147
MHD equations 175
MHD simulations 169, 170
MHD with the Hall term 195
Minor ions 212
Moment approach 474
Moment implicit method 74
Moment method 14
Momentum 115
  Equations 88

Flow 128
Flux tensor 128
Monte Carlo 5
Monte Carlo method 135, 137
Morning side aurora 342, 343
Multidimensional particle codes 103
Multidimensional particle simulations 187
Multiple wave components 258
Mutual impedance probe 280, 284

Near earth region 181
Nearest grid point 6
Neutral sheet 216
Newton's laws 4
Nightside auroral observations 341
Nightside precipitation 339
Non-dipolar geomagnetic field 460
Non-dipolar model 462
Nonlinear phenomena 3
Nonlinear plus inhomogeneity (NPI) effect 442
Nonlinear three wave interactions 444, 447
Non-periodic emissions 358
Nonresonant acceleration 529
Number densities 284
Numerical models 187
Numerical oscillations 177
Numerical plasma simulation 41
Numerical scheme accuracy 178
Numerical schemes 170, 171, 174, 178
Numerical simulation of ICWs 531

Oblique shocks 56, 57
One dimensional electrostatic plasma simulation code 6
O^+ transport 533
Outer plasmasphere 288
Oxygen cyclotron frequency 363
Waves 532, 533

Parallel shocks 58
Parameter initialization 149
Partial donor cell method 177
Particle advancement 20
Particle codes 146, 148, 150
Particle distributions and currents 378
Particle distribution functions 339
Particle dynamics 515, 517, 522
Particle equations 70
Particle fluid simulations 83, 84
Particle fluxes 519
Particle-grid interpolation spline 68
Particle-in-cell 55, 67
Simulation 3
Particle loading 8
Particle models 18
Particle momentum density 128
Particle move 146
Particle movers 150
Particle simulations 17, 90
Particle trapping 430
Particles or fluid codes 53
Particles, sorting the 10
Pc1-2 358
Pc1 pulsation source 372
Pedersen conductivity 306
Periodic emissions 358
Perpendicular shocks 61
Phase bunching 439, 440, 441, 529, 530, 531
Phase-mixing 435, 436
Phase trapping 435, 436, 439
PIC electrostatic force 68
Piston 132
Pitch angle diffusion 392
Pitch angle scattering 342
Planes of symmetry 174
Plasma boundary 543
Plasma bulk velocity 216
Plasma convection 212
Plasma densities 217, 275, 282
Measurements 291
Plasma diagnostics 368
Plasma dispersion expression 264
Plasma dispersion relations 258
Plasma distributions 337
Plasma flow 280, 488
Plasma mantle 215
Plasma number density 279
Plasma parameters 213, 276, 285
Plasma profiles 115
Plasma sheet 218, 313
Boundary layer 219
-tail 313
Plasma simulation 3, 5, 67
Codes 145

Plasma temperature and ray tracing 459
Plasma trough 278, 284
DE-1 Plasma wave instrument 261
Plasma wave receiving system 258
Plasma waves 257
Plasmaspheric hiss 258
Plasmoid 218, 225, 231
Poisson's equation 6, 7, 8
Polar cap 228
  Electron precipitation 344
Polar cusp 214, 215
Polar disturbances 304
Polar rain 344
Polar wind 280, 281, 282, 350, 351
Polarization parameters 362
Polarization reversals 359, 364
Ponderomotive force 445, 447
Ponderomotive potential 446
Post-processing 114
Postprocessor 149, 150
Potential drops 346
Power spectra 362
Predictor-corrector 53, 54
  Code 60, 61
  Technique 57, 62
Propagation in a cold plasma 365
Propagation of waves 257
Proton cyclotron instability 367
Proton fluxes 524
Pulsating aurora 342
Pulsations 517

Quasi-linear diffusion 526
Quasi-linear simulations 477
Quasi-linear theory 473
Quasineutrality 54, 55
Quasiperpendicular collisionless shock 59, 60, 62

Ray paths 462, 464
  in the IGRF model 463
Ray tracing 257, 452, 456, 458, 460
  Techniques 449, 450
Reconnection 156, 165, 166, 172, 213, 214, 399, 400, 401, 402, 403, 407
Refilling processes 277
Refractive index 458
Relativistic dispersion 26
Relativistic effects 18
Relativistic electron beams 36
Relativistic momentum 20
Relaxation oscillations 476
Relaxation sounding 279
Reprogramming 147
Resistive code 60
Resistive method 62
Resistive Ohm's law 53, 55
Resonance effects 274
Resonant particles 439
Ring current 375, 383, 384, 385
  Decay 391
  Distributions 382
  Generation 390
Ring distributions 524
Rotational instabilities 57
Rusanov scheme 175
R-X mode 23

Sampling 103, 104
Satellite-ground correlations 369
Satellites 257
Saturn 246
Scaling laws 471
Scatter operation 10
Searchoil magnetic wave field antenna 261
Shear Alfvén waves 480
Shock 61
Shock front 132
Shocks and reconnection 472
Simulation speed 178
Slow mode shocks 232
Slow shocks 58, 493
Slow waves 57
Solar shocks 171
Solar wind 57, 60, 160, 212, 213, 214, 289, 479, 485, 541
Solar wind electrons 344
Solenoidal vectors 33
Solid state disks 151
Soliton 133, 138
Solution of the coupled field 70
Sounding rockets 257
Space and time differencing 6
Space plasma simulation 187
Spectral methods 106

Spectral slot 368
Stationary shock structure 134
Step frequency correlation 261
Stochastic heating 121
Strahl electrons 344
Structured pulsations 370
Substorm recovery 226
Substorms 226, 493, 494
Sun aligned polar arcs 345
Suprathermal electrons 528
Superthermal ion beams 288
Superthermal plasma 277, 280
Symmetry boundary conditions 180
Synchronous orbit 360
System of units 1

Tandem mirror 542, 543
  Equilibria 541
Tearing mode 54, 58
Temperature equations 89
Terrestrial ring current 225
Theory, modelling and simulations 261
Thermal plasma 277, 280, 523
Thermal velocity 289
Theta aurora 228
Three-dimensional MHD calculations of the magnetosphere 187
3D Tensor-pressure equilibria 543
Time centering 21
Timestep limitation 19
Tracer explorers 386
Trajectories of test particles 114, 119
Transport algorithm 87
Transpolar arc 228
Transverse energization of ions 348
Tridiagonal solver 8
Truncation error analysis 160
Two stream instability 468

ULF wave generation 534, 535
ULF wave propagation 534
Ultra-low-frequency (ULF) waves 357, 515, 524
Upflowing electron beams 346
Upflowing ion beams 350
Upgoing ions 316
Upper hybrid resonance frequency 258
Upstream 54, 59

Vector code 148
Vectorization 146, 147
Vector machines 146
Vector potential for the transverse current 124
Velocity distribution functions 114
Viscous transport 494
Vlasov equation 4, 5
Vlasov-Maxwells' equations 411-416
Voyager 242

Wave characteristics 517
Wave distribution function 257
Waveform capture receivers 269
Wave frequency 519
Wave growth 533
Wave normal angle 259
Wave normal direction 364
Wave observations 241
Wave-particle interaction 435
Wave-particle resonance condition 22
Wave polarization 257
Waves in Maxwellian plasmas 422-427
Wave structure 517
Wave tunnelling 369
Weakly-collisional plasma 68
Weak turbulence theory 3
Whistler mode 439, 440, 452
  Propagation 449
Whistler phenomenon 450, 451
Whistler propagation 450
Whistlers 455, 456
Wideband receiver 263
Work 115

AUTHOR INDEX

Aldrich, Charles H. 131

Barnes, Christopher W. 145
Boehm, Manfred H. 313
Brackbill, Jeremiah U. 153
Brecht, Stephen H. 169
Buneman, Oscar 103

Cattell, Cynthia A. 313
Coroniti, Ferdinand V. 399

Dawson, John M. 187
Decyk, Viktor K. 113
Denavit, Jacques 83
Dum, Christian T. 467

Fennell, Joseph F. 337
Forslund, David W. 3
Frank, Louis A. 211
Fraser, Brian J. 357

Gendrin, Roger E. 515
Goertz, Cristoph K. 499

Hewett, Dennis W. 29
Hultqvist, Bengt 275

Kimura, Iwane 449

Langdon, A. Bruce 67

Matsumoto, Hiroshi 429
McNamara, Brendan 541
Mozer, Forrest S. 313

Okuda, Hideo 41

Potemra, Thomas A. 295
Pearlstein, L. Donald 541
Pritchett, Philip L. 17

Rönnmark, Kjell 411

Sato, Tetsuya 485
Scarf, Frederick L. 241
Shawhan, Stanley D. 257

Temerin, Michael 313

Williams, Donald J. 375
Winske, Dan 53
Wygant, J. 313

ANNOUNCEMENT

Space Plasma Simulations

Editors: Maha Ashour-Abdalla and Daryl Ann Dutton

Please note that a hardbound edition of these special issues of *Space Science Reviews*, Vol. 42 Nos. 1–4 (October/December 1985), are available from the publishers.

ISBN 90-277-2108-4 Prices: Dfl. 195,– / \$69.00 / £54.25